A Practical Guide to Quantum Machine Learning and Quantum Optimization

Hands-on Approach to Modern Quantum Algorithms

Elías F. Combarro

Samuel González-Castillo

<packt>

BIRMINGHAM—MUMBAI

A Practical Guide to Quantum Machine Learning and Quantum Optimization

Copyright © 2023 Packt Publishing

All rights reserved. No part of this book may be reproduced, stored in a retrieval system, or transmitted in any form or by any means, without the prior written permission of the publisher, except in the case of brief quotations embedded in critical articles or reviews.

Every effort has been made in the preparation of this book to ensure the accuracy of the information presented. However, the information contained in this book is sold without warranty, either express or implied. Neither the authors, nor Packt Publishing or its dealers and distributors, will be held liable for any damages caused or alleged to have been caused directly or indirectly by this book.

Packt Publishing has endeavored to provide trademark information about all of the companies and products mentioned in this book by the appropriate use of capitals. However, Packt Publishing cannot guarantee the accuracy of this information.

Associate Group Product Manager: Gebin George
Publishing Product Manager: Kunal Sawant
Content Development Editor: Rosal Colaco
Technical Editor: Maran Fernandes
Copy Editor: Safis Editing
Project Coordinator: Manisha Singh
Proofreader: Safis Editing
Indexer: Hemangini Bari
Production Designer: Vijay Kamble
Business Development Executive: Kriti Sharma
Developer Relations Marketing Executive: Sonia Chauhan

First published: March 2023

Production reference: 1170323

Published by Packt Publishing Ltd.
Livery Place
35 Livery Street
Birmingham
B3 2PB, UK.

ISBN 978-1-80461-383-2

www.packtpub.com

Contributors

About the authors

Elías F. Combarro holds degrees from the University of Oviedo (Spain) in both mathematics (1997, award for second highest grades in the country) and computer science (2002, award for highest grades in the country). After carrying out several stays as a visiting researcher at the Novosibirsk State University (Russia), he obtained a PhD in mathematics (Oviedo, 2001) with a dissertation on the properties of some computable predicates under the supervision of Prof. Andrey Morozov and Prof. Consuelo Martínez.

Since 2009, Elías F. Combarro has been a tenured associate professor at the Computer Science Department of the University of Oviedo. He has published more than 50 research papers in international journals on topics such as quantum computing, computability theory, machine learning, fuzzy measures, and computational algebra. His current research focuses on the application of quantum computing to algebraic, optimization, and machine-learning problems.

In 2020 and 2022, he was a Cooperation Associate at CERN openlab. Currently, he is Spain's representative on the Advisory Board of the CERN Quantum Technology Initiative.

To Adela, Paula and Sergio. You are my reason to live.

Samuel González-Castillo holds degrees from the University of Oviedo (Spain) in both mathematics and physics (2021). He is currently a mathematics research student at Maynooth University, where he works as a graduate teaching assistant.

He completed his physics bachelor thesis under the supervision of Prof. Elías F. Combarro, Prof. Ignacio F. Rúa (University of Oviedo), and Dr. Sofia Vallecorsa (CERN). In it, he worked alongside other researchers from ETH Zürich on the application of quantum machine learning to classification problems in high energy physics. In 2021, he was a summer student at CERN developing a benchmarking framework for quantum simulators. He has contributed to several conferences on quantum computing and related fields.

About the reviewers

Francisco Orts is a researcher at the Institute of Data Science and Digital Technologies, Vilnius University (Lithuania). He holds a PhD in computer science from the University of Almería (Spain) and is a collaborator of the High-Performance Computing research group of this university. He has worked as a computer scientist at construction, stock exchange, and IT services companies, with more than 15 years of experience in the sector. His research interests are multidimensional scaling, quantum computing, and high-performance computing.

Guillermo Botella (Senior Member, IEEE) received an MSc degree in physics in 1998, an MSc degree in electronic engineering in 2001, and a PhD degree (in computer engineering) in 2007, all from the University of Granada, Spain. He was an EU research fellow working with the University of Granada, Spain, and University College London, UK. He is currently an associate professor with the Department of Computer Architecture and Automation, Complutense University of Madrid, Spain. He has performed research stays from 2008 to 2012 with the Department of Electrical and Computer Engineering, Florida State University, USA. His current research interests include signal processing for FPGAs, GPGPUs, and novel computing paradigms such as analog and quantum computing.

Foreword

"I know you, you were working with Elías Combarro when he gave that course about quantum computing at CERN. That course changed my life!"

As I welcome students and more seasoned researchers to the CERN IT Department, it is not unusual to hear this kind of comment about Elías' lectures. I've been working at CERN for 25 years on R&D projects in computing and data science for high-energy physics and still I have not seen this happening that often with other courses.

When I started building the CERN Quantum Technology Initiative in 2018, quantum computing and its applications had already started growing at an accelerated pace. We were looking for ways of understanding the potential benefits for physics and contributing to the ongoing research in theory, computing, sensing, and communication. A daunting task despite the incredible work that CERN has been doing since 1954 in physics and computing for physics. Where to start? How to build knowledge? How to identify and address realistic problems? Which tools and techniques should we focus on?

Prof. Combarro joined CERN in 2020 for a short sabbatical, and he immediately became a reference for the team, inspiring students and researchers and helping us lay solid foundations for the work we were doing with the LHC experiments and the theory groups at CERN. Samuel González-Castillo joined the team a year later during his summer internship, building from the ground up our first prototype of a quantum systems benchmark framework.

This book is the perfect image of how I saw them in action. It will initially guide you across complex concepts with rare clarity, building solid foundations to work comfortably with quantum optimization methods, quantum machine learning, and hybrid architectures without ever losing sight of the goal of providing a realistic, practical, usable approach.

Chapters 1 and 2 will introduce you to the basics of quantum computing, building a reference of mathematical concepts and notations and a first practical overview of the "tools of the trade," the frameworks and platforms used to interact with quantum devices. Once the foundations are set like a sort of teaser of what's to come, the rest of the book guides you through two complementary paths, quantum optimization methods in Chapters 3 to 7, and quantum machine learning, quantum neural networks, and hybrid architectures in Chapters 8 to 12.

The authors not only provide clear formal explanations at every step, but also practical instructions and examples on how to implement and execute algorithms and methods on freely accessible actual quantum computers. Exercises (with detailed answers) are given throughout the book to check the progress of the exploration and gently nudge you beyond your comfort zone, always keeping the interest alive.

Whether you are at the beginning of your discovery of quantum computing or are looking to understand its potential in your ongoing research, this book will be a trustworthy guide on an exciting journey. And I'm sure that the next time I meet you, you will say *"I know you, you wrote the foreword for Elias' and Samuel's book. That book changed my life!"*

Alberto Di Meglio, MEng, PhD
Head of Innovation – Coordinator CERN Quantum Technology Initiative
Information Technology Department
CERN – European Organization for Nuclear Research

Acknowledgements

I would maintain that thanks are the highest form of thought; and that gratitude is happiness doubled by wonder.

— G.K. Chesterton

There are many people who we are grateful to because their support, help, knowledge, and advice were instrumental in shaping this book. First of all, we would like to thank Tomás Fernández Marcos. He taught both of us — albeit in different millennia! — some of the mathematical notions that, in our future, would be indispensable for our study of quantum computing. Then, he introduced us to each other, because he had the prescient intuition that we would share many interesting projects. It is fair to say that, without him, this book would not exist.

We would also like to thank Alberto Di Meglio for writing such a wonderful foreword for us. The rest of the book can only go downhill from there!

Many parts of this book originate from courses on quantum computing taught over the years. Those courses would not have been possible without the help and trust of Enrique Arias, Alberto Di Meglio, Melissa Gaillard, Ester Martín Garzón and José Ranilla, among others.

We are also indebted to all the colleagues with whom we have discussed different topics on quantum computing and from whom we have learned a lot. We are particularly grateful to those who gave us very useful feedback and comments on preliminary versions of our lectures and of material for this book, including Vasilis Belis, Héctor García Morales, Miguel

Hernández-Cáceres, Carla Rieger, Ignacio F. Rúa, Bruno Santidrián Manzanedo, Daniel Setó, Erik Skibinsky Gitlin, and Sofia Vallecorsa. We also want to give heartfelt thanks to Ferdous Khan. He was the first to suggest that our lectures should be collected in book form. We cannot overstate how much we appreciate his constant encouragement and support.

Of course, we would also like to thank our team at Packt. They believed in our ability to write a whole book on quantum computing much more than we did ourselves! And they provided many useful suggestions and advice, making the complicated process of preparing a technical manuscript as smooth as possible.

We've been extremely lucky to have such wonderful technical reviewers as Guillermo Botella and Francisco Orts. They went well beyond the call of duty to check that everything was correct, they gave us invaluable feedback and suggestions, and they located quite a number of errata and typos that would have been very embarrassing had they made it to the print version. Obviously, all remaining errors are our sole responsibility.

And last but certainly not least, we would like to thank our friends and families. Writing a book takes a lot of time. And when we say "a lot," we really mean "an incredibly awful lot." Sadly, we had to steal part of that time from them and, moreover, they had to listen to our complaints and worries when we went through rough spots in the writing process. If it weren't for them, we could not have made it. This is all because of them and dedicated to them.

Elías F. Combarro, Samuel González-Castillo
Oviedo/Maynooth
February 2023

Table of Contents

Foreword vii

Acknowledgements ix

Preface xxi

Part 1: I, for One, Welcome our New Quantum Overlords 1

Chapter 1: Foundations of Quantum Computing 3

1.1 Quantum computing: the big picture 4

1.2 The basics of the quantum circuit model 7

1.3 Working with one qubit and the Bloch sphere 9

1.3.1 What is a qubit? 9

1.3.2 Dirac notation and inner products 11

1.3.3 One-qubit quantum gates 13

1.3.4 The Bloch sphere and rotations 18

1.3.5 Hello, quantum world! 22

1.4 Working with two qubits and entanglement 23

1.4.1 Two-qubit states 24

1.4.2 Two-qubit gates: tensor products 27

1.4.3 The CNOT gate 28

1.4.4 Entanglement 30

1.4.5 The no-cloning theorem 32
1.4.6 Controlled gates 33
1.4.7 Hello, entangled world! 35
1.5 Working with multiple qubits and universality 36
1.5.1 Multi-qubit systems 36
1.5.2 Multi-qubit gates 39
1.5.3 Universal gates in quantum computing 40
Summary 42

Chapter 2: The Tools of the Trade in Quantum Computing 43

2.1 Tools for quantum computing: a non-exhaustive overview 44
2.1.1 A non-exhaustive survey of frameworks and platforms 44
2.1.2 Qiskit, PennyLane, and Ocean 48
2.2 Working with Qiskit 50
2.2.1 An overview of the Qiskit framework 50
2.2.2 Using Qiskit Terra to build quantum circuits 52
Initializing circuits 53
Quantum gates 54
Measurements 57
2.2.3 Using Qiskit Aer to simulate quantum circuits 58
2.2.4 Let's get real: using IBM Quantum 64
2.3 Working with PennyLane 69
2.3.1 Circuit engineering 101 70
2.3.2 PennyLane's interoperability 77
Love is in the Aer 78
Connecting to IBMQ 79
Summary 80

Part 2: When Time is Gold: Tools for Quantum Optimization 81

Chapter 3: Working with Quadratic Unconstrained Binary Optimization Problems 83

3.1 The Max-Cut problem and the Ising model **84**
3.1.1 Graphs and cuts 85
3.1.2 Formulating the problem 86
3.1.3 The Ising model 89
3.2 Enter quantum: formulating optimization problems the quantum way **91**
3.2.1 From classical variables to qubits 91
3.2.2 Computing expectation values with Qiskit 97
3.3 Moving from Ising to QUBO and back **105**
3.4 Combinatorial optimization problems with the QUBO model **109**
3.4.1 Binary linear programming 110
3.4.2 The Knapsack problem 114
3.4.3 Graph coloring 116
3.4.4 The Traveling Salesperson Problem 119
3.4.5 Other problems and other formulations 122
Summary **123**

Chapter 4: Adiabatic Quantum Computing and Quantum Annealing 125

4.1 Adiabatic quantum computing **126**
4.2 Quantum annealing **129**
4.3 Using Ocean to formulate and transform optimization problems **135**
4.3.1 Constrained quadratic models in Ocean 135
4.3.2 Solving constrained quadratic models with dimod 138
4.3.3 Running constrained problems on quantum annealers 141
4.4 Solving optimization problems on quantum annealers with Leap **146**
4.4.1 The Leap annealers 146
4.4.2 Embeddings and annealer topologies 150

4.4.3 Controlling annealing parameters 155
4.4.4 The importance of coupling strengths 160
4.4.5 Classical and hybrid samplers 164
Classical solvers 165
Hybrid solvers 167
Summary 169

Chapter 5: QAOA: Quantum Approximate Optimization Algorithm 171

5.1 From adiabatic computing to QAOA 172
5.1.1 Discretizing adiabatic quantum computing 172
5.1.2 QAOA: The algorithm 175
5.1.3 Circuits for QAOA 178
5.1.4 Estimating the energy 181
5.1.5 QUBO and HOBO 183
5.2 Using QAOA with Qiskit 188
5.2.1 Using QAOA with Hamiltonians 188
5.2.2 Solving QUBO problems with QAOA in Qiskit 195
5.3 Using QAOA with PennyLane 203
Summary 209

Chapter 6: GAS: Grover Adaptive Search 211

6.1 Grover's algorithm 212
6.1.1 Quantum oracles 214
6.1.2 Grover's circuits 216
6.1.3 Probability of finding a marked element 219
6.1.4 Finding minima with Grover's algorithm 223
6.2 Quantum oracles for combinatorial optimization 224
6.2.1 The quantum Fourier transform 225
6.2.2 Encoding and adding integer numbers 227
6.2.3 Computing the whole polynomial 232

6.2.4 Constructing the oracle 234

6.3 Using GAS with Qiskit **236**

Summary **242**

Chapter 7: VQE: Variational Quantum Eigensolver 245

7.1 Hamiltonians, observables, and their expectation values **246**

7.1.1 Observables 248

7.1.2 Estimating the expectation values of observables 253

7.2 Introducing VQE **260**

7.2.1 Getting excited with VQE 263

7.3 Using VQE with Qiskit **267**

7.3.1 Defining a molecular problem in Qiskit 267

7.3.2 Using VQE with Hamiltonians 270

7.3.3 Finding excited states with Qiskit 276

7.3.4 Using VQE with molecular problems 278

7.3.5 Simulations with noise 282

7.3.6 Running VQE on quantum computers 288

7.3.7 The shape of things to come: the future of Qiskit 291

7.4 Using VQE with PennyLane **294**

7.4.1 Defining a molecular problem in PennyLane 294

7.4.2 Implementing and running VQE 296

7.4.3 Running VQE on real quantum devices 298

Summary **300**

Part 3: A Match Made in Heaven: Quantum Machine Learning 303

Chapter 8: What Is Quantum Machine Learning? 305

8.1 The basics of machine learning **306**

8.1.1 The ingredients for machine learning 307

8.1.2 Types of machine learning 317

8.2 Do you wanna train a model? **319**
8.2.1 Picking a model 322
8.2.2 Understanding loss functions 327
8.2.3 Gradient descent 328
8.2.4 Getting in the (Tensor)Flow 332
8.2.5 Training the model 335
8.2.6 Binary classifier performance 340
8.3 Quantum-classical models **347**
Summary **350**

Chapter 9: Quantum Support Vector Machines 351

9.1 Support vector machines **352**
9.1.1 The simplest classifier you could think of 352
9.1.2 How to train support vector machines: the hard-margin case 356
9.1.3 Soft-margin training 361
9.1.4 The kernel trick 364
9.2 Going quantum **366**
9.2.1 The general idea behind quantum support vector machines 366
9.2.2 Feature maps 368
9.3 Quantum support vector machines in PennyLane **372**
9.3.1 Setting the scene for training a QSVM 372
9.3.2 PennyLane and scikit-learn go on their first date 375
9.3.3 Reducing the dimensionality of a dataset 377
9.3.4 Implementing and using custom feature maps 380
9.4 Quantum support vector machines in Qiskit **382**
9.4.1 QSVMs on Qiskit Aer 382
9.4.2 QSVMs on IBM quantum computers 385
Summary **386**

Chapter 10: Quantum Neural Networks 389

10.1 Building and training a quantum neural network **390**
10.1.1 A journey from classical neural networks to quantum neural networks 391
10.1.2 Variational forms 393
10.1.3 A word about measurements 398
10.1.4 Gradient computation and the parameter shift rule 401
10.1.5 Practical usage of quantum neural networks 403
10.2 Quantum neural networks in PennyLane **405**
10.2.1 Preparing data for a QNN 406
10.2.2 Building the network 410
10.2.3 Using TensorFlow with PennyLane 413
10.2.4 Gradient computation in PennyLane 420
10.3 Quantum neural networks in Qiskit: a commentary **426**
Summary **429**

Chapter 11: The Best of Both Worlds: Hybrid Architectures 431

11.1 The what and why of hybrid architectures **432**
11.2 Hybrid architectures in PennyLane **436**
11.2.1 Setting things up 436
11.2.2 A binary classification problem 438
11.2.3 Training models in the real world 442
11.2.4 A multi-class classification problem 449
A general perspective on multi-class classification tasks 450
Implementing a QNN for a ternary classification problem 452
11.3 Hybrid architectures in Qiskit **458**
11.3.1 Nice to meet you, PyTorch! 458
Setting up a model in PyTorch 460
Training a model in PyTorch 463
11.3.2 Building a hybrid binary classifier with Qiskit 472
11.3.3 Training Qiskit QNNs with Runtime 477
11.3.4 A glimpse into the future 481

Summary 482

Chapter 12: Quantum Generative Adversarial Networks 483

12.1 GANs and their quantum counterparts 484
12.1.1 A seemingly unrelated story about money 484
12.1.2 What actually is a GAN? 486
12.1.3 Some technicalities about GANs 491
12.1.4 Quantum GANs 493
12.2 Quantum GANs in PennyLane 495
12.2.1 Preparing a QGAN model 497
The training process 502
12.3 Quantum GANs in Qiskit 510
Summary 518

Afterword and Appendices 519

Chapter 13: Afterword: The Future of Quantum Computing 521

Appendix A: Complex Numbers 527

Appendix B: Basic Linear Algebra 531

Vector spaces 531
Bases and coordinates 533
Linear maps and eigenstuff 535
Inner products and adjoint operators 537
Matrix exponentiation 539
A crash course in modular arithmetic 540

Appendix C: Computational Complexity 543

A few words on Turing machines 544

Measuring computational time 546
Asymptotic complexity 547
P and NP 549
Hardness, completeness, and reductions 554
A very brief introduction to quantum computational complexity 558

Appendix D: Installing the Tools 565

Getting Python 565
Installing the libraries 568
Accessing IBM's quantum computers 571
Accessing D-Wave quantum annealers 572
Using GPUs to accelerate simulations in Google Colab 573

Appendix E: Production Notes 577

Assessments 579

Chapter 1, Foundations of Quantum Computing 579
Chapter 2, The Tools of the Trade in Quantum Computing 588
Chapter 3, Working with Quadratic Unconstrained Binary Optimization Problems 591
Chapter 4, Adiabatic Quantum Computing and Quantum Annealing 593
Chapter 5, QAOA: Quantum Approximate Optimization Algorithm 595
Chapter 6, GAS: Grover Adaptative Search 598
Chapter 7, VQE: Variational Quantum Eigensolver 600
Chapter 8, What is Quantum machine Learning? 606
Chapter 9, Quantum Support Vector Machines 608
Chapter 10, Quantum Neural Networks 609
Chapter 11, The Best of Both Worlds: Hybrid Architectures 610
Chapter 12, Quantum Generative Adversarial Networks 614

Bibliography 617

Index 627

Other Books You Might Enjoy 642

Preface

See that the imagination of nature is far, far greater than the imagination of man.

— Richard Feynman

Many people believe that quantum computing is very difficult to learn, assuming that it requires knowledge of arcane and obscure branches of mathematics, and that it can only be mastered with a strong background in physics. We couldn't disagree more. In fact, the set of prerequisites that you need for a journey into the depths of quantum computing is surprisingly small: just some basic linear algebra, a few notions from probability, and some familiarity with computer programming. All of these are acquired by students of mathematics, physics, computer science, and engineering in their first year of college.

For this reason, a few years ago, one of us started designing quantum computing courses that focused only on the essentials, trying to demystify the subject and make it accessible to as many students as possible. Prominent among these courses is the series of lectures titled "A Practical Introduction to Quantum Computing: From Qubits to Quantum Machine Learning and Beyond" taught online from CERN in 2020 (you can find the materials and recordings at `https://indico.cern.ch/event/970903/`).

The seed of this book was already present in that series of lectures (and in others taught at the University of Oviedo, the University of Castilla-La Mancha and the University of Almería, among other places). Some of the material (greatly adapted and expanded) comes from what was prepared for the CERN course and, more importantly, the main guiding principles behind the design of the lectures (discussed in detail in [1]) have remained unchanged in the preparation of the book that you are now reading.

The first of those principles is our firm belief that quantum computing should be about computing. Let us clarify that a little bit. This means that our end goal will be to enable you to run quantum algorithms on quantum computers and solve problems with them. To achieve that, you will need to know how to write code that implements some quantum algorithms. And you cannot do that if you only study quantum computing from a purely theoretical perspective. Indeed, you need to get your hands dirty, learn (at least!) one quantum programming language, and be able to translate abstract quantum operations into executable instructions.

This is why a big part of this book is devoted to introducing different quantum programming frameworks (Qiskit, PennyLane, D-Wave's Ocean) and explaining how to run different quantum algorithms with them. Contrary to other (excellent!) quantum computing books out there, here you will find code. Lots of code. Code that you will be able to run straightaway on simulators and actual quantum computers. Code that you can modify, experiment with, and adapt to your own problems and projects.

But this is not just a quantum programming book. Our goal is not to give you recipes for solving particular instances of particular problems. Our goal is for you to *understand* quantum computing. So our second guiding principle is a commitment to discuss all the mathematics behind every quantum algorithm covered in the book and behind each line of code in our examples — at least in a reasonable amount of detail.

This is important for a couple of reasons. On the one hand, quantum computing is still a young field. New, improved algorithms are proposed in scientific publications every day and some of them will become standard in the short or medium term. Eventually, you will want to understand and use those algorithms. But you cannot do it if you do not fully understand the algorithms they come from.

On the other hand, successfully programming and running an algorithm is the ultimate test to know if you really understand its principles. Computers are merciless. They do not tolerate imprecision or ambiguities. You really need to know each and every detail of an algorithm in order to implement it. We completely agree with Donald Knuth when he says that "a person does not really understand something until he can teach it to a computer."

So those are the two main pillars upon which we have built this book: code and the mathematics behind it. Enough mathematics to understand the code and enough code to make the mathematics clear and useful. We won't lie to you: it was difficult to find a balance between the two. Sometimes you may feel that our mathematical explanations run for too long. But just be patient and we promise that it will pay off when you see the formulas coming alive in the examples.

These two pillars were already present in the CERN lectures. But there is a difference between the topics covered there and the ones that we have selected for this book. We have decided not to include basic methods such as quantum teleportation [2], [3] and quantum key distribution with the BB84 protocol [4], or canonical algorithms like the ones by Deutsch and Jozsa [5] and Shor [6], for instance.

Fortunately, there is much more quantum computing material out there than four or five years ago, when we started developing our introductory courses. We think that there is no longer an urgent need to explain methods that are perfectly discussed in other books such as the (highly recommended) one by Sutor [7].

However, we do feel that there exists a need for a unified, detailed and practice-oriented explanation of many algorithms that are central to modern quantum computing and that are difficult to find together in a sole source. This includes a lot of methods that have been developed to solve optimization problems with quantum computers, and most algorithms (especially the ones based on variational circuits) from the field of quantum machine learning.

Many of these algorithms have been proposed fairly recently and have been designed to run on the kind of quantum computers available today (small, not fully connected, and susceptible to noise) as opposed to on idealized, fault-tolerant quantum processors. For this reason, these algorithms are currently the subject of intense research, because their true capabilities are not yet completely understood. There is some evidence that they may surpass classical algorithms in certain tasks, but this is still not so well-established as with other, older quantum algorithms such as Shor's.

Does this mean that this book is advanced or only for people already experienced in quantum computing? Not at all! It is true that, traditionally, one used to start studying quantum computing by going through protocols with just a few qubits, and then learning about Deutsch-Jozsa's, Simon's [8], and Bernstein-Vazirani's [5] algorithms, climbing all the way up to Shor's and Grover's [9] methods. If you know about those algorithms, that knowledge will certainly be useful, but it is not, by any means, necessary or expected to understand the topics that we will cover.

With this book, we want to provide you with a solid understanding of the principles behind modern quantum algorithms that have been proposed for the fields of optimization and machine learning, as well as to show you how to implement them and run them on quantum simulators and on real quantum hardware. This will allow you to start experimenting on problems of your own right away. We strongly believe that this is the perfect moment to start searching for use cases with current quantum computers. The algorithms that we present in this book are strong candidates to be among the first to be applied in practical situations in the near future, because most of them need much fewer resources than other earlier quantum algorithms (such as Shor's) and do not require error correction. Moreover, they can be understood and used without the need to know about previous developments in the field.

In fact, we have designed this book assuming that you have had no previous experience with quantum computing at all (we do assume that you have a working knowledge of complex numbers and linear algebra, although we also provide a refresher of both topics in the appendices).

The style of our exposition is mainly informal, without following the usual structure of definition-theorem-proof-corollary of many mathematical texts, but without sacrificing rigor at any point in the book. Whenever possible, we give detailed derivations that justify the mathematical properties that we use in our developments and analyses (or, at least, we provide an argument that may be extended to a full proof by just adding some small technical details). In the cases that proving a particular fact is beyond the scope of the book, we provide references in which a full treatment can be found.

Throughout all the text, we propose exercises that will help you understand important concepts and develop practical skills for manipulating formulas and writing your own quantum code. They are intended to be readily solved (we try to give useful hints for those exercises that are a lit bit more challenging), but, at the end of the book, we provide full, detailed solutions so that you check your understanding of the subject.

Quantum computing is a field in constant evolution, so we feel that it is especially important to give pointers to new developments, to variants of the algorithms that we present in the book, and to alternative approaches to solve the kind of problems that we study. We do this by including numerous boxes with the label "To learn more…". You can skip these boxes if you wish, as they are not necessary to follow the main text. However, we strongly recommend reading them, since they help to situate in a wider context the topics under study. Other boxes that we use throughout the book serve to highlight important facts, to give warnings about subtle points, or to remind you of central definitions and formulas. These should not be skipped. They are labeled "Important note" for a reason!

We've had a great time writing this book and we hope that it shows. But, above all, we hope that you find it useful. If it helps you in understanding the fascinating field of quantum computers a little bit better, we will consider our mission to be fulfilled.

Who this book is for

This book is for professionals from a wide variety of backgrounds, including computer scientists and programmers, engineers, physicists, chemists, and mathematicians. Basic knowledge of linear algebra and some programming skills (for instance, in Python) are assumed, although all mathematical prerequisites will be covered in the appendices.

What this book covers

This book is organized in three parts, an afterword, and some appendices as follows:

Part 1, I, for One, Welcome our New Quantum Overlords

Chapter 1, Foundations of Quantum Computing, briefly reviews the key ideas behind the quantum circuit model, fixing the notation that we will use throughout the book. It explores core ideas and notions, discussing quantum states, quantum gates and measurements; all starting completely from scratch. It also makes the book more self-contained and accessible for readers with different backgrounds.

Chapter 2, The Tools of the Trade in Quantum Computing, presents different quantum programming libraries that you can use to implement and run quantum methods, focusing especially on Qiskit and PennyLane. This chapter will guide you through the process of implementing quantum circuits and running them on simulators and actual quantum computers.

Part 2, When Time is Gold: Tools for Quantum Optimization

Chapter 3, Working with Quadratic Unconstrained Binary Optimization Problems, introduces a mathematical framework that will help us formulate combinatorial optimization problems in a way that will allow us to solve them with quantum algorithms. It also provides many examples of how to use this formalism in practice.

Chapter 4, Adiabatic Quantum Computing and Quantum Annealing, is devoted to quantum annealing, our first quantum optimization method. It starts by explaining all the mathematical details behind this algorithm and then covers all the practical aspects of using it to solve optimization problems. It also introduces Ocean, the library that we use to run programs on quantum annealers.

Chapter 5, QAOA: Quantum Approximate Optimization Algorithm, shows how to adapt the ideas behind quantum annealing to the quantum circuit model. It introduces QAOA, one of the most popular modern quantum algorithms, and studies its mathematical properties. This chapter also explains in detail how to use this algorithm with both Qiskit and PennyLane.

Chapter 6, GAS: Grover Adaptive Search, introduces Grover's algorithm and explains how to use it solve optimization problems. It focuses on designing oracles for optimization problems and on running the method with Qiskit.

Chapter 7, VQE: Variational Quantum Eigensolver, expands the applicability of the quantum optimization methods studied in the previous chapters to problems that are not combinatorial, including tasks from fields such as physics or quantum chemistry. It also shows how to run VQE with Qiskit and PennyLane, including how to use important techniques such as noise simulation and error mitigation.

Part 3, A Match Made in Heaven: Quantum Machine Learning

Chapter 8, What is Quantum Machine Learning?, gives a self-contained introduction to (classical) machine learning. It also explains the ways in which quantum computing can be used to define new machine learning methods.

Chapter 9, Quantum Support Vector Machines, studies our first quantum machine learning model: a quantum version of the famous Support Vector Machines. It explains how they can be derived mathematically and shows how to use them to solve classification problems with Qiskit and PennyLane.

Chapter 10, Quantum Neural Networks, shows how to construct quantum versions of neural networks by using variational circuits with different roles in the model. It also provides detailed examples of how to define and run these models using PennyLane and Qiskit.

Chapter 11, The Best of Both Worlds: Hybrid Architectures, is a very practical and hands-on chapter. It shows how to mix quantum and classical neural networks to create hybrid models. It also guides you through all the steps needed to implement these architectures in both PennyLane and Qiskit.

Chapter 12, Quantum Generative Adversarial Networks, shows how to create quantum generative models that are the quantum version of classical Generative Adversarial Networks (or GANs). In addition to explaining the architecture of the model, it also provides detailed examples, both in PennyLane and Qiskit, of how to use them in practice.

Afterword and Appendices

Chapter 13, Afterword: The Future of Quantum Computing, wraps up everything discussed in the book and hints at some possible developments for quantum computing in the short and medium term.

Appendix A, Complex Numbers, gives a quick recap of the most relevant properties of complex numbers and how to operate with them.

Appendix B, Basic Linear Algebra, is a refresher on the fundamentals of linear algebra, including vectors and matrices, important notions such as bases and eigenvalues, and even some concepts from modular arithmetic.

Appendix C, Computational Complexity, serves as a quick introduction to measuring the resources needed to solve problems with algorithms. It defines important concepts such as big O notation and reductions, and complexity classes such as the famous *P* and *NP*.

Appendix D, Installing the Tools, guides you through the process of installing the libraries needed in order to run the source code included in this book.

Appendix E, Production Notes, gives a glimpse of the process of writing a technical book like this one, including the software used to typeset formulas and create figures.

Assessments contains the solutions to all the exercises proposed in the main text.

Parts 2 and *3* are mostly self-contained and independent of each other (although, in the main text, we point out the connections between them whenever they exist). They can be used for self-study of quantum optimization and quantum machine learning, or to teach two short, independent courses on these topics or one full course on modern quantum algorithms. The strongest dependencies between chapters are shown in *Figure 1,* so you can know which chapters you may skip without losing track of the explanation.

To get the most out of this book

The concepts explained in this book are better understood by implementing algorithms that solve practical problems and by running them on simulators and actual quantum

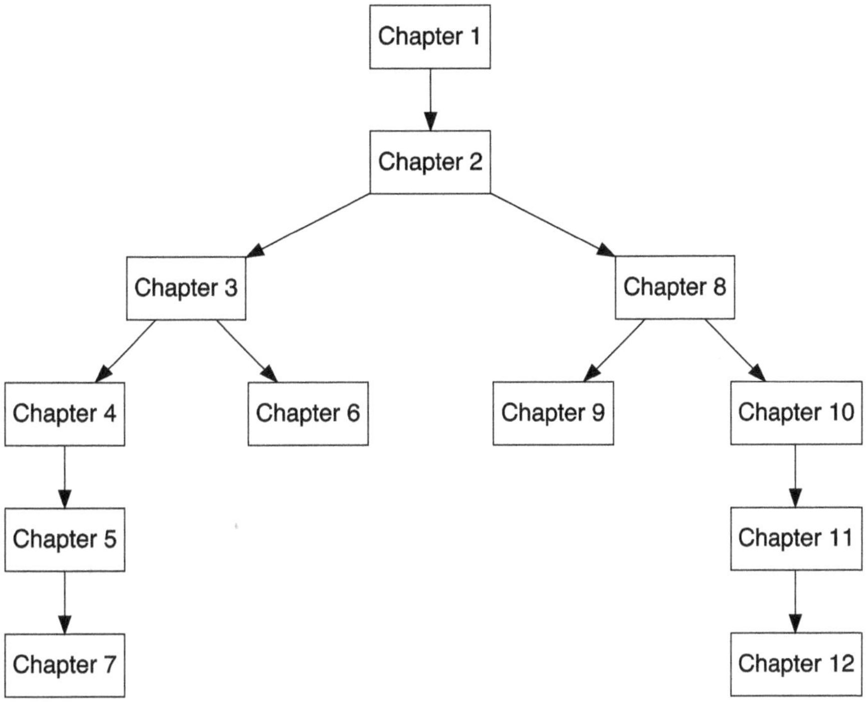

Figure 1: Strongest dependencies between chapters

computers. You will learn how to do all that starting from the very beginning of the book, but in order to run the code you will need to install some tools.

We recommend that you download the Jupyter notebooks from the link provided in the following section and that you follow the instructions given in *Appendix D, Installing the Tools,* to get your environment ready to rock!

Download the example code files

The code bundle for the book is also hosted on GitHub at `https://github.com/PacktPublishing/A-Practical-Guide-to-Quantum-Machine-Learning-and-Quantum-Optimization`. In case there's an update to the code, it will be updated on the existing GitHub repository.

We also have other code bundles from our rich catalog of books and videos available at `https://github.com/PacktPublishing/`. Check them out!

Download the color images

We also provide a PDF file that has color images of the screenshots/diagrams used in this book. You can download it here: `https://packt.link/FtU9t`.

Conventions used

There are a number of text conventions used throughout this book.

`CodeInText`: Indicates code words in text, database table names, folder names, filenames, file extensions, pathnames, dummy URLs, and user input. Here is an example: "We could create a `GroverOptimizer` object and directly use its `solve` method with `qp`."

A block of code is set as follows:

```
import dimod
J = {(0,1):1, (0,2):1}
h = {}
problem = dimod.BinaryQuadraticModel(h, J, 0.0, dimod.SPIN)
print("The problem we are going to solve is:")
print(problem)
```

Any command-line input or output is written as follows:

```
$ python3 script.py
```

Important ideas are highlighted in boxes like the following:

Important note

I am a box. I feel important. That's because I am important.

We sometimes include material for those of you who want to learn more. We format it as follows:

To learn more...

You don't have to read me if you don't want to.

There are a few exercises in the text, which are displayed as follows:

Exercise 0.1

Prove that every even number greater than two can be written as the sum of two prime numbers.

Get in touch

Feedback from our readers is always welcome.

General feedback: If you have questions about any aspect of this book, mention the book title in the subject of your message and email us at customercare@packtpub.com.

Errata: Although we have taken every care to ensure the accuracy of our content, mistakes do happen. If you have found a mistake in this book, we would be grateful if you would report this to us. Please visit `www.packtpub.com/support/errata` and fill in the form.

Piracy: If you come across any illegal copies of our works in any form on the Internet, we would be grateful if you would provide us with the location address or website name. Please contact us at copyright@packtpub.com with a link to the material.

If you are interested in becoming an author: If there is a topic that you have expertise in and you are interested in either writing or contributing to a book, please visit `authors.packtpub.com`.

Share your thoughts

Once you've read *A Practical Guide to Quantum Machine Learning and Quantum Optimization*, we'd love to hear your thoughts! Scan the QR code below to go straight to the Amazon review page for this book and share your feedback.

`https://packt.link/r/1-804-61383-5`

Your review is important to us and the tech community and will help us make sure we're delivering excellent quality content.

Download a free PDF copy of this book

Thanks for purchasing this book!

Do you like to read on the go but are unable to carry your print books everywhere? Is your eBook purchase not compatible with the device of your choice?

Don't worry, now with every Packt book you get a DRM-free PDF version of that book at no cost.

Read anywhere, any place, on any device. Search, copy, and paste code from your favorite technical books directly into your application.

The perks don't stop there, you can get exclusive access to discounts, newsletters, and great free content in your inbox daily.

Follow these simple steps to get the benefits:

1. Scan the QR code or visit the link below:

   `https://packt.link/free-ebook/9781804613832`

2. Submit your proof of purchase.
3. That's it! We'll send your free PDF and other benefits to your email directly.

Part 1

I, for One, Welcome our New Quantum Overlords

This part introduces the main concepts behind the quantum circuit model. You will learn how qubits store information, how to operate on that information with quantum gates, and how to obtain results with quantum measurements. You will also learn about some of the most important tools currently used to program quantum computers. In particular, we will discuss how to implement and execute quantum circuits with Qiskit and PennyLane.

This part includes the following chapters:

- *Chapter 1, Foundations of Quantum Computing*
- *Chapter 2, The Tools of the Trade in Quantum Computing*

1

Foundations of Quantum Computing

The beginning is always today.

— Mary Shelley

You may have heard that the mathematics needed to understand quantum computing is arcane, mysterious and difficult…but we utterly disagree! In fact, in this chapter, we will introduce all the concepts that you will need in order to follow the quantum algorithms that we will be studying in the rest of the book. Actually, you may be surprised to see that we will only rely on some linear algebra and a bit of (extremely simple) trigonometry.

We shall start by giving a quick overview of what quantum computing is, what the current state of the art is, and what the main applications are expected to be. After that, we will introduce the **model of quantum circuits**. There are several computational models for quantum computing, but this is the most popular one and, moreover, it's the one that we will be using throughout most of the book. Then, we will describe in detail what qubits are, how we can operate on them by using quantum gates, and how we can retrieve results by performing measurements. We will start with the simplest possible case — just a humble qubit! Then, we will steadily build upon that until we learn how to work with as many qubits as we want.

This chapter will cover the following topics:

- Quantum computing: the big picture
- The basics of the quantum circuit model
- Working with one qubit and the Bloch sphere
- Working with two qubits and entanglement
- Working with multiple qubits and universality

After reading this chapter, you will have acquired a solid understanding of the fundamentals of quantum computing and you will be more than ready to learn how practical quantum algorithms are developed.

1.1 Quantum computing: the big picture

In October 2019, an announcement made by a team of researchers from Google took the scientific world by storm. For the first time ever, a practical demonstration of quantum computational advantage had been shown. The results, published in the prestigious Nature journal [10], reported that a quantum computer had solved, in just a few minutes, a problem that would have taken the most powerful classical supercomputer in the world thousands of years.

Although the task solved by the quantum computer has no direct practical applications and it was later claimed that the computing time with classical resources had been overestimated (see [11] and, also, [12]), this feat remains a milestone in the history of computing and has fueled interest in quantum computing all over the world. So, what can these mysterious quantum computers do? How do they work in order to achieve these mind-blowing speed-ups?

We could define quantum computing as the study of the application of properties of quantum systems (such as superposition, entanglement, and interference) to accelerate some computational tasks. These properties do not manifest in our macroscopic world and, although they are present at the fundamental level in our computing devices, they are not explicitly used in the traditional computing models that we employ to build our microprocessors and to design our algorithms. For this reason, quantum computers behave in a radically different way to classical computers, making it possible to solve some tasks much more efficiently than with traditional computing devices.

The most famous problem for which quantum algorithms offer a huge advantage over classical methods is finding prime factors of big integers. The best known classical algorithm for this task requires an amount of time that grows almost exponentially with the length of the number (see *Appendix C, Computational Complexity*, for all the concepts referred to computational complexity, including exponential growth). Thus, factoring numbers that are several thousand bits long becomes infeasible with classical computers, and this inefficiency is the basis for some widely used cryptographic protocols, such as RSA, proposed by Rivest, Shamir, and Adleman [13].

Nevertheless, more than twenty years ago, the mathematician Peter Shor proved in a celebrated paper [6] that a quantum computer could factor numbers taking an amount of time that no longer grows exponentially with the size of the input, but only polynomially. Other examples in which quantum algorithms outperform classical ones include finding elements satisfying a given condition from an unsorted list (with Grover's algorithm [9]) or sampling from the solutions of systems of linear equations (using the famous HHL algorithm [14]).

Wonderful as the properties of these quantum algorithms are, they require quantum computers that are fault tolerant and more powerful than those available today. This is why, in the last few years, many researchers have focused on studying quantum algorithms that try to obtain some advantage with the noisy intermediate-scale quantum computers, also known as **NISQ devices**, that are at our disposal now. The **NISQ** name was coined by

John Preskill in a greatly enjoyable article [15] and has been widely adopted to describe the *evolutionary stage* in which quantum hardware currently is.

Machine learning and optimization are two of the fields that are being actively explored in this NISQ era. In these areas, many interesting algorithms have been proposed in recent years; some examples are the **Quantum Approximate Optimization Algorithm** (**QAOA**), the **Variational Quantum Eigensolver** (**VQE**), or different quantum flavors of machine learning models, including **Quantum Support Vector Machines** (**QSVMs**) and **Quantum Neural Networks** (**QNNs**).

Since these algorithms are fairly new, we still lack a complete understanding of their full capabilities. However, some partial theoretical results show some evidence that these approaches can offer some advantages over what is possible with classical computers, for instance, by giving us better approximations to the solutions of hard **combinatorial optimization problems** or by showing better performance when learning from particular **datasets**.

Exploring the real possibilities of these NISQ computers and the algorithms designed to take advantage of them will be crucial in the short and medium term, and it may very likely pave the way for the first practical applications of quantum computing to real-world problems.

We believe that you can be part of the exciting task of making quantum computing applications a reality and we would like to help you on that journey. But, for that, we need to start by setting in place the tools that we will be using throughout the book.

If you are already familiar with the quantum circuit model, you can skip the rest of this chapter. However, we recommend that you at least skim through the following sections so that you can get familiar with the conventions and choices of notation that we will use in this book.

1.2 The basics of the quantum circuit model

We have mentioned that quantum computing relies on quantum phenomena such as **superposition**, **entanglement**, and **interference** to perform computations. But what does this really mean? To make this explicit, we need to define a particular computational model that allow us to describe mathematically how to take advantage of all these properties.

There are many such models, including **quantum Turing machines**, **measurement-based quantum computing** (also known as **one-way quantum computing**), or **adiabatic quantum computing**, and all of them are equivalent in power. However, the most popular one — and the one that we will be using for the most part in the book — is the **quantum circuit model**.

To learn more...

In addition to the quantum circuit model, sometimes we will also use the adiabatic model. All the necessary concepts will be introduced in *Chapter 4, Quantum Adiabatic Computing and Quantum Annealing.*

Every computation has three elements: **data**, **operations**, and **output**. In the quantum circuit model, these correspond to some concepts that you may have already heard about: **qubits**, **quantum gates**, and **measurements**. Through the remainder of this chapter, we will briefly review all of them, highlighting some special details that will be of particular importance when talking about quantum machine learning and quantum optimization algorithms; at the same time, we will show the notation that will be used throughout the book. But before committing to that, let us have a quick overview of what a **quantum circuit** is.

Let's have a look at *Figure 1.1*. It shows a simple quantum circuit. The three horizontal lines that you see are sometimes called **wires**, and they represent the qubits that we are working with. Thus, in this case, we have three qubits. The circuit is meant to be read from left to right, and it represents all the different operations that are performed on the qubits. It is customary to assume that, at the very beginning, all the qubits are in state $|0\rangle$. You do not need to worry yet about what $|0\rangle$ means, but please notice how we have indicated that this is indeed the initial state of all the wires by writing $|0\rangle$ to the left of each of them.

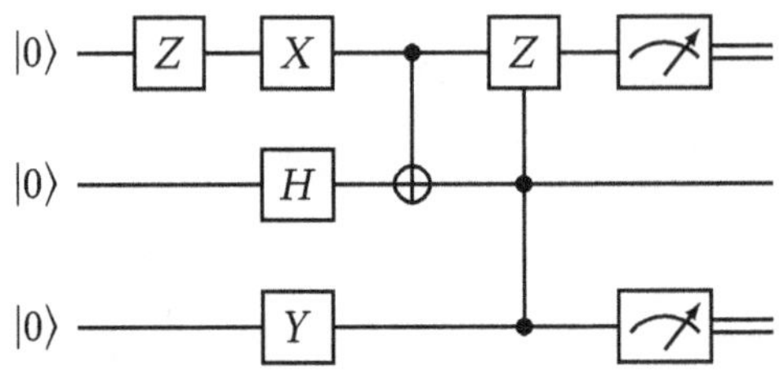

Figure 1.1: An example of a simple quantum circuit.

In that circuit, we start by applying an operation called a Z gate on the top qubit; we will explain in the next section what all of these operations do, but note that we represent them with little boxes with the name of the operation inside. After that initial Z gate, we apply individual gates X, H, and Y on the top, middle, and bottom qubits and, then, a two-qubit gate on the top and middle qubits followed by a three-qubit gate, which acts on all the qubits at the same time. Finally, we measure the top and bottom qubits (we will get to measurements in the next section, don't worry), and we represent this in the circuit using the **gauge symbol**. Notice that, after these measurements, the wires are represented with double lines, to indicate that we have obtained a result – technically, we say that the state of the qubit has **collapsed** to a classical value. This means that, from this point on, we do not have quantum data anymore, only classical bits. This collapse may seem a little bit mysterious (it is!), but don't worry. In the next section, we will explain in detail the process by which quantum information (qubits) is transformed into classical data (bits).

As you may have noticed, quantum circuits are somewhat similar to digital ones, in which we have wires representing bits and different logical gates such as **AND**, **OR**, and **NOT** acting on them. However, our qubits, quantum gates, and measurements obey the rules of quantum mechanics and show some properties that are not found in classical circuits. The rest of this chapter is devoted to explaining all of this in detail, starting with the simplest of cases, that of a single qubit, but growing all the way up to fully-fledged quantum circuits that can use as many qubits and gates as desired.

Ready? Let's start, then!

1.3 Working with one qubit and the Bloch sphere

One of the advantages of using a computational model is that you can forget about the particularities of the physical implementation of your computer and focus instead on the properties of the elements on which you store information and the operations you can perform on them. For instance, we could define a qubit as a (physical) quantum system that is capable of being in two different states. In practice, it could be a photon with two possible polarizations, a particle with two possible values for its spin, or a superconducting circuit, whose current can be flowing in one of two directions. When using the quantum circuit model, we can forget about those implementation details and just define a qubit... as a mathematical vector!

1.3.1 What is a qubit?

In fact, a **qubit** (short for **quantum bit**, sometimes also written as **qbit**, **Qbit** or even **q-bit**) is the minimal information unit in quantum computing. In the same way that a **bit** (short for **binary digit**) can be in state 0 or in state 1, a qubit can be in state $|0\rangle$ or in state $|1\rangle$. Here, we are using the so-called **Dirac notation**, where these funny-looking symbols surrounding 0 and 1 are called **kets** and are used to indicate that we are dealing with vectors instead of regular numbers. In fact, $|0\rangle$ and $|1\rangle$ are not the only possibilities for the state of a qubit and, in general, it could be in a **superposition** of the form

$$a\left|0\right\rangle + b\left|1\right\rangle,$$

where a and b are complex numbers, called **amplitudes**, such that $|a|^2 + |b|^2 = 1$. The quantity $\sqrt{|a|^2 + |b|^2}$ is called the **norm** or **length** of the state and, when it is equal to 1, we say that the state is **normalized**.

> **To learn more...**
>
> If you need a refresher on complex numbers or vector spaces, please check *Appendix A, Complex Numbers*, and *Appendix B, Basic Linear Algebra.*

All these possible values for the state of a single qubit are vectors that live in a complex vector space of dimension 2 (in fact, they live in what is called a **Hilbert space**, but since we will be working only with finite dimensions, there is no real difference). Thus we shall fix the vectors $|0\rangle$ and $|1\rangle$ as elements of a special **basis**, which we will refer to as the **computational basis**. We will represent these vectors, constituents of the computational basis, as the column vectors

$$|0\rangle = \begin{pmatrix} 1 \\ 0 \end{pmatrix}, \qquad |1\rangle = \begin{pmatrix} 0 \\ 1 \end{pmatrix}$$

and hence

$$a|0\rangle + b|1\rangle = a\begin{pmatrix} 1 \\ 0 \end{pmatrix} + b\begin{pmatrix} 0 \\ 1 \end{pmatrix} = \begin{pmatrix} a \\ b \end{pmatrix}.$$

If we are given a qubit and we want to determine or, rather, estimate its state, all we can do is perform a measurement and get one of two possible results: 0 or 1. We have nonetheless seen how a qubit can be in infinitely many states, so how does the state of a qubit determine the outcome of a measurement? As you likely already know, in quantum physics, these measurements are not deterministic, but probabilistic. In particular, given any qubit $a|0\rangle + b|1\rangle$, the probability of getting 0 upon a measurement is $|a|^2$, while that of getting 1 is $|b|^2$. Naturally, these two probabilities must add up to 1, hence the need for the **normalization condition** $|a|^2 + |b|^2 = 1$.

If upon measuring a qubit we get, let's say, 0, we then know that, after the measurement, the state of the qubit is $|0\rangle$, and we say that the qubit has **collapsed** into that state. If we obtain 1, the state collapses to $|1\rangle$. Since we are obtaining results that correspond to $|0\rangle$ and $|1\rangle$, we say that we are **measuring in the computational basis**.

Exercise 1.1

What is the probability of measuring 0 if the state of a qubit is $\sqrt{1/2}\,|0\rangle + \sqrt{1/2}\,|1\rangle$? And the probability of measuring 1? What if the state of the qubit is $\sqrt{1/3}\,|0\rangle + \sqrt{2/3}\,|1\rangle$? And if it is $\sqrt{1/2}\,|0\rangle - \sqrt{1/2}\,|1\rangle$?

So a qubit is, mathematically, just a 2-dimensional vector that satisfies a normalization condition. Who could have known? But the surprises do not end here. In the next subsection, we will see how we can use those funny-looking kets to compute inner products in a very easy way.

1.3.2 Dirac notation and inner products

Dirac notation can not only be used for column vectors, but also for row vectors. In that case, we talk of **bras**, which, together with kets, can be used to form **bra-kets**. This name is a pun, because, as we are about to show, bra-kets are, in fact, inner products that are written — you guessed it — between brackets. To be more mathematically precise, with each ket we can associate a bra that is its **adjoint** or **conjugate transpose** or **Hermitian transpose**. In order to obtain this adjoint, we take the ket's column vector, we transpose it and conjugate each of its coordinates (which are, as we already know, complex numbers). We use $\langle 0|$ to denote the bra associated with $|0\rangle$ and $\langle 1|$ to denote the bra associated with $|1\rangle$, so we have

$$\langle 0| = |0\rangle^{\dagger} = \begin{pmatrix} 1 \\ 0 \end{pmatrix}^{\dagger} = \begin{pmatrix} 1 & 0 \end{pmatrix}, \qquad \langle 1| = |1\rangle^{\dagger} = \begin{pmatrix} 0 \\ 1 \end{pmatrix}^{\dagger} = \begin{pmatrix} 0 & 1 \end{pmatrix}$$

and, in general,

$$a\langle 0| + b\langle 1| = a|0\rangle^\dagger + b|1\rangle^\dagger = a\begin{pmatrix} 1 & 0 \end{pmatrix} + b\begin{pmatrix} 0 & 1 \end{pmatrix} = \begin{pmatrix} a & b \end{pmatrix},$$

where, as it is customary, we use the dagger symbol (†) for the adjoint.

Important note

When finding the adjoint, do not forget to conjugate the complex numbers! For instance, it holds that

$$\begin{pmatrix} \frac{1-i}{2} \\ \frac{i}{\sqrt{2}} \end{pmatrix}^\dagger = \begin{pmatrix} \frac{1+i}{2} & \frac{-i}{\sqrt{2}} \end{pmatrix}.$$

One of the reasons why Dirac notation is so popular for working with quantum systems is that, by using it, we can easily compute the inner products of kets and bras. For instance, we can readily show that

$$\langle 0|0\rangle = \begin{pmatrix} 1 & 0 \end{pmatrix}\begin{pmatrix} 1 \\ 0 \end{pmatrix} = 1, \qquad \langle 0|1\rangle = \begin{pmatrix} 1 & 0 \end{pmatrix}\begin{pmatrix} 0 \\ 1 \end{pmatrix} = 0,$$

$$\langle 1|0\rangle = \begin{pmatrix} 0 & 1 \end{pmatrix}\begin{pmatrix} 1 \\ 0 \end{pmatrix} = 0, \qquad \langle 1|1\rangle = \begin{pmatrix} 0 & 1 \end{pmatrix}\begin{pmatrix} 0 \\ 1 \end{pmatrix} = 1.$$

This proves that $|0\rangle$ and $|1\rangle$ are not just elements of any basis but of an **orthonormal** one, since $|0\rangle$ and $|1\rangle$ are orthogonal and of length 1. Thus, we can compute the inner product of two states $|\psi_1\rangle = a|0\rangle + b|1\rangle$ and $|\psi_2\rangle = c|0\rangle + d|1\rangle$ using Dirac notation by noting that

$$\begin{aligned}\langle \psi_1|\psi_2\rangle &= (a^*\langle 0| + b^*\langle 1|)(c|0\rangle + d|1\rangle) \\ &= a^*c\langle 0|0\rangle + a^*d\langle 0|1\rangle + b^*c\langle 1|0\rangle + b^*d\langle 1|1\rangle \\ &= a^*c + b^*d,\end{aligned}$$

where a^* and b^* are the complex conjugates of a and b.

Exercise 1.2

What is the inner product of $\sqrt{1/2}\,|0\rangle + \sqrt{1/2}\,|1\rangle$ and $\sqrt{1/3}\,|0\rangle + \sqrt{2/3}\,|1\rangle$? And the inner product of $\sqrt{1/2}\,|0\rangle + \sqrt{1/2}\,|1\rangle$ and $\sqrt{1/2}\,|0\rangle - \sqrt{1/2}\,|1\rangle$?

To learn more...

Notice that, if $|\psi\rangle = a\,|0\rangle + b\,|1\rangle$, then $|\langle 0|\psi\rangle|^2 = |a|^2$, which is the probability of measuring 0 if the state is $|\psi\rangle$. This is not accidental. In *Chapter 7, VQE: Variational Quantum Eigensolver*, for example, we will use measurements in orthonormal bases other than the computational one, and we will see how, in that case, the probability of measuring the result associated to an element $|\varphi\rangle$ of a given orthonormal basis is exactly $|\langle\varphi|\psi\rangle|^2$.

We now know what qubits are, how to measure them, and even how to benefit from Dirac notation for some useful computations. The only thing remaining is to study how to operate on qubits. Are you ready? It is time for us to get you introduced to the mighty quantum gates!

1.3.3 One-qubit quantum gates

So far, we have focused on how a qubit stores information in its state and on how we can access (part of) that information with measurements. But in order to develop useful algorithms, we also need some way of manipulating the state of qubits to perform computations.

Since a qubit is, fundamentally, a quantum system, its evolution follows the laws of quantum mechanics. More precisely, if we suppose that our system is isolated from its environment, it obeys the famous **Schrödinger equation**.

To learn more...

The time-independent Schrödinger equation can be written as

$$H\left|\psi(t)\right\rangle = i\hbar\frac{\partial}{\partial t}\left|\psi(t)\right\rangle,$$

where H is the **Hamiltonian** of the system, $\left|\psi(t)\right\rangle$ is the state vector of the system at time t, i is the imaginary unit, and $\hbar$ is the reduced Planck constant.

We will talk more about Hamiltonians in *Chapter 3, QUBO: Quadratic Unconstrained Binary Optimization, Chapter 4, Quantum Adiabatic Computing and Quantum Annealing,* and *Chapter 7, VQE: Variational Quantum Eigensolver.*

Don't panic! To program a quantum computer, you don't need to know how to solve Schrödinger's equation. In fact, the only thing that you need to know is that its solutions are always a special type of linear transformations. For the purposes of the quantum circuit model, since we are working in finite-dimensional spaces and we have fixed a basis, the operations can be described by matrices that are applied to the vectors that represent the states of the qubits.

But not any kind of matrix does the trick. According to quantum mechanics, the only matrices that we can use are the so-called **unitary** matrices, which are the matrices U such that

$$U^{\dagger}U = UU^{\dagger} = I,$$

where I is the identity matrix and $U^{\dagger}$ is the adjoint of U, that is, the matrix obtained by transposing U and replacing each element by its complex conjugate. This means that any unitary matrix U is invertible and its inverse is given by $U^{\dagger}$. In the context of the quantum circuit model, the operations represented by these matrices are called quantum gates.

To learn more...

It is relatively easy to check that unitary matrices preserve vector lengths (see, for instance, Section *5.7.5* in *Dancing with Qubits*, by Robert Sutor [7]). That is, if U is a unitary matrix and $|\psi\rangle$ is a quantum state (and, hence, its norm is 1, as we already know) then $U|\psi\rangle$ also is a valid quantum state because its norm is still 1. For this reason, we can safely apply unitary matrices to our quantum states and rest assured that the resulting states will satisfy the normalization condition.

When we have just one qubit, our unitary matrices need to be of size 2×2 because the state vector is of dimension 2. Thus, the simplest example of a quantum gate is the identity matrix of dimension 2, which transforms the state of the qubit by... well, by not transforming it at all. A less boring example is the X gate, whose matrix is given by

$$X = \begin{pmatrix} 0 & 1 \\ 1 & 0 \end{pmatrix}.$$

The X gate is also called the **NOT** gate, because its action on the elements of the computational basis is

$$X|0\rangle = \begin{pmatrix} 0 & 1 \\ 1 & 0 \end{pmatrix}\begin{pmatrix} 1 \\ 0 \end{pmatrix} = \begin{pmatrix} 0 \\ 1 \end{pmatrix} = |1\rangle, \qquad X|1\rangle = \begin{pmatrix} 0 & 1 \\ 1 & 0 \end{pmatrix}\begin{pmatrix} 0 \\ 1 \end{pmatrix} = \begin{pmatrix} 1 \\ 0 \end{pmatrix} = |0\rangle,$$

which is exactly what the NOT gate does in classical digital circuits.

Exercise 1.3

Check that the gate X matrix is, indeed, unitary. What is the inverse of X? What is the action of X on a general qubit in a state of the form $a|0\rangle + b|1\rangle$?

A quantum gate with no classical analog is the **Hadamard** or H gate, given by

$$H = \begin{pmatrix} \frac{1}{\sqrt{2}} & \frac{1}{\sqrt{2}} \\ \frac{1}{\sqrt{2}} & -\frac{1}{\sqrt{2}} \end{pmatrix} = \frac{1}{\sqrt{2}} \begin{pmatrix} 1 & 1 \\ 1 & -1 \end{pmatrix}.$$

This gate is extremely useful in quantum computing, for it can create superposition. To be precise, if we apply the H gate on a qubit in state $|0\rangle$, we obtain

$$H|0\rangle = \frac{1}{\sqrt{2}}|0\rangle + \frac{1}{\sqrt{2}}|1\rangle = \frac{1}{\sqrt{2}}(|0\rangle + |1\rangle).$$

This state is so important that it has its own name and symbol. It is called the **plus** state and it is denoted by $|+\rangle$. In a similar way, we have that

$$H|1\rangle = \frac{1}{\sqrt{2}}(|0\rangle - |1\rangle)$$

and, as you probably guessed, this state is called the **minus** state and it is denoted by $|-\rangle$.

Exercise 1.4

Check that the gate H matrix is, indeed, unitary. What is the action of H on $|+\rangle$ and $|-\rangle$? What is the action of X on $|+\rangle$ and $|-\rangle$?

Of course, we can apply several gates to the same qubit one after the other. For instance, consider the following circuit:

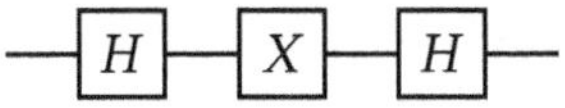

We read gates from left to right, so in the preceding circuit we would first apply an H gate, then an X gate and, finally, another H gate. You can easily check that, if the initial state of the qubit is $|0\rangle$, it would end up again in state $|0\rangle$. But were its initial state $|1\rangle$, the final state would become $-|1\rangle$.

It turns out that this operation is also very important, and, of course, it has its own name: we call it the Z gate. From its action on $|0\rangle$ and $|1\rangle$, we can tell that its matrix will be

$$Z = \begin{pmatrix} 1 & 0 \\ 0 & -1 \end{pmatrix},$$

something that we could have also deduced by multiplying the matrices of the gates H, X, and H one after the other.

Exercise 1.5

Check that $Z|0\rangle = |0\rangle$ and that $Z|1\rangle = -|1\rangle$ in two different ways. First, use Dirac notation and the actions of H and X (remember that we have defined Z as HXH). Then, derive the same result by performing the matrix multiplication

$$\begin{pmatrix} \frac{1}{\sqrt{2}} & \frac{1}{\sqrt{2}} \\ \frac{1}{\sqrt{2}} & -\frac{1}{\sqrt{2}} \end{pmatrix}\begin{pmatrix} 0 & 1 \\ 1 & 0 \end{pmatrix}\begin{pmatrix} \frac{1}{\sqrt{2}} & \frac{1}{\sqrt{2}} \\ \frac{1}{\sqrt{2}} & -\frac{1}{\sqrt{2}} \end{pmatrix}.$$

Since there are X and Z gates, you may be wondering if there is also a Y gate. Indeed, there is one, given by matrix

$$Y = \begin{pmatrix} 0 & -i \\ i & 0 \end{pmatrix}.$$

To learn more...

The set $\{I, X, Y, Z\}$, known as the set of **Pauli matrices**, is of great importance in quantum computing. One of its many interesting properties is that it constitutes a basis of the vector space of 2×2 complex matrices. We will work with it in *Chapter 7, VQE: Variational Quantum Eigensolver*, for instance.

Other important one-qubit gates include the S and T gates, whose matrices are

$$S = \begin{pmatrix} 1 & 0 \\ 0 & e^{i\frac{\pi}{2}} \end{pmatrix}, \qquad T = \begin{pmatrix} 1 & 0 \\ 0 & e^{i\frac{\pi}{4}} \end{pmatrix}.$$

But, of course, there is an (uncountably!) infinite number of 2-dimensional unitary matrices and we cannot just list them all here. What we will do instead is introduce a beautiful geometrical representation of single-qubit states, and, with it, we will explain how all one-qubit quantum gates can, in fact, be understood as certain kinds of rotations. Enter the Bloch sphere!

Exercise 1.6

Check that $T^2 = S$. Then, use the most beautiful formula ever (i.e., Euler's identity $e^{i\pi} + 1 = 0$) to check that $S^2 = Z$. Check also that S and T are unitary. Express $S^\dagger$ and $T^\dagger$ as powers of S and T.

1.3.4 The Bloch sphere and rotations

The general state of a qubit is described with two complex numbers. Since each of those numbers has two real components, it would be natural to think that we would need a four-dimensional real space in order to represent the state of a qubit. Surprisingly enough, all the possible states of a qubit can be drawn on the surface of an old-school sphere, which is a two-dimensional object!

To show how it can be accomplished, we need to remember that a complex number z can be written in polar coordinates as

$$z = re^{i\alpha},$$

where $r = |z|$ is a non-negative real number and α is an angle in $[0, 2\pi]$. Consider, then, a qubit in a state $|\psi\rangle = a|0\rangle + b|1\rangle$ and write a and b in polar coordinates as

$$a = r_1 e^{i\alpha_1}, \qquad b = r_2 e^{i\alpha_2}.$$

We know that $r_1^2 + r_2^2 = |a|^2 + |b|^2 = 1$ and, since $0 \leq r_1, r_2 \leq 1$, there must exist an angle θ in $[0, \pi]$ such that $\cos(\theta/2) = r_1$ and $\sin(\theta/2) = r_2$. The reason for considering $\theta/2$ instead of θ in the cosine and sine will be apparent in a moment. Notice that, by now, we have

$$|\psi\rangle = \cos\frac{\theta}{2} e^{i\alpha_1} |0\rangle + \sin\frac{\theta}{2} e^{i\alpha_2} |1\rangle .$$

Another crucial observation is that we can multiply $|\psi\rangle$ by a complex number c with absolute value 1 without changing its state. Indeed, it is easy to see that c does not affect the probabilities of obtaining 0 and 1 when measuring in the computational basis (check it!) and, by linearity, it comes out when applying a quantum gate U (that is, $U(c|\psi\rangle) = cU|\psi\rangle$). Thus, there is no operation — either unitary transformation or measurement — that allows us to distinguish $|\psi\rangle$ from $c|\psi\rangle$. We call c a **global phase** and we have just shown that it is physically irrelevant.

Important note

Notice, however, that **relative** phases are, unlike global ones, really relevant! For instance, $|+\rangle = \frac{1}{\sqrt{2}}(|0\rangle + |1\rangle)$ and $|-\rangle = \frac{1}{\sqrt{2}}(|0\rangle - |1\rangle)$ differ just in the phase of $|1\rangle$, but we can easily distinguish between them by first applying H to those states and then measuring them in the computational basis.

We can, thus, multiply $|\psi\rangle$ by $e^{-i\alpha_1}$ to obtain an equivalent representation

$$|\psi\rangle = \cos\frac{\theta}{2} |0\rangle + \sin\frac{\theta}{2} e^{i\varphi} |1\rangle ,$$

where we have defined $\varphi = \alpha_2 - \alpha_1$.

In this way, we can describe the state of any qubit with just two numbers $\theta \in [0, \pi]$ and $\varphi \in [0, 2\pi]$ that we can interpret as a polar angle and an azimuthal angle, respectively (that is, we are using what are known as **spherical coordinates**). This gives us a three-dimensional point

$$(\sin\theta\cos\varphi, \sin\theta\sin\varphi, \cos\theta)$$

that locates the state of the qubit on the surface of a sphere, called the **Bloch sphere** (see *Figure 1.2*).

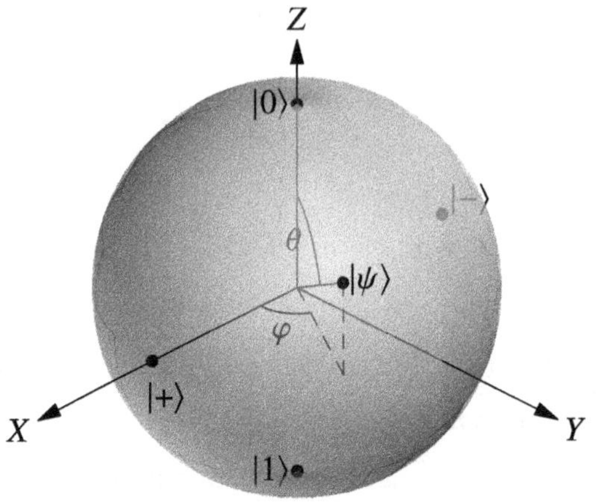

Figure 1.2: Qubit state $|\psi\rangle$ represented on the Bloch sphere.

Notice that θ runs from 0 to π to cover the whole range from the top to the bottom of the sphere. This is why we used $\theta/2$ in the representation of our preceding qubit. We only needed to get up to $\pi/2$ for our angles in the sines and cosines!

In the Bloch sphere, $|0\rangle$ is mapped to the North pole and $|1\rangle$ to the South pole. In general, states that are orthogonal with respect to the inner product are antipodal on the sphere. For instance, $|+\rangle$ and $|-\rangle$ both lie on the equator, but on opposite points of the sphere. As we already know, the X gate takes $|0\rangle$ to $|1\rangle$ and $|1\rangle$ to $|0\rangle$, but leaves $|+\rangle$ and $|-\rangle$ unchanged, at least up to an irrelevant global phase. In fact, this means that the X gate acts like a rotation of π radians around the X axis of the Bloch sphere..., so now you know why we use that name for the gate! In the same manner, Z and Y are rotations of π radians around the Z and Y axes, respectively.

We can generalize this behavior to obtain rotations of any angle around any axis of the Bloch sphere. For instance, for the X, Y, and Z axes we may define

$$R_X(\theta) = e^{-i\frac{\theta}{2}X} = \cos\frac{\theta}{2}I - i\sin\frac{\theta}{2}X = \begin{pmatrix} \cos\frac{\theta}{2} & -i\sin\frac{\theta}{2} \\ -i\sin\frac{\theta}{2} & \cos\frac{\theta}{2} \end{pmatrix},$$

$$R_Y(\theta) = e^{-i\frac{\theta}{2}Y} = \cos\frac{\theta}{2}I - i\sin\frac{\theta}{2}Y = \begin{pmatrix} \cos\frac{\theta}{2} & -\sin\frac{\theta}{2} \\ \sin\frac{\theta}{2} & \cos\frac{\theta}{2} \end{pmatrix},$$

$$R_Z(\theta) = e^{-i\frac{\theta}{2}Z} = \cos\frac{\theta}{2}I - i\sin\frac{\theta}{2}Z = \begin{pmatrix} e^{-i\frac{\theta}{2}} & 0 \\ 0 & e^{i\frac{\theta}{2}} \end{pmatrix} \equiv \begin{pmatrix} 1 & 0 \\ 0 & e^{i\theta} \end{pmatrix},$$

where we use the $\equiv$ symbol for equivalent action up to a global phase. Notice that $R_X(\pi) \equiv X$, $R_Y(\pi) \equiv Y$, $R_Z(\pi) \equiv Z$, $R_Z(\frac{\pi}{2}) \equiv S$, and $R_Z(\frac{\pi}{4}) \equiv T$.

Exercise 1.7

Check these equivalences by substituting the angles in the definitions of R_X, R_Y, and R_Z.

In fact, it can be proved (see, for instance, the book by Nielsen and Chuang [16]) that for any one-qubit gate U there exists a unit vector $r = (r_x, r_y, r_z)$ and an angle θ such that

$$U \equiv \cos\frac{\theta}{2}I - i\sin\frac{\theta}{2}(r_xX + r_yY + r_zZ).$$

For example, choosing $\theta = \pi$ and $r = (1/\sqrt{2}, 0, 1/\sqrt{2})$ we can obtain the Hadamard gate, for it holds that

$$H \equiv -i\frac{1}{\sqrt{2}}(X + Z).$$

Additionally, it can also be proved that, again for any one-qubit gate U, there exist three angles α, β, and γ such that

$$U \equiv R_Z(\alpha)R_Y(\beta)R_Z(\gamma).$$

In fact, you can obtain such a decomposition for any two rotation axes as long as they are not parallel, not just for Y and Z.

Moreover, in some quantum computing architectures (including the ones used by companies such as IBM), it is common to use a **universal** one-qubit gate, called the U-**gate**, that depends on three angles and is able to generate any other one-qubit gate. Its matrix is

$$U(\theta, \varphi, \lambda) = \begin{pmatrix} \cos\frac{\theta}{2} & -e^{i\lambda}\sin\frac{\theta}{2} \\ e^{i\varphi}\sin\frac{\theta}{2} & e^{i(\varphi+\lambda)}\cos\frac{\theta}{2} \end{pmatrix}.$$

Exercise 1.8

Show that $U(\theta, \varphi, \lambda)$ is unitary. Check that $R_X(\theta) = U(\theta, -\pi/2, \pi/2)$, that $R_Y(\theta) = U(\theta, 0, 0)$ and that, up to a global phase, $U(\theta) = R_Z(0, 0, \theta)$.

All these observations about how to construct one-qubit gates from rotations and parametrized families will be very important when we talk about variational forms and feature maps in *Chapter 9, Quantum Support Vector Machines*, and *Chapter 10, Quantum Neural Networks*, and also to construct controlled gates later in this chapter.

1.3.5 Hello, quantum world!

To put together everything that we have learned, we are going to create our very first complete quantum circuit. It looks like this:

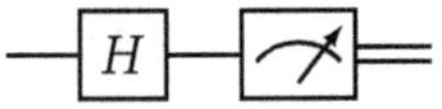

It doesn't seem very impressive, but let's analyze it part by part. As you know, following convention, the initial state of our qubit is assumed to be $|0\rangle$, so that's what we have before we do anything. Then we apply the H gate, so the state changes to $\sqrt{1/2}\,|0\rangle + \sqrt{1/2}\,|1\rangle$. Finally, we measure the qubit. The probability of obtaining 0 will be $\left|\sqrt{1/2}\right|^2 = 1/2$, and that of getting 1 will also be $1/2$, so we have created a circuit that — at least in theory — generates random bits following a perfectly uniform distribution.

To learn more...

Unbiased uniform bit distributions are of great relevance for multiple applications in simulation, cryptography, and even online gambling games. As we will learn in *Chapter 2, The Tools of the Trade*, current quantum computers deviate from this equilibrium because they are affected by noise and gate and measurement errors. However, protocols to extract perfect random bits even with noisy quantum computers have been proposed and could become one of the first practical applications of quantum computing (see, for instance, the paper by Acín and Masanes [17]).

We can modify the previous circuit to obtain any distribution over 0 and 1 that we desire. If we want the probability of measuring 0 to be $p \in [0, 1]$, we just need to consider $\theta = 2 \arccos \sqrt{p}$ and the following circuit:

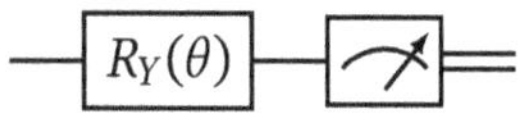

Exercise 1.9

Check that, with the preceding circuit, the state before measurement is $\sqrt{p}\,|0\rangle + \sqrt{1-p}\,|1\rangle$ and, hence, the probability of measuring 0 is p and that of measuring 1 is $1 - p$.

For now, this is all that we need to know about one-qubit states, gates, and measurements. Let us move on to two-qubit systems, where the mysteries of entanglement are awaiting to be revealed!

1.4 Working with two qubits and entanglement

Now that we have mastered the inner workings of solitary qubits, we are ready to up the ante. In this section, we will learn about systems of two qubits and how they can become entangled. We will first define the mathematical representation of two-qubit systems and how we can measure them. After that, we will study different quantum gates that can act on two qubits at once and we will have a look at some of their very interesting and slightly puzzling properties. We will conclude with a simple but enlightening example of a two-qubit circuit. We promise that the ride is going to be amazing!

1.4.1 Two-qubit states

So far, we have worked with qubits in isolation. But the real power of quantum computing cannot be unleashed unless qubits can talk to each other. We will start by considering the simplest case of quantum systems in which there is qubit interaction: two-qubit systems.

Of course, in a two-qubit system, each of the qubits can be in state $|0\rangle$ or in state $|1\rangle$. Thus, for the two qubits, we have four possible combinations: both are in state $|0\rangle$, the first one is in state $|0\rangle$ and the second one in state $|1\rangle$, the first one is in state $|1\rangle$ and the second one in state $|0\rangle$, or both are in state $|1\rangle$. These four possibilities form a basis (called the **computational basis**) of a 4-dimensional space and we denote them, respectively, by

$$|0\rangle \otimes |0\rangle, \ |0\rangle \otimes |1\rangle, \ |1\rangle \otimes |0\rangle, \ |1\rangle \otimes |1\rangle.$$

Here, $\otimes$ is the symbol for the **tensor product**. The tensor product of two column vectors is defined by

$$\begin{pmatrix} a_1 \\ a_2 \\ \vdots \\ a_n \end{pmatrix} \otimes \begin{pmatrix} b_1 \\ b_2 \\ \vdots \\ b_m \end{pmatrix} = \begin{pmatrix} a_1 \begin{pmatrix} b_1 \\ b_2 \\ \vdots \\ b_m \end{pmatrix} \\ a_2 \begin{pmatrix} b_1 \\ b_2 \\ \vdots \\ b_m \end{pmatrix} \\ \vdots \\ a_n \begin{pmatrix} b_1 \\ b_2 \\ \vdots \\ b_m \end{pmatrix} \end{pmatrix} = \begin{pmatrix} a_1 b_1 \\ a_1 b_2 \\ \vdots \\ a_1 b_m \\ a_2 b_1 \\ a_2 b_2 \\ \vdots \\ a_2 b_m \\ \vdots \\ a_n b_1 \\ a_n b_2 \\ \vdots \\ a_n b_m \end{pmatrix}.$$

Hence, the four basis states can be represented by four-dimensional column vectors given by

$$|0\rangle \otimes |0\rangle = \begin{pmatrix} 1 \\ 0 \\ 0 \\ 0 \end{pmatrix}, \qquad |0\rangle \otimes |1\rangle = \begin{pmatrix} 0 \\ 1 \\ 0 \\ 0 \end{pmatrix}, \qquad |1\rangle \otimes |0\rangle = \begin{pmatrix} 0 \\ 0 \\ 1 \\ 0 \end{pmatrix}, \qquad |1\rangle \otimes |1\rangle = \begin{pmatrix} 0 \\ 0 \\ 0 \\ 1 \end{pmatrix}.$$

Usually, we omit the $\otimes$ symbol and just write

$$|0\rangle\,|0\rangle\,,\ |0\rangle\,|1\rangle\,,\ |1\rangle\,|0\rangle\,,\ |1\rangle\,|1\rangle$$

or

$$|00\rangle\,,\ |01\rangle\,,\ |10\rangle\,,\ |11\rangle$$

or even

$$|0\rangle\,,\ |1\rangle\,,\ |2\rangle\,,\ |3\rangle\,.$$

Obviously, in this last case, the number of qubits that we are using must be clear from the context in order not to mistake the state $|0\rangle$ of a one-qubit system with the state $|0\rangle$ of a two-qubit system — or, as we will see soon, any other multi-qubit system!

As we have mentioned, these four states constitute a basis of the vector space of possible states for a two-qubit system. The general expression for the state of such a system is

$$|\psi\rangle = a_{00}\,|00\rangle + a_{01}\,|01\rangle + a_{10}\,|10\rangle + a_{11}\,|11\rangle$$

where a_{00}, a_{01}, a_{10}, and a_{11} are complex numbers (called amplitudes, remember?) such that $\sum_{x,y=0}^{1} |a_{xy}|^2 = 1$.

If we measure in the computational basis both qubits at this generic state that we are considering, we will obtain 00 with probability $|a_{00}|^2$, 01 with probability $|a_{01}|^2$, 10 with probability $|a_{10}|^2$, and 11 with probability $|a_{11}|^2$. In all those cases, the state will collapse to the state corresponding to the outcome of the measurement, just as with one-qubit systems.

Let's now say that we only measure one of the qubits. What happens then? Suppose that we measure the first qubit. Then, the probability of obtaining 0 will be $|a_{00}|^2 + |a_{01}|^2$, which is the sum of the probabilities of all the outcomes in which the first qubit can be 0. If we measure the first qubit and the result turns out to be 0, the system will not collapse completely, but it will remain in the state

$$\frac{a_{00}\,|00\rangle + a_{01}\,|01\rangle}{\sqrt{|a_{00}|^2 + |a_{01}|^2}},$$

where we have divided by $\sqrt{|a_{00}|^2 + |a_{01}|^2}$ to keep the state normalized. The situation in which the result of the measurement is 1 is analogous.

Exercise 1.10

Derive the formulas for the probability of measuring 1 on the first qubit in a general two-qubit state and for the state of the system after the measurement.

Dirac notation is also useful to compute inner products of two-qubit states. We only need to notice that

$$(\langle\psi_1| \otimes \langle\psi_2|)(|\varphi_1\rangle \otimes |\varphi_2\rangle) = \langle\psi_1|\varphi_1\rangle\,\langle\psi_2|\varphi_2\rangle\,,$$

apply distributivity and remember to conjugate the complex coefficients when obtaining a bra from a ket.

Then, for instance, we can notice that the inner product of $\frac{4}{5}\,|01\rangle + \frac{3i}{5}\,|11\rangle$ and $\frac{1}{\sqrt{2}}\,|00\rangle + \frac{1}{\sqrt{2}}\,|11\rangle$ is

$$\begin{aligned}
&\left(\frac{4}{5}\,\langle 01| - \frac{3i}{5}\,\langle 11|\right)\left(\frac{1}{\sqrt{2}}\,|00\rangle + \frac{1}{\sqrt{2}}\,|11\rangle\right) = \\
&\frac{4}{5\sqrt{2}}\,\langle 01|00\rangle + \frac{4}{5\sqrt{2}}\,\langle 01|11\rangle - \frac{3i}{5\sqrt{2}}\,\langle 11|00\rangle - \frac{3i}{5\sqrt{2}}\,\langle 11|11\rangle = \\
&\frac{4}{5\sqrt{2}}\,\langle 0|0\rangle\,\langle 1|0\rangle + \frac{4}{5\sqrt{2}}\,\langle 0|1\rangle\,\langle 1|1\rangle - \frac{3i}{5\sqrt{2}}\,\langle 1|0\rangle\,\langle 1|0\rangle - \frac{3i}{5\sqrt{2}}\,\langle 1|1\rangle\,\langle 1|1\rangle = -\frac{3i}{5\sqrt{2}},
\end{aligned}$$

since $\langle 0|1\rangle = \langle 1|0\rangle = 0$ and $\langle 0|0\rangle = \langle 1|1\rangle = 1$.

1.4.2 Two-qubit gates: tensor products

Of course, the operations that we can conduct on two-qubit systems need to be unitary. Thus, two-qubit quantum gates are 4×4 unitary matrices that act on 4-dimensional column vectors. The simplest way to construct such matrices is by taking the tensor product of two one-qubit quantum gates. Namely, if we consider two one-qubit gates U_1 and U_2 and two one-qubit states $|\psi_1\rangle$ and $|\psi_2\rangle$, we can form a two-qubit gate $U_1 \otimes U_2$ that acts on $|\psi_1\rangle \otimes |\psi_2\rangle$ as

$$(U_1 \otimes U_2)(|\psi_1\rangle \otimes |\psi_2\rangle) = (U_1 |\psi_1\rangle) \otimes (U_2 |\psi_2\rangle).$$

By linearity, we can extend $U_1 \otimes U_2$ to any combination of two-qubit states and we can associate a matrix to $U_1 \otimes U_2$. In fact, said matrix is given by the tensor product of the matrices associated to U_1 and U_2. More concretely, the expression for the tensor product, $A \otimes B$, of the 2×2 matrices A and B is

$$\begin{pmatrix} a_{11} & a_{12} \\ a_{21} & a_{22} \end{pmatrix} \otimes \begin{pmatrix} b_{11} & b_{12} \\ b_{21} & b_{22} \end{pmatrix} = \begin{pmatrix} a_{11}\begin{pmatrix} b_{11} & b_{12} \\ b_{21} & b_{22} \end{pmatrix} & a_{12}\begin{pmatrix} b_{11} & b_{12} \\ b_{21} & b_{22} \end{pmatrix} \\ a_{21}\begin{pmatrix} b_{11} & b_{12} \\ b_{21} & b_{22} \end{pmatrix} & a_{22}\begin{pmatrix} b_{11} & b_{12} \\ b_{21} & b_{22} \end{pmatrix} \end{pmatrix}$$
$$= \begin{pmatrix} a_{11}b_{11} & a_{11}b_{12} & a_{12}b_{11} & a_{12}b_{12} \\ a_{11}b_{21} & a_{11}b_{22} & a_{12}b_{21} & a_{12}b_{22} \\ a_{21}b_{11} & a_{21}b_{12} & a_{22}b_{11} & a_{22}b_{12} \\ a_{21}b_{21} & a_{21}b_{22} & a_{22}b_{21} & a_{22}b_{22} \end{pmatrix}.$$

Now it is easy to verify that this operation is indeed unitary and, hence, deserves the name of quantum gate.

Exercise 1.11

Check that, given any pair of unitary matrices U_1 and U_2, the inverse of $U_1 \otimes U_2$ is $U_1^\dagger \otimes U_2^\dagger$ and that $(U_1 \otimes U_2)^\dagger = U_1^\dagger \otimes U_2^\dagger$.

Tensor products of gates occur naturally when we have circuits with two qubits and pairs of individual one-qubit gates are acting on each of them. For instance, in the following circuit, the gate $X \otimes X$ acts on the two qubits and then it is followed by the gate $H \otimes I$, where I is the identity gate:

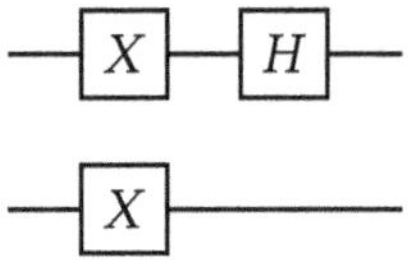

Exercise 1.12

Explicitly compute the matrices for the gates $X \otimes X$ and $H \otimes I$.

You may complain that we haven't done anything new so far. And you would be right! In fact, quantum gates that are obtained as the tensor product of one-qubit gates can be seen as operations on isolated qubits that just happen to be applied at the same time. But wait and see! In the next subsection, we will introduce a completely different way of acting on two-qubit systems.

1.4.3 The CNOT gate

By taking tensor products of one-qubit gates, we can only obtain operations that act on each qubit individually. But this just leaves us with a (rather boring) subset of all the possible two-qubit gates. There are many unitary matrices that cannot be written as the tensor product of other simple matrices. In the two-qubit case, probably the most important one is the **controlled-NOT** (or **controlled-X**) gate, usually called the **CNOT gate**, given by the unitary matrix

$$\begin{pmatrix} 1 & 0 & 0 & 0 \\ 0 & 1 & 0 & 0 \\ 0 & 0 & 0 & 1 \\ 0 & 0 & 1 & 0 \end{pmatrix}.$$

It is illuminating to see how this gate acts on the elements of the two-qubit computational basis. As you can easily check, we get

$$\mathrm{CNOT}\left|00\right\rangle = \left|00\right\rangle, \qquad \mathrm{CNOT}\left|01\right\rangle = \left|01\right\rangle, \qquad \mathrm{CNOT}\left|10\right\rangle = \left|11\right\rangle, \qquad \mathrm{CNOT}\left|11\right\rangle = \left|10\right\rangle.$$

This means that the value of the second qubit is flipped if and only if the value of the first qubit is 1. Or, to put it in other words, the application of a NOT gate on the second qubit (that we call the **target**) is **controlled** by the first qubit. Now the name of this gate makes much more sense, doesn't it?

In a quantum circuit, the CNOT gate is represented as follows:

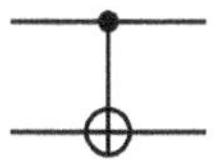

Notice that the control qubit is indicated by a solid black circle and the target qubit is indicated by the $\oplus$ symbol (the symbol for an X gate can also be used instead of $\oplus$).

Sometimes, technical difficulties restrict the number of CNOT gates that can be actually implemented on a quantum computer. For instance, on a certain quantum chip you may have the possibility of applying a CNOT gate targeting qubit 1 and controlled by qubit 0, but not the other way around. If you find yourself in such a situation, there's no need to panic. If you use the circuit

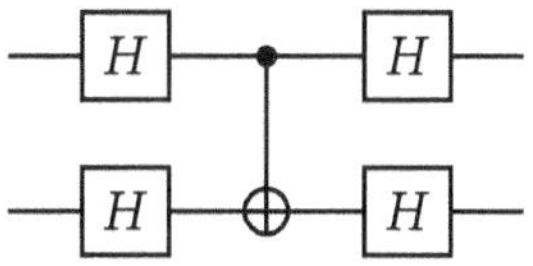

you are effectively applying a CNOT gate with target in the top qubit and control in the bottom one. And that's how you can save the day!

The CNOT gate can also be used to interchange or **swap** the states of two qubits, by using the following circuit:

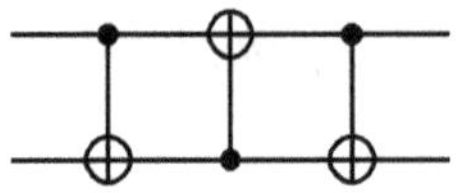

Exercise 1.13

Check these equivalences in two different ways: by computing the matrices of the circuits and by obtaining the result of using them with qubits in states $|00\rangle$, $|01\rangle$, $|10\rangle$, and $|11\rangle$.

In any case, the most prominent use of the CNOT gate is, without a doubt, the ability to create entanglement, an intriguing property of quantum systems that we will study next.

1.4.4 Entanglement

Oddly enough, in order to define when a quantum system is entangled, we first need to define when it is **not** entangled. We say that a state $|\psi\rangle$ is a **product state** if it can be written as the tensor product of two other states $|\psi_1\rangle$ and $|\psi_2\rangle$, each of at least one qubit, as in

$$|\psi\rangle = |\psi_1\rangle \otimes |\psi_2\rangle .$$

If $|\psi\rangle$ is not a product state, we say that it is **entangled.**

For example, $|01\rangle$ is a product state, because we know that it is just another way of writing $|0\rangle \otimes |1\rangle$. Also, $\sqrt{1/2}(|00\rangle + |10\rangle)$ is a product state, because we can factor $|0\rangle$ on the second qubit to obtain

$$\frac{1}{\sqrt{2}}(|00\rangle + |10\rangle) = \left(\frac{1}{\sqrt{2}}(|0\rangle + |1\rangle)\right)|0\rangle .$$

On the other hand, $\sqrt{1/2}\,(|00\rangle + |11\rangle)$ is an entangled state. No matter how hard you try, it is impossible to write it as a product of two one-qubit states. Suppose, for sake of contradiction, that it were possible. Then, you would have

$$\frac{1}{\sqrt{2}}\left(|00\rangle + |11\rangle\right) = \left(a\,|0\rangle + b\,|1\rangle\right)\left(c\,|0\rangle + d\,|1\rangle\right)$$
$$= ac\,|00\rangle + ad\,|01\rangle + bc\,|01\rangle + bd\,|11\rangle\,.$$

But this forces ad to be 0, because we have no $|01\rangle$ component in $\sqrt{1/2}(|00\rangle + |11\rangle)$. Then, either $a = 0$, in which case ac is 0, or $d = 0$, from which $bd = 0$ follows. In both cases, it is impossible to reach the equality that we needed. Thus, it follows that the state is entangled.

Exercise 1.14

Is $\sqrt{1/3}(|00\rangle + |01\rangle + |11\rangle)$ entangled? And what about $\frac{1}{2}(|00\rangle + |01\rangle + |10\rangle + |11\rangle)$?

When measured, entangled states can show correlations that go beyond what can be explained with classical physics. For instance, if we have the entangled state $\sqrt{1/2}\left(|00\rangle + |11\rangle\right)$ and we measure the first qubit, we can obtain 0 or 1, each with probability $1/2$. However, if we measure the second qubit afterwards, the result will be completely determined by the value obtained when measuring the first qubit and, in fact, will be exactly the same. If we invert the order and measure first the second qubit, then the result will be 0 or 1, with equal probability. But, in this case, the result of a subsequent measurement of the first qubit will be completely determined!

This still happens even if we separate the two qubits thousands of light years apart, as if one qubit could somehow know what the result of measuring the other qubit was. This curious behavior haunted many physicists during the 20th century, including Albert Einstein, who called it a *"spooky action at a distance"* (see [18]). Nevertheless, the effects of entanglement have been repeatedly demonstrated in uncountable experiments (in fact, the Nobel Prize in Physics 2022 was awarded to Alain Aspect, John F. Clauser, and Anton Zeilinger, pioneers in studying and testing this phenomenon in practice [3], [19]–[21]). And, very importantly for us, entanglement is one of the most powerful resources available in quantum computing.

But entanglement is, by no means, the only puzzling feature of qubit systems. In the next subsection, we are going to mathematically prove that copying quantum information, an operation that you may have taken for granted, is not possible in general. These qubits are, indeed, full of surprises!

1.4.5 The no-cloning theorem

Another peculiar property of quantum systems is that, in general, they don't allow us to **copy information**. Surprising as this may seem, it is just an easy consequence of the linearity of quantum gates. To show why, let us be more precise about what we would need in order to copy information, for instance with just two qubits. We would like to have a two-qubit quantum gate U that will be able to copy the first qubit into the second. That is, for any given quantum state $|\psi\rangle$, we would need

$$U|\psi\rangle|0\rangle = |\psi\rangle|\psi\rangle.$$

Then, $U|00\rangle = |00\rangle$ and $U|10\rangle = |11\rangle$ and, by linearity,

$$U\left(\frac{1}{\sqrt{2}}(|00\rangle + |10\rangle)\right) = \frac{1}{\sqrt{2}}\left(U|00\rangle + U|10\rangle\right) = \frac{1}{\sqrt{2}}\left(|00\rangle + |11\rangle\right).$$

We should highlight that the state that we have obtained is entangled, as we proved in the previous subsection.

Nevertheless, notice that, in our original state, we can factor the second $|0\rangle$ out to obtain

$$\frac{1}{\sqrt{2}}(|00\rangle + |10\rangle) = \left(\frac{|0\rangle + |1\rangle}{\sqrt{2}}\right)|0\rangle.$$

Then, in virtue of the action of U, we should have

$$U\left(\frac{1}{\sqrt{2}}(|00\rangle + |10\rangle)\right) = U\left(\left(\frac{|0\rangle + |1\rangle}{\sqrt{2}}\right)|0\rangle\right) = \frac{(|0\rangle + |1\rangle)}{\sqrt{2}}\frac{(|0\rangle + |1\rangle)}{\sqrt{2}},$$

which is a product state. However, we had obtained earlier that $U(\sqrt{1/2}(|00\rangle + |10\rangle)) = \sqrt{1/2}(|00\rangle + |11\rangle)$, which is entangled! This contradiction implies that, alas, no such U exists.

This remarkable result is called the **no-cloning theorem** and we should explain its meaning in a little more detail. On the one hand, notice that this does not imply that we cannot copy classical information. In fact, if $|\psi\rangle$ is just $|0\rangle$ or $|1\rangle$, we can easily achieve $U|\psi\rangle|0\rangle = |\psi\rangle|\psi\rangle$ by taking U to be the CNOT gate. On the other hand, the theorem applies to unknown states $|\psi\rangle$. If we know what $|\psi\rangle$ is – that is, if we know a circuit that prepares $|\psi\rangle$ starting from $|0\rangle$ – then, of course, we can create as many independent copies of it as we want. However, if $|\psi\rangle$ is handed to us without any additional information about its state, the no-cloning theorem shows that we cannot replicate its state in general.

To learn more...

The no-cloning theorem plays an important role in the security of quantum key distribution protocols such as the famous **BB84**, introduced in 1984 by Bennett and Brassard [4].

After this brief detour, let's return to our study of two-qubit quantum gates. In the next subsection, we will show how to construct many interesting two-qubit unitary operations whose action is controlled by one of their inputs.

1.4.6 Controlled gates

You may be wondering if, in addition to a controlled-X (or CNOT) gate, there are also **controlled-Y**, **controlled-Z**, or **controlled-H** gates. The answer is a resounding yes and, in fact, for any quantum gate U, it is possible to define a **controlled-U** (or, simply, **CU**) gate whose action on the computational basis is

$$CU|00\rangle = |00\rangle, \qquad CU|01\rangle = |01\rangle, \qquad CU|10\rangle = |1\rangle U|0\rangle, \qquad CU|11\rangle = |1\rangle U|1\rangle.$$

Exercise 1.15

Check that the matrix of CU is

$$\begin{pmatrix} 1 & 0 & 0 & 0 \\ 0 & 1 & 0 & 0 \\ 0 & 0 & u_{11} & u_{12} \\ 0 & 0 & u_{21} & u_{22} \end{pmatrix},$$

where $(u_{ij})_{i,j=1}^{2}$ is the matrix of U. Check also that CU is unitary. What is the adjoint of CU?

The circuit representation of a CU gate is similar to the one that we use for the CNOT gate, namely

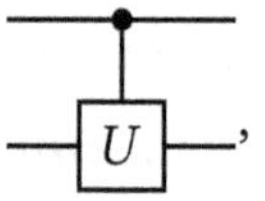

where the solid black circle indicates the control and the box with U inside indicates the target.

Constructing a controlled gate is simpler than it seems, provided your quantum computer already implements rotation gates and the two-qubit CNOT gate. In fact, from the decomposition in rotations that we mentioned at the end of *Section 1.3.4*, it can be proved (see the book by Nielsen and Chuang [16, Corollary 4.2]) that any one-qubit quantum gate U can be written in the form

$$U = e^{i\theta} AXBXC$$

for some angle θ and gates A, B, and C such that $ABC = I$. Then, the following circuit implements CU:

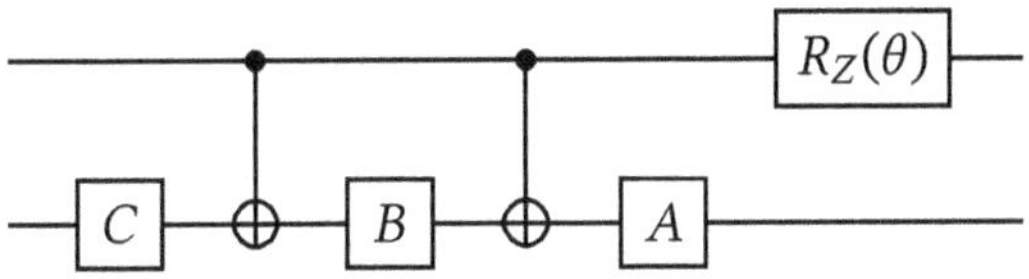

Sometimes, though, constructing a controlled gate is much easier. For instance, it can be shown that a controlled-Z gate can be obtained from a controlled-X and two H gates, as shown in the identity of the following circuits:

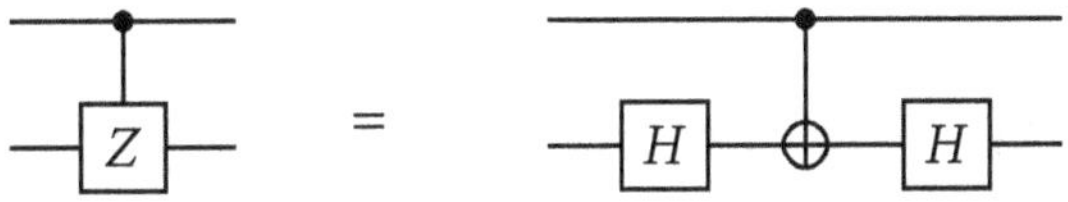

> **Exercise 1.16**
>
> Prove the preceding equivalence.

We now have everything we need in order to construct our first two-qubit quantum circuit. Let's get those qubits entangled!

1.4.7 Hello, entangled world!

To finish up with our study of two-qubit systems, let us show how to create entangled states with the help of the CNOT gate. Consider the following circuit:

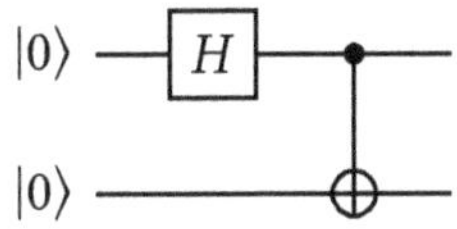

Initially, the state of the system is $|00\rangle$. After we apply the H gate, we get into the state $\sqrt{1/2}(|00\rangle + |10\rangle)$. Finally, when we apply the CNOT gate, the state changes to $\sqrt{1/2}(|00\rangle + |11\rangle)$, which, as we proved in *Section 1.4.4*, is indeed an entangled state.

The state $\sqrt{1/2}\,(|00\rangle + |11\rangle)$ is known as a **Bell state**, of which there are four. The other three are $\sqrt{1/2}\,(|00\rangle - |11\rangle)$, $\sqrt{1/2}(|10\rangle + |01\rangle)$, and $\sqrt{1/2}(|10\rangle - |01\rangle)$. All of them are entangled, and they can be prepared with circuits similar to the preceding one.

Exercise 1.17

Show that all four Bell states are entangled. Obtain circuits to prepare them. *Hint*: you can use Z and X gates after the CNOT in the preceding circuit.

We are now ready for the big moment. In the next section, we will finally learn how to work with not just one or two qubits, but with as many as we can get in our quantum computers.

1.5 Working with multiple qubits and universality

Now that we have mastered working with two-qubit systems, it will be fairly straightforward to generalize all the notions that we have been studying to the case in which the number of qubits in our circuits is arbitrarily big. You know the drill: we will start by defining, mathematically, what a multi-qubit system is, we will then learn how to measure it and, finally, we will introduce quantum gates that act on many qubits at the same time.

1.5.1 Multi-qubit systems

With all that we have learned so far, it will now be very easy to understand how to work with **multi-qubit systems**.

As you may have already deduced, if we have n qubits, the states that constitute the computational basis are

$$\begin{gathered}
\left|0\right\rangle \otimes \left|0\right\rangle \otimes \cdots \otimes \left|0\right\rangle, \\
\left|0\right\rangle \otimes \left|0\right\rangle \otimes \cdots \otimes \left|1\right\rangle, \\
\vdots \\
\left|1\right\rangle \otimes \left|1\right\rangle \otimes \cdots \otimes \left|1\right\rangle.
\end{gathered}$$

We usually omit the ⊗ symbol to write

$$|0\rangle\,|0\rangle \cdots |0\rangle\,,$$
$$|0\rangle\,|0\rangle \cdots |1\rangle\,,$$
$$|1\rangle\,|1\rangle \cdots |1\rangle$$

or

$$|00\cdots 0\rangle\,, |00\cdots 1\rangle\,, \ldots, |11\cdots 1\rangle$$

or simply

$$|0\rangle\,, |1\rangle\,, \ldots, |2^n - 1\rangle\,.$$

Important note

When using the $|0\rangle, |1\rangle, \ldots, |2^n - 1\rangle$ notation for basis states, the total number of qubits must be clear from context. Otherwise, a state like, for example, $|2\rangle$ might mean either $|10\rangle$, $|010\rangle$, $|0010\rangle$, or any string with leading zeroes and ending in 10…and that would be an intolerable ambiguity!

Of course, a generic state of the system will then be of the form

$$|\psi\rangle = a_0\,|0\rangle + a_1\,|1\rangle + \ldots + a_{2^n-1}\,|2^n - 1\rangle$$

subject to the only condition that the amplitudes a_i should be complex numbers such that $\sum_{l=0}^{2^n-1} |a_l|^2 = 1$. Our dear old friend, the normalization condition!

To learn more…

Notice that the number of parameters describing the general state of an n-qubit system is exponential in n. For highly entangled states, we do not know how to represent all this information in a more succinct way and it is strongly suspected that it is not possible. Part of the power of quantum computing comes from this possibility of implicitly working with 2^n complex numbers by manipulating just n qubits.

Exercise 1.18

Check that the basis state $|j\rangle$ is represented by a 2^n-dimensional column vector whose j-th component is 1, while the rest are 0 (*Hint*: Use, repeatedly, the expression for the tensor product of column vectors that we discussed in *Section 1.4.1* and the fact that the tensor product is associative). Deduce that any n-qubit state can be represented by a 2^n-dimensional column vector with unit length.

If we decide to measure all the qubits of the system in the computational basis, we will obtain m with probability $|a_m|^2$. If that is the case, then the state will collapse to $|m\rangle$. But if we only measure one of the qubits, say the j-th one, then we will obtain 0 with probability

$$\sum_{l \in J_0} |a_l|^2,$$

where J_0 is the set of numbers whose j-th bit is 0. In this scenario, the state of the system after measuring 0 would be

$$\frac{\sum_{l \in J_0} a_l |l\rangle}{\sqrt{\sum_{l \in J_0} |a_i|^2}}.$$

Exercise 1.19

Derive the formulas for the case in which the result of the measurement is 1.

Exercise 1.20

What is the probability of getting 0 when we measure the second qubit of $(1/2)|100\rangle + (1/2)|010\rangle + \sqrt{1/2}|001\rangle$? What will the state be after the measurement if we indeed get 0?

Computing inner products of n-qubit systems in Dirac notation is very similar to doing it with two-qubit systems. The procedure is analogous to the one we showed in *Section 1.4.1*, but taking into account that

$$(\langle\psi_1| \otimes \ldots \otimes \langle\psi_n|)(|\varphi_1\rangle \otimes \ldots \otimes |\varphi_n\rangle) = \langle\psi_1|\varphi_1\rangle \ldots \langle\psi_n|\varphi_n\rangle .$$

Exercise 1.21

Compute the inner product of $|x\rangle$ and $|y\rangle$, where x and y are both binary strings of length n. Use your result to prove that $\{|x\rangle\}_{x\in\{0,1\}^n}$ is, indeed, an orthonormal basis.

Exercise 1.22

Compute the inner product of the states $\sqrt{1/2}\,(|000\rangle + |111\rangle)$ and $1/2\,(|000\rangle + |011\rangle + |101\rangle + |110\rangle)$.

We can now turn to the question of how to operate on many qubits at once. Let's define multi-qubit gates!

1.5.2 Multi-qubit gates

Since n-qubit states are represented by 2^n-dimensional column vectors, n-qubit gates can be identified with $2^n \times 2^n$ unitary matrices. Similar to the two-qubit case, we can construct n-qubit gates by taking the tensor product of gates on a smaller number of qubits. Namely, if U_1 is an n_1-qubit gate and U_2 is an n_2-qubit gate, then $U_1 \otimes U_2$ is an $(n_1 + n_2)$-qubit gate and its matrix is given by the tensor product of the matrices U_1 and U_2.

To learn more...

The expression for the tensor product of two matrices A and B is

$$\begin{pmatrix} a_{11} & \ldots & a_{1q} \\ \vdots & \ddots & \vdots \\ a_{p1} & \ldots & a_{pq} \end{pmatrix} \otimes B = \begin{pmatrix} a_{11}B & \ldots & a_{1q}B \\ \vdots & \ddots & \vdots \\ a_{p1}B & \ldots & a_{pq}B \end{pmatrix}.$$

However, there are n-qubit gates that cannot be constructed as tensor products of smaller gates. One such example is the **Toffoli** or **CCNOT** gate, a three-qubit gate that acts on the computational basis as

$$\mathrm{CCNOT}\left|x\right\rangle\left|y\right\rangle\left|z\right\rangle = \left|x\right\rangle\left|y\right\rangle\left|z \oplus (x \wedge y)\right\rangle,$$

where $\oplus$ is the **XOR** function and $\wedge$ is the symbol for the AND Boolean function. Thus, CCNOT applies a doubly controlled (in this case, by the first two qubits) NOT gate to the third qubit — hence the name!

Exercise 1.23

Obtain the matrix for the CCNOT gate and verify that it is unitary.

The Toffoli gate is important because, using it and with the help of auxiliary qubits, we can construct any classical Boolean operator. For instance, $\mathrm{CCNOT}\left|1\right\rangle\left|1\right\rangle\left|z\right\rangle = \left|1\right\rangle\left|1\right\rangle\left|\neg z\right\rangle$ (where $\neg z$ is the negation of z) and $\mathrm{CCNOT}\left|x\right\rangle\left|y\right\rangle\left|0\right\rangle = \left|x\right\rangle\left|y\right\rangle\left|x \wedge y\right\rangle$. This shows that, with quantum circuits, we can simulate the behavior of any classical digital circuit at the cost of using some additional ancillary qubits, since any Boolean function can be built with just negations and conjunctions. This is somewhat surprising, because we know that all quantum gates are invertible, while not all Boolean functions are. It then follows that we could make all of our digital circuits reversible just by implementing a classical version of the Toffoli gate!

We will not be studying any other concrete examples of gates that act on three (or more!) qubits because, in fact, we can simulate their behavior with circuits that only use one- and two-qubit gates. Keep on reading to know how!

1.5.3 Universal gates in quantum computing

Current quantum computers can't implement every possible quantum gate. Instead, they rely on **universality results** that show how any unitary operation can be decomposed as a circuit that uses a reduced set of **primitive** gates. In previous sections, we mentioned, for instance, that any one-qubit gate can be obtained by using just R_Z and R_Y rotations. It turns out that similar results exist for the general case of n-qubit quantum gates.

To us, it will be important to know that, for any unitary operation, we can construct a circuit that implements it using only one-qubit gates and the CNOT gate. For this reason, we say that those gates are **universal** – in the same sense that, for example, negation and conjunction are universal for Boolean logic. This fact will be crucial for our study of **feature maps** and **variational forms** in connection to **quantum neural networks** and other quantum machine learning models.

To learn more...

In addition to one-qubit gates plus CNOT, there are many other sets of universal gates. For instance, it can be shown that the three gates H, T, and CNOT can be used to approximate any unitary operation to any desired accuracy – and they are universal in that sense. See Section *4.5* of the book by Nielsen and Chuang [16] for proofs of these facts and for more examples of universal gate sets.

To illustrate how CNOT and one-qubit gates can be used to implement any other quantum gate, the following circuit shows a possible decomposition of the Toffoli gate targeting the top qubit:

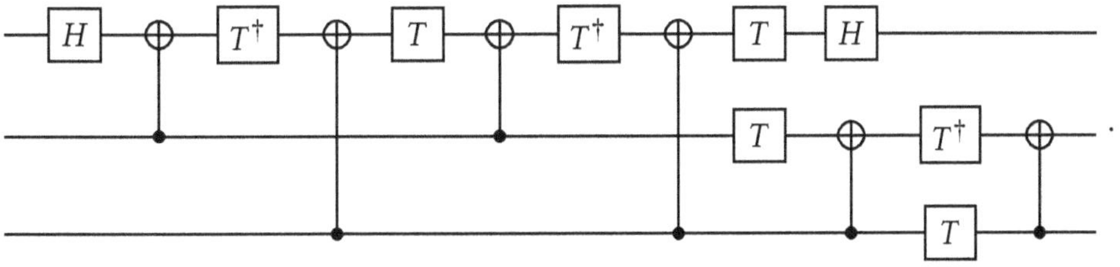

Exercise 1.24

Verify that the preceding circuit implements the Toffoli gate by checking its action on the states of the computational basis.

This concludes our review of the fundamentals of quantum computing. We've come a long way since the beginning of this chapter, but by now we have mastered all the mathematics that we will need in order to study quantum machine learning and quantum optimization algorithms. Soon, we will see all these concepts in action!

Summary

In this chapter, we have introduced the quantum circuit model and the main concepts that it relies on: qubits, gates, and measurements. We have started by studying the most humble circuits, those that only have one or two qubits, but we have used our experience with them to grow all the way up to multi-qubit systems. In the process, we have discovered some powerful properties, such as superposition and entanglement, and we have mastered the mathematics – mainly some linear algebra – needed to work with them.

These notions will be extremely valuable to us, because they make up the language in which we will be describing the quantum algorithms for machine learning and optimization that we will study in the rest of the book. Soon, all the pieces will come together to form a beautiful structure. And we will be able to appreciate it and understand it fully because of the solid foundations that we have acquired by now.

In the next chapter, we will start applying all that we have learned by implementing and running quantum circuits on quantum simulators and on actual quantum computers. We don't know about you, but we are pretty excited!

2

The Tools of the Trade in Quantum Computing

Give us the tools, and we will finish the job.

— Winston Churchill

We are all very much looking forward to having a "Q1 Pro" quantum chip in our laptops, but — much to our regret — the technology is not there just yet. Nevertheless, we do have some actual quantum computers that, with their limitations, are able to execute quantum algorithms. And, furthermore, our good old classical computers can actually do a very decent job at simulating ideal quantum computers, at least for a low number of qubits.

In this chapter, we will explore the tools that allow us to implement quantum algorithms using the quantum circuit model and run them on simulators or on real quantum hardware.

We will begin by going through some of the most widely-used quantum software frameworks and platforms out there. Then, we will see how to work with the two software frameworks that we are going to use more extensively throughout this book: Qiskit and PennyLane.

We'll cover the following topics in this chapter:

- Tools for quantum computing: a non-exhaustive overview
- Working with Qiskit
- Working with PennyLane

After reading this chapter, you will have a broad perspective on the range of software tools and platforms available for quantum computing. Moreover, you will know how to implement and execute quantum algorithms — both on simulators and real quantum hardware — using Qiskit and PennyLane.

2.1 Tools for quantum computing: a non-exhaustive overview

In this book, we will work mostly with two quantum frameworks: **Qiskit** and **PennyLane**. These frameworks are powerful, very widely used, and are backed by strong user communities, but they are by no means the only interesting options available. There is currently a plethora of wonderful software frameworks for quantum computing, so much so that it can sometimes feel overwhelming!

2.1.1 A non-exhaustive survey of frameworks and platforms

In this section, we will briefly go through some of the most popular frameworks out there. Most of these frameworks are free, both as in *free beer* and as in *free speech.*

- **Quirk**: We can begin with a simple yet powerful simulator of quantum circuits: Quirk (`https://algassert.com/quirk`). Unlike all the other frameworks that we will discuss, this one does not work with code, but with a graphical user interface that runs as a web application. This makes it ideal for running demonstrations of algorithms or for quick prototyping.

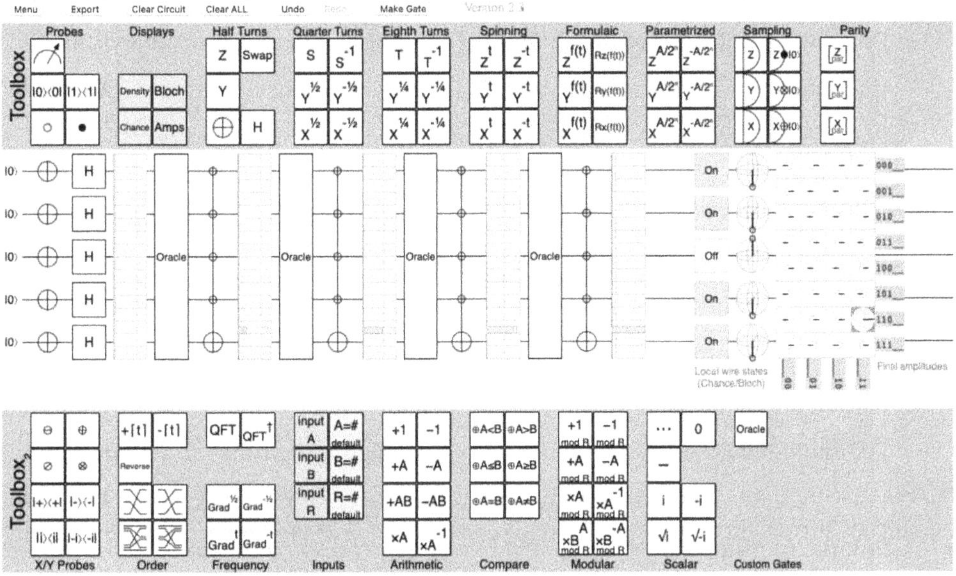

Figure 2.1: Quirk demonstrating the Grover search algorithm

With Quirk, you can build quantum circuits arranging gates with a **drag-and-drop interface**. It includes the most common quantum gates, in addition to a handful of custom gates that are used for the demonstration of some algorithms. You can have a look at *Figure 2.1* to see Quirk in action.

- **Q# and Microsoft's QDK**: Most quantum software frameworks rely on a **host** classical programming language (usually **Python**). **Q#** (read as "*Q sharp*") is an exception to this rule: it is a purpose-built programming language for quantum computing developed at Microsoft. This language is used in Microsoft's **Quantum Development Kit** (**QDK**) (`https://azure.microsoft.com/en-us/resources/development-kit/quantum-computing/`), which includes several simulators for running quantum algorithms and assessing their performance. With the QDK, you can also send your quantum algorithms to real quantum computers using **Azure Quantum**.

In addition to this, Microsoft provides solutions that enable Q# to run interactively and to be inter-operable with other programming languages such as Python. Moreover, the Azure Quantum service, which allows you to run quantum algorithms on real hardware, also works with other quantum software frameworks.

- **QuEST**. The **Quantum Exact Simulation Toolkit (QuEST)** (`https://quest.qtechtheory.org/`) is a simulation framework written in **C++** and built with performance in mind. With it, you can write a single implementation of a quantum algorithm that can be compiled into binary code, yielding a program that will simulate the algorithm. Using this framework, you can be *closer to the metal* and make sure that all the hardware resources available are used in an optimal way. This makes QuEST a very interesting choice for hardware-intensive simulations with a large number of qubits. In other words, if you ever wanted to test how many qubits your computer was able to handle, QuEST might be the way to go (you might want to have a fire extinguisher at hand though, just in case things get hot!).

  The source code for QuEST is freely available online: (`https://github.com/quest-kit/QuEST`). It is mostly C and C++ code that you can use on any device.

- **Cirq** (`https://quantumai.google/cirq`) is a quantum software framework developed at **Google** that uses Python as a host language. It can be used to design quantum algorithms and to simulate them on a classical device or send them to real quantum hardware. Furthermore, Cirq is integrated in TensorFlow Quantum (`https://www.tensorflow.org/quantum`), Google's framework for quantum machine learning.

- **Qiskit** (`https://qiskit.org/`) is **IBM**'s quantum framework. It relies on Python as a host language and provides a wide range of simulators as well as allowing the submission of algorithms to IBM's real quantum hardware. In addition to this, Qiskit provides an extensive library of quantum circuits and algorithms, many of which we will use in this book!

In particular, when it comes to machine learning, Qiskit includes some interesting models with the tools necessary to train and execute them; and it also provides a **PyTorch** interface for the training of quantum machine learning models (we will explore Qiskit and all its secrets in detail in the following section).

- **PennyLane** (`https://pennylane.ai/`) is a quantum framework built specifically for quantum machine learning, but it can perfectly be used as a general-purpose quantum computing framework. It is quite a newcomer to the quantum programming scene and it is being developed at **Xanadu**. Like Qiskit, it uses Python as a host language. Any quantum algorithms written in PennyLane can be sent to real quantum computers and executed in a broad collection of simulators.

  PennyLane is one of the best frameworks out there when it comes to inter-operability. Thanks to a wide collection of **plugins**, you can export PennyLane circuits to other frameworks and execute them there – taking advantage of some of the features that these other frameworks may have.

  When it comes to machine learning, PennyLane provides some built-in tools, but it is also highly inter-operable with classical machine learning frameworks such as **scikit-learn**, **Keras**, **TensorFlow**, and **PyTorch**. We will discuss this framework in detail later in this chapter.

- **Ocean** (`https://docs.ocean.dwavesys.com/en/stable/`) is a Python library developed by the Canadian company **D-Wave**. In contrast with the other software packages that we have mentioned so far, Ocean's goal is not the implementation and execution of quantum circuits. Instead, this library allows you to define instances of combinatorial optimization problems of different types and to solve them both with classical algorithms and on D-Wave's **quantum annealers** (special quantum computers that are not general-purpose but oriented to solving optimization problems).

  Starting in *Chapter 3, QUBO: Quadratic Unconstrained Binary Optimization*, we will introduce the concepts that we will need in order to understand how to define problems with Ocean. And in *Chapter 4, Quantum Adiabatic Computing and Quantum Annealing*, we will learn how to use Ocean in all its glory, both to solve these optimization problems with classical algorithms and, of course, with quantum algorithms in actual quantum computers!

To learn more...

The source code (and binaries, if any) for these frameworks can be downloaded from their official websites. For detailed installation instructions for Qiskit, Pennylane, and Ocean, you can refer to *Appendix D, Installing the Tools*.

- **Amazon Braket**: Amazon Web Services offers Amazon Braket (`https://aws.amazon.com/braket/`), a paid cloud service that makes it possible to use a wide range of implementations of real quantum computers. In order to execute code on these computers, they provide their own *device-agnostic SDK*, but they also fully support PennyLane and Qiskit, and there are even plugins to work with Ocean (`https://amazon-braket-ocean-plugin-python.readthedocs.io/`).

2.1.2 Qiskit, PennyLane, and Ocean

As we have just seen, there is an abundant range of choices when it comes to software frameworks for quantum computing, and this might make you wonder why we are going to stick with Qiskit, PennyLane, and Ocean. Of course, this is not a choice made at random; we have a bunch of good reasons!

Regarding Qiskit, it is just massive: the level of built-in algorithms and functionalities that it includes is simply unmatched. And not only that, but it also has a very strong community of users and it is supported by most quantum hardware providers out there. In other words, Qiskit could easily be considered a *lingua franca* of quantum computing.

PennyLane, on the contrary, is not as widely used a framework as Qiskit (at least for now), but we believe that it is one of the most promising newcomers to the world of quantum computing.

In particular, when it comes to quantum machine learning, it is very hard to say that there is anything better than PennyLane. On the one hand, PennyLane simply runs very smoothly and is beautifully documented and, on the other, its interoperability with other quantum and **Machine Learning** (**ML**) frameworks knows no rival.

This is why we believe that Qiskit and PennyLane are better choices than, for instance, **Q#** or **Cirq** (which, of course, are also great frameworks on their own). Regarding QuEST, it is true that the performance of the simulators provided by Qiskit and PennyLane may not be as good as the performance that QuEST would yield. But we should also take into account that QuEST is not nearly as user-friendly as Qiskit or PennyLane, and it lacks many of their features; for instance, QuEST does not have any built-in tools or interfaces for training quantum machine learning models. In any case, we should remark that while running circuits on the simulators bundled in Qiskit and PennyLane may not be as efficient as running them in QuEST, for our purposes, the performance that we can get with them is more than good enough! Nevertheless, if you are still eager to get the performance boost that QuEST could give you, you should know that there is a community plugin that allows PennyLane to work with the QuEST simulator.

Finally, we have chosen Ocean because it is completely unique in that it probably is the only software package that lets you work with **quantum annealers**, both to define problems and to run them on actual quantum hardware. It is also very easy to learn...at least once you understand how to define combinatorial optimization problems in the **Ising** and QUBO models. But don't worry; we will extensively study those frameworks in *Chapter 3, QUBO: Quadratic Unconstrained Binary Optimization*, and, by *Chapter 4, Quantum Adiabatic Computing and Quantum Annealing*, we will be more than ready to write our very first programs using Ocean.

At this point, we have a good global understanding of the current landscape of tools for quantum computing. In the upcoming sections, we will take our first steps in using them, and we shall begin with Qiskit.

2.2 Working with Qiskit

In this section, we will learn how to work with the Qiskit framework. We will first discuss the general structure of Qiskit, and then we will study how to implement quantum circuits in Qiskit using quantum gates and measurements. Then, we will explore how to run these circuits using the simulators provided by Qiksit and also real quantum computers available for free thanks to IBM. This section is key, for we will use Qiskit extensively in this book.

> **Important note**
>
> Quantum computing is a rapidly-evolving field... and so are its software frameworks! We are going to work with **version 0.39.2** of Qiskit. Keep in mind that, if you are using a different version, things may have changed. In case of doubt, you should always refer to the documentation (`https://qiskit.org/documentation/`).

2.2.1 An overview of the Qiskit framework

The Qiskit framework [22] consists of the components depicted in *Figure 2.2*. At the very foundation of Qiskit lies **Qiskit Terra**. This package is responsible for handling quantum circuits and providing the necessary tools for constructing them. It also includes a basic Python-based simulator (**BasicAer**) and it can work with the **IBM Quantum provider** to execute circuits on IBM's quantum hardware.

Qiskit Aer is built on top of Qiskit Terra, and it provides a suite of high-performance quantum simulators written in **C++** and designed to use hardware resources more efficiently.

We can think of Qiskit Aer and Terra as the core of the Qiskit framework; they are included when you do a simple installation of Qiskit. In addition to these components, however, there are still a few more:

- **Qiskit Machine Learning** implements some well-known quantum machine learning algorithms that are suitable for **NISQ** devices. We will extensively work with this package in *Part 3, A Match Made in Heaven: Quantum Machine Learning*, of this book.

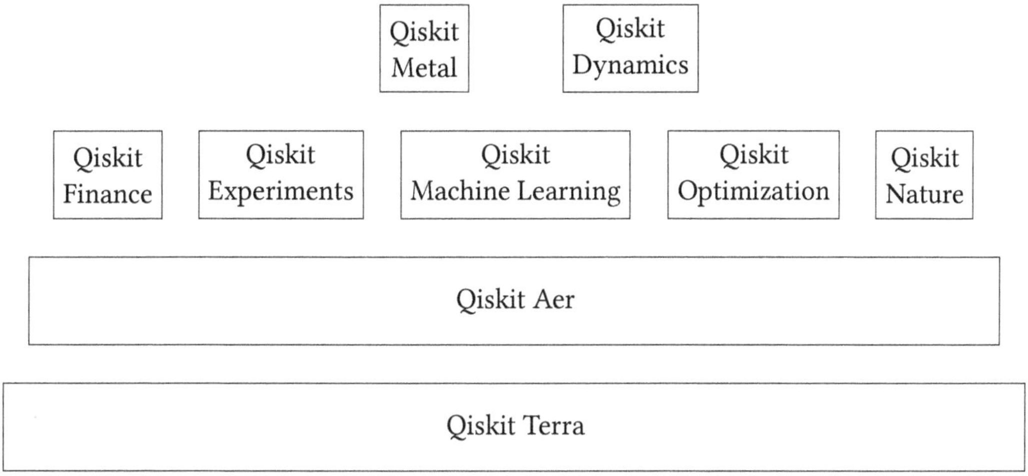

Figure 2.2: Components of the Qiskit framework

This package also provides an optional interface with **PyTorch** that can be utilized in the training of quantum machine learning models.

- **Qiskit Optimization** implements some quantum optimization algorithms. We will work with them in *Part 2, When Time is Gold: Tools for Quantum Optimization*, of this book.
- This book focuses on quantum machine learning and quantum optimization, but quantum computing has more exciting applications in other specific fields. Two good examples are the natural sciences, in particular, quantum physics and chemistry, and finance. You can find some algorithms related to problems in these fields in the **Qiskit Nature** and **Qiskit Finance** packages. As an interesting fact, you should know that the possibility of doing more efficient computations for quantum-mechanical systems was one of the initial motivations for exploring the idea of quantum computing in the first place. We will briefly explore some of these applications in *Chapter 7, VQE: Variational Quantum Eigensolver*.
- **Qiskit Experiments** provides a range of tools for working with noisy quantum computers, that is, current quantum devices that are subject to different types of errors and external noise, to characterize, benchmark, and calibrate them.

- Lastly, **Qiskit Metal** and **Qiskit Dynamics** are the most recent additions to Qiskit. Qiskit Metal can be used to design real quantum devices, while Qiskit Dynamics provides tools to work with models of quantum systems.

Exercise 2.1

Follow the instructions in *Appendix D, Installing the Tools*, to do a full installation of **version 0.39.2** of the Qiskit package.

Once you have installed Qiskit, you can load it in Python running `import qiskit`. If you want to check the version of Qiskit that you are running, you might be tempted to look for it in `qiskit.__version__`, but that would give you the version of the Qiskit Terra package, not that of Qiskit itself! If you want to find the version of the Qiskit framework, you will have to access `qiskit.__qiskit_version__`. That will give you a dictionary with the versions of all the components of Qiskit (including Qiskit itself). Hence, the version of Qiskit will be `qiskit.__qiskit_version__['qiskit']`.

To learn more...

Qiskit is updated quite frequently. To stay up to date with the new features, we recommend you visit `https://qiskit.org/documentation/release_notes.html`.

Now that we are all set up, it is time for us to build some circuits with our feet on the Terra!

2.2.2 Using Qiskit Terra to build quantum circuits

In order to get started, let us, first of all, import Qiskit as follows:

```
from qiskit import *
```

Notice how in the preceding subsection we imported Qiskit with `import qiskit` to be able to check its version number. For the remainder of this chapter, we will assume that Qiskit has been imported as `from qiskit import *`.

We will now explore how to implement quantum algorithms (in the form of quantum circuits) using Qiskit.

Initializing circuits

In Qiskit, circuits are represented as objects of the `QuantumCircuit` class. When we initialize such an object, we may give some optional arguments depending on how many qubits and bits we want our circuit to have. For example, if we want our circuit to have `n` qubits, we may invoke `QuantumCircuit(n)`. If we also want it to have `m` classical bits to store the results of measuring our qubits, we can run `QuantumCircuit(n, m)`. Once we have a quantum circuit object, we can get an ASCII representation of it in the terminal by calling the `draw` method. For instance, if we executed `QuantumCircuit(2,2).draw()`, we would get the following:

```
q_0:


q_1:


c_0:


c_1:
```

Of course, in this representation we can only see the names of the qubits and the bits that we have created, for that is all that we have in our circuit so far.

These ASCII representations are fine, but we can all agree that they are not necessarily very stylish. If you want to get something more fancy, you can pass the optional argument `'mpl'` (short for **matplotlib**) to `draw`. Keep in mind that, if you use Python on your terminal rather than in a **Jupyter notebook** (old school for the win!), you might also have to use `draw('mpl', interactive = True)`. However, this might not work if your environment doesn't support graphical user interfaces.

Qubits and classical bits in Qiskit are grouped in quantum and classical **registers**. By default, when you create a circuit `QuantumCircuit(n, m)`, Qiskit groups your qubits in a quantum register `q` and your bits in a classical register `c`. You may however want to have a different arrangement of registers or you may want to give them different names. In order to do this, you can create your own registers, which will be objects of the `QuantumRegister` and `ClassicalRegister` classes. When initializing these registers, you are free to specify some `size` and `name` parameters. Once you have created some quantum and classical registers `reg_1`,...,`reg_n`, you can stack them in a circuit with a call of the form `QuantumCircuit(reg_1,...,reg_n)`. In this way, we could execute the following code:

```
qreg1 = QuantumRegister(size = 2, name = "qrg1")
qreg2 = QuantumRegister(1, "qrg2")
creg = ClassicalRegister(1, "oldschool")

qc = QuantumCircuit(qreg1, creg, qreg2)
```

And we would get the following result if we ran `qc.draw()`:

```
   qrg1_0:

   qrg1_1:

     qrg2:

oldschool:
```

Quantum gates

We now have a circuit with a bunch of qubits — not a bad way to get started! By default, all those qubits will be initialized to a state, $|0\rangle$, but, of course, if we want to get some computing done, we better be able to bring some quantum gates to the table.

The way you add quantum gates to a circuit `qc` is by executing methods of that circuit. For instance, if you want to apply an X gate on the first qubit of the circuit `qc`, you can just run `qc.x(0)`.

Important note

As is often the case in Python, quantum bits in a circuit are 0-indexed! This means that the first qubit will be labeled as 0, the second as 1, and so on.

When we have different quantum registers in a circuit, we can still refer to qubits by their indices. The qubits of the first register that we added will have the first indices, the subsequent indices will correspond to the qubits of the second register, and so on and so forth. Classical registers and quantum registers are fully independent in this matter.

This, however, can be somewhat inconvenient if we have many registers, but don't worry, Qiskit has got you covered. While referring to gates by their index can be convenient, we can also refer to them directly! Let us say that we have a setup like the previous one, with the circuit `qc` with quantum registers `qreg1` and `qreg2`. Running `qc.x(2)` would have the same effect as executing `qc.x(qreg2[0])`. Moreover, if we called `qc.x(qreg1)`, that would be the same as applying both `qc.x(0)` and `qc.x(1)` one after the other.

The following are the Qiskit methods for applying some of the most common one-qubit gates (the ones that we studied in *Sections 1.3.3* and *1.3.4*) on a qubit `q0`:

- In order to apply one of the Pauli gates, X, Y, or Z, we can call `x(q0)`, `y(q0)`, or `z(q0)` respectively.
- The method `h(q0)` can be used to apply a Hadamard gate.
- We can apply rotation gates R_X, R_Y, or R_Z parametrized by `theta` with the `rx(theta,q0)`, `ry(theta,q0)`, or `rz(theta,q0)` methods respectively.
- We can apply the universal one-qubit gate $U(\theta, \varphi, \lambda)$ parametrized by `theta`, `phi`, and `lambd` as `u(theta, phi, lambd, q0)`.

Of course, we also have methods for multi-qubit gates. Most notably, a controlled X, Y, Z, or H gate with control qubit `q0` on a target `qt` can be applied with the `cx(q0, qt)`, `cy(q0, qt)`, `cz(q0, qt)`, and `ch(q0, qt)` methods respectively. In full analogy, a controlled rotation gate R_X, R_Y, or R_Z parametrized by a value `theta` can be added with the `crx(theta, q0, qt)`, `cry(theta, q0, qt)`, and `crz(theta, q0, qt)` methods, where, as before, `q0` represents the control qubit and `qt` the target one.

Important note

Remember that the controlled X gate is the famous **CNOT**. We love entangling qubits, so we will certainly be using that `cx` method a lot!

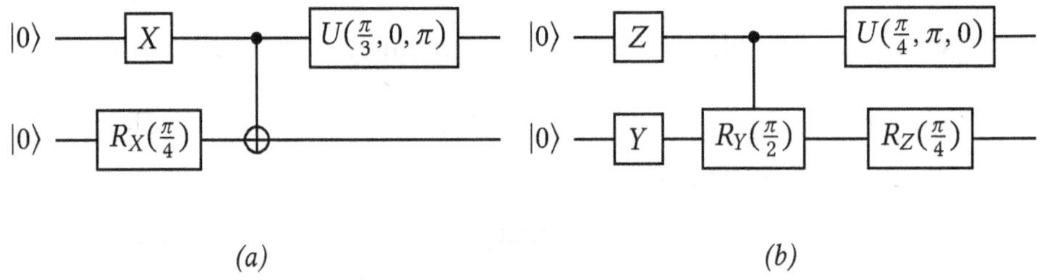

Figure 2.3: Sample quantum circuits that we can construct in Qiskit.

Now might be a good moment to step back for a second and see all that we have done come alive. For example, let us try to construct the circuit depicted in *Figure 2.3a*. Using all that we have learned, we could implement this circuit in Qiskit as follows:

```
import numpy as np

qc = QuantumCircuit(2) # Initialise the circuit.

# We can now apply the gates sequentially.
qc.x(0)
qc.rx(np.pi/4, 1)
qc.cx(0, 1)
qc.u(np.pi/3, 0, np.pi, 0)
```

And now if we run `qc.draw("mpl")` to verify that our implementation is correct, we will get the output shown in *Figure 2.4*.

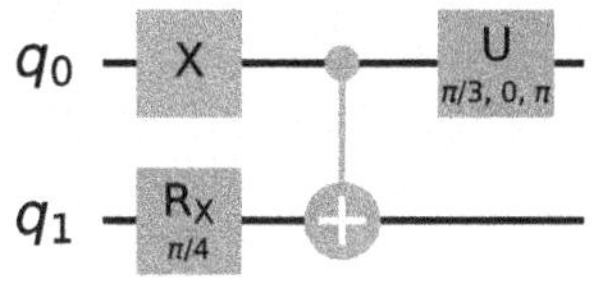

Figure 2.4: Qiskit output for the circuit in Figure 2.3a.

Exercise 2.2

Construct the circuit in *Figure 2.3b*. Draw the result and use the output to verify whether your circuit implementation is correct.

Measurements

We now know how to add quantum gates to a circuit, so there is only one ingredient that we are missing: measurement operators. It turns out that this couldn't be easier. If you want to perform a measurement (in the computational basis) at any point in a circuit, you can do so by calling the `measure(qbits,bits)` method, where `qbits` should be a list with all the qubits that you want to measure and `bits` should be a list with all the classical bits on which you want the measurements to be stored. Of course, the lists must be of the same length.

If you just want to measure all the qubits and do not want to bother with creating a classical register of the appropriate size, you can just call the `measure_all` method. This will add as many bits to your circuit as qubits it has, and it will measure each qubit and send the results to these bits. If you have already added classical bits to store the measurement results, you can still use them with the `measure_all` method: all you have to do is set the `add_bits` parameter to `False`.

Exercise 2.3

Implement your own version of the `measure_all` method. You may need to use the `add_register` method of the `QuantumCircuit` class, which takes some register objects as arguments and appends them to the circuit.

So now we can construct our own quantum circuits, but we still need to find a way to run them using Qiskit. We will do that in the next subsection. Let's fly in the Aer!

2.2.3 Using Qiskit Aer to simulate quantum circuits

As we mentioned when we introduced the `Qiskit` framework, the `Terra` package includes a Python-based simulator, **BasicAer**. While this simulator is good enough for most basic tasks, it is largely out-powered by the simulators included in the `Aer` package, so we will only discuss these here.

If we want to use the Aer simulator, it will not suffice to import Qiskit. This time, we will also have to run the following:

```
from qiskit.providers.aer import AerSimulator
```

Once we have done the necessary imports, we can create an Aer simulator object in one of the following ways, depending on whether or not we have configured our system to use a GPU:

```
sim = AerSimulator()
sim_GPU = AerSimulator(device = 'GPU')
```

If you have a GPU and have configured your Qiskit installation properly (see *Appendix D, Installing the Tools*, for some instructions), using the GPU-powered simulator will yield much better results for demanding simulation tasks. Nevertheless, you should keep in mind that for less resource-intensive simulations, using the GPU may actually lead to worse performance because of the communication overhead. For the remainder of this section, we will use the simulator without a GPU. If we were to use `sim_GPU`, everything would be fully analogous.

To learn more...

To simulate circuits using a GPU, you will need the `qiskit-aer-gpu` package. This package is written using CUDA. For this reason, it only supports NVIDIA GPUs.

As we already know, when we measure a quantum state the result is probabilistic. For that reason, we usually run several executions or **shots** of a given circuit and then compute some statistics on the results. If we wanted to simulate the execution of `nshots` shots of a circuit `qc`, we would have to run `job = execute(qc, sim, shots = nshots)`, and we could retrieve a *result* object by calling `result = job.result()`; by the way, the default value for `shots` is `1024`. With this result object, we could get the simulated frequency counts using `result.get_counts()`, which would give us a dictionary with the absolute frequencies of each outcome.

Let's try to make this more clear with an example. We will consider a very simple two-qubit circuit with a single Hadamard gate in the top qubit. We will then measure both qubits in the circuit and simulate 1024 shots:

```
qc = QuantumCircuit(2, 2)
qc.h(0)
qc.measure(range(2), range(2))

job = execute(qc, sim, shots = 1024)
result = job.result()
counts = result.get_counts()
print(counts)
```

To learn more...

If you have a job running, in an object called `job`, and you want to check its status, you can import `job_monitor` from `qiskit.providers.ibmq.job` and run `job_monitor(job)`.

Important note

When getting results for measurements in Qiskit, you need to keep in mind that the top qubit becomes the least significant bit, and so on. This is, if you have two qubits and, when measured, the top one (qubit 0) has value 0 and the bottom one (qubit 1) has value 1, the result will be interpreted as `10`, not as `01`.

This is the opposite of what we have been doing so far – and also the opposite of what most of the world agrees on. Thus, what we call state 10, Qiskit will call `01`.

For most practical purposes, we can simply ignore this issue and assume that when we have to access qubit q in a circuit with n qubits, we need to use the index $n-q-1$ (remember that we start counting qubits at 0). This is what we will do implicitly when we use Qiskit in practice.

Theoretically, we know that the state before the measurement is $\sqrt{1/2}(|00\rangle + |10\rangle)$, so we would expect an even distribution of frequencies for the (Qiskit) outcomes 01 and 00, and we should not see any occurrences of other results. Indeed, when we ran the code, we got the following:

```
{'01': 519, '00': 505}
```

Needless to say, you will not get the same results! But you will certainly get something with the same flavor.

Notice that, in the preceding measurement, we have labeled as 10 the state with 1 in the first qubit (qubit 0) (which is consistent with the notation that we have been using). Nonetheless, Qiskit, being consistent with its own notation, has labeled its corresponding outcome as 01.

To learn more...

If you want to obtain reproducible results when executing circuits in Qiskit, you need to use two parameters of the `execute` function: `seed_transpiler` and `seed_simulator`. They are used to set the initial values for the pseudo-random number generators that are used in the process of transpiling the circuit – we will talk about this later in this section – and when sampling from the results of the measurements. If you use some fixed seeds, you will always get the same results. This can be useful, for instance, for debugging purposes.

All these numbers are good, but we all know that a picture is worth a thousand words. Thankfully, the folks at IBM agree, and they have been thoughtful enough to bundle some fancy visualization tools right into Qiskit. For instance, we could run these instructions:

```
from qiskit.visualization import *
plot_histogram(counts)
```

And we would get the plot shown in *Figure 2.5*. This function admits the optional argument `filename`, which, if provided, will lead to the figure being saved with the given string as the filename.

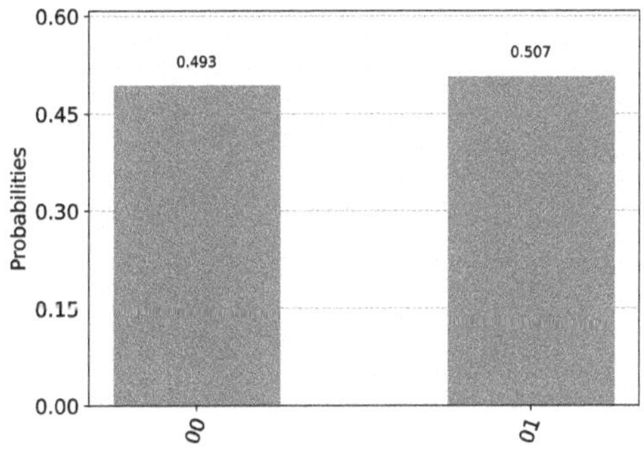

Figure 2.5: Histogram generated by Qiskit

As an interesting fact, you should know that the Aer simulator can simulate the execution of circuits using different methods. Nevertheless, unless we ask otherwise, our circuits will always be simulated using the **statevector** method, which — as the name suggests — computes the exact quantum state (or state vector) of the system through the circuit in order to generate results.

Now, if the simulator does compute the quantum state of the system, why settle for some simulated samples when we could be getting the actual state? Of course, the state of a circuit is something we don't have access to when working with real quantum computers (we can only obtain results by performing measurements), but, hey, simulating circuits should have its own perks too!

If we want to access the state vector at any point in a quantum circuit qc, all we have to do is call `qc.save_statevector()`, just as if we were adding another gate. Then, once the circuit has been simulated and we have gotten our results from the execution job, we can get the state vector using the `get_statevector` method, just as we would use `get_counts`. Actually, if our circuit also has measurements, we could do both at the same time. For instance, we can consider this example:

```
qc = QuantumCircuit(2, 2)
qc.h(0)
qc.save_statevector()
qc.measure(0,0)
qc.measure(1,1)

result = execute(qc, sim, shots = 1024).result()
sv = result.get_statevector()
print(sv)
counts = result.get_counts()
print(counts)
```

Notice how, in this piece of code, we have measured the two qubits in the circuit using two individual instructions instead of just calling `qc.measure(range(2), range(2)`. When we run that, we get the following output:

```
Statevector([0.70710678+0.j, 0.70710678+0.j, 0.        +0.j,
             0.        +0.j],
            dims=(2, 2))
{'00': 486, '01': 538}
```

That is exactly what we should have expected. We just need to remember that $1/\sqrt{2} \approx 0.7071\ldots$! In the output given by Qiskit, the first element of the state vector array is the amplitude of the basis state $|00\rangle$, the second element is the amplitude of $|10\rangle$ (remember the Qiskit convention for naming basis states, so for Qiskit, the label of this state would be `01`), and the following ones are the amplitudes of $|01\rangle$ and $|11\rangle$ respectively.

To learn more...

It is possible to save multiple state vectors in order to retrieve them later. For this, one needs to pass the optional argument `label` to `save_statevector`, specifying a label that uniquely identifies the state vector in the circuit. Then, the state vectors can be extracted as a dictionary from the results object `result` using `result.data()`.

Another possibility that the Aer simulator offers us is computing the unitary matrix that would represent, up to any given point, the transformations that have been performed by the circuit. In order to get this matrix, we could use the `save_unitary` and `get_unitary` methods, which would work in full analogy to `save_statevector` and `get_statevector`. As wonderful as this may seem, there is a small caveat, which is that these matrices cannot be computed with the statevector method; instead, one needs to use the **unitary** method, which does not support measurements and does not allow access to the state vector of the circuit. In any case, this is not a big deal, for one can always combine different methods of simulation provided the simulated circuits are tweaked accordingly.

To see this in action, let us run the following example:

```
sim_u = AerSimulator(method = 'unitary')
qc = QuantumCircuit(1)
qc.h(0)
qc.save_unitary()

result = execute(qc, sim_u).result()
U = result.get_unitary(decimals = 4)
print(U)
```

When we execute this code, this is the output that we get:

```
Operator([[ 0.7071+0.j,  0.7071-0.j],
          [ 0.7071+0.j, -0.7071+0.j]],
         input_dims=(2,), output_dims=(2,))
```

That is, just as it should be, the matrix of the Hadamard gate.

Notice, by the way, how we have used the optional argument `decimals` to limit the precision of the output. This can also be used in the `get_statevector` method.

We are now able to use Qiskit to construct and simulate circuits, but there's something we are missing: how to actually run them on real quantum hardware. That's what the next subsection is all about.

2.2.4 Let's get real: using IBM Quantum

Now we know how to use the tools provided by Qiskit Aer in order to perform ideal simulations of quantum circuits, but we know that real quantum computers, even if with limitations, exist and are actually accessible, so why not give them a shot (pun intended)?

IBM provides, for free, access to some of its real quantum computers. In order to gain access, all you have to do is sign up for a free IBM ID account. With it, you can log into the IBM Quantum website (`https://quantum-computing.ibm.com/`) and get your API token (refer to *Appendix D, Installing the Tools*, for more details).

Once you have your token, the next thing you should do is head to your local environment and execute the instruction `IBMQ.save_account("TOKEN")`, where, of course, you should replace `TOKEN` with your actual token. With that out of the way, we may run the following piece of code:

```
provider = IBMQ.load_account()
print(provider.backends(simulator = False))
```

This will allow us to load our account details and get a list of all the available real quantum devices. If you have an ordinary free account, you could expect to get something like this as output:

```
[<IBMQBackend('ibmq_lima') from IBMQ(hub='ibm-q',
    group='open', project='main')>,
<IBMQBackend('ibmq_belem') from IBMQ(hub='ibm-q',
    group='open', project='main')>,
<IBMQBackend('ibmq_quito') from IBMQ(hub='ibm-q',
    group='open', project='main')>,
<IBMQBackend('ibmq_manila') from IBMQ(hub='ibm-q',
    group='open', project='main')>,
<IBMQBackend('ibm_nairobi') from IBMQ(hub='ibm-q',
    group='open', project='main')>,
<IBMQBackend('ibm_oslo') from IBMQ(hub='ibm-q',
    group='open', project='main')>]
```

If we used the argument `simulator = True`, we would get a list of all the available cloud simulators. Their main advantage is that some of them are capable of running circuits with many more qubits than an ordinary computer can handle.

A naive way of picking any of these providers would be just choosing one element in the list, for instance, taking `dev = provider.backends(simulator = False)[0]`. Alternatively, if you knew the name of the device that you want to use (let it be `ibmq_lima`, for instance), you could simply run `dev = provider.get_backend('ibmq_lima')`. Once you have chosen a device, that is, a backend object, you can get some of its configuration details by calling the `configuration` method (with no arguments). This will return an object with information about the device. For instance, in order to know how many qubits a provider `dev` has, we could just access `dev.configuration().n_qubits`.

Nevertheless, rather than picking a device at random or by a fancy location name, we can try to do some filtering first. When calling `get_backend`, we can pass an optional `filters` parameter. This should be a one-argument function that would only return `True` for the devices that we want to pick. For instance, if we wanted to get a list of all the real devices with at least 5 qubits, we could use the following:

```
dev_list = provider.backends(
    filters = lambda x: x.configuration().n_qubits >= 5,
    simulator = False)
```

Out of all of these devices, maybe it is wise to just use the one that is the least busy. For this, we can simply execute the following:

```
from qiskit.providers.ibmq import *
dev = least_busy(dev_list)
```

To learn more...

The `least_busy` accepts an optional parameter called `reservation_lookahead`. This is the number of minutes for which a device needs to be free of reservations to be considered as a candidate for being the least busy one. The default value of the parameter is 60. So if it ever happens to you that `least_busy` does not return a suitable device, you may set `reservation_lookahead=None` to also consider computers that are under reservation.

And now, running a circuit on the device that we have selected with a certain number of shots will be completely analogous to running it on a simulator. Actually, we can run it on both and compare the results!

```
from qiskit.providers.ibmq.job import job_monitor

# Let us set up a simple circuit.
qc = QuantumCircuit(2)
qc.h(0)
qc.cx(0,1)
qc.measure_all()
```

```
# First, we run the circuit using the statevector simulator.
sim = AerSimulator()
result = execute(qc, sim, shots = 1024).result()
counts_sim = result.get_counts()

# Now we run it on the real device that we selected before.
job = execute(qc, dev, shots = 1024)
job_monitor(job)

result = job.result()
counts_dev = result.get_counts()
```

It will probably take a while to get the results (you will get status updates thanks to the `job_monitor(job)` instruction). In fact, sometimes, you can experience quite long waiting times because many users are submitting jobs at the same time. But with a little bit of patience, the results will eventually come! Once the execution has finished, we can print the results, and we could get something like this:

```
print(counts_sim)
print(counts_dev)
```

```
{'11': 506, '00': 518}
{'00': 431, '01': 48, '10': 26, '11': 519}
```

That is close, but far from ideal! We can see how in the execution on real hardware, we get some outputs – namely, `10` and `01` – that should not even be allowed in the first place. This is the effect of the **noise** of real quantum computers, which makes them deviate from perfect mathematical simulations.

To learn more...

Here, we have only worked with ideal simulations. You can also perform noisy simulations in Qiskit, which can more faithfully resemble the behavior of the quantum computers that are at our disposal today. Moreover, you could configure these simulations to use the same noise parameters as those measured in the real quantum devices owned by IBM. We will learn how to do this in *Chapter 7, VQE: Variational Quantum Eigensolver.*

Important note

When executing quantum circuits on actual quantum hardware, you have to be aware of the fact that real quantum systems only implement certain gates, and thus some of the gates that make up the circuit may have to be decomposed using the gates available. For instance, it is typical to decompose multi-qubit gates into qubits that act on just one or two qubits or to simulate CNOT gates between qubits that are not directly connected in the quantum computer by first swapping the qubits, then applying an actually existing CNOT gate, and then swapping back the qubits.

This process is called **transpilation**, and, using the code that we have considered, we have let Qiskit take care of all its details automatically. However, it is possible to dive deeper into this and to hack it as much as one could want! For instance, you can use the `transpile` method to manually define a transpilation, specifying the gates that are present in the computer or the qubits that are actually connected, among other things. See the documentation at `https://qiskit.org/documentation/stubs/qiskit.compiler.transpile.html` for more details.

In this section, we have gained a good understanding of the general structure of the Qiskit framework, we have learned how to implement circuits in it, and how to simulate them and run them on real hardware through IBM Quantum. In the following section, we will do the same for another very interesting framework: PennyLane. Let's get to it!

2.3 Working with PennyLane

The structure of PennyLane [23] is more simple than that of Qiskit. PennyLane mainly consists of a core software package, which comes with all the features that you would expect: it allows you to implement quantum circuits, it comes with some wonderful built-in simulators, and it also allows you to train quantum machine learning models (both with native tools and with a **TensorFlow** interface).

In addition to this core package, PennyLane can be extended with a wide selection of plugins that provide interfaces to other quantum computing frameworks and platforms. At the time of writing, these include **Qiskit**, **Amazon Braket**, the **Microsoft QDK**, and **Cirq**, among many others that we have not mentioned in our introduction. In addition, there is a community plugin, **PyQuest**, that makes PennyLane interoperable with the QuEST simulator (`https://github.com/johannesjmeyer/pennylane-pyquest`).

In short, with PennyLane, it's not that you get the best of both worlds. You truly can get the best of any world!

Important note

We are going to work with **version 0.26** of PennyLane. If you are using a different version, some things may be different. If in doubt, you should always check the documentation (`https://pennylane.readthedocs.io/en/stable/`).

Exercise 2.4

Follow the instructions in *Appendix D, Installing the Tools*, to install **version 0.26** of PennyLane and its Qiskit plugin.

Once you have installed PennyLane, you can import it. Following the conventions set out in PennyLane's documentation, we will do it as follows:

```
import pennylane as qml
```

After running this instruction, you can check which version of PennyLane you are running by printing the string `qml.__version__`.

Now that we are all set up, let's build our first circuit, shall we?

2.3.1 Circuit engineering 101

The way quantum circuits are built in PennyLane is fundamentally different to the way they are constructed in Qiskit.

In Qiskit, if we wanted to implement a quantum circuit, we would initialize a `QuantumCircuit` object and manipulate it with some methods; some of these methods would be used to add gates to the circuit, some to perform measurements, and some to specify where we wanted to extract information about the state of the circuit.

In PennyLane, on the other hand, if you want to run a circuit, you need two elements: a `Device` object and a function that specifies the circuit.

To put it in simple terms, a `Device` object is PennyLane's virtual analog of a quantum device. It is an object with methods that allow it to run any circuit that it is given (through a simulator, through an interface with other platforms, or however it may be!). For example, if we have a circuit and we want to run it on the `default.qubit` simulator (more on that later in this section) using two qubits, we will need to use this device:

```
dev = qml.device('default.qubit', wires = 2)
```

Notice, by the way, how the number of qubits available is a property of the device object itself.

Now that we have a device, we need to define the specification of our circuit. As we mentioned earlier, that is as easy as defining a function. In this function, we will execute instructions that will correspond to the actions of the quantum gates that we want to use. Lastly, the output of the function will be whichever information we want to get out of the circuit – whether it be the state of the circuit, some measurement samples, or whatever it may be. Of course, the output that we can get will depend on the device that we are using.

Let us illustrate this with an example:

```
def qc():
    qml.PauliX(wires = 0)
    qml.Hadamard(wires = 0)
    return qml.state()
```

Here we have a very basic circuit specification. In this circuit, we get the state vector (with `qml.state()`), after we first apply an X gate on the first qubit and then an H gate on the first qubit too. We do this by calling, in sequence, `qml.PauliX` and `qml.Hadamard`, specifying the wires on which we want the gates to act. In most non-parametrized gates, `wires` is the first positional argument, and it does not have a default value, so you need to provide one. In the case of single-qubit gates, this value must be an integer representing the qubit on which the gate is meant to act. Analogously, for multi-qubit gates, `wires` must be a list of integers.

You may have noticed that the naming conventions for gate classes in PennyLane differ from those for gate methods in Qiskit. The functions for the X, Y, and Z Pauli gates are, respectively, `qml.PauliX`, `qml.PauliY`, and `qml.PauliZ`. Also, as we have just seen, the function for the Hadamard gate is `qml.Hadamard`.

In regard to rotation gates, we can apply R_X, R_Y, and R_Z parametrized by `theta` on a wire `w` using the instructions `qml.RX(phi=theta, wires=w)`, `qml.RY(phi=theta, wires=w)`, and `qml.RZ(phi=theta, wires=w)` respectively. In addition, the universal single-qubit gate $U(\theta, \varphi, \lambda)$ can be applied on a wire `w` calling `qml.U3(theta, phi, lambd, w)`.

Lastly, the controlled Pauli gates can be applied on a pair of qubits `w = [w0, w1]` using the instructions `qml.CNOT(w)`, `qml.CY(w)` and `qml.CZ(w)`. The first wire, `w0`, is meant to be the control qubit, while the second one must be the target. Controlled X, Y, and Z rotations parametrized by an angle `theta` can be added with the instructions `qml.CRX(theta, w)`, `qml.CRY(theta, w)`, and `qml.CRZ(theta, w)` respectively.

In any case, we now have a two-qubit device `dev` and we have a circuit function `qc`. How do we assemble these two together and run the circuit? Easy, all we have to do is execute the following:

```
qcirc = qml.QNode(qc, dev) # Assemble the circuit & the device.
qcirc() # Run it!
```

If we run this, we will get the following result,

```
tensor([ 0.70710678+0.j,  0.        +0.j, -0.70710678+0.j,
         0.        +0.j], requires_grad=True)
```

which makes perfect sense, for we know that

$$(HX \otimes I)|00\rangle = (H \otimes I)|10\rangle = \frac{1}{\sqrt{2}}(|00\rangle - |10\rangle) \approx (0.7071\ldots)(|00\rangle - |10\rangle).$$

As a fun fact, the result of assembling a circuit function and a device is known, in PennyLane jargon, as a **Quantum Node** (or **QNode**, for short).

Important note

> PennyLane, unlike Qiskit, labels state like most people do: assigning to the first qubit the most significant bit. Thus, a PennyLane output of `10` corresponds to the state $|10\rangle$.

Notice how, consistent with PennyLane's convention for labeling states, the state vector is returned as a list with the amplitudes of the states in the computational basis. The first element corresponds to the amplitude state $|0 \cdots 0\rangle$, the second one to that of $|0 \cdots 01\rangle$, and so on.

In the preceding example, we should remark that at no point in the definition of the function `qc` did we specify the number of qubits of the circuit — we left that to the device. When we create a QNode, PennyLane assumes that the device has enough qubits to execute the circuit specification. If that isn't the case, we will encounter a `WireError` exception when executing the corresponding QNode.

If you — like most of us! — are lazy, all this process of defining a function and assembling it with a device might seem overwhelmingly exhausting. Thankfully, the folks at PennyLane were kind enough to provide a shortcut. If you have a device `dev` and want to define a circuit for it, you could just do the following:

```
@qml.qnode(dev) # We add this decorator to use the device dev.
def qcirc():
    qml.PauliX(wires = 0)
    qml.Hadamard(wires = 0)
    return qml.state()

# Now qcirc is already a QNode. We can just run it!
qcirc()
```

Now that is much cuter! By placing the `@qml.qnode(dev)` decorator before the definition of our circuit function, it automatically became a QNode without us having to do anything else.

We have seen how circuits in PennyLane are implemented as simple functions, and this begs the question: are we allowed then to use parameters in these functions? The answer is a resounding yes. Let us say that we want to construct a one-qubit circuit, parametrized by a certain `theta`, which performs an X-rotation by this parameter. Doing so is as easy as this:

```
dev = qml.device('default.qubit', wires = 1)
@qml.qnode(dev)
def qcirc(theta):
    qml.RX(theta, wires = 0)
    return qml.state()
```

And, with this, for any value `theta` of our choice, we can run `qcirc(theta)` and get our result. This way of handling parameters is very handy and convenient. Of course, you can use loops and conditionals dependent on circuit parameters within the definition of a circuit. The possibilities are endless!

If at any point you need to draw a circuit in PennyLane, that is not an issue: it is fairly straightforward. Once you have a quantum node `qcirc`, you can pass this node to the `qml.draw` function. This will itself return a function, `qml.draw(qcirc)`, which will take the same arguments as `qcirc` and will give you a string that *draws* the circuit for each choice of those arguments. We may see this more clearly with an example. Let us execute the following piece of code to draw the `qcirc` circuit that we've just considered for $\theta = 2$:

```
print(qml.draw(qcirc)(theta = 2))
```

Upon running that, we get the following representation of the circuit:

```
0: --RX(2.00)--|  State
```

So far, we have only performed simulations that return the state vector of the circuit at the end of its executions, but, naturally, that is just one of the many options that PennyLane provides. These are some, but not all, of the return values that we can have in a circuit function:

- If we want to get the state of the circuit at the end of its execution, we can, as we have seen, return `qml.state()`.
- If we wish to get a list with the probabilities of each state in the computational basis of a list of wires `w`, we can return `qml.probs(wires = w)`.
- We can get a sample of measurements in the computational basis of some wires `w` by returning `qml.sample(wires = w)`; the `wires` argument is optional (if no value is provided, all qubits are measured). When we get a sample, we have to specify its size by either setting a `shots` argument when invoking the device or by setting it when calling the QNode.

We will explore some additional possibilities for return values in *Chapter 10, Quantum Neural Networks.* We already know how to get the state of the circuit. In order to illustrate the other return values that we may use, let us execute the following piece of code:

```
dev = qml.device('default.qubit', wires = 3)

# Get probabilities
@qml.qnode(dev)
def qcirc():
    qml.Hadamard(wires = 1)
    return qml.probs(wires = [1, 2]) # Only the last 2 wires.
prob = qcirc()
print("Probs. wires [1, 2] with H in wire 1:", prob)

# Get a sample, not having specified shots in the device.
@qml.qnode(dev)
def qcirc():
    qml.Hadamard(wires = 0)
    return qml.sample(wires = 0) # Only the first wire.
s1 = qcirc(shots = 4) # We specify the shots here.
print("Sample 1 after H:", s1)

# Get a sample with shots in the device.
dev = qml.device('default.qubit', wires = 2, shots = 4)
@qml.qnode(dev)
def qcirc():
    qml.Hadamard(wires=0)
    return qml.sample() # Will sample all wires.
s2 = qcirc()
print("Sample 2 after H x I:", s2)
```

The output we got with this execution is the following (the samples returned in your case will probably be different):

```
Probs. wires [1, 2] with H in wire 1: [0.5 0.  0.5 0. ]
Sample 1 after H: [0 1 0 0]
Sample 2 after H x I: [[1 0], [0 0], [0 0], [1 0]]
```

There might be a bit to unpack here. So, first of all, we are returned a list of probabilities; these are, in accordance with PennyLane's conventions, the probabilities of getting 00, 01, 10, and 11. In these possible outcomes, the first (leftmost) bit represents the outcome of the first measured qubit: in our case, since we are measuring wires `[1, 2]`, the second wire of the circuit, wire 1. The second (rightmost) bit represents the outcome of the second measured qubit: in our case, the third wire of the circuit. For example, the first number in the list of probabilities represents the probability of getting 00 (that is, 0 in both wires). The second number in the list would be the probability of getting 01 (0 in wire 1 and 1 in wire 2). And so on.

Lastly, in the next two examples, we are getting some measurement samples. In the first case, we specify that we want to measure only the first qubit (wire 0), and, when we call the QNode, we ask for 4 shots; since we hadn't specified a default number of shots when defining the device, we need to do it in the execution. And with that, we have a sample of the first qubit. In our case, the results were first 0, then 1, and then two more zeros.

In the last example, we define a two-qubit circuit and we measure all the wires. We already specified a default number of shots (4) when we defined the device, so we don't need to do it when calling the QNode. And, upon execution, we are given a sample of measurements. Each item in the list corresponds to a sample. Within each sample, the first element gives the result of measuring the first qubit of the circuit, the second element the result of measuring the second qubit, and so on it would go. For example, in our case, we see that in the first measurement we obtained 1 on the first qubit and 0 on the second one.

Exercise 2.5

Implement the circuits in *Figure 2.3* and verify that you get the same state vector that we get when simulating with Qiskit Aer.

Keep in mind that, as we have mentioned before, Qiskit and PennyLane use different conventions when naming basis states. Be careful with that!

To learn more...

If you want to get reproducible results using PennyLane's simulator, you can set a seed `s` using the instruction `np.random.seed(s)` after importing the `numpy` package as `np`.

So far, we have been working with devices based on the `default.qubit` simulator, which is a Python-based simulator with some basic functionalities. We will introduce more simulators when we dive into the world of quantum machine learning. For now, however, you should at least know about the existence of the `lightning.qubit` simulator, which relies on a C++ backend and provides a significant boost in performance, especially for circuits with a large number of qubits. Its usage is analogous to that of the `default.qubit` simulator. Furthermore, there is a `lightning.gpu` simulator that can enable the Lightning simulator to rely on your GPU. It can be installed as a plugin. As in the case of Qiskit, at the time of writing this book, it only supports NVIDIA GPUs (and mainly just quite modern ones!).

2.3.2 PennyLane's interoperability

We have mentioned plenty of times how one of PennyLane's virtues is its ability to communicate with other quantum frameworks. We will now try to demonstrate this with PennyLane's Qiskit interface. Parlez-vous Qiskit?

When you install the Qiskit plugin for PennyLane, you gain access to a new set of devices: most notably, a `qiskit.aer` device that allows you to use the Aer simulator directly from

PennyLane, and a qiskit.ibmq device that enables you to run circuits on the real quantum computers available from IBM Quantum.

Love is in the Aer

If we want to simulate a circuit in PennyLane using Aer, all we have to do is use a device with the qiskit.aer simulator – and, of course, having installed the appropriate plugin (refer to *Appendix D, Installing the Tools*). This will allow us to get measurement samples and also measurement probabilities (through qml.sample and qml.probs respectively). Actually, the measurement probabilities returned by these Aer devices are approximations of the exact probabilities: they are obtained by sampling and returning the empirical probabilities. By default, in an Aer device, the number of shots is fixed at 1024, following Qiskit's conventions. Of course, the number of shots can be adjusted as with any other PennyLane device.

We can see a qiskit.aer device in action with the following code sample:

```
dev = qml.device('qiskit.aer', wires = 2)
@qml.qnode(dev)
def qcirc():
    qml.Hadamard(wires = 0)
    return qml.probs(wires = 0)
s = qcirc()
print("The probabilities are", s)
```

When we run this, we can get something like the following output:

```
The probabilities are [0.48535156 0.51464844]
```

This shows that, indeed, the result is not analytical, but empirical and extracted from a sample. If you want to obtain the state vector, you need to use the following instruction when creating the device:

```
dev = qml.device('qiskit.aer', wires = 2,
    backend='aer_simulator_statevector', shots = None)
```

This will allow you to use `qml.state()` to retrieve the state amplitudes, as we did with the PennyLane devices. Moreover, if you try to get probabilities using `qml.probs` with this device object, you will now get analytical results. For example, if you run the previous example on this device, you will always obtain `[0.5, 0.5]`.

Connecting to IBMQ

The ability to (partially) use the Aer simulator is probably not the most appealing feature of PennyLane's Qiskit interface. Nevertheless, being able to connect to IBM's quantum computers is a more exciting possibility.

In order to connect to an IBM Quantum device, we will first load Qiskit and get the name of the least busy hardware backend, just as we did in the previous section:

```
from qiskit import *
from qiskit.providers.ibmq import *

# Save our token if we haven't already.
IBMQ.save_account('TOKEN')
# Load the account and get the name of the least busy backend.
prov = IBMQ.load_account()
bck = least_busy(prov.backends(simulator = False)).name()
# Invoke the PennyLane IBMQ device.
dev = qml.device('qiskit.ibmq', wires = 1,
    backend = bck, provider = prov)

# Send a circuit and get some results!
@qml.qnode(dev)
def qcirc():
    qml.Hadamard(wires = 0)
    return qml.probs(wires = 0)
print(qcirc())
```

Upon executing the preceding code, we get our desired result:

```
[0.51660156 0.48339844]
```

And that's how you can send jobs to IBM Quantum using PennyLane!

To learn more...

Of course, you can use any quantum device that is accessible with your IBM account, not just the least busy one. You only need to replace the definition of the backend in the previous code with a direct specification of a particular computer as we did in the previous section.

With this introduction to the workings of Qiskit and PennyLane and all the mathematical concepts that we studied in *Chapter 1, Foundations of Quantum Computing*, we are now ready to begin solving problems with actual quantum algorithms. And that is exactly what we will do, starting in the next chapter. Let the quantum games begin — and may the shots be ever in your favor!

Summary

In this chapter, we have explored some of the frameworks and platforms that can enable us to implement, simulate, and run quantum algorithms. We have also learned how to work with two of these frameworks: Qiskit and PennyLane, which are very widely used. In addition to this, we have learned how to use the IBM Quantum platform to execute quantum circuits on real hardware, sending them from either Qiskit or PennyLane.

With the skills that you have gained in this chapter, you are now able to implement and execute your own circuits. Moreover, you are now well-prepared to read the rest of the book, since we will be using Qiskit and PennyLane extensively.

In the next chapter, we will take our first steps in putting all this knowledge into practice. We shall dive into the world of quantum optimization!

Part 2

When Time is Gold: Tools for Quantum Optimization

This part focuses on the use of quantum algorithms to solve optimization problems. You will learn about **Quadratic Unconstrained Binary Optimization** (**QUBO**) problems and how to solve them with quantum annealers and digital quantum computers. You will also learn about more general optimization problems and about the Variational Quantum Eigensolver.

The chapters included in this part are the following:

- *Chapter 3, Working with Quadratic Unconstrained Binary Optimization Problems*
- *Chapter 4, Adiabatic Quantum Computing and Quantum Annealing*
- *Chapter 5, QAOA: Quantum Approximate Optimization Algorithm*
- *Chapter 6, GAS: Grover Adaptive Search*
- *Chapter 7, VQE: Variational Quantum Eigensolver,*

3

Working with Quadratic Unconstrained Binary Optimization Problems

The universe cannot be read until we have learned the language and become familiar with the characters in which it is written.

— Galileo Galilei

Starting with this chapter, we will be studying different algorithms that have been proposed to solve optimization problems with quantum computers. We will work both with **quantum annealers** and with computers that implement the **quantum circuit model**. We will use methods such as the **Quantum Approximate Optimization Algorithm** (QAOA), **Grover's Adaptive Search** (GAS), and the **Variational Quantum Eigensolver** (VQE). We will also learn how to adapt these algorithms to different types of problems, and how to run them on simulators and actual quantum computers.

But before we can do all that, we need a language in which we can state problems in a manner that makes it possible for a quantum computer to solve them. In this regard, with the **Quadratic Unconstrained Binary Optimization** (**QUBO**) framework, we can formulate many different optimization problems in a way that maps directly into the quantum setting, allowing us to use a plethora of quantum algorithms to try to find solutions that are optimal or, at least, close to optimal.

This chapter will introduce all the tools that we need to work with QUBO formulations. We will start by studying the **maximum cut** (or **Max-Cut**) problem in graphs, probably the simplest problem that can be formulated in the QUBO framework, and we will work our way up from there.

We'll cover the following topics in this chapter:

- The Max-Cut problem and the Ising model
- Enter quantum: formulating optimization problems the quantum way
- Moving from Ising to QUBO and back
- Combinatorial optimization problems with the QUBO model

After reading this chapter, you will be ready to write your own optimization problems in a format that will allow you to solve them using quantum computers.

3.1 The Max-Cut problem and the Ising model

In order for us to understand how to use quantum computers to solve optimization problems, we need to get used to some abstractions and techniques that we will develop throughout this chapter. To get started, we will consider the problem of finding what we call **maximum cuts** in a mathematical structure called a **graph**. This is possibly the simplest problem that can be written in the formalism that we will be using in the following chapters. It will help us in gaining intuition and it will provide a solid foundation for formulating more complicated problems later on.

3.1.1 Graphs and cuts

When you are given a graph, you are essentially given some *elements*, which we will refer to as **vertices**, and some *connections* between pairs of these vertices, which we will call **edges**. See *Figure 3.1* for an example of a graph with five vertices and six edges.

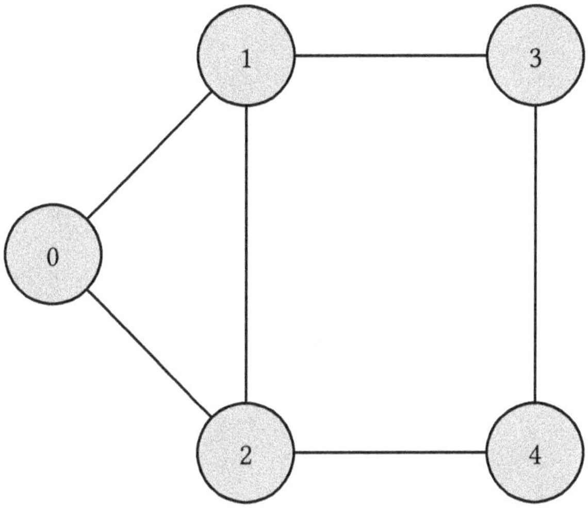

Figure 3.1: Example of a graph

Given a graph, the **Max-Cut problem** consists in finding a **maximum cut** of it. That is, we want to divide the vertices of the graph into two sets — that's what we call **cutting** the graph into two parts — such that the number of edges with extremes in different sets of the cut is the maximum possible. We call the number of such edges the **size of the cut**, and we say that these edges are **cut**. You can imagine that, for instance, the vertices represent workers of a company, edges have been added between people who don't get along that well, and you need to form two teams trying to minimize the number of conflicts by putting potential enemies in different teams.

Figure 3.2 presents two different cuts for the graph in *Figure 3.1*, using different colors for vertices that go in different sets and using dashed lines for edges that have extremes in different parts of the cut. As you can see, the cut in *Figure 3.2a* has size 5, while the cut in *Figure 3.2b* is of size 4. In fact, it is easy to check that no cut of this graph can have a size bigger than 5 since vertices 0, 1, and 2 can't all go in different sets and, hence, at least one of the edges (0, 1), (0, 2), or (1, 2) will not be cut. The cut in *Figure 3.2a* is, then, a **maximum** or **optimal** cut.

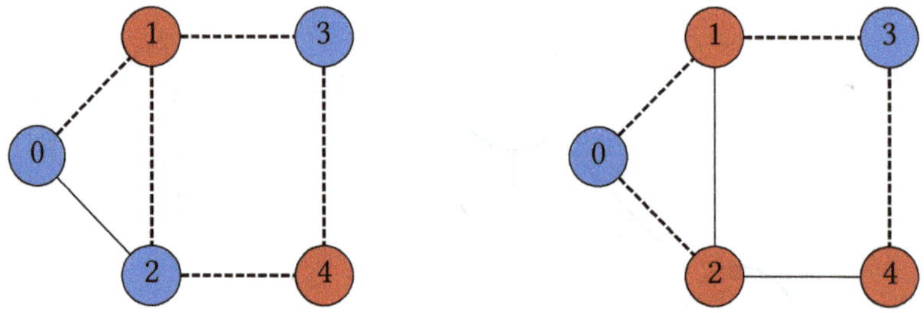

(a) Optimal cut. *(b) Non-optimal cut.*

Figure 3.2: Two different cuts of the same graph

Exercise 3.1

The maximum cut in a graph doesn't need to be unique. Find a maximum cut for the graph in *Figure 3.1* in which vertices 0 and 1 are in the same set.

So now we know what the Max-Cut problem is about. But how do we formulate it mathematically? We will learn exactly that in the next subsection.

3.1.2 Formulating the problem

Surprisingly enough, we can formulate the Max-Cut problem as a combinatorial optimization problem with no reference whatsoever to graphs, edges, or vertices. In order to do that, we associate a variable z_i to each vertex $i = 0, \ldots, n-1$ of the graph. Variables z_i will take value 1 or -1. Each assignment of values to the variables determines a cut: vertices whose variables take value 1 will be in one set and vertices whose variables take value -1 will be in the other one. For instance, for the cut of *Figure 3.2a* we could have $z_0 = z_2 = z_3 = 1$ and $z_1 = z_4 = -1$. Notice that, for our purposes, we could also represent that cut with the assignment $z_0 = z_2 = z_3 = -1$, $z_1 = z_4 = 1$.

The key observation to formulate Max-Cut as a combinatorial optimization problem is to notice that, if there is an edge between two vertices j and k, then that edge is cut if and only if $z_j z_k = -1$. This is because if the two vertices are in the same set, then either $z_j = z_k = 1$ or $z_j = z_k = -1$ and, consequently, $z_j z_k = 1$. However, if they are in different sets, then either $z_j = 1$ and $z_k = -1$, or $z_j = -1$ and $z_k = 1$, yielding $z_j z_k = -1$.

Thus, our problem can be written as

$$\begin{array}{ll} \text{Minimize} & \sum_{(j,k)\in E} z_j z_k \\ \text{subject to} & z_j \in \{-1, 1\}, \qquad j = 0, \ldots, n-1 \end{array}$$

where E is the set of edges in the graph and the vertices are $\{0, \ldots, n-1\}$. For instance, for the graph in *Figure 3.1*, we would have the following formulation:

$$\begin{array}{ll} \text{Minimize} & z_0 z_1 + z_0 z_2 + z_1 z_2 + z_1 z_3 + z_2 z_4 + z_3 z_4 \\ \text{subject to} & z_j \in \{-1, 1\}, \qquad j = 0, \ldots, 4. \end{array}$$

Note that the cut $z_0 = z_2 = z_3 = 1$, $z_1 = z_4 = -1$ (which is the one in *Figure 3.2a*), attains a value of -4 in the function to be minimized, which is the minimum possible value for this particular case – but notice that it does not coincide with the number of edges that are cut! The cut $z_0 = z_3 = -1$, $z_1 = z_2 = z_4 = 1$, on the other hand, achieves a value of -2, showing once again that the cut on *Figure 3.2b* is not optimal.

Exercise 3.2

Write the Max-Cut problem for the graph in *Figure 3.3* as an optimization problem. What is the value of the function to be minimized when $z_0 = z_1 = z_2 = 1$ and $z_3 = z_4 = z_5 = -1$. Is it an optimal cut?

At first sight, solving the Max-Cut problem might seem easy enough. However, it is an **NP-hard** problem (refer to *Appendix C, Computational Complexity*, for more details on this

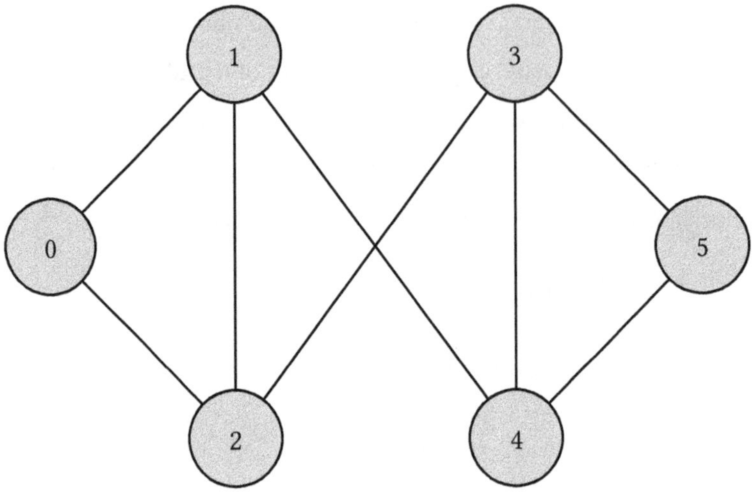

Figure 3.3: Another example of a graph

kind of problem). That means that if we were able to solve it efficiently with a classical algorithm, we would have $P = NP$, something the scientific community strongly believes not to be true. This would be the case even if we could find a classical algorithm that approximates the optimal cut within a factor of $16/17$, as was proved by Håstad in a paper published in 2001 [24]. So, even if we resort to looking for precise enough approximations, the problem is indeed hard!

To learn more...

If you want to learn more about P, NP, and NP-hard problems, please check *Appendix C, Computational Complexity.* We will be discussing the **ratio of approximation** that quantum algorithms can achieve for the Max-Cut problem in *Chapter 5, QAOA: Quantum Approximate Optimization Algorithm.*

We are now able to formulate Max-Cut as a minimization problem in which the variables take values 1 and -1. Is this just accidental or are there more problems that can be written in a similar way? Keep on reading and you will learn the answer in the next subsection.

3.1.3 The Ising model

The Max-Cut problem, as formulated in the previous pages, can be seen as just a particular case of a seemingly unrelated problem in statistical physics: finding the state of minimum **energy** of an instance of the **Ising model**. For the physics geeks out there, this is a mathematical model for the ferromagnetic interaction of particles with **spin**, usually arranged in a lattice (see *Figure 3.4* and refer to the book by Gallavotti [25] for more details). The particle spins are represented by variables z_j that can take values 1 (spin up) or -1 (spin down) – sounds familiar, doesn't it?

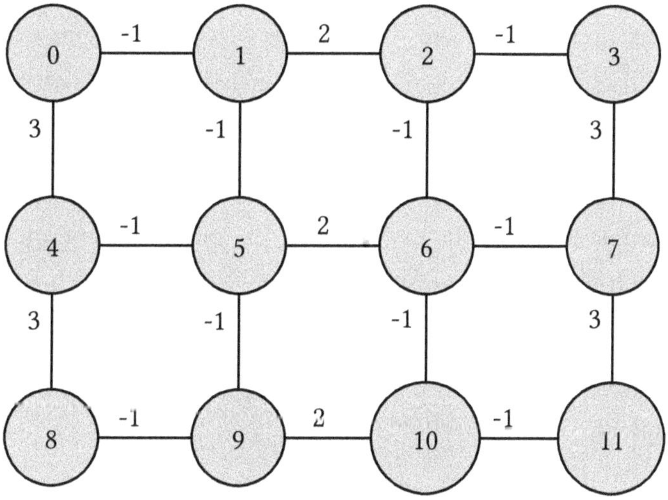

Figure 3.4: Example of the Ising model

The total energy of the system is given by a quantity called the **Hamiltonian** function (more about this later in this chapter) defined by

$$-\sum_{j,k} J_{jk} z_j z_k - \sum_j h_j z_j$$

where the coefficients J_{jk} represent the interaction between particles j and k (usually, only non-zero for adjacent particles) and the coefficients h_j represent the influence of an external magnetic field on particle j.

Finding the state of minimum energy of the system consists in obtaining a spin configuration for which the Hamiltonian function attains its minimum value. As you can easily check yourself, when all the J_{jk} coefficients are -1 and all the h_j coefficients are 0, the problem is exactly the same as getting the maximum cut in a graph — although in a completely different context! Of course, this makes the problem of finding the state of minimum energy of a given Ising model an NP-hard problem.

To learn more...

The quantum annealers that we will be using in *Chapter 4, Quantum Adiabatic Computing and Quantum Annealing*, are quantum computers constructed with the specific purpose of sampling from states of low energy of systems whose behavior can be described with the Ising model. We will use this property to try to approximate solutions to Max-Cut and many other related problems.

Let's give an example of the problem of finding the minimum energy state of an Ising model. Imagine that we have particles arranged as in *Figure 3.4*, where the numbers on the edges represent the coefficients J_{jk} and we assume that the external magnetic field is homogeneous and that all coefficients h_j are equal to 1. Then, the problem can be formulated as follows:

$$\begin{array}{ll}\text{Minimize} & z_0z_1 - 2z_1z_2 + z_2z_3 - 3z_0z_4 + z_1z_5 + z_2z_6 - 3z_3z_7 \\ & + z_4z_5 - 2z_5z_6 + z_6z_7 - 3z_4z_8 + z_5z_9 + z_6z_{10} - 3z_7z_{11} \\ & + z_8z_9 - 2z_9z_{10} + z_{10}z_{11} - z_0 - z_1 - z_2 - z_3 - z_4 - z_5 \\ & - z_6 - z_7 - z_8 - z_9 - z_{10} - z_{11} \\ \text{subject to} & z_j \in \{-1, 1\}, \qquad j = 0, \ldots, 11.\end{array}$$

This seems a little more involved than the formulations of the Max-Cut problem that we have seen so far, but it clearly follows the same pattern. However, you could be wondering what all this has to with quantum computing, since all these formulas only involve classical variables. Fair point! It is now time to use our knowledge of qubits and quantum gates to try to see all these problems under a different, quantum light.

3.2 Enter quantum: formulating optimization problems the quantum way

In this section, we will unveil how all the work that we have done so far in this chapter has followed a secret plan! Was the choice of z as the name for the variables in our problems completely arbitrary? Of course not! If it made you think of those lovely Z quantum gates and matrices that we introduced back in *Chapter 1, Foundations of Quantum Computing*, you were on the right track. It will be the key to introducing the *quantum factor* into our problems, as we will begin to see in the next subsection.

3.2.1 From classical variables to qubits

So far, the formulations that we have considered for the Max-Cut problem and for the Ising model are purely classical. They do not mention quantum elements such as qubits, quantum gates, or measurements. But we are closer than you might think to being able to give a quantum formulation for these problems. We will start with a very simple instance of the Max-Cut problem and show how we can easily transform it into *quantum form*. Consider the graph in *Figure 3.5*. We already know that the corresponding Max-Cut problem can be written as follows:

$$\begin{array}{lll} \text{Minimize} & z_0 z_1 + z_0 z_2 & \\ \text{subject to} & z_j \in \{-1, 1\}, & j = 0, 1, 2. \end{array}$$

The crucial observation that we need to make in order to transform this formulation into a quantum one is that our beloved Z matrix can be used to evaluate the different terms in

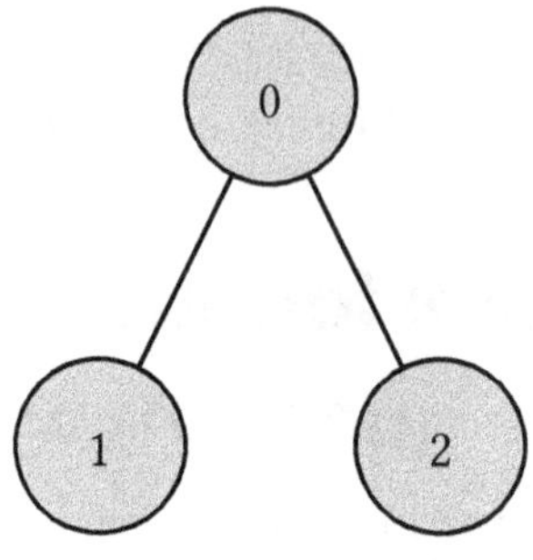

Figure 3.5: A very simple Max-Cut problem

the function that we need to minimize. Namely, it is easy to check that

$$\langle 0| Z |0\rangle = \begin{pmatrix} 1 & 0 \end{pmatrix}\begin{pmatrix} 1 & 0 \\ 0 & -1 \end{pmatrix}\begin{pmatrix} 1 \\ 0 \end{pmatrix} = 1, \qquad \langle 1| Z |1\rangle = \begin{pmatrix} 0 & 1 \end{pmatrix}\begin{pmatrix} 1 & 0 \\ 0 & -1 \end{pmatrix}\begin{pmatrix} 0 \\ 1 \end{pmatrix} = -1.$$

Now, consider the tensor product $Z \otimes Z \otimes I$ and basis state $|010\rangle$. We know from *Section 1.5.1* that

$$\begin{aligned}\langle 010| Z \otimes Z \otimes I |010\rangle &= \langle 010| (Z|0\rangle \otimes Z|1\rangle \otimes I|0\rangle) \\ &= \langle 0| Z |0\rangle \langle 1| Z |1\rangle \langle 0| I |0\rangle = 1 \cdot (-1) \cdot 1 = -1.\end{aligned}$$

We interpret $|010\rangle$ as representing a cut in which vertices 0 and 2 are assigned to one set (because the value of qubits 0 and 2 in $|010\rangle$ is 0) and vertex 1 is assigned to the other (because qubit 1 has value 1 in $|010\rangle$). Then, the fact that the product $\langle 010| Z \otimes Z \otimes I |010\rangle$ evaluates to -1 means that edge $(0, 1)$ has extremes in different sets of the cut; that is because we have used $Z \otimes Z \otimes I$, having Z operators acting on qubits 0 and 1. This behavior is analogous to the one that we had with term $z_0 z_1$ in the function to minimize our classical formulation of the problem.

In fact, $Z \otimes Z \otimes I$ is usually denoted by just $Z_0 Z_1$ (the subindices indicate the positions of each Z gate; the other positions are assumed to be the identity) and, following this convention, we would have, for instance,

$$\langle 010| Z_0 Z_2 |010\rangle = \langle 0| Z |0\rangle \langle 1| I |1\rangle \langle 0| Z |0\rangle = 1 \cdot 1 \cdot 1 = 1$$

because the edge $(0, 2)$ is not cut with this particular assignment.

Of course, this is analogous for any basis state $|x\rangle$ with $x \in \{000, 001, \ldots, 111\}$, so $\langle x| Z_j Z_k |x\rangle$ will be -1 if the edge (j, k) is cut under an assignment x and it will be 1 otherwise. We only need to notice that if j and k are in different parts of the cut, then their qubits will have different values and the product will be -1.

Furthermore, by linearity it holds that

$$\langle x| (Z_0 Z_1 + Z_0 Z_2) |x\rangle = \langle x| Z_0 Z_1 |x\rangle + \langle x| Z_0 Z_2 |x\rangle .$$

Thus, we can rewrite our problem as finding a basis state $|x\rangle$ for which $\langle x| (Z_0 Z_1 + Z_0 Z_2) |x\rangle$ attains a minimum.

Exercise 3.3

Compute $\langle 010| (Z_0 Z_1 + Z_0 Z_2) |010\rangle$ and $\langle 100| (Z_0 Z_1 + Z_0 Z_2) |100\rangle$. Does any of those states minimize $\langle x| (Z_0 Z_1 + Z_0 Z_2) |x\rangle$?

But that is not the end of the story. For any basis state $|x\rangle$, it holds that either $Z_j Z_k |x\rangle = |x\rangle$ or $Z_j Z_k |x\rangle = -|x\rangle$, as you can easily check. Notice that this proves that each $|x\rangle$ is an **eigenvector** of $Z_j Z_k$ with **eigenvalue** either 1 or -1 (refer to *Appendix B, Basic Linear Algebra*, for more information about eigenvectors and eigenvalues). Thus, for $x \neq y$ we will have

$$\langle y| Z_j Z_k |x\rangle = \pm \langle y|x\rangle = 0,$$

because $\langle y|x\rangle = 0$ whenever $x \neq y$, as we proved in *Section 1.5.1*.

Consequently, since a general state $|\psi\rangle$ can always be written as $|\psi\rangle = \sum_x a_x |x\rangle$, it follows by linearity that

$$\begin{aligned}\langle\psi| Z_j Z_k |\psi\rangle &= \left(\sum_y a_y^* \langle y|\right) Z_j Z_k \left(\sum_x a_x |x\rangle\right) = \sum_y \sum_x a_y^* a_x \langle y| Z_j Z_k |x\rangle \\ &= \sum_x |a_x|^2 \langle x| Z_j Z_k |x\rangle,\end{aligned}$$

where we have used $a_x^* a_x = |a_x|^2$.

Hence, again by linearity, it is true that

$$\begin{aligned}\langle\psi| (Z_0 Z_1 + Z_0 Z_2) |\psi\rangle &= \langle\psi| Z_0 Z_1 |\psi\rangle + \langle\psi| Z_0 Z_2 |\psi\rangle \\ &= \sum_x |a_x|^2 \langle x| Z_0 Z_1 |x\rangle + \sum_x |a_x|^2 \langle x| Z_0 Z_2 |x\rangle \\ &= \sum_x |a_x|^2 \langle x| (Z_0 Z_1 + Z_0 Z_2) |x\rangle.\end{aligned}$$

We know that $\sum_x |a_x|^2 = 1$ and that every $|a_x|^2$ is non-negative, so it holds that

$$\begin{aligned}\sum_x |a_x|^2 \langle x| (Z_0 Z_1 + Z_0 Z_2) |x\rangle &\geq \sum_x |a_x|^2 \langle x_{\min}| (Z_0 Z_1 + Z_0 Z_2) |x_{\min}\rangle \\ &= \langle x_{\min}| (Z_0 Z_1 + Z_0 Z_2) |x_{\min}\rangle \sum_x |a_x|^2 \\ &= \langle x_{\min}| (Z_0 Z_1 + Z_0 Z_2) |x_{\min}\rangle,\end{aligned}$$

where $|x_{\min}\rangle$ is a basis state (there could be more than one) for which $\langle x| (Z_0 Z_1 + Z_0 Z_2) |x\rangle$ is minimum and, hence, $x_{\min}$ represents a maximum cut.

This may all seem a little bit too abstract. But what we have proved is simply that the minimum over all possible quantum states is always reached on one of the basis states — which are the only ones that we can directly interpret as representing cuts. Then, we can rewrite the problem of finding a maximum cut for the graph of *Figure 3.5* as follows:

$$\text{Minimize} \quad \langle\psi|\left(Z_0Z_1 + Z_0Z_2\right)|\psi\rangle = \langle\psi|Z_0Z_1|\psi\rangle + \langle\psi|Z_0Z_2|\psi\rangle,$$

where $|\psi\rangle$ is taken from the set of quantum states on 3 qubits.

Notice the change that we have introduced. In our previous formulation, we were only minimizing over basis states, but now that we know that the minimum over all possible states is reached on a basis state, we are minimizing over all possible quantum states. This will make our life easier in future chapters when we introduce quantum algorithms to solve this kind of problem, because we will be justified in using any quantum state instead of just constraining ourselves to those that come from the basis.

Important note

Although the minimum energy is always achieved on one basis state, it could also be the case that it is also achieved on a non-basis state. In fact, if two different basis states $|x\rangle$ and $|y\rangle$ achieve the minimum energy, then any superposition $a|x\rangle + b|y\rangle$ is of minimum energy as well. That is the case, for example, for Z_0Z_1 for which both $|01\rangle$ and $|10\rangle$ have energy -1. Then, any superposition $a|01\rangle + b|10\rangle$ also achieves energy -1, which is the minimum possible for Z_0Z_1.

It can be easily checked that our preceding argument holds for any number of qubits and any sum of tensor products Z_jZ_k, so if we have a graph with set of vertices V, of size n, and set of edges E, we can rewrite the Max-Cut problem for the graph as follows:

$$\text{Minimize} \quad \sum_{(j,k)\in E} \langle\psi|Z_jZ_k|\psi\rangle,$$

where $|\psi\rangle$ is taken from the set of quantum states on n qubits.

Important note

Let's take a step back and examine what we have proved. First, notice that matrices such as

$$\sum_{(j,k)\in E} Z_j Z_k$$

are **Hermitian** or **self-adjoint**. This means that they are equal to their conjugate transposes, as you can easily verify, and they have particular properties such as having real eigenvalues and being able to form an orthonormal basis with their eigenvectors (refer to *Appendix B, Basic Linear Algebra*, for more details). In our case, we have proved that the computational basis *is* such an orthonormal basis of eigenvectors. Furthermore, the quantity

$$\langle\psi|\left(\sum_{(j,k)\in E} Z_j Z_k\right)|\psi\rangle = \sum_{(j,k)\in E} \langle\psi| Z_j Z_k |\psi\rangle,$$

which is usually called the **expectation value** of $\sum_{(j,k)\in E} Z_j Z_k$, attains its minimum value on one of those eigenvectors, called the **ground state**.

This result is known as the **variational principle**, and we will revisit it in a more general form in *Chapter 7, VQE: Variational Quantum Eigensolver.*

For the Ising model, the situation is exactly the same. We can go through an analogous reasoning, only this time also involving terms of the form Z_j. Each Z_j is a tensor product with all the factors equal to the identity matrix except for the one in the j-th position, which is Z. Then, finding the state of minimum energy of an Ising model with n particles and coefficients J_{jk} and h_j is equivalent to the following problem:

$$\text{Minimize} \qquad -\sum_{(j,k)\in E} J_{jk} \langle\psi| Z_j Z_k |\psi\rangle - \sum_j h_j \langle\psi| Z_j |\psi\rangle,$$

where $|\psi\rangle$ is taken from the set of quantum states on n qubits.

So, we have been able to cast several combinatorial optimization problems into a *quantum form*. More concretely, we have rewritten our problems as instances of finding the ground state of a self-adjoint matrix called the **Hamiltonian** of the system. Notice, however, that we do not really need to obtain the exact ground state. If we can prepare a state $|\psi\rangle$ such that the amplitude $a_{x_{\min}} = \langle x_{\min}|\psi\rangle$ is big in absolute value, then we will have a high probability of finding $x_{\min}$ when we measure $|\psi\rangle$. This approach will be behind the algorithms that we will introduce in *Chapters 4* through *7*.

In the following sections, we will see that the possibility of rewriting combinatorial optimization problems as instances of ground state problems is not just a happy coincidence, but rather the norm, and we will show how to write many other important problems in this form. But, before we turn to that, let us write some code to work with those tensor products of Z matrices and to compute their expectation values.

3.2.2 Computing expectation values with Qiskit

In *Chapter 2, The Tools of the Trade in Quantum Computing*, we introduced the main ways in which Qiskit can be used to work with quantum circuits and to execute them on simulators and real quantum computers. But Qiskit also allows us to work with quantum states and Hamiltonians, combining them with tensor products and computing their expectation values, something that can be useful when dealing with optimization problems, as we have just seen. Learning how to perform these computations will help make the concepts that we have introduced more concrete. Moreover, in *Chapter 5, QAOA: Quantum Approximate Optimization Algorithm*, we will be working extensively with Hamiltonians in Qiskit, so we will need to know how to initialize and manipulate them.

Let's start by showing, for example, how to define in Qiskit a basis state of three qubits such as $|100\rangle$. We can do this in several different ways. For instance, we can first define one-qubit states $|0\rangle$ and $|1\rangle$ and compute their tensor products. There are several possible approaches to achieve this. The first one is to directly use the amplitudes to initialize a `Statevector` object. To do that, we need to import the class and then call its constructor with input `[1,0]` (the amplitudes of $|0\rangle$) as shown in the following fragment of code:

```
from qiskit.quantum_info import Statevector
zero = Statevector([1,0])
print("zero is", zero)
```

The output that you will get if you run this code is

```
zero is Statevector([1.+0.j, 0.+0.j],
            dims=(2,))
```

which shows that, indeed, we have created a quantum state and set it to $|0\rangle$. Of course, to initialize a quantum state to $|1\rangle$, we can run

```
one = Statevector([0,1])
print("one is",one)
```

obtaining this output:

```
one is Statevector([0.+0.j, 1.+0.j],
            dims=(2,))
```

An alternative, probably more convenient way of achieving the same result is to initialize the `Statevector` object from an integer such as `0` or `1`. We will use the `from_int` method and it is important to also use the `dims` parameter to indicate the size of the statevector. Otherwise, `0` could be interpreted to be $|0\rangle$ or $|00\rangle$ or $|000\rangle$ or... (as we mentioned in *Section 1.4.1*). In our case, we set `dims = 2`, but in general, we will have to set `dims` to 2^n, where n is the number of qubits, because that is the number of amplitudes on an n-qubit system. Then, we can run

```
zero = Statevector.from_int(0, dims = 2)
one = Statevector.from_int(1, dims = 2)
print("zero is",zero)
print("one is",one)
```

which results in the following output, as expected:

```
zero is Statevector([1.+0.j, 0.+0.j],
            dims=(2,))
one is Statevector([0.+0.j, 1.+0.j],
            dims=(2,))
```

In either case, we can now construct states with a higher number of qubits by computing tensor products with the `tensor` method, as shown in the following lines:

```
psi = one.tensor(zero.tensor(zero))
print("psi is",psi)
```

After running them, we will get the following output:

```
psi is Statevector([0.+0.j, 0.+0.j, 0.+0.j, 0.+0.j, 1.+0.j, 0.+0.j,
             0.+0.j, 0.+0.j],
            dims=(2, 2, 2))
```

Notice that the amplitude whose value is 1 is in the fifth position. It corresponds to $|100\rangle$, because, in binary, 100 is 4 and we start counting on 0.

As you can imagine, both the way in which we compute the tensor product and the representation as an amplitude vector can become difficult to parse when we are working with many qubits. The following lines show a more concise way of using tensor products and a much more beautiful way of presenting states, but they achieve exactly the same result as the code shown previously:

```
psi = one^zero^zero
psi.draw("latex")
```

In this case, the output will be just $|100\rangle$. More readable, right?

A faster way of constructing the $|100\rangle$ state is using, again, the `from_int` method, as in

```
psi = Statevector.from_int(4, dims = 8)
```

where we specify that we are working with three qubits by setting `dims = 8` (because we need 8 amplitudes to define a three-qubit state).

So, we now know a bunch of ways of creating basis states. What about states that are in superposition? Well, it couldn't be easier, because, in Qiskit, you can simply multiply basis states by amplitudes and then add them together. For instance, the instructions

```
from numpy import sqrt
ghz = 1/sqrt(2)*(zero^zero^zero) + 1/sqrt(2)*(one^one^one)
```

create the state $1/\sqrt{2}|000\rangle + 1/\sqrt{2}|111\rangle$.

Important note

It may seem that we have included some unnecessary parenthesis in the previous code. However, if you remove them, you will not get the expected result. Qiskit overloads the ^ operator to be used as the tensor product operation. But, in Python, ^ has a lower precedence than `+`, so we need the parenthesis for the operations to be performed in the desired order.

To learn more...

An additional, indirect way of setting the values of a quantum state is creating a quantum circuit that prepares the state and running it to obtain the state vector with `get_stavector` as we learned to do in *Chapter 2, The Tools of the Trade in Quantum Computing*; or you can even just pass the quantum circuit to the `Statevector` constructor. For instance, to create basis states, you would only need a circuit with X gates on the qubits that you need to be set to 1. If you use this method, however, you need to be careful to remember that qubit 0 in Qiskit circuits is represented as the rightmost one in kets. Thus, if you have a three-qubit `QuantumCircuit` called `qc` and you use `qc.x(0)`, you will obtain $|001\rangle$!

In order to compute expectation values, quantum states are not enough. We also need to create Hamiltonians. For now, we will learn how to work with tensor products of Z gates, like the ones we used in the previous section, starting with simple ones that can be stored in Qiskit `Pauli` objects. Qiskit offers several ways to initialize them, like in the case of `Statevector` objects. The first one is to use a string to specify the positions of Z and I matrices in the product. For instance, if we are working with three qubits and we want to create Z_0Z_1 (which is, as you surely remember, the tensor product $Z \otimes Z \otimes I$), we can use the following instructions:

```
from qiskit.quantum_info import Pauli
Z0Z1 = Pauli("ZZI")
print("Z0Z1 is",Z0Z1)
print("And its matrix is")
print(Z0Z1.to_matrix())
```

They give the following output:

```
Z0Z1 is ZZI
And its matrix is
[[ 1.+0.j  0.+0.j  0.+0.j  0.+0.j  0.+0.j  0.+0.j  0.+0.j  0.+0.j]
 [ 0.+0.j  1.+0.j  0.+0.j  0.+0.j  0.+0.j  0.+0.j  0.+0.j  0.+0.j]
 [ 0.+0.j  0.+0.j -1.+0.j  0.+0.j  0.+0.j  0.+0.j  0.+0.j  0.+0.j]
 [ 0.+0.j  0.+0.j  0.+0.j -1.+0.j  0.+0.j  0.+0.j  0.+0.j  0.+0.j]
 [ 0.+0.j  0.+0.j  0.+0.j  0.+0.j -1.+0.j  0.+0.j  0.+0.j  0.+0.j]
 [ 0.+0.j  0.+0.j  0.+0.j  0.+0.j  0.+0.j -1.+0.j  0.+0.j  0.+0.j]
 [ 0.+0.j  0.+0.j  0.+0.j  0.+0.j  0.+0.j  0.+0.j  1.+0.j  0.+0.j]
 [ 0.+0.j  0.+0.j  0.+0.j  0.+0.j  0.+0.j  0.+0.j  0.+0.j  1.+0.j]]
```

The matrix representing Z_0Z_1 is of size 8×8 and, as you can see, it can be hard to read. Fortunately, we can use the fact that tensor products of diagonal matrices are always diagonal, and print only the non-zero coefficients with the following instructions:

```
print("The sparse representation of Z0Z1 is")
print(Z0Z1.to_matrix(sparse=True))
```

They will give us:

```
Z0Z1 is ZZI
The sparse representation of Z0Z1 is
  (0, 0)    (1+0j)
  (1, 1)    (1+0j)
  (2, 2)    (-1+0j)
  (3, 3)    (-1+0j)
  (4, 4)    (-1+0j)
  (5, 5)    (-1+0j)
  (6, 6)    (1+0j)
  (7, 7)    (1+0j)
```

To learn more...

When constructing a `Pauli` object, we can also specify which positions of the tensor product are Z matrices, passing them as a vector of ones (indicating the presence of Z) and zeroes (indicating the absence of Z or, equivalently, the presence of I). Since the construction method is more general and it can be used to create other tensor products, we would need to specify another vector with positions of X matrices, which we will set to all zeroes for the moment.

For instance, you can run something like `Z0Z1 = Pauli(([0,1,1],[0,0,0]))` in order to obtain $Z \otimes Z \otimes I$. Notice that, because of the convention of qubit numbering in Qiskit, we need to use `[0,1,1]` for the vector of Z positions instead of `[1,1,0]`.

The main drawback of working with `Pauli` objects is that you cannot add them or multiply them by scalars. To get something like $Z_0Z_1 + Z_1Z_2$, we need first to convert the `Pauli` objects to `PauliOp`, which we can then add together as shown in the following code:

```
from qiskit.opflow.primitive_ops import PauliOp
H_cut = PauliOp(Pauli("ZZI")) + PauliOp(Pauli("ZIZ"))
print("H_cut is")
```

```
print(H_cut)
print("The sparse representation of H_cut is")
print(H_cut.to_spmatrix())
```

The output, in this case, is:

```
H_cut is
1.0 * ZZI
+ 1.0 * ZIZ
The sparse representation of H_cut is
  (0, 0)	(2+0j)
  (3, 3)	(-2+0j)
  (4, 4)	(-2+0j)
  (7, 7)	(2+0j)
```

Since the sum of diagonal matrices is diagonal, we have used the sparse representation to more compactly show the non-zero terms of `H_cut`. Notice that even some of the diagonal terms are zero, because some elements of Z_0Z_1 cancel with those of Z_0Z_2.

A more compact way of obtaining the same Hamiltonian is:

```
from qiskit.opflow import I, Z
H_cut = (Z^Z^I) + (Z^I^Z)
print("H_cut is")
print(H_cut)
```

This evaluates to:

```
H_cut is
1.0 * ZZI
+ 1.0 * ZIZ
```

Notice that we have used ^ to compute tensor products and parenthesis to get the operation priorities right.

Of course, more complicated Hamiltonians, even including coefficients, can be constructed.

For example,

```
H_ising = -0.5*(Z^Z^I) + 2*(Z^I^Z) -(I^Z^Z) + (I^Z^I) -5*(I^I^Z)
```

defines the Hamiltonian $-1/2Z_0Z_1 + 2Z_0Z_2 - Z_1Z_2 + Z_1 - 5Z_2$.

Now we are ready to compute expectation values. Thanks to the code that we have written and executed so far, `psi` stores $|100\rangle$ and `H_cut` stores $Z_0Z_1 + Z_1Z_2$. Then, computing $\langle 100| (Z_0Z_1 + Z_1Z_2) |100\rangle$ is as easy as running the following instruction:

```
print("The expectation value is", psi.expectation_value(H_cut))
```

This will give us the following output:

```
The expectation value is (-2+0j)
```

Since $Z_0Z_1 + Z_0Z_2$ is the Hamiltonian for the Max-Cut problem of the graph in *Figure 3.5*, this indicates that the assignment represented by $|100\rangle$ (vertex 0 in one set and 1 and 2 in the other) cuts the two edges of the graph and is, therefore, an optimal solution. Notice how the output is represented as a complex number because inner products can, in general, have imaginary parts. However, these expectation values will always be real, and the coefficients that will go with the imaginary unit — represented in Python as `j` — will just be 0.

Exercise 3.4

Write code to compute the expectation value of all the possible cuts of the graph in *Figure 3.5*. How many optimal solutions are there?

If you want to evaluate expressions such as $\langle\psi| H_{\text{cut}} |\psi\rangle$ step by step, you can also use Qiskit to first compute $H_{\text{cut}} |\psi\rangle$ and, then, the inner product of $|\psi\rangle$ with that. This can be achieved with the following instruction:

```
print("The expectation value is", psi.inner(psi.evolve(H_cut)))
```

Here, the `evolve` method is used to compute the matrix-vector multiplication, and `inner` is, obviously, used for the inner product.

Important note

We must stress that all these operations are numerical and not something that we can run on actual quantum computers. In fact, as you already know, on real devices we have no access to the full state vector: this is something that we can only do when we run circuits on simulators. In any case, we know that state vectors grow exponentially in size with the number of qubits, so simulations can easily become unfeasible in many scenarios. But don't worry. In *Chapter 5, QAOA: Quantum Approximate Optimization Algorithm*, we will learn how to use quantum computers to estimate expectation values of tensor products of Z matrices. In *Chapter 7, VQE: Variational Quantum Eigensolver*, we will do the same with more general tensor products. In fact, the procedure that we will use will clarify why we call these quantities *expectation values*!

But enough of tensor products and expectation values for now. Instead, in the next section, we will introduce a new formalism that will allow us to formulate some optimization problems more naturally than with the Ising model.

3.3 Moving from Ising to QUBO and back

Consider the following problem. Let's say that you are given a set of integers S and a target integer value T, and you are asked whether there is any subset of S whose sum is T. For instance, if $S = \{1, 3, 4, 7, -4\}$ and $T = 6$, then the answer is affirmative, because $3 + 7 - 4 = 6$. However, if $S = \{2, -2, 4, 8, -12\}$ and $T = 1$, the answer is negative because all the numbers in the set are even and they cannot add up to an odd number.

This problem, called the **Subset Sum** problem, is known to be NP-**complete** (see, for instance, *Section 7.5* in the book by Sipser [26] for a proof). It turns out that we can **reduce** the Subset Sum problem to finding a spin configuration of minimal energy for an Ising model, (which is an NP-hard problem – see *Section 3.1.3*). This means that we can rewrite any instance of Subset Sum as an Ising ground state problem (check *Appendix C, Computational Complexity*, for a refresher on **reductions**).

However, it may not be directly evident how to do so.

In fact, it is much simpler to pose the Subset Sum problem as a minimization problem by using binary variables instead of variables that take 1 or -1 values. Indeed, if we are given $S = \{a_0, \ldots, a_m\}$ and an integer T, we can define binary variables x_j, $j = 0, \ldots, m$, and consider

$$c(x_0, x_1, \ldots, x_m) = (a_0 x_0 + a_1 x_1 + \ldots + a_m x_m - T)^2 .$$

Clearly, the Subset Sum problem has a positive answer if and only if we can find binary values x_j, $j = 0, \ldots, m$, such that $c(x_0, x_1, \ldots, x_m) = 0$. In that case, the variables x_j that are equal to 1 will indicate which numbers from the set are selected for the sum. But $c(x_0, x_1, \ldots, x_m)$ is always non-negative, so we have reduced the Subset Sum problem to finding the minimum of $c(x_0, x_1, \ldots, x_m)$: if the minimum is 0, the Subset Sum has a positive solution; otherwise, it doesn't.

For example, for the case of $S = \{1, 4, -2\}$ and $T = 2$ that we considered previously, the problem would be

$$\begin{array}{ll} \text{Minimize} & x_0^2 + 8x_0x_1 - 4x_0x_2 - 4x_0 + 16x_1^2 - 16x_1x_2 - 16x_1 + 4x_2^2 + 8x_2 + 4 \\ \text{subject to} & x_j \in \{0, 1\}, \qquad j = 0, \ldots, m \end{array}$$

where we have expanded $(x_0 + 4x_1 - 2x_2 - 2)^2$ to obtain the expression to be optimized. If you wish, you can simplify it a little by taking into account that $x_j^2 = x_j$ always holds for binary variables. In any case, $x_0 = 0, x_1 = x_2 = 1$ would be an optimal solution for this problem.

Notice that, in all of these cases, the function $c(x_0, x_1, \ldots, x_m)$ that we need to minimize is a polynomial of degree 2 on the binary variables x_j. We can thus generalize this setting and define **Quadratic Unconstrained Binary Optimization** (**QUBO**) problems, which are of the form

$$\begin{aligned} &\text{Minimize} \quad q(x_0, \dots, x_m) \\ &\text{subject to} \quad x_j \in \{0, 1\}, \qquad j = 0, \dots, m \end{aligned}$$

where $q(x_0, \dots, x_m)$ is a quadratic polynomial on the x_j variables. The reason why these problems are called QUBO should now be clear: we are minimizing quadratic expressions over binary variables with no restrictions (because every combination of zeroes and ones is acceptable).

From the preceding reduction of the Subset Sum problem, it follows that QUBO problems are NP-hard. Indeed, the QUBO model is very flexible, and it enables us to formulate many optimization problems in a natural way. For example, it is quite easy to recast any Ising minimization problem as a QUBO instance. If you need to minimize

$$-\sum_{j,k} J_{jk} z_j z_k - \sum_j h_j z_j$$

with some variables z_j, $j = 0, \dots, m$, taking values 1 or -1, you can define new variables $x_j = (1 - z_j)/2$. Obviously, x_j will be 0 when z_j is 1, and 1 when z_j is -1. Furthermore, if you make the substitutions $z_j = 1 - 2x_j$, you obtain a quadratic polynomial in the binary variables x_j that takes exactly the same values as the energy function of the original Ising model. If you minimize the polynomial for the variables x_j, you can then recover the spin values z_j that achieve the minimal energy.

In case you were wondering, yes, you can also use the substitution $z_j = 2x_j - 1$ to transform Ising problems into QUBO formalism. In that case, values of z_j equal to -1 would be taken to values of x_j equal to 0 and values of z_j equal to 1 would be taken to 1 in x_j. However, we will stick to the transformation $z_j = 1 - 2x_j$ for the rest of the book.

For instance, if the Ising energy is given by $(-1/2)z_0z_1 + z_2$, then, under the transformation $z_j = 1 - 2x_j$, the corresponding QUBO problem will be the following:

$$\begin{array}{ll} \text{Minimize} & -2x_0x_1 + x_0 + x_1 - 2x_2 + \frac{1}{2} \\ \text{subject to} & x_j \in \{0,1\}, \qquad j = 0,1,2. \end{array}$$

You can also go from a QUBO problem to an Ising model instance by using the $x_j = (1-z_j)/2$ substitution. However, you will need to pay attention to a couple of details. Let's illustrate them with an example. Suppose that your QUBO problem is asking to minimize $x_0^2+2x_0x_1-3$. Then, when you substitute the x_j variables, you obtain

$$\frac{z_0^2}{4} + \frac{z_0z_1}{2} - z_0 - \frac{z_1}{2} - \frac{9}{4}.$$

But the Ising model does not allow squared variables or independent terms! Fixing these problems is not difficult, though. Regarding the squared variables, we can simply notice that it always holds that $z_j^2 = 1$, because z_j is either 1 or -1. Thus, we replace each squared variable with the constant value 1. In our case, we would get

$$\frac{1}{4} + \frac{z_0z_1}{2} - z_0 - \frac{z_1}{2} - \frac{9}{4} = \frac{z_0z_1}{2} - z_0 - \frac{z_1}{2} - 2.$$

Then, we can simply drop the independent term, because we are dealing with a minimization problem and it won't influence the choice of optimal variables (however, you should add it back when you want to recover the original value of the function to minimize). In the preceding example, the equivalent Ising minimization problem would then be the following:

$$\begin{array}{ll} \text{Minimize} & \frac{z_0z_1}{2} - z_0 - \frac{z_1}{2} \\ \text{subject to} & z_j \in \{1,-1\}, \qquad j = 0,1,2. \end{array}$$

It is easy to check that this problem has two optimal solutions: $z_0 = z_1 = 1$ and $z_0 = 1, z_1 = -1$, both attaining the -1 value. If we add back the independent term of value -2 that we had dropped, we obtain an optimal cost of -3 in the QUBO problem. These solutions correspond to $x_0 = x_1 = 0$ and $x_0 = 0, x_1 = 1$, respectively, which indeed evaluate to -3 and are optimal for the original problem.

Exercise 3.5

Write the Subset Sum problem for $S = \{1, -2, 3, -4\}$ and $T = 0$ as a QUBO problem and transform it into an instance of the Ising model.

So now we know how to go from QUBO problems to Ising energy minimization problems and back, and we can use either formalism — whichever is more convenient at any given moment. In fact, as we will learn in *Chapters 4* and *5*, the Ising model is the preferred formulation when solving combinatorial optimization problems with quantum computers. Also, the software tools that we will be using (Qiskit and D-Wave's Ocean) will help us in rewriting our QUBO problems in the Ising formalism by using transformations like the ones we have described in this section.

We now have all the mathematical tools that we need in order to work with combinatorial optimization problems if we want to solve them with quantum computers. Let's play with our new, shiny toys and use them to write some important problems in QUBO formalism.

3.4 Combinatorial optimization problems with the QUBO model

In this final section of the chapter, we are going to introduce some techniques that will allow us to write many important optimization problems as QUBO and Ising instances, so we can later solve them with different quantum algorithms. These examples will also help you understand how to formulate your own optimization problems under these models, which is the first step in order to be able to use quantum computers to solve them.

3.4.1 Binary linear programming

Binary linear programming problems involve optimizing a linear function on binary variables subject to linear constraints. Thus, the general form is

$$\begin{array}{ll} \text{Minimize} & c_0x_0 + c_1x_1 + \ldots + c_mx_m \\ \text{subject to} & Ax \leq b, \\ & x_j \in \{0,1\}, \qquad j = 0, \ldots, m, \end{array}$$

where c_j are integer coefficients, A is an integer matrix, x is the transpose of $(x_0, \ldots, x_m)$, and b is an integer column vector.

An example of this type of problem could be the following:

$$\begin{array}{ll} \text{Minimize} & -5x_0 + 3x_1 - 2x_2 \\ \text{subject to} & x_0 + x_2 \leq 1, \\ & 3x_0 - x_1 + 3x_2 \leq 4 \\ & x_j \in \{0,1\}, \qquad j = 0, 1, 2. \end{array}$$

Binary linear programming (also known as **zero-one linear programming**) is NP-hard. In fact, the decision version in which the goal is to determine if there is any assignment of zeroes and ones that satisfies the linear constraints (with no actual optimization performed) was one of Richard M. Karp's original 21 NP-complete problems published in his famous paper on reducibility [27]. Assignments that satisfy the constraints are called **feasible.**

To write a binary linear program in QUBO formalism, we need to perform some transformations. The first one is to convert the inequality constraints into equality constraints by adding **slack variables**. This is better understood with an example. In the preceding problem, we have two constraints: $x_0 + x_2 \leq 1$ and $3x_0 - x_1 + 3x_2 \leq 4$. In the first one, the minimum value of the left-hand side is 0, attained when both x_0 and x_2 are 0. Thus, if we add a new binary slack variable y_0 to that left-hand side and substitute $\leq$ with $=$, we have

$$x_0 + x_2 + y_0 = 1,$$

which can be satisfied if and only if $x_0 + x_2 \leq 1$ can be satisfied. Indeed, if $x_0 = x_2 = 0$, then we can take $y_0 = 1$; and, if $x_0 = 0$ and $x_2 = 1$, or $x_0 = 1$ and $x_2 = 0$, we can take $y_0 = 0$. If $x_0 = x_2 = 1$, it is not possible to satisfy the constraint. That's why we can replace $x_0 + x_2 \leq 1$ with $x_0 + x_2 + y_0 = 1$ without changing the set of feasible solutions.

In the same way, the minimum value of $3x_0 - x_1 + 3x_2$ is -1, and it is achieved when $x_0 = 0$, $x_1 = 1$, and $x_2 = 0$.

Important note

Notice how the general rule to minimize these linear expressions on binary variables within some constraints is to set the variables with positive coefficients to 0 and those with negative coefficients to 1.

Then, in order for $3x_0 - x_1 + 3x_2$ to reach up to 4, which is the right-hand side of the constraint, we may need to add a number as big as 5. But to write non-negative numbers up to 5 we need only three bits, so we can add three new binary variables, y_1, y_2, and y_3, and consider

$$3x_0 - x_1 + 3x_2 + y_1 + 2y_2 + 4y_3 = 4,$$

which can be satisfied if and only if $3x_0 - x_1 + 3x_2 \leq 4$ can be satisfied.

To learn more...

In fact, notice that $y_1 + 2y_2 + 4y_3$ may go up to 7, but we only need to go up to 5. Thus, we could also use

$$3x_0 - x_1 + 3x_2 + y_1 + 2y_2 + 2y_3 = 4$$

as a replacement for $3x_0 - x_1 + 3x_2 \leq 4$.

Putting it all together, our original problem is equivalent to the following one:

$$\begin{aligned}
\text{Minimize} \quad & -5x_0 + 3x_1 - 2x_2 \\
\text{subject to} \quad & x_0 + x_2 + y_0 = 1, \\
& 3x_0 - x_1 + 3x_2 + y_1 + 2y_2 + 2y_3 = 4 \\
& x_j \in \{0, 1\}, \qquad j = 0, 1, 2, \\
& y_j \in \{0, 1\}, \qquad j = 0, 1, 2, 3.
\end{aligned}$$

Now, we are ready to write the problem as a QUBO instance. The only thing that we need to do is to incorporate the constraints as **penalty terms** in the expression that we are trying to minimize. For that, we use an integer B (for which we will select a concrete value later on) and consider the problem

$$\begin{aligned}
\text{Minimize} \quad & -5x_0 + 3x_1 - 2x_2 + B(x_0 + x_2 + y_0 - 1)^2 \\
& + B(3x_0 - x_1 + 3x_2 + y_1 + 2y_2 + 2y_3 - 4)^2 \\
\text{subject to} \quad & x_j \in \{0, 1\}, \qquad j = 0, 1, 2, \\
& y_j, \in \{0, 1\}, \qquad j = 0, 1, 2, 3,
\end{aligned}$$

which is already in QUBO form.

Since the new problem is unconstrained, we need to set B big enough so that violating the constraints does not *pay off*. If one of the original constraints is violated, the terms that are multiplied by B will be greater than 0. Moreover, the expression that we wanted to minimize in the original formulation of the problem was $-5x_0 + 3x_1 - 2x_2$, which can reach a minimum value of -7 (when $x_0 = x_2 = 1$ and $x_1 = 0$) and a maximum value of 3 (when $x_0 = x_2 = 0$ and $x_1 = 1$). Thus, if we choose, for instance, $B = 11$, any assignment that violates the constraints will achieve a value greater than at least 4 and will never be selected as the optimal solution to the QUBO problem if there is at least one feasible solution (which is the case for this particular problem).

In this way, a QUBO problem whose optimal solution is the same as the optimal solution of the original one is the following one:

$$
\begin{aligned}
\text{Minimize} \quad & -5x_0 + 3x_1 - 2x_2 + 11(x_0 + x_2 + y_0 - 1)^2 \\
& + 11(3x_0 - x_1 + 3x_2 + y_1 + 2y_2 + 2y_3 - 4)^2 \\
\text{subject to} \quad & x_j \in \{0,1\}, \qquad j = 0,1,2, \\
& y_j, \in \{0,1\}, \qquad j = 0,1,2,3.
\end{aligned}
$$

If you expand the expression to minimize, you will obtain a quadratic polynomial in the x_j variables, exactly what we need in the QUBO formulation.

To learn more...

Integer linear programming is a generalization of binary linear programming in which non-negative integer variables are used instead of binary ones. In some instances of that kind of problem, the constraints allow us to deduce that the integer variables are bounded. For instance, if you have the constraint

$$2a_0 + 3a_1 \leq 10$$

then you can deduce that $a_0 \leq 5$ and $a_1 \leq 3$. Since both a_0 and a_1 are non-negative, we can replace them with expressions in binary variables in the same way that we introduced slack variables for binary integer programs. In this case, for instance, we can replace a_0 with $x_0 + 2x_1 + 4x_2$ and a_1 with $x_3 + 2x_4$. In that manner, the integer linear program is transformed into an equivalent binary linear program that, in turn, can be written as a QUBO problem.

The procedure that we have studied in this section can be applied to transform any binary linear program into a QUBO problem. You only need to first introduce slack variables and then add penalty terms that substitute the original constraints. This is quite useful, since many important problems can be written directly as binary linear programs. In the next subsection, we will give a prominent example.

3.4.2 The Knapsack problem

In the famous **Knapsack problem**, you are given a list of objects $j = 0, \ldots, m$, each of them with a weight w_j and a value c_j. You are also given a maximum weight W and the goal is to find a collection of objects that maximizes the total value without going over the maximum weight allowed. Think of it as if you were going on a journey and you want to pack as many valuable objects as possible without getting a knapsack that is too heavy to carry.

For instance, you can have objects with values $c_0 = 5$, $c_1 = 3$, and $c_2 = 4$ and weights $w_0 = 3$, $w_1 = 1$, and $w_2 = 1$. If the maximum weight is 4, then the optimal solution would be to choose objects 0 and 2 for a total value of 9. However, that solution is unfeasible if the maximum weight is 3. In that case, we should choose objects 1 and 2 to obtain a total value of 7.

Although, at first sight, this problem may seem easy to solve, the fact is that (surprise, surprise!) it is NP-hard. In fact, if we consider a decision version of the problem in which we are also given a value V and we are asked if there is a selection of objects with a value of at least V that also satisfies the weight constraint, then the problem is NP-complete.

To learn more...

Proving that the decision version of the Knapsack problem is NP-complete is easy, because we already know that the Subset Sum problem is NP-complete. Suppose, then, that you are given an instance of the Subset Sum problem with set $S = \{a_0, \ldots, a_m\}$ and target sum T. Then, you can recast this as an instance of the Knapsack problem by considering objects $j = 0, \ldots, m$ with values $c_j = a_j$ and weights $w_j = a_j$, maximum weight $W = T$, and minimum total value $V = T$. Then, a solution to the Knapsack decision problem will give you a selection of objects $j_0, \ldots, j_k$ such that $a_{j_0} + \ldots + a_{j_k} \leq W = T$ because of the weight constraint and such that $a_{j_0} + \ldots + a_{j_k} \geq V = T$ because of the minimum value condition. Obviously, that selection of objects would also be a solution to the Subset Sum problem.

It is straightforward to write the Knapsack problem as a binary linear program. We only need to define binary variables x_j, $j = 0, \ldots, m$, that indicate whether we choose object j (if $x_j = 1$) or not (if $x_j = 0$) and consider

$$\begin{aligned} \text{Minimize} \quad & -c_0x_0 - c_1x_1 - \ldots - c_mx_m \\ \text{subject to} \quad & w_0x_0 + w_1x_1 + \ldots + w_mx_m \leq W, \\ & x_j \in \{0,1\}, \qquad j = 0, \ldots, m, \end{aligned}$$

where c_j are the object values, w_j are their weights, and W is the maximum weight of the knapsack. Notice that, since the original problem was asking to maximize the value, we are now minimizing the negative value, which is completely equivalent.

For instance, in the example that we considered previously with object values $5, 3$, and 4, weights $3, 1$, and 1, and maximum weight 3, the problem would be the following:

$$\begin{aligned} \text{Minimize} \quad & -5x_0 - 3x_1 - 4x_2 \\ \text{subject to} \quad & 3x_0 + x_1 + x_2 \leq 3, \\ & x_j \in \{0,1\}, \qquad j = 0,1,2. \end{aligned}$$

Of course, then we can add slack variables and introduce penalty terms, just as we did in the previous subsection, to rewrite the program as a QUBO problem. And that is exactly what we will need to do in order solve these problems with our quantum algorithms!

Exercise 3.6

Consider objects with values $3, 1, 7, 7$ and weights $2, 1, 5, 4$. Write the Knapsack problem for the case in which the maximum weight is 8 as a binary linear program.

To learn more...

A variant of the Knapsack problem allows us to choose several copies of the same object to put into the knapsack. In that case, we should use integer variables instead of binary ones to represent the number of times that each object is chosen. However, notice that, once we know the maximum weight allowed, each integer variable is bounded. Thus, we can use the technique that we explained at the end of the previous subsection in order to replace the integer variables with binary variables. And, afterwards, of course, we can rewrite the problem using the QUBO formalism.

For our next examples of optimization problems, we shall go back to working with graphs. In fact, in the next subsection, we will deal with a very colorful problem!

3.4.3 Graph coloring

In this subsection and the next, we will study some problems that are related to graphs but have many applications in different fields. The first one is graph coloring, in which we are given a graph and we are asked to assign a color to each vertex in such a way that vertices that are connected by an edge (also called **adjacent** vertices) receive different colors. Usually, we are asked to do this using the minimum possible number of colors or using no more than a given number of different colors. If we can color a graph with k colors, we say that it is k-**colorable**. The minimum number of colors needed to color a graph is called its **chromatic number**.

In *Figure 3.6*, we present three color assignments for the same graph. The one in *Figure 3.6a* is not a valid coloring, because there are adjacent vertices that share the same color. The one in *Figure 3.6b* is valid, but not optimal, because we do not need more than three colors for this graph, as *Figure 3.6c* proves.

The graph coloring problem may look like a children's game. However, many very relevant practical problems can be written as instances of graph coloring. For example, imagine that your company has several projects and you need to assign supervisors to each of them, but some projects are incompatible because of time overlaps or other restrictions. You can

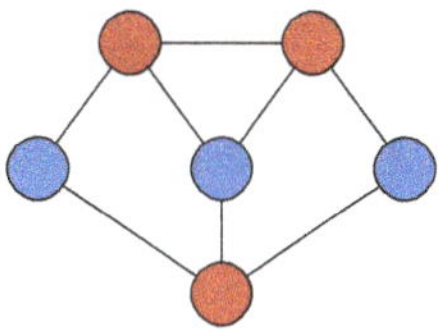

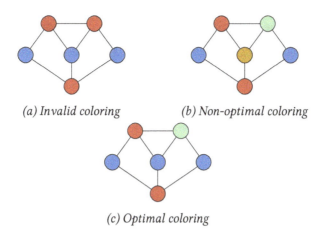

(a) Invalid coloring

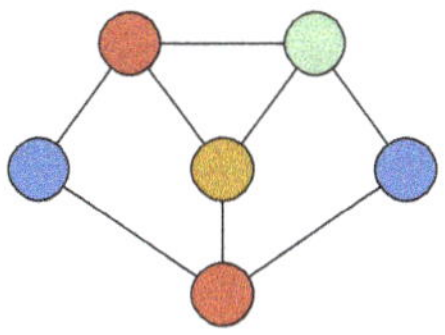

(b) Non-optimal coloring

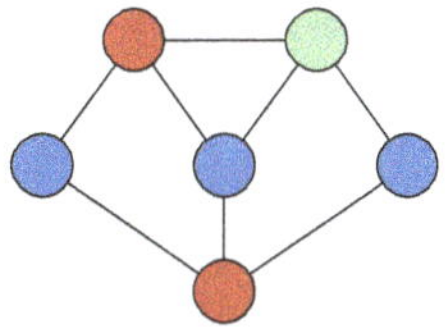

(c) Optimal coloring

Figure 3.6: Different colorings of a graph

create a graph in which the projects are the vertices and two projects are connected by an edge if and only if they are incompatible. Then, finding the chromatic number of the graph is equivalent to finding the minimum number of project leaders that you need to assign. Furthermore, finding a coloring will give you a way of assigning the supervisors while satisfying the constraints.

To learn more...

The history of graph coloring dates back to the middle of the 19th century and it is full of surprising plot twists. It originated from a seemingly simple problem: finding the minimum number of colors needed in order to color a geographical map in such a way that any pair of neighboring countries receive different colors. But, in spite of its *humble* origins, it evolved to even become a philosophical debate about the validity of computer-assisted mathematical proofs!

A very enjoyable popular recounting of this long and winding process can be found in *Four Colors Suffice*, by Robin Wilson [28].

Deciding whether a graph is 2-colorable or not is relatively easy. Indeed, notice that the vertices of a 2-colorable graph can be assigned to two disjoint sets depending on the color they receive and such that there are no edges among vertices of the same set — that's why these graphs are said to be **bipartite graphs**. But it is a well-known fact (originally proved by König in 1936) that a graph is bipartite if and only if it has no **cycles** of odd length (refer to *Section 1.6* in the book by Diestel [29]) and we can check for the presence of cycles by, for instance, computing the powers of the adjacency matrix of the graph (cf. *Section 10.4.7* in Rosen's book on discrete mathematics [30]). However, checking if a graph is k-colorable is NP-complete for any $k \geq 3$ (see the paper by Garey, Johnson, and Stockmeyer [31]) and, thus, computing the chromatic number of a graph is NP-hard.

Suppose we have a graph with vertices $0, \ldots, m$. In order to write the problem of determining if a graph is k-colorable using the QUBO framework, we will define some binary variables x_{jl} with $j = 0, \ldots, m$ and $l = 0, \ldots, k-1$. The variable x_{jl} will get value 1 if the vertex j receives the l-th color (for simplicity, colors are usually identified with numbers) and 0 otherwise. Then, the condition that vertex j receives exactly one color can be algebraically written as

$$\sum_{l=0}^{k-1} x_{jl} = 1.$$

For this condition to hold, there must exist l such that $x_{jl} = 1$ and such that $x_{jh} = 0$ for any $h \neq l$, exactly as we need.

On the other hand, we need to impose the constraint that adjacent vertices are not assigned the same color. Notice that in the case that two vertices j and h receive the same color l, then we would have $x_{jl}x_{hl} = 1$. Thus, for adjacent vertices j and h we need to impose

$$\sum_{l=0}^{k-1} x_{jl}x_{hl} = 0.$$

We can write these constraints as penalty terms in the expression to minimize in our QUBO problem to get

$$\begin{aligned} \text{Minimize} \quad & \sum_{j=0}^{m}\left(\sum_{l=0}^{k-1} x_{jl} - 1\right)^2 + \sum_{(j,h)\in E}\sum_{l=0}^{k-1} x_{jl}x_{hl} \\ \text{subject to} \quad & x_{jl} \in \{0,1\}, \qquad j = 0, \ldots, m, l = 0, \ldots, k-1, \end{aligned}$$

where E is the set of edges of the graph. Notice that we do not need to square the terms $\sum_{l=0}^{k-1} x_{jl}x_{hl}$ because they are always non-negative. If we find that the optimal solution of the problem is 0, then the graph is k-colorable. Otherwise, it is not.

Exercise 3.7

Consider a graph with vertices $0, 1, 2$, and 3, and edges $(0, 1)$, $(0, 2)$, $(1, 3)$, and $(2, 3)$. Write the QUBO version of the problem of checking whether the graph is 2-colorable.

In the next subsection, we will study another optimization problem on graphs. Do you like traveling? Then, prepare yourself to optimize your travel plans with the help of QUBO formalism.

3.4.4 The Traveling Salesperson Problem

The **Traveling Salesperson Problem** (or, simply, **TSP**) is one of the most famous problems in combinatorial optimization. The goal of the problem is very simple to state: you need to find a route that goes through each of the cities in a given set once and only once while minimizing some global quantity (distance traveled, time spent, total cost...).

We can formulate the problem mathematically using graphs. In this formulation, we would be given a set of vertices $j = 0, \ldots, m$ representing the cities, and, for each pair of vertices j and l, we would also be given the cost w_{jl} of traveling from j to l (this cost does not need to be the same, in general, as w_{lj}). We then need to find a **path** in the graph (that is, a set of edges such that the end of an edge is the beginning of the next one) that visits each vertex once and only once and that minimizes the sum of the costs of all the edges used.

To learn more...

As you have probably guessed, the TSP is NP-hard (refer to *Chapter 15* in the book on combinatorial optimization by Korte and Vygen [32] for more details). In fact, given a graph, a set of costs for the edges and a value C, it is NP-complete to decide whether there is a path that visits all the cities and has a cost less than or equal to C.

For instance, in *Figure 3.7* we can see a TSP instance with four cities. The numbers that appear to label the edges are their costs. For simplicity, we have assumed that for every pair of vertices, the travel cost is, in this case, the same in both directions.

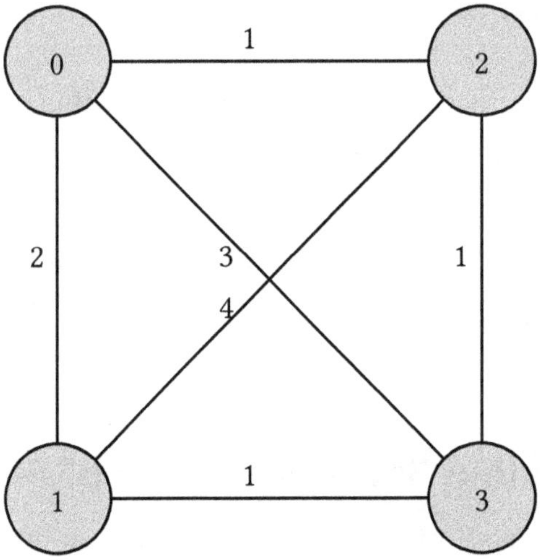

Figure 3.7: An example of the Traveling Salesperson Problem

To formulate the TSP in the QUBO framework, we will define binary variables x_{jl} that will indicate the order of visiting the different vertices. More concretely, if vertex j is the l-th in the tour, then x_{jl} will be 1 and x_{jh} will be 0 for $h \neq l$. Thus, for every vertex j, we need to impose the constraint

$$\sum_{l=0}^{m} x_{jl} = 1,$$

because every vertex needs to be visited exactly once. But we also need to impose

$$\sum_{j=0}^{m} x_{jl} = 1$$

for every position l, because we can only visit one city at a time.

If those two constraints are met, we will have a path that visits every vertex once and only once. However, that is not enough. We also want to minimize the total cost of the path, so we need an expression that gives us that cost in terms of the x_{jl} variables. Notice that an edge (j, k) is used if and only if the vertices j and k are consecutive in the path. That is, if and only if there exists an l such that j is visited in position l and k is visited in position $l + 1$. In that case, the cost of using the edge will be given by $w_{jk}x_{jl}x_{kl+1}$, because $x_{jl}x_{kl+1} = 1$. But if j and k are not consecutive in the path, then $x_{jl}x_{kl+1} = 0$ for every l, which is also the cost of that path in our route – we are not using it, so we don't pay for it!

Thus, the total cost of the tour is given by

$$\sum_{l=0}^{m-1}\sum_{j=0}^{m}\sum_{k=0}^{m} w_{jk}x_{jl}x_{kl+1},$$

where we are assuming that $w_{jj} = 0$ for $j = 0, \dots, m$ – staying in the same place costs nothing!

Then, we can incorporate the constraints as penalty terms in the function to minimize and write the TSP problem as

$$\begin{array}{ll} \text{Minimize} & \displaystyle\sum_{l=0}^{m-1}\sum_{j=0}^{m}\sum_{k=0}^{m} w_{jk}x_{jl}x_{kl+1} + B\left(\sum_{l=0}^{m} x_{jl} - 1\right)^2 + B\left(\sum_{j=0}^{m} x_{jl} - 1\right)^2 \\ \text{subject to} & x_{jl} \in \{0, 1\}, \qquad j, l = 0, \dots, m, \end{array}$$

where B is chosen so that unfeasible solutions never achieve an optimal value. For instance, if we select

$$B = 1 + \sum_{j,k=0}^{m} w_{jk},$$

then those solutions that violate the constraints will get a penalty that is bigger than the cost of any valid tour and will not be selected as optimal.

Exercise 3.8

Obtain the expression for the route cost in the TSP problem with the graph in *Figure 3.7*.

We have shown how to formulate several important problems in the QUBO formalism. But the ones that we have focused on are not, by any means, the only ones that these techniques can address. In the next subsection, we will give you a couple of hints of where to look for more.

3.4.5 Other problems and other formulations

In this chapter, we have introduced the Ising and QUBO models and we have shown how to use them to formulate combinatorial optimization problems. In fact, in the last part of the chapter, we have studied several famous problems, including binary linear programming and the Traveling Salesperson Problem, and we have given QUBO formulations for them.

The possibility of using these frameworks to formulate optimization problems is not reduced to the examples that we have worked with. Other important problems that can be readily written as QUBO and Ising instances include finding cliques in graphs, determining whether a logic formula is satisfiable, and scheduling jobs under constraints. Using the techniques described in this chapter, you are now equipped to write your own formulations for these and other problems.

However, it is useful to have some references to problems that have already been formulated as QUBO instances to be used out of the box or to serve as inspiration if your problem doesn't exactly fit any of them. A good survey of such formulations is the one compiled by Lucas [33], which includes all 21 of Karp's NP-complete problems and more.

An important thing that you should always keep in mind when using the QUBO framework is that, usually, there is more than just one way of formulating a problem. For example, sometimes it is straightforward to state a problem as a binary linear program and then use the transformations that we have studied to obtain a QUBO and, eventually, an Ising version of the problem. Nevertheless, it could be possible that with a different approach, you may find a more compact formulation that reduces, for instance, the number of variables or the length of the expression to minimize.

In recent years, the comparison of alternative QUBO formulations of important combinatorial optimization problems has become a very active research area. It is good advice to keep an open mind (and an eye on the scientific literature) because, in many cases, choosing the right formulation can be a crucial factor to obtain better results when using quantum computers to solve optimization problems.

To learn more...

A recent paper by Salehi, Glos, and Miszczak [34] addresses the task of representing the TSP and some of its variants with QUBO formalism and studies how different formulations can affect the performance of quantum optimization algorithms.

Our next stop will be using actual quantum devices to solve the type of problem that we have been focusing on in this chapter. Get ready to learn how to use quantum annealers!

Summary

This chapter has been devoted to introducing two different mathematical frameworks, the Ising model and the QUBO formalism, which allow us to write combinatorial optimization problems in a way that we will later be able to use to find approximate solutions with the help of quantum computers. We started with some simple examples and worked our way up to some famous problems such as graph coloring and the Traveling Salesperson Problem.

In order to achieve that, we studied different techniques that find wider applications in the process of writing optimization problems for quantum computers. We saw, for example, how to use slack variables and how to replace constraints with penalty terms. We also learned how to transform integer variables into a series of binary ones.

After all that we have covered in this chapter, you are now prepared to write your own problems in the languages required by optimization algorithms that can run on quantum computers. The rest of the chapters in this part of the book will be devoted to learning how to implement and run those quantum optimization algorithms. In fact, in the next chapter, we will explain how to use a type of quantum computer called a **quantum annealer** to solve QUBO and Ising problems.

4

Adiabatic Quantum Computing and Quantum Annealing

Love's a different sort of thing, hot enough to make you flow into something, interflow, cool and anneal and be a weld stronger than what you started with.

— Theodore Sturgeon

In the previous chapter, we studied how to formulate different combinatorial optimization problems as QUBO instances that, in turn, could be rewritten as the optimization problem of finding a state with minimum energy in an Ising model system. In this chapter, we will use this fact to introduce a way of using **quantum annealers** — a special type of quantum computer — to try to find (approximate) solutions to those combinatorial optimization problems.

But, in order to do that, we first need to talk a little bit more about Hamiltonians and their ground states, as well as the central role they play in adiabatic quantum computing.

The topics that we will cover in this chapter are as follows:

- Adiabatic quantum computing
- Quantum annealing
- Using Ocean to formulate and transform optimization problems
- Solving optimization problems on quantum annealers with Leap

Here we go!

4.1 Adiabatic quantum computing

In *Chapter 1, Foundations of Quantum Computing*, we focused mainly on quantum circuits but we briefly mentioned that there were other equivalent quantum computing models. One of them is adiabatic quantum computing, introduced in 2000 by Farhi, Goldstone, Gutmann, and Sipser in a widely influential paper [35].

When using quantum circuits, we apply operations (our beloved quantum gates) through discrete, sequential steps. However, adiabatic quantum computing relies on the use of continuous transformations. Namely, we will use a Hamiltonian $H(t)$ that will vary with time and that will be the driving force to change the state of our qubits according to the time-dependent Schrödinger equation:

$$H(t)\left|\psi(t)\right\rangle = i\hbar\frac{\partial}{\partial t}\left|\psi(t)\right\rangle.$$

To learn more...

As you may remember, in *Chapter 1, Foundations of Quantum Computing*, we talked about the **time-independent** Schrödinger equation. In that case, the Hamiltonian – which you can think of as a mathematical object that can describe the energy of the system — remained unchanged throughout the process. Now, we'll consider situations in which this energy can vary with time. This is the case, for instance, if you are applying an electromagnetic pulse to your qubits and you change its intensity or its frequency.

The terms in this equation are the time-dependent Hamiltonian $H(t)$, the state vector of the system $|\psi(t)\rangle$, the imaginary unit i (defined by $i^2 = -1$), and the reduced Planck's constant $\hbar$.

In addition to using time-dependent Hamiltonians, there is another ingredient that we need for our new quantum computing model: the idea of **adiabatic evolution**. Roughly speaking, an adiabatic process is one in which the "energy configuration" of the system changes "very gently" (there are quite a few quotation marks here, aren't there?). But... what does this have to do with quantum computing and how does it help us in finding solutions to our problems?

The key observation is that we will be considering problems whose optimal solutions will correspond to minimum-energy or ground states of some Hamiltonian of an Ising model. So, if we start with our system in the ground state (for some Hamiltonian) and we evolve it adiabatically, we know that it will remain in a ground state through the whole process. We won't be adding enough energy for the system to "jump" to the next energy level: this is, in more "physical" terms, to go from the ground state to an **excited state**. And we can use that to our advantage, because if we engineer the procedure so that the final Hamiltonian of the system is the one whose ground state will yield the solution to our problem, then we only need to measure the system to get the solution we are looking for.

Important note

To put it in a nutshell, the idea behind adiabatic quantum computing is to start with a simple Hamiltonian, one for which we can easily obtain — and prepare! — the ground state, and evolve it "carefully." We do this so that we remain in the ground state all the time, slowly changing our system until the ground state of its Hamiltonian is the solution to our problem. And then, bang, we perform a measurement and get our result!

Of course, the crucial thing here is how to perform the evolution to ensure that it is, indeed, adiabatic. But don't worry, the **adiabatic theorem** has got our backs there. This result, originally proved by Max Born and Vladimir Fock [36], two of the fathers of quantum mechanics, says that for your process to be adiabatic, it should be slow enough. You may ask: how slow? Well, the total time should be inversely proportional to the square of the **spectral gap**, which is the minimum difference in energy between the ground state and the first excited state of the Hamiltonian during the whole evolution.

This makes perfect intuitive sense. If there is always a big difference in energy between the ground state and the first excited state, then you can speed things up a little bit – you won't risk jumping to the next energy level. However, if the difference is small, you'd better be careful, lest you accidentally go up a step (or several!) on the energy ladder.

Now that we have a clear understanding of the ideas behind adiabatic quantum computing, let's make things a little bit more formal. Suppose that you have a problem for which H_1 is the Hamiltonian whose ground state encodes the result that you want to find. For instance, H_1 could be an Ising Hamiltonian that you obtained from transforming a QUBO problem. Now, imagine that your system is in the ground state of some initial Hamiltonian H_0. We will soon discuss how to choose H_0, but for now just think that you can prepare its ground state easily enough so that it is a natural choice for you.

Suppose that we run the process for total time T. The time-dependent Hamiltonian that we will consider will be of the form

$$H(t) = A(t)H_0 + B(t)H_1,$$

where A and B are real-valued functions that accept inputs over the interval $[0, T]$ such that $A(0) = B(T) = 1$ and $A(T) = B(0) = 0$. Notice that it holds that $H(0) = H_0$ and $H(T) = H_1$, exactly as we desired. A common choice for the functions A and B is to set $A(t) = 1 - t/T$ and $B(t) = t/T$. Nonetheless, as we will see later in this chapter, sometimes we also use other options, under the requirement that they satisfy the aforementioned boundary conditions.

Adiabatic quantum computing is polynomially equivalent to other quantum computing models, as proved by Aharonov et al. [37], including the quantum circuit model. This means that anything that is efficiently computable in one of these models is also efficiently computable in adiabatic quantum computing, and vice versa. Consequently, you can choose to use any of these models depending on the particulars of your problem or, as we will see in the next section, on the kind of quantum computer that you have access to.

4.2 Quantum annealing

Although we have just seen that adiabatic quantum computing is, theoretically, a perfectly viable alternative to the quantum circuit model, in its practical incarnation it is usually implemented in a restricted version called **quantum annealing**.

Quantum annealing relies on the same core idea as adiabatic quantum computing: it takes an initial Hamiltonian H_0, a final Hamiltonian H_1 whose ground state encodes the solution to the problem of interest, and it gradually changes the acting Hamiltonian from the initial to the final one by using some functions A and B (as described in the previous section) to decrease the action of H_0 and to increase the action of H_1. However, quantum annealing deviates from full adiabatic quantum computing in two ways. First of all, in practical implementations of quantum annealing, the final Hamiltonian H_1 that can be realized cannot be chosen completely at will, but has to be selected from a certain, restricted class. A typical option is an Ising Hamiltonian of the form

$$-\sum_{j,k} J_{jk} Z_j Z_k - \sum_j h_j Z_j,$$

which is the quantum version of the one we introduced in *Section 3.1.3*. In this case, the user has the freedom of selecting the J_{jk} and h_j coefficients within certain ranges. Due to this restriction in the choice of the final Hamiltonian, quantum annealing, unlike adiabatic quantum computing, is not universal and can only be used to solve a specific (but still very important!) type of problem. On the bright side, physical quantum devices based on quantum annealing are simpler to construct, making it possible to scale the size of these **quantum annealers** up to hundreds or even thousands of qubits.

The initial Hamiltonian in the quantum annealing setup is also usually fixed to be $H_0 = -\sum_{j=0}^{n-1} X_j$, where n is the number of qubits, and X_j stands for the tensor product in which the X matrix is acting on qubit j with the rest of positions occupied by I, the identity matrix. The ground state of H_0 is easily seen to be $\bigotimes_{i=0}^{n-1} |+\rangle$, the tensor product of n copies of the plus state, which is relatively easy to prepare because it is completely unentangled.

Exercise 4.1

Prove that $|\psi_0\rangle = \bigotimes_{i=0}^{n-1} |+\rangle$ has the minimum possible energy for $H_0 = -\sum_{j=0}^{n-1} X_j$ by first showing that, for each j and each state $|\psi\rangle$, it holds that $\langle\psi| X_j |\psi\rangle \leq 1$ and then showing that $\langle\psi_0| X_j |\psi_0\rangle = 1$ for each j.

Thus, the Hamiltonian used in quantum annealing is given by

$$H(t) = -A(t)\sum_{j=0}^{n-1} X_j - B(t)\sum_{j,k} J_{jk}Z_jZ_k - B(t)\sum_{j} h_jZ_j,$$

where J_{jk} and h_j are some adjustable coefficients, and A and B are functions such that $A(0) = B(T) = 1$ and $A(T) = B(0) = 0$, with T being the total **annealing time**. In this context, A and B are called the **annealing schedule**.

The other important deviation from the adiabatic quantum computing model is that, in quantum annealing, evolution is no longer guaranteed to be adiabatic. There are two main reasons for this decision. As you surely remember, the spectral gap is the minimum of the difference between the ground state and the first excited state of $H(t)$ for $t \in [0, T]$.

Computing this spectral gap can be very difficult. Actually, it can be even harder than finding the ground state that we are looking for, as proved by Cubitt et al. [38]. The second reason is that, even if we are able to compute the time that we need for the process to be adiabatic, it can be so big that it wouldn't be practical – or even possible! – to run the system evolution for so long.

Thus, in quantum annealing, we run the evolution for a certain amount of time that need not satisfy the conditions for adiabaticity, and hope to still be able to find good approximations of the optimal solution to our problem. In fact, we don't strictly need to remain in the ground state of $H(t)$. Since, at the end, we are going to measure the state, it would be enough if the amplitude of an optimal or sufficiently good solution in our final state were big enough. That's because, then, the probability of obtaining a useful result will still be high. And, of course, we can always repeat the process several times and keep the best of all measurements!

In 2011, the Canadian company D-Wave was the first to ever commercialize a quantum device that implemented quantum annealing as we have just described it. That quantum annealer, called D-Wave One, had 128 qubits, while one of D-Wave's most recent quantum devices, the Advantage, has more than 5000 qubits, and it's available for you to use online!

We need to keep in mind that, with these quantum computers, the evolution process will not be adiabatic in general, so there is no guarantee that the exact solution will be found in all cases. But, all over the world, many research teams and prominent companies – from sectors as diverse as finance, logistics, and aircraft manufacturing – are actively exploring the practical applications of quantum annealers. We will devote the rest of this chapter to showing you how you can also try them for your own optimization problems.

Using D-Wave's quantum annealers is much easier than you may think. First of all, you need to follow the instructions in *Appendix D, Installing the Tools*, to install Ocean, which is D-Wave's quantum annealing Python library, and to create a free account on D-Wave Leap, a cloud service where you can get one minute per month of free computing time on D-Wave's quantum annealers. This may not seem like much, but you will see that it is enough to run quite a number of experiments.

To learn more...

If one minute per month proves not to be enough for your annealing necessities, both D-Wave Leap and Amazon Braket offer paid access to quantum annealers. Obviously, the pricing of these services varies from time to time, so please check their websites to check the current rates and conditions.

Once you have everything set up, you can access quantum annealers to find an approximation of a solution to any combinatorial optimization problem that you may have written as either an instance of finding the ground state of an Ising model or as a QUBO problem. For instance, let's try to solve the MaxCut problem for the graph in *Figure 3.5*. As you surely remember, we can pose it as finding the ground state of

$$Z_0Z_1 + Z_0Z_2,$$

which is, of course, an Ising Hamiltonian in which $J_{01} = J_{02} = 1$ and the rest of the coefficients are 0.

All we need to tell the quantum annealer is that those are the coefficients we want to use, and then we can perform the annealing multiple times to obtain some results that will hopefully solve our problem. To specify the problem, we can use the `dimod` package, included in the Ocean library, as follows:

```
import dimod
J = {(0,1):1, (0,2):1}
h = {}
problem = dimod.BinaryQuadraticModel(h, J, 0.0, dimod.SPIN)
print("The problem we are going to solve is:")
print(problem)
```

The output will be the following:

```
The problem we are going to solve is:
BinaryQuadraticModel({0: 0.0, 1: 0.0, 2: 0.0},
    {(1, 0): 1.0, (2, 0): 1.0}, 0.0, 'SPIN')
```

There are a couple of things to notice here. First, we have used `J` for the coefficients of the degree 2 terms — `(0,1):1` sets the J_{01} coefficient to 1 and `(0,2):1` sets $J_{02} = 1$ — and `h` for the linear ones. Those coefficients that we do not specify are automatically set to 0 by the `BinaryQuadraticModel` constructor, but we still need to pass both the `J` and the `h` parameters (even in our case, where the latter is empty). Notice that in the output we get `(1, 0): 1.0, (2, 0): 1.0`, which seems to be the reverse of what we used. But they are exactly the same, because $Z_0Z_1 = Z_1Z_0$ and, thus, the situation is symmetrical. Second, we have used `0.0` as the value for the **offset**, which is a constant term that can be added to the Hamiltonian. Finally, we have used the `dimod.SPIN` parameter because we are working with an Ising Hamiltonian and, thus, the values of our variables are 1 and -1. In just a minute, we will see how to use binary variables instead. But, before that, let's use the following code to run the annealing process on one of the quantum annealers:

```python
from dwave.system import DWaveSampler
from dwave.system import EmbeddingComposite
sampler = EmbeddingComposite(DWaveSampler())
result = sampler.sample(problem, num_reads=10)
print("The solutions that we have obtained are")
print(result)
```

What we are doing here is, first, importing `DWaveSampler`, which will give us access to the quantum annealers, and then `EmbeddingComposite`, which will allow us to **map** or **embed** our problem into the actual qubits of the annealer — don't worry, we will explain this in detail later. For now, you can think of this as an automatic way of selecting a few qubits in the computer that will be used to represent our variables. After that, we create an object `sampler` that we then use to obtain 10 samples or possible solutions to our problem. This is where the actual execution on the actual quantum annealer happens. After that, we just print the result, which will vary from execution to execution. This is because we are using an actual quantum computer, which is, as you know, essentially probabilistic. In our case, we obtained the following:

```
The solutions that we have obtained are
   0  1  2 energy num_oc. chain_.
0 +1 -1 -1   -2.0       6     0.0
1 -1 +1 +1   -2.0       4     0.0
['SPIN', 2 rows, 10 samples, 3 variables]
```

This means that we obtained two different solutions: $z_0 = 1$, $z_1 = -1$, and $z_2 = -1$, and $z_0 = -1$, $z_1 = 1$, and $z_2 = 1$, both with energy -2; the first one was measured in 6 of the executions and the second in the remaining 4 — we will explain what the `chain_.` data means later in the chapter. But meanwhile, we can rejoice. These two solutions are, indeed, maximum cuts in our graph, as you can easily check!

We can get some additional information from the `result` variable. In fact, we can access the best solution through `result.first` and the total time that we used the quantum annealer for, with `result.info['timing']['qpu_access_time']`. This is the amount that will be subtracted from your monthly 60 seconds... or that you will be charged for if you have a paying plan. In our case, the time that we used the annealer for was 15 832.16, which may look like a huge number if you don't realize that it is actually measured in microseconds. So for the 10 samples we used about 0.016 seconds. That minute of access doesn't seem so short anymore, right?

We can also use `dimod` to work with QUBO problems. We will need to specify the coefficients of the degree 2 terms, the linear coefficients — remember that, in QUBO, we are using binary variables, so expressions like x_3^2 can be simplified to x_3 — and the independent coefficient, exactly as in the Ising case. The only change is that we will use the `dimod.BINARY` parameter when creating our problem with the `BinaryQuadraticModel` constructor.

Exercise 4.2

Create an instance of a simple QUBO problem and solve it with an annealer. Notice that the values for the variables in the solution will be 0 and 1 instead of 1 and -1.

This is just the simplest kind of execution that we can run on a quantum annealer, in which we have used all the default parameters. But the Ocean software implements many other functionalities that allow us, for instance, to work more comfortably with optimization problems and to control the settings of our experiments more precisely, including the annealing time and other important values. The rest of this chapter will guide you through the most important features and options to help you get the most of your time with annealers, starting with how to use Ocean to work with optimization problems.

4.3 Using Ocean to formulate and transform optimization problems

As we have just seen, the `BinaryQuadraticModel` class can be used to define both Ising and QUBO problems. But `dimod` also offers other models and utilities that will make our lives a little bit easier. Let's start by studying how we can conveniently define problems with linear restrictions.

4.3.1 Constrained quadratic models in Ocean

You surely remember that a problem like

$$\begin{aligned} \text{Minimize} \quad & -5x_0 + 3x_1 - 2x_2 \\ \text{subject to} \quad & x_0 + x_2 \leq 1, \\ & 3x_0 - x_1 + 3x_2 \leq 4 \\ & x_j \in \{0,1\}, \qquad j = 0,1,2 \end{aligned}$$

is an instance of binary linear programming. In *Section 3.4.1*, we studied this family of problems in detail and we showed that they can be transformed into the QUBO and Ising models by using slack variables and penalty terms.

So, imagine that you want to solve the preceding problem in a quantum annealer. Do you need to perform all those boring transformations in order to obtain the QUBO coefficients and then use them to define a `BinaryQuadraticModel` object? No! Fortunately, `dimod` provides the `ConstrainedQuadraticModel` class, which simplifies the process of working with problems that involve linear constraints.

In order to instantiate our binary linear program as a `ConstrainedQuadraticModel` object, the first thing that we need to do is to define the variables that we want to use and their types. In our case, we have three binary variables that we can define with the following piece of code:

```
x0 = dimod.Binary("x0")
x1 = dimod.Binary("x1")
x2 = dimod.Binary("x2")
```

With these instructions, we have simply created three binary variables and we have labeled them so that we can use them in mathematical expressions and easily identify them when we print them.

To learn more...

If you have used symbolic mathematics libraries (for instance, SymPy), you will recognize that the principles at work here are very similar.

Now, we are going to define a `ConstrainedQuadraticModel` object and we are going to set the **objective** (the function that we seek to minimize) and also fix the constraints of the problem. For that, we will use the variables that we have just created. This can be achieved with the following instructions:

```
blp = dimod.ConstrainedQuadraticModel()
blp.set_objective(-5*x0+3*x1-2*x2)
```

```
blp.add_constraint(x0 + x2 <= 1, "First constraint")
blp.add_constraint(3*x0 -x1 + 3*x2 <= 4, "Second constraint")
```

Setting the objective or adding constraints automatically adds all the variables involved to the problem object. Notice also that we have provided labels to identify the constraints. If you prefer not to do it, then `dimod` will randomly assign an alphanumeric string to each constraint and it will be used as its name, should you need it. If, later on, you want to rename any of them, you can use the `relabel_constraints` method.

We can inspect the elements of `blp` by accessing its `variables`, `objective`, and `constraints` attributes. Thus, we can execute these instructions:

```
print("Our variables are:")
print(blp.variables)
print("Our objective is:")
print(blp.objective)
print("Our constraints are:")
print(blp.constraints)
```

And we will obtain something like this:

```
Our variables are:
Variables(['x0', 'x1', 'x2'])
Our objective is:
ObjectiveView({'x0': -5.0, 'x1': 3.0, 'x2': -2.0}, {}, 0.0,
    {'x0': 'BINARY', 'x1': 'BINARY', 'x2': 'BINARY'})
Our constraints are:
{'First constraint': Le(ConstraintView({'x0': 1.0, 'x2': 1.0}, {}, 0.0,
    {'x0': 'BINARY', 'x2': 'BINARY'}), 1.0), 'Second constraint':
    Le(ConstraintView({'x0': 3.0, 'x1': -1.0, 'x2': 3.0}, {}, 0.0,
    {'x0': 'BINARY', 'x1': 'BINARY', 'x2': 'BINARY'}), 4.0)}
```

Notice that both the objective and the constraints are internally represented as quadratic functions and, therefore, they formally have quadratic terms, linear terms, and an offset or independent term. In our case, only the linear part of the constraints is non-empty, and the offset is 0 in both cases.

To learn more...

As you can see from the output, the constraints that we have created are instances of the `dimod.sym.Le` class, where `Le` stands for *less than or equal to*. You can also create equality constraints, which will belong to the `dimod.sym.Eq` class or inequality constraints with $\geq$, which will be `dimod.sym.Ge` objects. Of course, an equality constraint is equivalent to one `Le` constraint plus one `Ge` constraint with the same left- and right-hand sides. And we can transform `Le` constraints into `Ge` constraints — and the other way around — by multiplying everything by -1.

Now, we know how to construct problems with constraints using `dimod`. In the next subsection, we will learn how to use the problems that we have defined to compute the cost of different value assignments, check if those assignments satisfy the constraints, and to also find the optimal solution to the problem.

4.3.2 Solving constrained quadratic models with dimod

The `dimod` package provides many tools to work with the constrained quadratic problems that we have just introduced. For instance, we can define an assignment of values to the variables, check if it is feasible, and compute its cost for the problem defined in the previous subsection by using the following instructions:

```
sample1 = {"x0":1, "x1":1, "x2":1}
print("The assignment is", sample1)
print("Its cost is", blp.objective.energy(sample1))
print("Is it feasible?",blp.check_feasible(sample1))
print("The violations of the constraints are")
print(blp.violations(sample1))
```

We are using the assignment $x_0 = x_1 = x_2 = 1$, so when we execute the code we obtain the following output:

```
The assignment is {'x0': 1, 'x1': 1, 'x2': 1}
Its cost is -4.0
Is it feasible? False
The violations of the constraints are
{'First constraint': 1.0, 'Second constraint': 1.0}
```

This tells us that the assignment is not feasible, and the `violations` method gives us the amount by which the left-hand side of each inequality is bigger than the right-hand side.

If, on the other hand, we want to try the $x_0 = x_1 = 0, x_2 = 1$ assignment, we can use the following code:

```
sample2 = {"x0":0, "x1":0, "x2":1}
print("The assignment is", sample2)
print("Its cost is", blp.objective.energy(sample2))
print("Is it feasible?",blp.check_feasible(sample2))
print("The violations of the constraints are")
print(blp.violations(sample2))
```

The result that we obtain is the following:

```
The assignment is {'x0': 0, 'x1': 0, 'x2': 1}
Its cost is -2.0
Is it feasible? True
The violations of the constraints are
{'First constraint': 0.0, 'Second constraint': -1.0}
```

In this case, the assignment is feasible and, therefore, no violation term is positive.

The dimod package also provides a brute-force solver that tries all possible assignments and sorts them according to their cost, from lowest to highest. Using it with our example is as simple as running

```
solver = dimod.ExactCQMSolver()
solution = solver.sample_cqm(blp)
print("The list of assignments is")
print(solution)
```

to obtain

```
The list of assignments is
  x0 x1 x2 energy num_oc. is_sat. is_fea.
6  1  0  1   -7.0       1 arra...   False
2  1  0  0   -5.0       1 arra...    True
7  1  1  1   -4.0       1 arra...   False
3  1  1  0   -2.0       1 arra...    True
4  0  0  1   -2.0       1 arra...    True
0  0  0  0    0.0       1 arra...    True
5  0  1  1    1.0       1 arra...    True
1  0  1  0    3.0       1 arra...    True
['INTEGER', 8 rows, 8 samples, 3 variables]
```

The first number is just an identifier of the assignment. It is followed by the values given to the variables. Then, we find the cost of the assignment – or, rather, its energy, if interpreted in terms of the Hamiltonian. After that, comes the times this solution has been found, which will always be 1 with this solver. Finally, we find information about which constraints are satisfied and whether the solution is feasible or not. It is very important to notice that the assignments are ordered by cost, but some of them may be unfeasible, even the first one, as in this case.

In fact, if we execute solution.first, we will obtain this output:

```
Sample(sample={'x0': 1, 'x1': 0, 'x2': 1}, energy=-7.0,
    num_occurrences=1, is_satisfied=array([False, False]),
    is_feasible=False)
```

where we can see that this assignment does not satisfy either of the two constraints of our problem. If you want the optimal solution to the problem, you should always remove the unfeasible solutions first with the **filter** method, using an instruction like the following:

```
feasible_sols = solution.filter(lambda s: s.is_feasible)
```

Then, if you access `feasible_sols.first`, you will get

```
Sample(sample={'x0': 1, 'x1': 0, 'x2': 0}, energy=-5.0, num_occurrences=1,
    is_satisfied=array([ True,  True]), is_feasible=True)
```

which is, indeed, the optimal solution to our binary linear program.

Of course, all these computations are done with a (very inefficient) classical algorithm. In the next subsection, we explain how to use actual quantum annealers to try to solve the problems that we have defined.

4.3.3 Running constrained problems on quantum annealers

As useful as the `ConstrainedQuadraticModel` class is, we cannot use it to define problems that can be run on quantum annealers. In order to do that, we first need to eliminate the constraints and create a `BinaryQuadraticModel` object that we can later execute on actual quantum hardware as we did in *Section 4.2*. Fortunately, the process is really simple thanks to the utilities provided in the Ocean library. Let's see how this works with an example.

To illustrate the general procedure, let's define a simple constrained problem with the following code:

```
y0, y1 = dimod.Binaries(["y0", "y1"])
cqm = dimod.ConstrainedQuadraticModel()
```

```
cqm.set_objective(-2*y0-3*y1)
cqm.add_constraint(y0 + 2*y1 <= 2)
```

We can transform this constrained problem into an unconstrained one by using the `cqm_to_bqm` method as follows:

```
qubo, invert = dimod.cqm_to_bqm(cqm, lagrange_multiplier = 5)
print(qubo)
```

In a moment, we will explain what `invert` is and how it is used, but for now let's focus on the output of those instructions, which will be something similar to the following:

```
BinaryQuadraticModel({'y0': -17.0, 'y1': -23.0,
    'slack_03b79fa9-3faa-410c-800b-65cfaf281cdf_0': -15.0,
    'slack_03b79fa9-3faa-410c-800b-65cfaf281cdf_1': -15.0},
    {('y1', 'y0'): 20.0,
    ('slack_03b79fa9-3faa-410c-800b-65cfaf281cdf_0', 'y0'): 10.0,
    ('slack_03b79fa9-3faa-410c-800b-65cfaf281cdf_0', 'y1'): 20.0,
    ('slack_03b79fa9-3faa-410c-800b-65cfaf281cdf_1', 'y0'): 10.0,
    ('slack_03b79fa9-3faa-410c-800b-65cfaf281cdf_1', 'y1'): 20.0,
    ('slack_03b79fa9-3faa-410c-800b-65cfaf281cdf_1',
    'slack_03b79fa9-3faa-410c-800b-65cfaf281cdf_0'): 10.0},
    20.0, 'BINARY')
```

That is quite a mouthful, but we promise that it is not nearly as complicated as it seems. In fact, with what you already know from *Chapter 3, QUBO: Quadratic Unconstrained Binary Optimization*, you could have computed a similar output yourself! Let's unpack it.

Since this is an unconstrained problem, what we are seeing is the specification of the cost function. First, we have the linear part, which starts with `'y0': -17.0`. It tells us that, in the objective function, y_0 has coefficient -17, y_1 has coefficient -23, and the two other variables have coefficient -15. Then comes the quadratic part, with coefficient 20 for the $y_0 y_1$ term, something that we deduce from the `('y1', 'y0'): 20.0` value, and with coefficients 10 and 20 for the other products of two variables. Finally, 20 is the independent term or offset, and we are also told that all the variables are binary.

But, where do all these coefficients come from? What our good friend `dimod` is doing here is nothing but applying the transformations that we studied in *Section 3.4.1*. First, two slack variables – with quite ugly random names – are introduced to transform the inequality constraint into an equality one. Then, the equality constraint is incorporated into the cost function as a penalty term with a penalty coefficient (the `lagrange_multiplier` parameter), which equals 5. And that's all! It wasn't that mysterious after all, was it?

Exercise 4.3

Check that the QUBO problem returned by `cqm_to_bqm` coincides with what you would obtain should you apply the transformations explained in *Section 3.4.1*. Don't forget the offset!

We can now use a quantum annealer to solve the problem defined in the `qubo` object just as we did in *Section 4.2*. For instance, we can run the following code:

```
sampler = EmbeddingComposite(DWaveSampler())
result = sampler.sample(qubo, num_reads=10)
print("The solutions that we have obtained are")
print(result)
```

Do not forget to import `EmbeddingComposite` and `DWaveSampler` if you haven't done it yet. If you run these instructions, you will obtain an output similar to the following:

```
  slack_03b79fa9-3faa-410c-800b-65cfaf281cdf_0 ... y1 energy num_oc. ...
0                                             0 ...  1   -3.0       5 ...
1                                             1 ...  0   -2.0       3 ...
2                                             0 ...  0   -2.0       1 ...
3                                             0 ...  1    0.0       1 ...
['BINARY', 4 rows, 10 samples, 4 variables]
```

We can all agree that this is not very informative. The problem here is that we are looking at solutions to the transformed problem, which include the slack variables with all those long, cryptic names. Of course, we don't really care about the slack variable values — we only introduced them in order to write our problem without any constraints. So, what can we do? Here's where the `invert` object comes to our rescue! It allows us to retrieve the solutions to the original problem from the solutions to the transformed one. Thus, we can run the following instructions:

```
samples = []
occurrences = []
for s in result.data():
    samples.append(invert(s.sample))
    occurrences.append(s.num_occurrences)
sampleset = dimod.SampleSet.from_samples_cqm(samples,cqm,
    num_occurrences=occurrences)
print("The solutions to the original problem are")
print(sampleset)
```

and obtain the following output, which now only shows the original variables:

```
The solutions to the original problem are
  y0 y1 energy num_oc. is_sat. is_fea.
3  1  1   -5.0       1 arra...   False
0  0  1   -3.0       5 arra...    True
1  1  0   -2.0       3 arra...    True
2  1  0   -2.0       1 arra...    True
['INTEGER', 4 rows, 10 samples, 2 variables]
```

Here, we have created a new `SampleSet` object — the type of structure in which `dimod` stores the results of solvers or samplers — from the samples obtained with the transformed problem. Notice that we use `invert` to eliminate the slack variables and that, by passing the `cqm` problem to the `from_samples_cqm` method, the energy without the penalties is computed, as well as the feasibility status of each assignment. In fact, notice that when printing the solutions that we sampled for the transformed problem, we obtained a solution with 0 energy. It corresponds to the assignment $y_0 = 1$ and $y_1 = 1$, which, on the original problem, had energy -5. The difference in the two energies comes from the fact that this assignment is unfeasible, and in the unconstrained problem it receives a penalty for it. Notice that we have also used the number of occurrences to keep track of how many times each solution is sampled.

So, in this way, we have recovered some solutions to the original problem, but there are a couple of details that we still need to fix. The first one is that if we want to only retain the feasible solutions, we need to use the **filter** method as we did in the previous subsection when using `ExactCMQSolver`. The second has to do with the repetition — which we can observe in the last two outputs — of the solution that sets $y_0 = 1$ and $y_0 = 0$. These two solutions come from two different assignments in the transformed problem, but they only differed in the values given to the slack variables. So, when those slack variables are eliminated, they produce exactly the same assignment. If we want them to be considered together, as they should be, we can use the `aggregate` method. Putting it all together, we can execute this code:

```
final_sols = sampleset.filter(lambda s: s.is_feasible)
final_sols = final_sols.aggregate()
print("The final solutions are")
print(final_sols)
```

This will print

```
  y0 y1 energy num_oc. is_sat. is_fea.
0  0  1   -3.0       5 arra...    True
1  1  0   -2.0       4 arra...    True
['INTEGER', 2 rows, 9 samples, 2 variables]
```

which is something that we can indeed use to solve our problem.

In this section, we have learned how to work with constrained problems and how to solve them both by brute force and with quantum annealers, transforming them into something that the quantum computer can use and then getting back to the original formulation. In the next section, we will study how to have more control over what the quantum annealer is doing in order to find the ground state of our Hamiltonians. Let's fiddle a little bit with the inner workings of those shiny quantum computers!

4.4 Solving optimization problems on quantum annealers with Leap

So far, we have run a couple of different optimization problems on actual quantum annealers. However, we have always used the default parameters and we do not even know the characteristics of the quantum computers that we are using. In this section, we shall remedy that. We will explain the different types of annealers that we can access through D-Wave Leap. We will also explore several hyperparameters that we can tweak when we are using these devices, and we will explain how to adjust the way in which our problems are embedded in the physical qubits – we will finally learn what that mysterious `EmbeddingComposite` object is used for!

4.4.1 The Leap annealers

You can list the devices to which you have access with your Leap account by using the `get_solvers` method in this way:

```python
from dwave.cloud import Client
for solver in Client.from_config().get_solvers():
    print(solver)
```

The results will depend on your actual access privileges, but for a typical free account you will see something like this:

```
BQMSolver(id='hybrid_binary_quadratic_model_version2')
DQMSolver(id='hybrid_discrete_quadratic_model_version1')
```

```
CQMSolver(id='hybrid_constrained_quadratic_model_version1')
StructuredSolver(id='Advantage_system6.1')
StructuredSolver(id='Advantage2_prototype1.1')
StructuredSolver(id='DW_2000Q_6')
StructuredSolver(id='Advantage_system4.1')
```

In this case, there are seven different solvers in total, of three different types. First, we have those with the word *hybrid* in their identifier. We will talk about them later in the chapter, but, for now, it suffices to know that they combine classical and quantum resources to solve problems. The other four, called `DW_2000Q_6`, `Advantage_system4.1`, `Advantage_system6.1`, and `Advantage2_prototype1.1`, are pure quantum annealers. These are the devices that are selected when we use `DWaveSampler` to solve a problem, as we have been doing so far in this chapter. Let's explore their properties in more detail.

We can select a particular annealer by using the `solver` parameter in the `DWaveSampler` constructor and then access the properties of the device with the `properties` attribute. For instance, for the `DW_2000Q_6` annealer, we can run the following instructions

```
from dwave.system import DWaveSampler
sampler=DWaveSampler(solver='DW_2000Q_6')
print("Name:",sampler.properties["chip_id"])
print("Number of qubits:",sampler.properties["num_qubits"])
print("Category:",sampler.properties["category"])
print("Supported problems:",sampler.properties["supported_problem_types"])
print("Topology:",sampler.properties["topology"])
print("Range of reads:",sampler.properties["num_reads_range"])
```

to obtain this output:

```
Name: DW_2000Q_6
Number of qubits: 2048
Category: qpu
```

```
Supported problems: ['ising', 'qubo']
Topology: {'type': 'chimera', 'shape': [16, 16, 4]}
Range of reads: [1, 10000]
```

If we do the same but with the `Advantage_system4.1` solver, we obtain:

```
Name: Advantage_system4.1
Number of qubits: 5760
Category: qpu
Supported problems: ['ising', 'qubo']
Topology: {'type': 'pegasus', 'shape': [16]}
Range of reads: [1, 10000]
```

The properties of the `Advantage_system6.1` solver will be exactly the same — except for the name, of course.

Finally, for the `Advantage2_prototype1.1` solver, we obtain the following output:

```
Name: Advantage2_prototype1.1
Number of qubits: 576
Category: qpu
Supported problems: ['ising', 'qubo']
Topology: {'type': 'zephyr', 'shape': [4, 4]}
Range of reads: [1, 10000]
```

To learn more...

The solvers have many other properties that we haven't yet discussed; we will study some of the most relevant ones in *Sections 4.4.2*, *4.4.3*, and *4.4.4*. You can nonetheless access all of them through the `properties` dictionary, printing it directly. But be careful: some of them can be huge to print, such as `properties["qubits"]`, which contains information about all the potentially thousands of qubits in a device!

Some properties are the same for the four devices. For instance, as we can see, all are of type **qpu**, which means that they are **quantum processing units** or quantum annealers. Also, all of them accept problems in the QUBO or Ising formats — but not constrained problems; that is why we had to transform them before running them in the previous section — and all can be used to obtain between 1 and 10 000 samples at a time. The main difference, other than the number of qubits — which is notably bigger in the `Advantage_system4.1` and `Advantage_system6.1` devices — is the **topology**. This refers to the way in which the qubits are connected to each other in the machine, and determines which **couplings** — or connections between variables — can be used to define our problems...unless we use an embedding, which will help us in mapping our coefficients to actual qubit connections.

The `Advantage2_prototype1.1` solver is a little bit special. As can be inferred from the name, it is a prototype of a new family of annealers that D-Wave will introduce in 2023-2024 — that's why, for now, it has fewer qubits than the rest of the devices, but the full version has been announced to have more than 7000 qubits. It uses a new topology, called Zephyr, which is designed to increase connectivity and decrease errors. At the time of writing, the available device is not a final version. For this reason, we will not use it in the examples that we will be working with, nor will we describe its properties and topology in detail. Notice, however, that everything that we explain about how to work with the devices translates, with no changes, to this new annealer.

We have summarized some of the annealer properties in *Table 4.1.*

Annealer name	Number of qubits	Topology
`DW_2000Q_6`	2048	Chimera
`Advantage_system4.1`	5760	Pegasus
`Advantage_system6.1`	5760	Pegasus
`Advantage2_prototype1.1`	576	Zephyr

Table 4.1: Summary of annealer properties

In the next subsection, we shall explore in more detail the annealers' topologies and how we can embed our problems in them.

4.4.2 Embeddings and annealer topologies

In current quantum computers, be they annealers or gate-based devices, technological difficulties prevent qubits from being connected in an all-to-all way. In fact, each qubit is usually connected exclusively to some of its neighbours and we can only apply two-qubit gates or use couplings (that is, use non-zero coefficients in the Ising model) between those qubits that are actually linked. The particular way in which the qubits are connected in a certain quantum chip is called its **topology** and sometimes it is important to be aware of it when we design our algorithms or when we anneal our problems.

For instance, the topology of the `DW_2000Q_6` annealer is called, as we saw in the previous subsection, Chimera. It consists of cells of 8 qubits organized into two groups of 4 qubits each. All the qubits in one group are connected to all the qubits in the other group, but there are no connections inside each group. For graph connoisseurs, the connections follow a complete bipartite graph $K_{4,4}$, which is depicted in *Figure 4.1*.

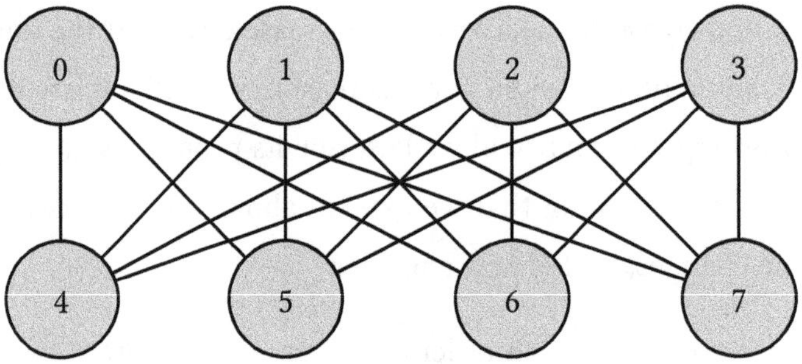

Figure 4.1: Qubit connections in a Chimera cell

The `DW_2000Q_6` computer has 256 of these cells, organized in a 16×16 grid, making for a total of $8 \cdot 16 \cdot 16 = 2048$ qubits, as expected. Each of the qubits that occupy the positions 0 to 3 in a cell is also connected to the qubits in the same position of the vertically adjacent cells. In a similar way, each qubit in positions 4 to 7 is also connected to the qubit in the same position of the horizontally adjacent cells. In total, each qubit will be connected to four other qubits in the same cell and to two other qubits (or one, if it lives in a border cell) from other cells.

We can obtain a list enumerating all the connections by using the `properties["couplers"]` attribute as follows:

```
sampler=DWaveSampler(solver='DW_2000Q_6')
print("Couplings:",sampler.properties["couplers"])
```

Running this, we will get a very long list that begins like this:

```
[[0, 4], [1, 4], [2, 4], [3, 4], [0, 5], [1, 5], [2, 5], [3, 5],
[0, 6], [1, 6], [2, 6], [3, 6], [0, 7], [1, 7], [2, 7], [3, 7],
[4, 12], [8, 12], [9, 12], [10, 12], [11, 12], [5, 13], [8, 13],
[9, 13], [10, 13], [11, 13], [6, 14], [8, 14], [9, 14], [10, 14],
[11, 14], [7, 15], [8, 15], [9, 15], [10, 15], [11, 15], [12, 20],
[16, 20], [17, 20], [18, 20], [19, 20], [13, 21], [16, 21],
[17, 21], [18, 21], [19, 21], [16, 22], [17, 22]...
```

A slightly more readable way of obtaining the same information is by using `sampler.adjacency`, which will we give us a dictionary indexed by qubit numbers and values that specify the qubits that are connected to the qubit in the key. In our case, it starts like this:

```
{0: {4, 5, 6, 7, 128}, 1: {4, 5, 6, 7, 129},
 2: {4, 5, 6, 7, 130}, 3: {4, 5, 6, 7, 131},
 4: {0, 1, 2, 3, 12}, 5: {0, 1, 2, 3, 13},
 6: {0, 1, 2, 3, 14}, 7: {0, 1, 2, 3, 15},
 8: {12, 13, 14, 15, 136}, ...
```

Exercise 4.4

Pick the qubits numbered from 0 to 7 and check that their connections correspond to the description of the Chimera topology that we have made in the text. Notice that they all lie in the top-left corner cell, so they will be connected to one cell on the right and one cell below.

To learn more...

The `Advantage_system4.1` and `Advantage_system6.1` annealers use a topology called **Pegasus**. It also groups qubits into cells, but their structure is more involved than that of the Chimera cells. Every qubit is connected to up to 15 qubits, compared to the maximum of 6 in `DW_2000Q_6`.

This topology also contains groups of 4 qubits that are all connected to each other, making it much easier to embed problems into it, as we will see later in the *Section 4.4.2*.

Describing the Pegasus topology in detail would get us too far out from our path, but you can find all the information in *Section 2.3* of the *QPU-Specific Physical Properties* document for the `Advantage_system4.1` and `Advantage_system6.1` annealers. You can download it, together with the corresponding documents for the rest of the D-Wave quantum computers, at `https://docs.dwavesys.com/docs/latest/doc_physical_properties.html`.

One important thing to note about the Chimera topology is that it does not contain triangles. That is, there are no three vertices all connected to each other. Thus, if our Ising Hamiltonian is something like $Z_0Z_1 + Z_0Z_2 + Z_1Z_2$, we cannot directly map it to qubits in the `DW_2000Q_6` annealer. What can we do then? Is it impossible to solve such a problem with this computer? Don't worry, embeddings are here to save the day!

An **embedding** is, essentially, a way of mapping the qubits in our problem Hamiltonian to the physical qubits in the annealer. The trick here is that this mapping need not be one to one. In fact, we can use several physical qubits (what we call a **chain**) to represent a single qubit from our problem. In that case, though, we want all the qubits in the same chain to have the same value when we measure them. To guarantee that, we need to use coupling strengths that are negative and big in absolute value.

For instance, if qubits 12 and 20 are part of the same chain, the coefficient for $(12, 20)$ could be, for instance, -15. Then, the term $-15Z_{12}Z_{20}$ will be part of the spin Hamiltonian that we want to minimize and it will make it very likely for Z_{12} and Z_{20} to be equal to each other, because that will make the total energy significantly lower.

Of course, the embedding needs to define the physical qubits (chains) used to represent each problem qubit, ensure that they can be connected correctly, and compute some appropriate coupling strengths for the chains. This may seem very complicated, but Ocean can compute embeddings automatically for us. Let's see, with an example, how to do it for a simple case. We could use the following code:

```
# Define the problem
J = {(0,1):1, (0,2):1, (1,2):1}
h = {}
triangle = dimod.BinaryQuadraticModel(h, J, 0.0, dimod.SPIN)
# Embed it and solve it on the DW_2000Q_6 annealer
sampler = EmbeddingComposite(DWaveSampler(solver = "DW_2000Q_6"))
result = sampler.sample(triangle, num_reads=10,
    return_embedding = True)
print("The samples obtained are")
print(result)
print("The embedding used was")
print(result.info["embedding_context"])
```

In these instructions, we first define a problem that requires three qubits to be connected together, something that we know is not directly possible with the annealer that we have selected. But, since we are using `EmbeddingComposite`, a way of embedding our graph in the actual annealer topology is automatically found for us, and we can run the annealing process and obtain some samples. By setting the `return_embedding` parameter to `True`, we also recover the embedding information. Let's see what the output of running this code may look like:

```
The samples obtained are
   0  1  2 energy num_oc. chain_.
```

```
0 +1 -1 -1   -1.0       3     0.0
1 +1 +1 -1   -1.0       2     0.0
2 +1 -1 +1   -1.0       2     0.0
3 -1 +1 +1   -1.0       3     0.0
['SPIN', 4 rows, 10 samples, 3 variables]
The embedding used was
{'embedding': {1: (1015, 1008), 0: (1011,), 2: (1012,)},
'chain_break_method': 'majority_vote', 'embedding_parameters': {},
'chain_strength': 1.9996979771955565}
```

As you can see, `EmbeddingComposite` performs the embedding in a way that is completely transparent for the user and, in fact, the samples returned only refer to the variables in the original problem. However, underneath the hood, variable 0 has been mapped to qubit 1011, variable 2 has been mapped to qubit 1012, and variable 1 is represented by the chain formed by qubits 1008 and 1015. The coupling strength for these two qubits was almost 2, which is bigger than the coefficients of the original problem in order to prevent the qubits in the chain from having different values. If, for whatever reason, these two qubits in the chain happen to receive different values, the chain is said to be **broken** and the method specified in `'chain_break_method'` would be used to assign a value to variable 1. In this case, that value would be a simple majority vote between the qubits in the chain.

To learn more...

Finding a suitable embedding is an *NP*-hard problem. However, the `minorminer` package included with Ocean provides heuristics for finding embeddings that usually work well in practice. These are used by `EmbeddingComposite`.

In addition to `EmbeddingComposite`, there are other classes in Ocean that allow you to find embeddings for your problems. For instance, `AutoEmbeddingComposite` first tries to run the problem on the annealer directly, not using an embedding, and only looks for one if it is needed; this can save some computing time in some cases. The `FixedEmbeddingComposite` class doesn't compute an embedding, but uses whichever one is passed as a parameter;

in this case, the embedding should be a Python dictionary with the format shown in the previous output. We also can use `LazyFixedEmbeddingComposite`, which only computes the embedding for a problem on the first call to the `sample` method, storing it for future calls; `EmbeddingComposite`, on the other hand, recomputes the embedding with each call to `sample`.

So, that should cover most of your needs for embedding problems into any annealer topology. But we're not done yet! There are still some additional parameters that we can control when running problems with Ocean on actual quantum devices. We will study some of the most important ones in the next subsection.

4.4.3 Controlling annealing parameters

You surely remember from the beginning of this chapter that for an evolution to be adiabatic (and, hence, for the system to remain in a state of minimal energy), it needs to be slow enough. However, this condition is difficult to meet in practice, so we just resort to running the evolution for a short period of time, resulting in what we call quantum annealing.

The question is: to what extent can we control the annealing process with D-Wave's quantum annealers? It turns out that there are quite a number of things that we can do in order to try to improve the results for our combinatorial optimization problems. The first (and most obvious) is changing the duration of the annealing process.

You can easily check the range of annealing times that a device supports, as well as its default annealing time, using instructions like the following:

```
sampler = DWaveSampler(solver = "Advantage_system4.1")
print("The default annealing time is",
    sampler.properties["default_annealing_time"],"microseconds")
print("The possible values for the annealing time (in microseconds)"\
    " lie in the range",sampler.properties["annealing_time_range"])
```

The output, in this case, will be as follows:

```
The default annealing time is 20.0 microseconds
```

```
The possible values for the annealing time (in microseconds)
    lie in the range [0.5, 2000.0]
```

Exercise 4.5

Check the default annealing time and the annealing time range for the `DW_2000Q_6` annealer.

Modifying the annealing time for a problem couldn't be easier. For instance, imagine that we want to increase it to 100 microseconds and sample from the triangle problem that we defined in the previous subsection. Then, the only modification that we would need to apply is adding the `annealing_time` parameter when calling the `sample` method. We could run, for instance, the following code:

```
J = {(0,1):1, (0,2):1, (1,2):1}
h = {}
triangle = dimod.BinaryQuadraticModel(h, J, 0.0, dimod.SPIN)
sampler = EmbeddingComposite(DWaveSampler(solver = "DW_2000Q_6"))
result = sampler.sample(triangle, num_reads=10, annealing_time = 100)
print("The samples obtained are")
print(result)
```

Important note

In order to try to obtain better and better solutions, you may be tempted to increase the annealing time to its maximum possible value. However, be warned that this may have two unwanted consequences. On the one hand, the longer you run the annealing process, the higher the possibility that external interactions will affect the system state and ruin your computation: you might get worse results instead of better ones! On the other hand, by increasing the annealing time, you will obviously spend more time using the quantum processing unit…and you will be charged accordingly!

With Ocean, the options to control the annealing process are not reduced to just modifying the annealing time. You can also tailor, to some extent, the annealing schedule itself. As we already know, this refers to the A and B functions in the expression

$$H(t) = -A(t)\sum_{j=0}^{n-1} X_j - B(t)\sum_{j,k} J_{jk}Z_jZ_k - B(t)\sum_j h_jZ_j,$$

which defines the Hamiltonian that we use in the annealing process.

You may remember that we only required A and B to satisfy that $A(0) = B(T) = 1$ and $A(T) = B(0) = 0$, where T is the total annealing time, but we did not restrict in any way how A and B should behave except for these boundary conditions. D-Wave's annealers have default schedules. You can find them at `https://docs.dwavesys.com/docs/latest/doc_physical_properties.html` and in the annealing schedule sections of the user manuals for the devices, which you can find on that same web page. We can modify those default schedules by specifying the values that we want the functions to take at some intermediate times.

We can define a custom annealing schedule through a list of pairs of real numbers. The first number of each pair needs to be a time value given in microseconds and the second one has to be a number between 0 and 1. This second number is called the **anneal fraction**, usually denoted by s. The higher the value of s is, the higher the value of B and the lower the value of A will be. As a consequence, when $s = 1$, we can interpret that B is 1 and A is 0; when $s = 0$, we can interpret that A is 1 and B is 0.

There are two types of annealing schedules that we can use. The first one is called **forward annealing** and corresponds to the usual annealing process that we have been studying since the beginning of this chapter. It starts with $(0, 0)$ and ends at $(T, 1)$, where T is the total annealing time — which, of course, must not be bigger than the maximum annealing time allowed by the device. In addition, the values of s must monotonically increase over the time points.

An example of a forward annealing schedule could be the following:

```
forward_schedule=[[0.0, 0.0], [5.0, 0.25], [25, 0.75], [30, 1.0]]
```

In this case, s starts at 0, gets value 0.25 at 5 microseconds, 0.75 at 25 microseconds and, finally, 1 at 30 microseconds, which is the end of the annealing process. The growth of s will be linear between the points specified in the schedule. To use this custom schedule in a device, you only need to pass it as the `anneal_schedule` parameter. For instance, you can do something like the following:

```
forward_schedule=[[0.0, 0.0], [5.0, 0.25], [25, 0.75], [30, 1.0]]
sampler = EmbeddingComposite(DWaveSampler())
result = sampler.sample(triangle, num_reads=10,
    anneal_schedule = forward_schedule)
```

Here, `triangle` is the problem that we defined in the previous code block.

To learn more...

Controlling the annealing schedule can be useful for certain problems, especially if you know that at some points the ground state and the first excited state are closer. In this case, you can use a custom schedule to slow the annealing process down on those "dangerous" regions, while allowing it to go faster on other, less problematic, time intervals.

In addition to forward annealing, we can also use **reverse annealing**. In reverse annealing, s starts at 1, decreases for some time, and then increases back to 1 at the end of the annealing process. An example of a reverse annealing schedule could be

```
reverse_schedule=[[0.0, 1.0], [10.0, 0.5], [20, 1.0]]
```

where, as in the case of forward annealing, the values of s are linearly interpolated between the points given in the list.

When using reverse annealing, you also need to specify an initial state. This is because now we do not start with a Hamiltonian whose ground state is known to us. You can do that with the `initial_state` parameter of the `sample` method. Reverse annealing is commonly used on an approximate solution that we already have in an attempt to find a better one. In this case, we take that solution to be the initial state, we decrease the intensity of the final Hamiltonian for some time, and then we increase it again in an attempt to obtain a new solution with a lower energy.

There are two different ways in which we can use reverse annealing. We can run several repetitions of the annealing process on the same initial state with the `reinitialize_state=True` option when calling `sample`. Alternatively, we can use the final (measured) state of one execution as the initial state of the next one by setting `reinitialize_state=False`.

Let's now look at an example in which we will apply reverse annealing to a simple problem. The following code, in which we use the `triangle` problem defined previously, is almost self-explanatory:

```
reverse_schedule=[[0.0, 1.0], [10.0, 0.5], [20, 1.0]]
initial_state = {0:1, 1:1, 2:1}
sampler = EmbeddingComposite(DWaveSampler())
result = sampler.sample(triangle, num_reads=10,
    anneal_schedule = reverse_schedule,
    reinitialize_state=False, initial_state = initial_state)
print("The samples obtained are")
print(result)
```

A possible output of these instructions could be the following:

```
The samples obtained are
   0  1  2 energy num_oc. chain_.
0 -1 +1 -1   -1.0       1     0.0
1 +1 +1 -1   -1.0       1     0.0
2 +1 +1 -1   -1.0       1     0.0
3 -1 +1 -1   -1.0       1     0.0
4 -1 -1 +1   -1.0       1     0.0
5 +1 -1 +1   -1.0       1     0.0
```

```
6 +1 +1 -1    -1.0        1     0.0
7 +1 +1 -1    -1.0        1     0.0
8 +1 -1 -1    -1.0        1     0.0
9 -1 +1 -1    -1.0        1     0.0
['SPIN', 10 rows, 10 samples, 3 variables]
```

To learn more...

Some researchers have found that reverse annealing can be more effective than forward annealing for some problems. For a very illuminating example, please check the paper by Carugno et al. [39].

Now, we know how to control both the annealing time and the schedule. In the next subsection, we will explain why it is important to set the coupling strengths and the penalty terms wisely, something that is easily overlooked, but that can greatly affect the results of our executions.

4.4.4 The importance of coupling strengths

You surely remember that there are a couple of situations in which we have to select values for some arbitrary constants that are used to set coupling strengths in the annealer. The first situation is having to introduce constraints as penalty terms in the objective function, using the `lagrange_multiplier` parameter in the `cqm_to_bqm` method of the `dimod` package. The second one is having to select the coupling strengths for the chains in a particular embedding, which is usually handled automatically by classes such as `EmbeddingComposite`.

It would be very natural to think that you would want these constants to be as big as possible. After all, you are not interested in solutions that do not satisfy the problem constraints and you do not want your chains to be broken. However, there is an important detail that makes choosing the values of these constants a little bit trickier than expected.

It turns out that the range of values that you use for qubit couplings in D-Wave's annealers is not arbitrarily large. For example, the following instructions allow us to check what the possible values are for the case of the `Advantage_system4.1` device – and, of course, if you change the solver name, you can get the values for any other annealers as well:

```
sampler = DWaveSampler("Advantage_system4.1")
print("The coupling strength range is", sampler.properties["h_range"])
```

The output that you will get if you run these instructions is the following:

```
The coupling strength range is [-4.0, 4.0]
```

This means that, if you set coupling strengths (that is, J coefficients) that in absolute value are bigger than 4, the largest one will be scaled down to 4. . . and the rest of the coefficients in your model will be scaled down accordingly. This can cause some of the values to be very close together, even closer than the resolution of the device, affecting the results of the annealing process. Let's illustrate it with an example.

The following code defines a constrained problem, converts it into an unconstrained model using the penalty constant $M = 10$, and then runs it on `Advantage_system4.1` taking 100 samples. Then, it converts the samples back to the variables of the original problem, aggregates the results, as we did in *Section 4.3.3*, and shows the frequency of each obtained solution:

```
sampler = EmbeddingComposite(DWaveSampler("Advantage_system4.1"))
# Define the problem
x0 = dimod.Binary("x0")
x1 = dimod.Binary("x1")
x2 = dimod.Binary("x2")
blp = dimod.ConstrainedQuadraticModel()
blp.set_objective(-5*x0+3*x1-2*x2)
blp.add_constraint(x0 + x2 <= 1, "First constraint")
blp.add_constraint(3*x0 -x1 + 3*x2 <= 4, "Second constraint")
```

```
# Convert the problem and run it
qubo, invert = dimod.cqm_to_bqm(blp, lagrange_multiplier = 10)
result = sampler.sample(qubo, num_reads=100)
# Aggregate and show the results
samples = []
occurrences = []
for s in result.data():
    samples.append(invert(s.sample))
    occurrences.append(s.num_occurrences)
sampleset = dimod.SampleSet.from_samples_cqm(samples,blp,
    num_occurrences=occurrences)
print("The solutions to the original problem are")
print(sampleset.filter(lambda s: s.is_feasible).aggregate())
```

When we ran this code, we obtained the following output:

```
The solutions to the original problem are
  x0 x1 x2 energy num_oc. is_sat. is_fea.
0  1  0  0   -5.0      21 arra...    True
1  1  1  0   -2.0      32 arra...    True
2  0  0  1   -2.0      11 arra...    True
3  0  0  0    0.0      17 arra...    True
4  0  1  1    1.0      10 arra...    True
5  0  1  0    3.0       9 arra...    True
['INTEGER', 6 rows, 100 samples, 3 variables]
```

So, in 21 out of 100 samples, we have obtained the optimal solution, and, in 43 more cases, we obtained solutions with the second lowest energy. Not too bad…but not very good either. The not-so-obvious problem behind this result is that the penalty constant (the `lagrange_multiplier` parameter) is too big compared to the range of energies of the objective function. In fact, if you use `ExactSolver` on the transformed problem, you can easily check that all the assignments that are unfeasible on the original problem get energy 16 or higher on the transformed one, while the feasible solutions always get energy 3 or lower. That is a huge gap!

But notice what happened when we ran the same code after reducing the penalty constant to 4. In that case, we obtained the following result:

```
The solutions to the original problem are
  x0 x1 x2 energy num_oc. is_sat. is_fea.
0  1  0  0   -5.0      30 arra...    True
1  1  1  0   -2.0      31 arra...    True
2  0  0  1   -2.0      16 arra...    True
3  0  0  0    0.0       8 arra...    True
4  0  1  1    1.0      10 arra...    True
5  0  1  0    3.0       3 arra...    True
['INTEGER', 6 rows, 98 samples, 3 variables]
```

As you can see, the frequency of the optimal solution has increased to 30 and the two solutions with the second lowest energy appear, more or less, the same number of times as in the experiment with `lagrange_multiplier=10`. In this case (check it by using `ExactSolver`), the unfeasible solutions all have energy that is at least 4 in the transformed problem, so all the feasible solutions have lower energy. Notice, though, that the gap is now much smaller and we only recovered 98 feasible solutions from the 100 samples.

We even tried a more extreme experiment, setting `lagrange_multiplier=1`. When we ran it, we obtained the following output:

```
The solutions to the original problem are
  x0 x1 x2 energy num_oc. is_sat. is_fea.
0  1  0  0   -5.0      76 arra...    True
1  0  0  1   -2.0       5 arra...    True
2  1  1  0   -2.0      11 arra...    True
3  0  0  0    0.0       1 arra...    True
```

```
4  0  1  1    1.0       1 arra...    True
['INTEGER', 5 rows, 94 samples, 3 variables]
```

The frequency of the optimal solution has dramatically improved, up to 76 out of 100 samples. However, we also "lost" 6 samples because they corresponded to unfeasible solutions. In this case, there are some unfeasible solutions with energy as low as -2 in the transformed problem. This is still bigger than the optimal energy, which is -5, but the low energy of these unfeasible solutions can fool the annealer into selecting them at times, as we have seen.

Setting a good penalty constant can be difficult, because it involves having some information about the energy distribution of the solutions to the problem. But let the examples here serve as a warning that you should not just use any value for `lagrange_multiplier`, because setting it too high can affect the quality of your solutions. In case of doubt, try some different options and keep the one that offers the best results.

To learn more...

Something similar may happen when the value of the coupling strength for chains in an embedding is too big. Fortunately, the methods used by `EmbeddingComposite` and its relatives take this into account and will try to keep the value as low as possible without risking breaking many chains. But should you need, for some reason, to create your own embedding, do not take the choice of the coupling strength lightly.

You now know how to adjust the most important parameters that govern quantum annealing and, more importantly, you understand the implications of such adjustments. But it turns out that D-Wave offers other ways of solving optimization problems beyond "pure" quantum annealing. Let us study them in the following subsection.

4.4.5 Classical and hybrid samplers

We have already seen that `dimod` provides a **classical** solver called `ExactSolver`. And it's not alone! In Ocean, we can also find solvers such as, for example, `SimulatedAnnealing` or `SteepestDescentSolver` that do not rely on any quantum resources whatsoever.

The purpose of including these classical solvers in a quantum optimization library is two-fold. On the one hand, it allows you to try and use different methods to solve your problems. On the other hand, they can be used in combination with quantum annealers in what D-Wave calls **hybrid solvers**. Let's briefly study these two aspects.

Classical solvers

Using classical solvers with Ocean couldn't be simpler. As long as you have a QUBO or Ising problem, you can use the `sample` method of any classical solver to get (approximate) solutions to it, exactly like you would do with a quantum annealer. In fact, you can also use the `num_reads` parameter to specify the number of samples that you want.

We will devote the rest of this subsection to describing the classical solvers included in Ocean at the time of writing.

`SteepestDescentSolver`

This is included in the `greedy` package and it is a discrete version of the gradient descent algorithm for continuous optimization (more on that in *Chapter 8, What is Quantum Machine Learning?*). At each step, it selects one direction (that is, one variable flip) in which the decrease in energy is bigger. We can use it as shown in the following piece of code, where we first define a simple Ising problem and then we sample from it:

```
import greedy
import dimod
J = {(0,1):1, (1,2):1, (2,3):1, (3,0):1}
h = {}
problem = dimod.BinaryQuadraticModel(h, J, 0.0, dimod.SPIN)
# Sample with SteepestDescentSolver
solver = greedy.SteepestDescentSolver()
solution = solver.sample(problem, num_reads = 10)
print(solution.aggregate())
```

The output of running these instructions will be similar to the following:

```
0  1  2  3 energy num_oc. num_st.
0 +1 -1 +1 -1   -4.0       5       1
2 -1 +1 -1 +1   -4.0       3       1
1 +1 -1 -1 +1    0.0       1       0
3 +1 +1 -1 -1    0.0       1       0
['SPIN', 4 rows, 10 samples, 4 variables]
```

As you can see, this is exactly the format the we already know and love from using quantum solvers.

To learn more...

In addition to the `num_reads` parameter, you can also set `initial_states` to specify the solutions from which the descent is going to start. If you don't use this parameter, then the initial states will be selected at random. In that case, you can use the `seed` argument should you want your results to be reproducible.

`TabuSolver`

This solver is included in the `tabu` package. It is an example of a **local search** algorithm. That is, it tries to improve a solution by exploring its neighbors — the solutions that can be obtained by flipping one variable, for instance. In this, the method is somewhat similar to the greedy descent algorithm implemented in `SteepestDescentSolver`, but it tries to avoid falling into local minima by sometimes accepting solutions with higher energy than the current one, and it also "remembers" solutions it has already visited in order not to explore them again — that is where the name *tabu* comes from.

Ocean implements the multistart tabu algorithm described in [40]. It can be used with the following instructions:

```
import tabu
solver = tabu.TabuSampler()
solution = solver.sample(problem, num_reads = 10)
print(solution.aggregate())
```

The tabu algorithm also accepts `initial_states` and `seed` parameters.

`SimulatedAnnealingSampler`

This is included in the `neal` package and it implements the heuristic algorithm known as **simulated annealing** [41]. It is another local search algorithm that explores the neighbourhood of the candidate solution that it is considering at a given moment. With that, it tries to move to solutions with lower energy. However, like tabu search, it can move to solutions with higher energy with some probability. This probability is bounded by a global "temperature" parameter that decreases with time, eventually reaching 0, inspired by the way in which metals become less malleable when they cool down during annealing – hence the name of the method. In fact, quantum annealing is seen by some people as a quantum version of simulated annealing. The analogy they make is that the intensity of the initial Hamiltonian H_0 can be understood as analogous to the temperature in simulated annealing: it allows the solutions to move or "tunnel" to some neighboring ones and it decreases over time. Simulated annealing can be used in Ocean as follows:

```
import neal
solver = neal.SimulatedAnnealingSampler()
solution = solver.sample(problem, num_reads = 10)
print(solution.aggregate())
```

As you surely guessed, the `initial_states` and `seed` parameters are also supported.

These samplers are all classical algorithms that do not use quantum resources. However, they can be combined with quantum annealers, as we show in the next subsection.

Hybrid solvers

In addition to quantum annealers and classical solvers, Ocean also provides the programmer with hybrid solvers that try to combine the best of both worlds. You may remember that, back in *Section 4.4.1*, these hybrid solvers were listed among the devices available through your Leap account. Finally, the time to learn how to use them has come!

Let's start with `LeapHybridSampler`. This sampler accepts QUBO and Ising problems and can scale up to a high number of variables because, internally, it divides the problem, assigns different parts to classical solvers and quantum annealers, and then reconstructs a global solution from the local ones. Its use is very similar to that of the samplers that we have studied so far. For instance, you can run the following instructions, with `problem` as defined in the previous subsection — or just any other QUBO or Ising problem:

```
import dwave.system
sampler = dwave.system.LeapHybridSampler()
solution = solver.sample(problem, num_reads = 10)
print(solution.aggregate())
```

One interesting property of `LeapHybridSampler` and the rest of the hybrid samplers is what is called the **quota conversion rate**. It can be checked through the following property:

```
sampler.properties["quota_conversion_rate"]
```

In the case of `LeapHybridSampler`, it is 20. This means that for each 20 microseconds that you use this hybrid sampler, you will get charged just 1 microsecond of quantum processor access because the quantum annealers are not used for the whole computation. Neat, right?

Ocean also provides `LeapHybridCQMSampler`, which is used similarly to `LeapHybridSampler`, but with constrained problems like the ones we defined in *Section 4.3.3*. Finally, there is also `LeapHybridDQMSampler`, which works with discrete quadratic problems defined as objects of the `DiscreteQuadraticModel` class.

To learn more...

We have not worked with the `DiscreteQuadraticModel` class, but it is very similar to `BinaryQuadraticModel`. The main difference is that it accepts variables that take a finite amount of different values instead of just 0 and 1. The problems defined through this class can be converted to binary quadratic problems by **one-hot encoding**; that is, each discrete variable is represented by a vector of n binary variables, where n is the total number of values that the original discrete variable

> can take. The restriction is that only one of those variables can take the value 1 at a given time. So, if binary variable number 3 is 1, this means that the original variable takes the value 3.

This ends our study of quantum annealing and its use in combinatorial optimization. But this kind of problem can also be solved with algorithms designed for quantum computers based on the quantum circuit model. That will be the topic of our next chapter.

Summary

In this chapter, you have learned about the adiabatic quantum computing model, which is equivalent to the quantum circuit model that we had already studied. Instead of discrete quantum gates, adiabatic quantum computing uses continuous evolution through a time-dependent Hamiltonian. You have learned how to select this Hamiltonian to encode combinatorial optimization problems and how, if the evolution is slow enough, the adiabatic theorem guarantees that we will measure the ground state at the end of the process.

You have also learned that, in practice, quantum annealing is used instead of adiabatic quantum computing, because adiabatic evolution can take too long for the process to be feasible. What is more, you now know how to use actual quantum annealers through D-Wave Leap to find approximate solutions to combinatorial optimization problems in several different ways.

You also know how to control several parameters of the annealing process, in order to improve the quality of the solutions that you can find with quantum annealers. Finally, you have also learned how to use hybrid solvers that divide big problems into smaller pieces and combine classical and quantum techniques to find a global solution to the original problem.

We will now turn to using quantum computers based on the quantum circuit model. But we will not forget about optimization problems and Hamiltonians. In fact, as you will soon see, the topic of our next chapter will be how to discretize quantum annealing so that it can be implemented with quantum gates.

5

QAOA: Quantum Approximate Optimization Algorithm

True optimization is the revolutionary contribution of modern research to decision processes.

— George Dantzig

The techniques that we have introduced in the two previous chapters already allow us to solve combinatorial optimization problems on quantum computers. Specifically, we have studied how to write problems using the QUBO formalism and how to use quantum annealers to sample approximate solutions. This is an important approach to quantum optimization, but it is not the only one.

In this chapter, we are going to show how the ideas that we have already explored can also be used on digital quantum computers. We will be using our beloved quantum circuits — with all their qubits, quantum gates, and measurements — to solve combinatorial optimization problems formulated in the QUBO framework.

More concretely, we will be studying the famous **Quantum Approximate Optimization Algorithm (QAOA)**, which is a gate-based algorithm that can be understood to be the counterpart to quantum annealing in the quantum circuit model. We will start by introducing all the theoretical concepts that are needed in order to understand this algorithm, then we will study the kind of circuits used in its implementation, and finally, we will explain how to run QAOA with both Qiskit and PennyLane.

After reading this chapter, you will understand how QAOA works, you will know how to design the circuits used in the algorithm, and you will be able to solve your own combinatorial optimization problems using QAOA in Qiskit and PennyLane.

The topics that we will cover in this chapter are the following:

- From adiabatic computing to QAOA
- Using QAOA with Qiskit
- Using QAOA with PennyLane

5.1 From adiabatic computing to QAOA

In this first section, we will introduce all the theoretical concepts that will allow us to understand QAOA in depth. But before that, we will give an intuitive idea of how QAOA works by studying its relationship with quantum annealing. Sounds interesting? Then keep on reading, because here we go!

5.1.1 Discretizing adiabatic quantum computing

In the previous chapter, we studied adiabatic quantum computing and its practical realization, quantum annealing, and we learned how to use them in order to obtain approximate solutions to combinatorial optimization problems. Both of these techniques relied on the adiabatic theorem. When we applied them, we used a time-dependent Hamiltonian that induced a continuous transformation of the state of a quantum system: from an initial state to a final state that – hopefully – has a big overlap with the solution to our problem.

A natural question to ask is whether there is any sort of analog to this way of solving optimization problems for circuit-based quantum computers. At first sight, there is an apparent difficulty in this idea, because in the quantum circuit model we apply *instantaneous* operations – quantum gates – that change the state vector in discrete steps. How can we resolve this "tension" between these discrete operations and the continuous evolution that we rely on for adiabatic quantum computing?

The answer is that we may **discretize** any continuous evolution, approximating it with a sequence of small, discrete changes. This process, sometimes called **Trotterization**, is the inspiration for the topic to which this chapter is devoted: the Quantum Approximate Optimization Algorithm – QAOA, for short.

QAOA was initially proposed [42] as a **discretization** or Trotterization of adiabatic quantum computing with the goal of approximating the optimal solutions to combinatorial optimization problems. As you surely remember, the Hamiltonian that is used in adiabatic quantum computing – and in quantum annealing – is of the form

$$H(t) = A(t)H_0 + B(t)H_1,$$

with H_0 and H_1 two fixed Hamiltonians and $A(t)$ and $B(t)$ functions satisfying $A(0) = B(T) = 1$ and $A(T) = B(0) = 0$, where T is the total time of the process. It turns out that the evolution of the quantum system is governed by the famous time-dependent Schrödinger equation, and if you can solve it, you will have an expression for the state vector of your system at any moment t between 0 and T.

However, to understand QAOA we don't need to learn how to solve the Schrödinger equation – that was close, but we managed to dodge the bullet! All that we need to know is that, applying discretization, we can express the solution as a product of operators of the form

$$e^{i\Delta t(A(t_c)H_0+B(t_c)H_1)}$$

applied to the initial state. Here, i is the imaginary unit, t_c is a fixed time point in $[0, T]$, and Δt is a small amount of time. The key idea is that in the interval $[t_c, t_c + \Delta t]$ we assume that the Hamiltonian is constant and equal to $H(t_c) = A(t_c)H_0 + B(t_c)H_1$. Of course, the smaller Δt is, the better this approximation will be. It is also important to notice that $e^{i\Delta t(A(t_c)H_0+B(t_c)H_1)}$ is a unitary transformation, just like the ones we studied in *Chapter 1, Foundations of Quantum Computing*. In fact, you surely remember that some of the quantum gates that we introduced in that chapter, such as R_X, R_Y, and R_Z, are also exponentials of some matrices. Using this discretization technique, if $|\psi_0\rangle$ is the initial state, then the final state can be approximated by

$$\left(\prod_{m=0}^{p} e^{i\Delta t(A(t_m)H_0+B(t_m)H_1)}\right)|\psi_0\rangle,$$

where $t_m = m\frac{\Delta t}{T}$ and $p = \frac{T}{\Delta t}$.

In order to compute this state with a quantum circuit, we just need an additional approximation. As you know, for any real numbers a and b, it holds that $e^{a+b} = e^a e^b$. The analogous identity for exponentials of matrices does not hold in general — unless the matrices commute. However, if Δt is small, then

$$e^{i\Delta t(A(t_c)H_0+B(t_c)H_1)} \approx e^{i\Delta t A(t_c)H_0} e^{i\Delta t B(t_c)H_1},$$

which is known as the **Lie-Trotter formula**.

Putting it all together, the final state of the adiabatic evolution can be approximated by

$$\prod_{m=0}^{p} e^{i\Delta t A(t_m)H_0} e^{i\Delta t B(t_m)H_1} |\psi_0\rangle,$$

which is the inspiration for QAOA, as we will see in the next subsection.

5.1.2 QAOA: The algorithm

The starting point and goal of QAOA are exactly the same as those of quantum annealing. We begin with a combinatorial optimization problem that we want to solve, and we encode it, as we learned in *Chapter 3, Working with Quadratic Unconstrained Binary Optimization Problems*, into an Ising Hamiltonian H_1. In order to find its ground state and solve our problem, we seek to apply a quantum state evolution similar to that of quantum annealing, but using a quantum circuit instead of a quantum annealer.

In light of the discretization of adiabatic evolution that we obtained at the end of the previous subsection, the idea behind QAOA is simple. In order to simulate with a quantum circuit the evolution of a state under a time-dependent Hamiltonian, you only need to take an initial state $|\psi_0\rangle$ and then alternate for p times the application of the operators $e^{i\gamma H_1}$ and $e^{i\beta H_0}$ for some values of γ and β. In the next subsection, by the way, we will see that the unitary transformations $e^{i\gamma H_1}$ and $e^{i\beta H_0}$ can be implemented with just one-qubit and two-qubit quantum gates.

What we are doing is, then, using a quantum circuit to prepare a state of the form

$$e^{i\beta_p H_0} e^{i\gamma_p H_1} \ldots e^{i\beta_2 H_0} e^{i\gamma_2 H_1} e^{i\beta_1 H_0} e^{i\gamma_1 H_1} |\psi_0\rangle ,$$

where $p \geq 1$. Usually, we collect all the coefficients in the exponents in two tuples $\boldsymbol{\beta} = (\beta_1, \ldots, \beta_p)$ and $\boldsymbol{\gamma} = (\gamma_1, \ldots, \gamma_p)$ and we denote the whole state by $|\boldsymbol{\beta}, \boldsymbol{\gamma}\rangle$.

In QAOA, we choose a fixed value of p and we have some values for $\boldsymbol{\beta}$ and $\boldsymbol{\gamma}$. Instead of thinking of the values for $\boldsymbol{\beta}$ and $\boldsymbol{\gamma}$ as small increments of time multiplied by intensity coefficients given by the functions A and B, as we did in the previous subsection, we'll just consider them to be "plain real numbers." And this is where the magic kicks in. Since we are now free to choose their values as we see fit...why not choose the *best* possible values for them?

But what does *best* mean here? Remember that we are just trying to find the ground state of H_1, so, for us, the lower the value of the energy $\langle \boldsymbol{\beta}, \boldsymbol{\gamma} | H_1 | \boldsymbol{\beta}, \boldsymbol{\gamma} \rangle$, the better. In this way, we have transformed our optimization problem into another one: finding the values $\boldsymbol{\beta}$ and $\boldsymbol{\gamma}$ that minimize

$$E(\boldsymbol{\beta}, \boldsymbol{\gamma}) = \left\langle \boldsymbol{\beta}, \boldsymbol{\gamma} \right| H_1 \left| \boldsymbol{\beta}, \boldsymbol{\gamma} \right\rangle .$$

Notice that, since the values $\boldsymbol{\beta}$ and $\boldsymbol{\gamma}$ are real and so is the energy $E(\boldsymbol{\beta}, \boldsymbol{\gamma})$, what we have in our hands is the old problem of finding a minimum for a real-valued function with real inputs. There are many algorithms that we can apply for this, for instance, the famous **gradient descent algorithm**, which we will be using to train machine learning models in *Part 3, A Match Made in Heaven: Quantum Machine Learning*, of this book. However, there is an important twist. As we know, the number of amplitudes needed to describe a state like $\left| \boldsymbol{\beta}, \boldsymbol{\gamma} \right\rangle$ is exponential in the number of qubits that we are using. Thus, computing $E(\boldsymbol{\beta}, \boldsymbol{\gamma})$ may be difficult with just a classical computer.

But it turns out that estimating values of $E(\boldsymbol{\beta}, \boldsymbol{\gamma})$ is something that we can do very efficiently with a quantum computer — at least when the number of terms in H_1 is polynomial in the number of qubits, something that is usually the case in the problems we are interested in. In the next subsection, we will explain in detail how to compute that kind of estimation, but for now just keep in mind that, given some values $\boldsymbol{\beta}$ and $\boldsymbol{\gamma}$, we can rely on the quantum computer to compute $E(\boldsymbol{\beta}, \boldsymbol{\gamma})$.

Then, we can take any classical algorithm for function minimization and, whenever it needs to compute a value of the E function, we use a quantum computer to estimate it and we give that value back to the classical algorithm until it needs another of value E. At that moment, we again use the quantum computer to obtain it and so on and so forth, all until we meet the stopping criteria of the classical algorithm. This is what we call a **hybrid algorithm**, one where the classical and the quantum computer work in tandem to solve a problem. We will see this kind of interaction many more times throughout the book.

Once we have obtained the optimal values $\boldsymbol{\beta}^*$ and $\boldsymbol{\gamma}^*$ for $\boldsymbol{\beta}$ and $\boldsymbol{\gamma}$ — or, at least, an estimation of them — we can use the quantum computer once more in order to prepare the state $\left| \boldsymbol{\beta}^*, \boldsymbol{\gamma}^* \right\rangle$. This state should have a sizeable overlap with the ground state of H_1, so when we measure it in the computational basis, we will have a good chance of obtaining a string of zeros and ones that is a good solution to our original problem – the one encoded in H_1 by using the techniques of *Chapter 3, Working with Quadratic Unconstrained Binary Optimization Problems.*

We now have all the pieces of the puzzle, so let's put them all together. The input to QAOA is an Ising Hamiltonian H_1, the ground state of which we wish to approximate because it encodes the solution to a certain combinatorial optimization problem. To that end, we consider the energy function $E(\boldsymbol{\beta}, \boldsymbol{\gamma})$ as defined before and we proceed to minimize it. For that, we choose $p \geq 1$ and some initial values $\boldsymbol{\beta}_0$ and $\boldsymbol{\gamma}_0$ that we shall use as the starting point for some classical minimization algorithm. Then, we run the minimization algorithm and, whenever it requests an evaluation of E on some points $\boldsymbol{\beta}$ and $\boldsymbol{\gamma}$, we use the quantum computer to prepare the state $\left|\boldsymbol{\beta}, \boldsymbol{\gamma}\right\rangle$ and estimate its energy, and we return the value to the classical algorithm. We continue this process until the classical minimization algorithm stops, returning some optimal values $\boldsymbol{\beta}^*$ and $\boldsymbol{\gamma}^*$. As a final step, we use the quantum computer to prepare $\left|\boldsymbol{\beta}^*, \boldsymbol{\gamma}^*\right\rangle$. When we measure it, we obtain a – hopefully, good – approximate solution to our combinatorial problem.

We have collected all these steps as pseudocode in the following algorithm. Notice that there are just two points at which the quantum computer is required.

Algorithm 5.1 (QAOA).

Choose a value for p
Choose a starting set of values $\boldsymbol{\beta} = (\beta_1, \ldots, \beta_p)$ and $\boldsymbol{\gamma} = (\gamma_1, \ldots, \gamma_p)$
while the stopping criteria are not met **do**
    Prepare state $\left|\boldsymbol{\beta}, \boldsymbol{\gamma}\right\rangle$ ▷ *This is done on the quantum computer!*
    From measurements of $\left|\boldsymbol{\beta}, \boldsymbol{\gamma}\right\rangle$, estimate $E(\boldsymbol{\beta}, \boldsymbol{\gamma})$
    Update $\boldsymbol{\beta}$ and $\boldsymbol{\gamma}$ according to the minimization algorithm
Obtain the optimal values $\boldsymbol{\beta}^*$ and $\boldsymbol{\gamma}^*$ returned by the minimization algorithm
Prepare state $\left|\boldsymbol{\beta}^*, \boldsymbol{\gamma}^*\right\rangle$ ▷ *This is done on the quantum computer!*
Measure the state to obtain an approximate solution

To learn more...

Just before bringing this subsection to an end, we thought this could be a good time for us to share a historical fact with you.

When it was introduced in a 2014 paper [42], QAOA provided a better ratio of approximation for the Max-Cut problem than any existing classical algorithm that would run on polynomial time. And we say that it *provided* because, soon after, this claim was challenged by a paper [43] that presented a classical algorithm that could beat QAOA.

What can we say? Sometimes classics refuse to die!

The description of QAOA that we have discussed may seem a little bit abstract. But don't worry. In the next subsection, we will make all of this much more concrete, because we will be studying in detail the quantum circuits that are needed to implement the parts of the algorithm that run on quantum computers.

5.1.3 Circuits for QAOA

As we have just seen, quantum computers are only used at certain steps in QAOA. And, in fact, those steps always involve the preparation of a state of the form

$$\left|\boldsymbol{\beta},\boldsymbol{\gamma}\right\rangle = e^{i\beta_p H_0} e^{i\gamma_p H_1} \ldots e^{i\beta_2 H_0} e^{i\gamma_2 H_1} e^{i\beta_1 H_0} e^{i\gamma_1 H_1} \left|\psi_0\right\rangle,$$

where $\left|\psi_0\right\rangle$ is the ground state of H_0. Of course, we need to prepare the state with adequate quantum gates on a quantum circuit, so let's analyze the operations that we need to perform. A crucial observation is that the Hamiltonians H_0 and H_1 take a very specific form. As we studied in the previous chapter, H_0 is usually taken to be $-\sum_{j=0}^{n-1} X_j$, while H_1 is an Ising Hamiltonian of the form

$$-\sum_{j,k} J_{jk} Z_j Z_k - \sum_j h_j Z_j,$$

where the coefficients J_{jk} and h_j are real numbers.

The ground state of H_0 is $\bigotimes_{i=0}^{n-1} |+\rangle$, as you proved in *Exercise 4.1*. This state can be easily prepared: starting from $|0\rangle$, you just need to use a Hadamard gate on each qubit of the circuit.

That was easy, so let's now focus on the operations of the form $e^{i\beta_k H_0}$, with β_j a real number. Notice that $H_0 = -\sum_{j=0}^{n-1} X_j$ and that all the X_j matrices commute with each other, so we can replace the exponential of the sum with the product of the exponentials. Therefore, it holds that

$$e^{i\beta_k H_0} = e^{-i\beta_k \sum_{j=0}^{n-1} X_j} = \prod_{j=0}^{n-1} e^{-i\beta_k X_j}.$$

But $e^{-i\beta X_j}$ is the expression for the rotation gate $R_X(2\beta)$, so this means that we just need to apply this gate to each of the qubits in our circuit. Neat, isn't it?

The last type of operation that we need to translate into quantum gates is $e^{i\gamma_l H_1}$ for any real coefficient γ_l. We know that H_1 is a sum of terms of the form $J_{jk}Z_jZ_k$ and h_jZ_j. Again, these matrices commute with each other, so we get

$$e^{i\gamma_l H_1} = e^{-i\gamma_l(\sum_{j,k} J_{jk}Z_jZ_k + \sum_j h_jZ_j)} = \prod_{j,k} e^{-i\gamma_l J_{jk}Z_jZ_k} \prod_j e^{-i\gamma_l h_jZ_j}.$$

Similar to the case of H_0, the operations of the form $e^{-i\gamma_l h_j Z_j}$ can be carried out with rotation gates R_Z. Thus, we only need to learn how to implement $e^{-i\gamma_l J_{jk} Z_j Z_k}$. To keep things simple, let's denote the real number $\gamma_l J_{jk}$ by a. Notice that $e^{-iaZ_jZ_k}$ is the exponential of a diagonal matrix, because Z_jZ_k is the tensor product of diagonal matrices. In fact, it holds that if $|x\rangle$ is a computational basis state in which qubits j and k have the same value, then

$$e^{-iaZ_jZ_k} |x\rangle = e^{-ia} |x\rangle .$$

On the other hand, if qubits j and k have different values, then

$$e^{-iaZ_jZ_k} |x\rangle = e^{ia} |x\rangle .$$

This unitary action is implemented by the circuit in *Figure 5.1*, where, for simplicity, we have only depicted qubits j and k – the action on the rest of the qubits would be the identity gate.

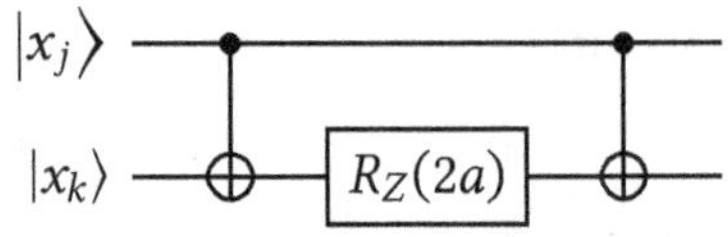

Figure 5.1: Implementation of $e^{-iaZ_jZ_k}$

We now have all the elements in place, so let's illustrate them with an example. Imagine that the Ising Hamiltonian of your problem is $3Z_0Z_2 - Z_1Z_2 + 2Z_0$. Then, the circuit used by QAOA to prepare $|\beta, \gamma\rangle$ with $p = 1$ is the one shown in *Figure 5.2*.

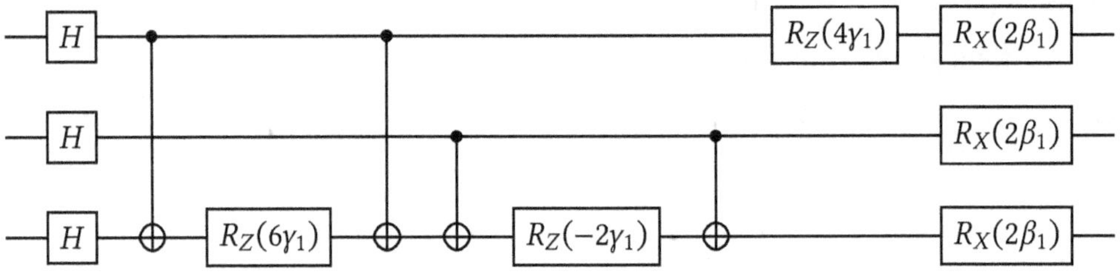

Figure 5.2: QAOA circuit with $p = 1$

Notice how we first prepare the ground state of H_0 with a column of Hadamard gates. Then, we have the implementation of $e^{-i3\gamma_1 Z_0 Z_2}$ with a CNOT gate between qubits 0 and 2, an R_Z gate on qubit 2, and another CNOT gate between qubits 0 and 2. The implementation of $e^{i\gamma_1 Z_1 Z_2}$ is similar, but on qubits 1 and 2. Then, we use an R_Z gate on qubit 0 to implement $e^{-i2\gamma_1 Z_0}$. Finally, a column of R_X gates implements $e^{-i\beta_1 \sum_j X_j}$. If we increased the number of **layers** p, the circuit would grow by repeating for another $p - 1$ times the very same circuit structure shown in *Figure 5.2* except for the initial Hadamard gates. Additionally, we would have to replace the parameters γ_1 and β_1 by γ_2 and β_2 in the second layer, by γ_3 and β_3 in the third, and so on and so forth.

Exercise 5.1

Obtain the QAOA circuit for $Z_1Z_3 + Z_0Z_2 - 2Z_1 + 3Z_2$ with $p = 1$.

Now that we know all the circuits that we need for QAOA, let's study how to use them in order to estimate the energy of the states $\left|\beta,\gamma\right\rangle$.

5.1.4 Estimating the energy

The circuits that we have just studied allow us to prepare any state of the form $\left|\beta,\gamma\right\rangle$. But we are not interested in the states themselves. What we need is their energy with respect to H_1, because that is the quantity that we want to minimize. That is, we need to evaluate $\left\langle\beta,\gamma\right| H_1 \left|\beta,\gamma\right\rangle$, but, of course, we don't have access to the state vector because we are preparing the state with a quantum computer. So, what can we do?

The key observation here is that we already know how to evaluate efficiently $\langle x| H_1 |x\rangle$ for any basis state $|x\rangle$. In fact, $\langle x| H_1 |x\rangle$ is the value of x in the cost function of our combinatorial optimization problem, because we derived H_1 from it. So, for instance, if we are trying to solve a Max-Cut problem, each $|x\rangle$ represents a cut and we can easily compute – with a classical computer – the cost of that cut, as we did in *Section 3.1.2*.

What is more, we can also evaluate $\langle x| H_1 |x\rangle$ directly from the expression of the Hamiltonian. We only need to notice that $\langle x| Z_j |x\rangle = 1$ if the j-th bit of x is 0 and that $\langle x| Z_j |x\rangle = -1$ otherwise. In a similar way, $\langle x| Z_jZ_k |x\rangle = 1$ if the j-th and k-th bits of x are equal and $\langle x| Z_jZ_k |x\rangle = -1$ if they are different.

Then, by linearity, we can easily evaluate $\langle x| H_1 |x\rangle$. For instance, if $H_1 = 3Z_0Z_2 - Z_1Z_2 + 2Z_0$, we will have

$$\langle 101| H_1 |101\rangle = 3\langle 101| Z_0Z_2 |101\rangle - \langle 101| Z_1Z_2 |101\rangle + 2\langle 101| Z_0 |101\rangle = 3 + 1 - 2 = 4.$$

Exercise 5.2

Evaluate $\langle 100| H_1 |100\rangle$ with $H_1 = 3Z_0Z_2 - Z_1Z_2 + 2Z_0$.

We also know that we can always write $\left|\beta,\gamma\right\rangle$ as a linear combination of basis states. Namely, we have

$$\left|\beta,\gamma\right\rangle = \sum_x a_x \left|x\right\rangle$$

for certain amplitudes a_x such that $\sum_x |a_x|^2 = 1$.

But then it holds that

$$\left\langle\beta,\gamma\right| H_1 \left|\beta,\gamma\right\rangle = \left(\sum_y a_y^* \left\langle y\right|\right) H_1 \left(\sum_x a_x \left|x\right\rangle\right) = \sum_y \sum_x a_y^* a_x \left\langle y\right| H_1 \left|x\right\rangle = \sum_x |a_x|^2 \left\langle x\right| H_1 \left|x\right\rangle,$$

because $H_1 \left|x\right\rangle$ is always a multiple of $\left|x\right\rangle$ (just notice that H_1 is a diagonal matrix because it is a sum of diagonal matrices), because $\langle y|x\rangle = 0$ when $y \neq x$, and because $a_x^* a_x = |a_x|^2$.

Now, since we can compute $\left\langle x\right| H_1 \left|x\right\rangle$ easily with the classical computer, we have reduced our problem to computing the values $|a_x|^2$. But $|a_x|^2$ is the probability of measuring $\left|x\right\rangle$ when the state $\left|\beta,\gamma\right\rangle$ is prepared — this is the reason why, back in *Chapter 3, Working with Quadratic Unconstrained Binary Optimization Problems*, we referred to expressions of the form $\left\langle\psi\right| H_1 \left|\psi\right\rangle$ as expectation values; they are indeed the expected or average energy under H_1 when we measure the state $\left|\psi\right\rangle$!

From this observation, it follows that we can use the quantum computer to prepare $\left|\beta,\gamma\right\rangle$ and measure it M times to make the estimation

$$\left\langle\beta,\gamma\right| H_1 \left|\beta,\gamma\right\rangle \approx \sum_x \frac{m_x}{M} \left\langle x\right| H_1 \left|x\right\rangle,$$

where m_x is the number of times that x was measured. Of course, the higher the value of M, the better this approximation will be.

Important note

In the process of estimating the energies of all the different states prepared with quantum computers, we will compute the cost, for our optimization problem, of many binary strings x. Of course, it would be wise for us to always keep the best x

seen during the optimization process. Occasionally, it might be even better than the ones we obtain when measuring the final state $|\beta^*, \gamma^*\rangle$.

We've now covered all that we needed to know about the inner workings of QAOA. Before we move on to show how to implement and use this algorithm with Qiskit and PennyLane, we will introduce a little perk that we get from the fact that we are no longer using quantum annealers, but universal quantum computers instead. This will help us in formulating some problems in a more natural way, as we will show in the next subsection.

5.1.5 QUBO and HOBO

Up to this point, we have only considered problems that can be written under the QUBO formalism. That is, minimization problems in which the cost function is a quadratic polynomial on binary variables that had no constraints on the values they could take. This is less restricting than it may seem, because QUBO is NP-hard and there are many important problems that we can rewrite via reductions, as we saw in *Section 3.4*.

However, consider a problem like the famous **satisfiability** or **SAT**. In it, we are given a Boolean formula on binary variables and we have to determine whether there is any assignment of values that makes the formula true. For example, we may receive

$$(x_0 \vee \neg x_1 \vee x_2) \wedge (\neg x_0 \vee x_1 \vee \neg x_2) \wedge (x_0 \vee x_1 \vee x_2),$$

which is **satisfiable** (by assigning *true* to all the variables, for instance). Or we can be given

$$x_0 \wedge \neg x_0,$$

which is clearly **unsatisfiable**.

SAT is easily seen to be in NP (and, in fact, it is NP-complete – see *Section 7.4* in Sipser's book [26]). Then, we know that there must be a way of rewriting any SAT instance in the QUBO formalism. But the task becomes much easier if we relax the conditions in the QUBO formulation by allowing binary polynomials of *any order*. Let's see why!

Let us consider, for the sake of an example, the formula $(x_0 \lor \neg x_1 \lor x_2) \land (\neg x_0 \lor x_1 \lor \neg x_2) \land (x_0 \lor x_1 \lor x_2)$. We'll show how it can be represented as the polynomial

$$p(x_0, x_1, x_2) = (1 - x_0)x_1(1 - x_2) + x_0(1 - x_1)x_2 + (1 - x_0)(1 - x_1)(1 - x_2),$$

on the binary variables x_0, x_1, and x_2. Let's say that we consider some assignment of truth values for the variables in the original formula, and we set $x_i = 1$ in the polynomial if x_i is true and $x_i = 0$ if x_i is false. It's easy to see that the original formula will be true under this assignment if and only if $p(x_0, x_1, x_2) = 0$, and that it will be false if and only if $p(x_0, x_1, x_2) > 0$. Thus, if the polynomial p is 0 for some values of its variables, then the original formula must be satisfiable. Otherwise, it has to be unsatisfiable.

Then, we can rewrite our original problem as follows:

$$\begin{array}{ll} \text{Minimize} & (1 - x_0)x_1(1 - x_2) + x_0(1 - x_1)x_2 + (1 - x_0)(1 - x_1)(1 - x_2) \\ \text{subject to} & x_j \in \{0, 1\}, \qquad j = 0, 1, 2. \end{array}$$

If the minimum of the polynomial is 0, then the formula will be satisfiable. Otherwise, the formula will be unsatisfiable.

With a simple transformation, we have been able to reformulate our problem as something that looks very much like a QUBO instance. But wait! This is *not* a QUBO problem. The reason is that the degree of the binary polynomial is 3 and not 2, as you can easily check by expanding its expression. These optimization problems in which we are asked to minimize a binary polynomial — of any degree — with no additional restrictions are called **Higher Order Binary Optimization (HOBO)** or **Polynomial Unconstrained Binary Optimization (PUBO)** problems, for obvious reasons.

The method that we have applied is quite general. In fact, it is easy to see that we can apply it to any Boolean formula that is given as conjunctions of disjunctions of variables and negations of variables. Something like, for instance, $(x_0 \vee \neg x_1 \vee x_2) \wedge (\neg x_0 \vee x_1 \vee \neg x_2) \wedge (x_0 \vee x_1 \vee x_2)$ or $(x_0 \vee x_1 \vee \neg x_2 \vee x_3) \wedge (\neg x_0 \vee x_1 \vee x_2 \vee x_3) \wedge (\neg x_0 \vee x_1 \vee \neg x_2 \vee \neg x_3)$. We say that these formulas are in **conjunctive normal form** or **CNF**. In this case, we can just obtain an associated polynomial consisting of a sum of products. Each product will correspond to one of the disjunctions of the formula. If a variable x appears negated in the disjunction, it will appear as x in the product. If it appears in positive form, it will appear as $1 - x$ in the product.

Exercise 5.3

Write the HOBO version of the SAT problem with the Boolean formula

$$(x_0 \vee x_1 \vee \neg x_2 \vee x_3) \wedge (\neg x_0 \vee x_1 \vee x_2 \vee x_3) \wedge (\neg x_0 \vee x_1 \vee \neg x_2 \vee \neg x_3).$$

And what about Boolean formulas that are not in CNF? In this case, we can apply a method, called the **Tseitin transformation**, that runs in polynomial time and gives us a formula in CNF that is satisfiable if and only if the original formula was satisfiable (see [44, Chapter 2] for more details). In fact, the resulting formula will be in **3-CNF**, meaning that the disjunctions will involve at most three variables or negations of variables. This is very convenient, because it guarantees that the process of expanding the polynomial to obtain the coefficients will be efficient.

But enough about satisfiability. Let's come back to HOBO problems. How can we solve them? One way of doing this is by transforming them into QUBO problems. There are different techniques for rewriting HOBO problems as QUBO instances by introducing auxiliary variables. For example, you can substitute products xy by a new binary variable z as long as you introduce a penalty term $xy - 2xz - 2yz + 3z$, which is 0 if and only if $xy = z$. In this way, you can reduce a term of order $m + 1$ such as $x_0 x_1 \cdots x_m$ to a term of order m of the form $z x_2 \cdots x_m$ and a quadratic penalty term on x_0, x_1, and z. By repeating this process as many times as needed, you can obtain an equivalent problem in which the objective function is a binary quadratic polynomial. Transformations of this kind are used in D-Wave's Ocean, where you can find `BinaryPolynomial` objects that you can reduce to polynomials of degree 2 with the `make_quadratic` function.

However, if you are using QAOA, you can deal with HOBO problems directly. We can consider a binary polynomial of any degree and transform it using the techniques of *Section 3.3*. We will end up having a Hamiltonian that is a sum of tensor products of Z_j matrices. The only difference is that, now, these products can involve more than just one or two Z_j matrices.

This implies that, when we set out to create a circuit for $e^{-i\gamma_l H_1}$, we may need to implement unitary operations of the form $e^{-iaZ_{j_1}Z_{j_2}\cdots Z_{j_m}}$, with $m > 2$. But that is not much more difficult than implementing $e^{-iaZ_jZ_k}$. In fact, we can almost repeat the argument in *Section 5.1.3*, because both $Z_{j_1}Z_{j_2}\cdots Z_{j_m}$ and $e^{-iaZ_{j_1}Z_{j_2}\cdots Z_{j_m}}$ are diagonal matrices. In fact, if $|x\rangle$ is a basis state, then

$$e^{-iaZ_{j_1}Z_{j_2}\cdots Z_{j_m}}|x\rangle = e^{-ia}|x\rangle$$

if the sum of the bits of x in positions $j_1, j_2, \ldots, j_m$ is even, and

$$e^{-iaZ_{j_1}Z_{j_2}\cdots Z_{j_m}}|x\rangle = e^{ia}|x\rangle$$

if the sum is odd.

This unitary action can be implemented by using consecutive CNOT gates with control qubits in $j_1, j_2, \ldots, j_{m-1}$ and targets in j_m, then an R_Z gate with parameter $2a$ on qubit j_m and, again, consecutive CNOT gates with control qubits in $j_{m-1}, j_{m-2}, \ldots, j_1$ and targets in j_m. *Figure 5.3* illustrates this procedure for the case of $e^{-iaZ_0Z_1Z_3}$, under the assumption that we only have four qubits.

Of course, this operation would be just one part in the implementation of $e^{-i\gamma_l H_1}$ and we would need to repeat a similar process for each term of the Hamiltonian.

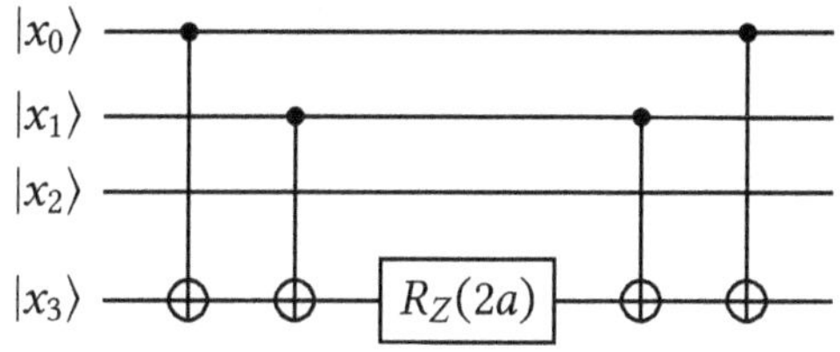

Figure 5.3: Implementation of $e^{-iaZ_0Z_1Z_3}$

Exercise 5.4

Implement, in a circuit with 5 qubits, the operation $e^{-i\frac{\pi}{4}Z_0Z_2Z_4}$.

Notice that we can also estimate the energy of a Hamiltonian H_1 that includes tensor products of Z matrices in a way that is very similar to the one that we explained in *Section 5.1.4*. The key fact is that, for any basis state $|x\rangle$, it holds that

$$\langle x| Z_{j_1}Z_{j_2}\cdots Z_{j_m}|x\rangle = 1$$

if the sum of the bits of x in positions $j_1, j_2, \ldots, j_m$ is even, and

$$\langle x| Z_{j_1}Z_{j_2}\cdots Z_{j_m}|x\rangle = -1$$

otherwise. By linearity, we can then evaluate $\langle x| H_1 |x\rangle$ and, from that, we can estimate $\langle\psi| H_1 |\psi\rangle$ by measuring $|\psi\rangle$ a number of times, exactly as we did in *Section 5.1.4*.

Exercise 5.5

Evaluate $\langle 100| H_1 |100\rangle$ with $H_1 = Z_0Z_1Z_2 + 3Z_0Z_2 - Z_1Z_2 + 2Z_0$.

We have now covered all the necessary concepts to understand QAOA in all its glory. In the next two sections, we will show how to implement and run this algorithm with both Qiskit and PennyLane.

5.2 Using QAOA with Qiskit

With everything that we have learned in the previous sections of this chapter and what we already know about Qiskit from *Chapter 2, The Tools of the Trade in Quantum Computing*, and *Section 3.2.2*, we could implement our own Qiskit version of QAOA. However, there is no need for that! As we shall show in this section, the Qiskit Optimization package provides all that is necessary to run QAOA on both quantum simulators and actual quantum computers. Moreover, it includes a set of tools to work directly with problems written under the QUBO formalism. As a matter of fact, in this section, we will also see how, underneath the hood, Qiskit uses the very same mathematical concepts that we have been studying.

Let's start by explaining how to work with QAOA in Qiskit when we already have the problem Hamiltonian.

5.2.1 Using QAOA with Hamiltonians

If we have the Hamiltonian H_1 that encodes our optimization problem, it is very easy to use Qiskit's QAOA implementation to approximate its ground state. Let's start with a simple example in which we have $H_1 = Z_0 Z_1$. We can create this Hamiltonian and prepare the corresponding QAOA circuit with the following lines of code:

```
from qiskit.opflow import Z
from qiskit.algorithms import QAOA

H1 = Z^Z # Define Z_0Z_1
qaoa = QAOA()
circuit = qaoa.construct_circuit([1,2],H1)[0]
circuit.draw(output="mpl")
```

As a result, we will obtain the circuit shown in *Figure 5.4*. We can see how it starts with two Hadamard gates, which are then followed by the exponential of H_1 and then the exponential of H_0 (because $H_0 = X_0X_1 = X_0I + IX_1$). This is exactly the circuit that we derived in the first part of the chapter.

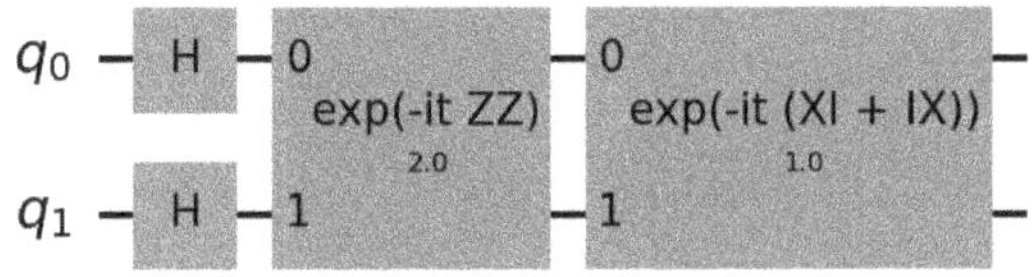

Figure 5.4: QAOA circuit for $H_1 = Z_0Z_1$

In order to create the circuit, in addition to `H1`, we've passed `[1,2]` to the method called `construct_circuit`. This list contains the $\boldsymbol{\beta}$ and $\boldsymbol{\gamma}$ parameters that we want to use. In *Figure 5.4*, this is indicated by the numbers below the exponentials in the gate boxes. Notice that this means that the first element in `[1,2]` is what we call β_1 and the second is γ_1. Also notice that we have used `[0]` after the call to `construct_circuit`. This is because this method, in general, returns a list of several circuits – but in this case, there is only one, which is the one that we pick.

We can visualize the circuit in more detail by decomposing the exponentials – that is, transforming them into simpler gates – a couple of times. For that, we may use

```
circuit.decompose().decompose().draw(output="mpl")
```

to get the circuit shown in *Figure 5.5*. The sequence of gates in that circuit is exactly the one that we would expect from our derivations earlier in this chapter, because $U(\pi/2, 0, \pi) = H$, as you can easily check from the definition that we gave in *Section 1.3.4* (U_3 is Qiskit's name for our U gate).

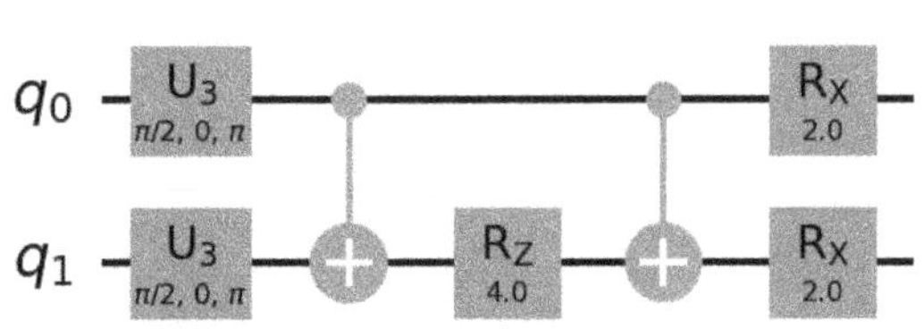

Figure 5.5: QAOA circuit for $H_1 = Z_0Z_1$*, more detailed*

By default, the value of p for QAOA in Qiskit is 1. However, we can change it by using the `reps` parameter when calling the class constructor. For instance, the following code can be used to obtain the QAOA circuit for Z_0Z_1 with $p = 2$, $\beta_1 = 1$, $\beta_2 = 2$, $\gamma_1 = 3$, and $\gamma_2 = 4$:

```
qaoa = QAOA(reps = 2)
circuit = qaoa.construct_circuit([1,2,3,4],H1)[0]
circuit.decompose().decompose().draw(output="mpl")
```

The result of the execution is the circuit shown in *Figure 5.6.*

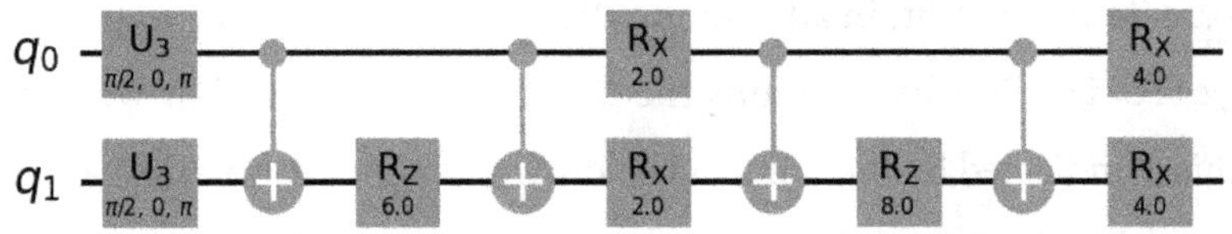

Figure 5.6: QAOA circuit for $H_1 = Z_0Z_1$ with $p = 2$

All this is well and good, but we haven't yet solved any optimization problems! For that, we need to pass two additional parameters when creating a `QAOA` object. The first one is a `QuantumInstance`. That is, some backend capable of executing the QAOA quantum circuit to evaluate the energy of the $\left|\boldsymbol{\beta}, \boldsymbol{\gamma}\right\rangle$ states. The second one is a classical optimizer, which will set initial values for $\boldsymbol{\beta}$ and $\boldsymbol{\gamma}$, use the `QuantumInstance` to evaluate state energies, and update the $\boldsymbol{\beta}$ and $\boldsymbol{\gamma}$ parameters in order to optimize them, until some stopping criterion is met.

To learn more...

The choice of classical optimizer can have a big impact on the execution time and quality of the solutions obtained with QAOA.

For some insights into this, you can refer to [45].

In the following piece of code, we give an example of how to create the quantum instance and the classical minimizer objects, and of how to use them with QAOA to solve a simple problem:

```
from qiskit.utils import algorithm_globals, QuantumInstance
from qiskit import Aer
from qiskit.algorithms.optimizers import COBYLA
seed = 1234
algorithm_globals.random_seed = seed
quantum_instance = QuantumInstance(Aer.get_backend("aer_simulator"),
                   seed_simulator=seed, seed_transpiler=seed,
                   shots = 10)
qaoa = QAOA(optimizer = COBYLA(), quantum_instance=quantum_instance)
result = qaoa.compute_minimum_eigenvalue(H1)
print(result)
```

Here, we are relying on the previous definition of `H1` as `Z^Z`, running the circuit on the `Aer` simulator with 10 shots, and using `COBYLA` as the classical optimizer — for an updated list of minimizers, please refer to Qiskit's documentation at `https://qiskit.org/documentation/stubs/qiskit.algorithms.optimizers.html`. We are also setting seeds for those processes that require random numbers, in order to obtain reproducible results.

If you run the preceding instructions, you will obtain the following output:

```
{   'aux_operator_eigenvalues': None,
    'cost_function_evals': 20,
    'eigenstate': {'01': 0.5477225575051661, '10': 0.8366600265340756},
    'eigenvalue': (-1+0j),
    'optimal_parameters': {
    ParameterVectorElement(γ[0]): -0.847240391875931,
    ParameterVectorElement(β[0]): 6.7647519845416655},
    'optimal_point': array([ 6.76475198, -0.84724039]),
    'optimal_value': -1.0,
    'optimizer_evals': None,
    'optimizer_time': 0.07764506340026855}
```

That is quite a lot of information! Let's try to explain the most relevant pieces. The first thing that we need to understand is that this result refers to the final state $|\boldsymbol{\beta}^*, \boldsymbol{\gamma}^*\rangle$ obtained by QAOA, not to the solutions that we would obtain if we measured it. In fact, this state is reconstructed from 10 measurements — because our simulator is using 10 shots — but those measurements are not given as part of the output. Instead, we get the `eigenstate` field, which shows that we have $|01\rangle$ with an amplitude roughly 0.5477 and $|10\rangle$ with an amplitude about 0.8367. These numbers are, in fact, $\sqrt{3/10}$ and $\sqrt{7/10}$, which means that, once the state with the optimal parameters found by the minimizer was prepared and measured, `01` was obtained 3 out of 10 times and `10` was obtained the remaining 7 times. This state can be prepared with the QAOA circuit by using the optimal parameters reported in the result: $\beta_1 \approx 6.7648$ and $\gamma_1 \approx -0.8472$.

Notice that, with respect to the Z_0Z_1 Hamiltonian, both $|01\rangle$ and $|10\rangle$ have an expected value of -1, which is the optimal energy. This means that we have been able to find an optimal solution — two of them, in fact — with QAOA! To get to this result, QAOA evaluated the energy function — by preparing a circuit with some values of β_1 and γ_1 and measuring it to estimate its expectation value — 20 times, as indicated by the `cost_function_evals` field, and it used about 0.08 seconds of computing time — your running time will most surely be different from ours, though.

All this has been done with the Aer simulator. If we wanted to use a real quantum computer, we could just replace the `backend` in the instantiation of the `QuantumInstance` object and use some of the quantum devices provided by IBM, as we showed in *Section 2.2.4*. However, this is not the best way to proceed. The problem with this straightforward approach is that you will be running the classical part of the algorithm locally. Then, each time that an energy estimation is required, a new job will be submitted to the quantum computer and you will have to wait on the queue if other users have also sent jobs to execute. This can be quite slow, not because of the process itself, but because of the queue waiting times.

Fortunately, Qiskit has recently introduced a new module called Runtime that allows us to reduce the execution time for hybrid algorithms such as QAOA. Instead of submitting each circuit individually, with Runtime you can submit a **program** that includes both the classical and the quantum part of the algorithm. The program is then queued just once, greatly speeding up the whole execution.

Using Runtime with QAOA is very easy. In fact, we just need to specify the same elements that we have used with the `QAOA` class, but in a slightly different way. The following piece of code shows an example of how to do this:

```
from qiskit import IBMQ
provider = IBMQ.load_account()
program_id = "qaoa"

H1 = Z^Z
opt = COBYLA()
reps = 1
shots = 1024

runtime_inputs = {
    "operator": H1,
    "reps": reps,
    "optimizer": opt,
    "initial_point": [0,0],
    "use_swap_strategies": False
}

options = {"backend_name": "ibmq_belem"}
job = provider.runtime.run(program_id=program_id,
      options=options, inputs=runtime_inputs)
```

This will create and run a QAOA Runtime job whose quantum part will be executed in the quantum computer that we have specified in the `options["backend_name"]` field – in our case, ibmq_belem. We have used the Z_0Z_1 Hamiltonian, COBYLA as the optimizer, a value of $p = 1$ (specified with the reps variable), and 1024 shots.

We have also chosen the initial values of β_1 and γ_1 to be 0 with the initial_point field. If this field is not used, the initial values are chosen at random.

Once the program has finished running (you can keep track of its execution using job.status()), the result can be retrieved with job.result(). We can access some of its parameters with the following instructions:

```
result = job.result()
print("Optimizer time", result['optimizer_time'])
print("Optimal value", result['optimal_value'])
print("Optimal point", result['optimal_point'])
print("Optimal state", result['eigenstate'])
```

In our case, running those lines of code offered the following result:

```
Optimizer time 88.11612486839294
Optimal value -0.84765625
Optimal point [0.42727683 2.39693691]
Optimal state {'00': 0.2576941016011038, '01': 0.691748238161833,
               '10': 0.6584783595532961, '11': 0.14657549249448218}
```

As you can appreciate, the results are slightly worse than with the simulator, due to the influence of noise. But the two optimal basis states — 01 and 10 — are still the most probable ones and, therefore, if we prepare and measure the final state several times, we will have a very high probability of finding an optimal solution.

So now we know how to solve problems with QAOA using Qiskit, both with simulators and with actual quantum computers. However, so far, we've had to prepare the Hamiltonian of the problem ourselves, and that is not ideal. As we learned in *Chapter 3, Working with Quadratic Unconstrained Binary Optimization Problems*, for many problems, it is more convenient to work with a QUBO formulation or even to write the problem as a binary linear program. Would it be possible to use those formalisms directly with QAOA in Qiskit? Absolutely! We'll show you how in the next subsection.

5.2.2 Solving QUBO problems with QAOA in Qiskit

Qiskit provides tools to work with quadratic problems, both with and without constraints, which are similar to the ones that we studied when working with Ocean in *Chapter 4, Quantum Adiabatic Computing and Quantum Annealing*. For example, we can define a simple binary program with the following piece of code:

```
from qiskit_optimization.problems import QuadraticProgram

qp = QuadraticProgram()
qp.binary_var('x')
qp.binary_var('y')
qp.binary_var('z')

qp.minimize(linear = {'y':-1}, quadratic = {('x','y'):2, ('z','y'):-4})
qp.linear_constraint(linear = {'x':1, 'y':2, 'z':3},
    sense ="<=", rhs = 5)

print(qp.export_as_lp_string())
```

As you can see, we are defining a quadratic problem with three binary variables, a function to minimize that has a linear and a quadratic part, and a linear constraint in the binary variables. When we run these instructions, we obtain the following output:

```
\ This file has been generated by DOcplex
\ ENCODING=ISO-8859-1
\Problem name: CPLEX

Minimize
 obj: - y + [ 4 x*y - 8 y*z ]/2
```

```
Subject To
 c0: x + 2 y + 3 z <= 5

Bounds
 0 <= x <= 1
 0 <= y <= 1
 0 <= z <= 1

Binaries
 x y z
End
```

The problem has exactly the elements that we specified. The only detail that may deserve a small explanation is why the quadratic part of the objective function is represented as $(4xy - 8yz)/2$ instead of $2xy - 4yz$. The reason for this seemingly odd choice is that, in this way, the matrix with the quadratic coefficients can be made symmetric. Instead of having 2 for the xy coefficient and 0 for the yx product, the value is duplicated and both terms will have 2 as their coefficient — but then we need to divide by 2 so that the total coefficient remains as in the original specification.

To learn more…

The internal representation of these quadratic problems is the one used by CPLEX, an IBM package that is used to solve optimization problems with classical methods. You can learn more about CPLEX on its web page: `https://www.ibm.com/products/ilog-cplex-optimization-studio`.

Once we have a `QuadraticProgram` object, we can solve it with one of the algorithms provided by Qiskit. To achieve this, we can use `MinimumEigenOptimizer` together with a concrete solver. For example, we can use a classical exact solver, which tries every possible solution and selects the optimal one. In Qiskit, this is as simple as using the following instructions:

```
from qiskit_optimization.algorithms import MinimumEigenOptimizer
from qiskit.algorithms import NumPyMinimumEigensolver
np_solver = NumPyMinimumEigensolver()
np_optimizer = MinimumEigenOptimizer(np_solver)
result = np_optimizer.solve(qp)
print(result)
```

The result of the execution is as follows:

```
fval=-5.0, x=0.0, y=1.0, z=1.0, status=SUCCESS
```

As you can see, we obtain the optimal assignment ($x = 0$, $y = 1$ and $z = 1$), the optimal value of the function (in this case, -5), and whether the assignment satisfies the constraints, indicated by the `SUCCESS` value, or not – if there were no assignments satisfying the constraints, we would obtain `INFEASIBLE` as the value for `status`.

In a similar way, we can use QAOA to solve the problem with the following instructions:

```
from qiskit import Aer
from qiskit.algorithms import QAOA
from qiskit.algorithms.optimizers import COBYLA
from qiskit.utils import QuantumInstance
quantum_instance = QuantumInstance(Aer.get_backend("aer_simulator"),
    shots = 1024)
qaoa = QAOA(optimizer = COBYLA(),
    quantum_instance=quantum_instance, reps = 1)
qaoa_optimizer = MinimumEigenOptimizer(qaoa)
result = qaoa_optimizer.solve(qp)
print(result)
```

In this case, the result will be the same one that we obtained with `NumPyMinimumEigensolver`. But we can also obtain additional information with the following instructions:

```
print('Variable order:', [var.name for var in result.variables])
for s in result.samples:
    print(s)
```

The result will be something like the following:

```
Variable order: ['x', 'y', 'z']
SolutionSample(x=array([0., 1., 1.]), fval=-5.0,
probability=0.11621093749999999, status=<OptimizationResultStatus.SUCCESS: 0>)
SolutionSample(x=array([0., 1., 0.]), fval=-1.0,
probability=0.107421875, status=<OptimizationResultStatus.SUCCESS: 0>)
SolutionSample(x=array([1., 0., 1.]), fval=0.0,
probability=0.1494140625, status=<OptimizationResultStatus.SUCCESS: 0>)
SolutionSample(x=array([0., 0., 1.]), fval=0.0,
probability=0.1103515625, status=<OptimizationResultStatus.SUCCESS: 0>)
SolutionSample(x=array([1., 0., 0.]), fval=0.0,
probability=0.103515625, status=<OptimizationResultStatus.SUCCESS: 0>)
SolutionSample(x=array([0., 0., 0.]), fval=0.0,
probability=0.1416015625, status=<OptimizationResultStatus.SUCCESS: 0>)
SolutionSample(x=array([1., 1., 0.]), fval=1.0,
probability=0.13769531249999997, status=<OptimizationResultStatus.SUCCESS: 0>)
SolutionSample(x=array([1., 1., 1.]), fval=-3.0,
probability=0.1337890625, status=<OptimizationResultStatus.INFEASIBLE: 2>)
```

First, we have printed the order of variables, to more easily interpret the assignments considered by the solver. Then, we have a listing of the different solutions that are part of the final, optimal state found by QAOA. Each item of the list includes the assignment, the energy or function value, the probability of obtaining the corresponding basis state when measuring the QAOA state, and whether the solution is feasible or not — `status=<OptimizationResultStatus.SUCCESS: 0>` indicates that the solution is feasible, while `status=<OptimizationResultStatus.INFEASIBLE: 2>` indicates that it is not.

Exercise 5.6

Modify the code that we have just run to make the results reproducible. *Hint*: you can set seeds in the same way that we did in *Section 5.2.1*.

We can also obtain full information about the QAOA execution by using the following:

```
print(result.min_eigen_solver_result)
```

We would obtain something like the following (where we have truncated part of the output):

```
{   'aux_operator_eigenvalues': None,
    'cost_function_evals': 32,
    'eigenstate': {   '000000': 0.09375,
                      '000001': 0.03125,
                      '000010': 0.05412658773652741,
                      [.....]
                      '111101': 0.11692679333668567,
                      '111110': 0.08838834764831845,
                      '111111': 0.07654655446197431},
    'eigenvalue': (-14.7548828125+0j),
    'optimal_parameters': {
    ParameterVectorElement(γ[0]): -5.087643335935586,
    ParameterVectorElement(β[0]): -0.24590437874189125},
    'optimal_point': array([-0.24590438, -5.08764334]),
    'optimal_value': -14.7548828125,
    'optimizer_evals': None,
    'optimizer_time': 0.6570718288421631}
```

Notice, however, that these assignments include the auxiliary variables used in the transformation from constrained to unconstrained problem, as in the procedure that we studied in *Chapter 3, Working with Quadratic Unconstrained Binary Optimization Problems*, and the function values are also the ones taken in the transformed problem. In fact, you can obtain the corresponding QUBO problem with the following code:

```
from qiskit_optimization.converters import QuadraticProgramToQubo
qp_to_qubo = QuadraticProgramToQubo()
qubo = qp_to_qubo.convert(qp)
print(qubo.export_as_lp_string())
```

The output will be the following:

```
\ This file has been generated by DOcplex
\ ENCODING=ISO-8859-1
\Problem name: CPLEX

Minimize
 obj: - 80 x - 161 y - 240 z - 80 c0@int_slack@0 - 160 c0@int_slack@1
      - 160 c0@int_slack@2 + [ 16 x^2 + 68 x*y + 96 x*z +
      32 x*c0@int_slack@0 + 64 x*c0@int_slack@1 + 64 x*c0@int_slack@2
      + 64 y^2 + 184 y*z + 64 y*c0@int_slack@0 + 128 y*c0@int_slack@1
      + 128 y*c0@int_slack@2 + 144 z^2 + 96 z*c0@int_slack@0
      + 192 z*c0@int_slack@1 + 192 z*c0@int_slack@2
      + 16 c0@int_slack@0^2 + 64 c0@int_slack@0*c0@int_slack@1
      + 64 c0@int_slack@0*c0@int_slack@2 + 64 c0@int_slack@1^2
      + 128 c0@int_slack@1*c0@int_slack@2 + 64 c0@int_slack@2^2 ]/2
      + 200
Subject To

Bounds
 0 <= x <= 1
 0 <= y <= 1
 0 <= z <= 1
```

```
 0 <= c0@int_slack@0 <= 1
 0 <= c0@int_slack@1 <= 1
 0 <= c0@int_slack@2 <= 1

Binaries
 x y z c0@int_slack@0 c0@int_slack@1 c0@int_slack@2
End
```

As you can see, this is now a QUBO problem in which slack variables and penalty terms have been introduced, exactly as we did in *Chapter 3, Working with Quadratic Unconstrained Binary Optimization Problems.*

To learn more...

In the `qiskit_optimization.converters` module, you can also find the functions `InequalityToEquality`, `IntegerToBinary`, and `LinearEqualityToPenalty`. The `QuadraticProgramToQubo` function calls them to convert quadratic programs with constraints into QUBO instances, by first introducing slack variables to transform inequalities into equalities, then transforming the integer slack variables into binary ones, and finally, replacing the equality constraints with penalty terms.

You may now be wondering how to use `MinimumEigenOptimizer` with a quantum computer instead of with a simulator. Of course, when defining the `quantum_instance` parameter to use with the `QAOA` object, you can simply declare a real quantum device. But, as we have already mentioned, that would imply entering the device queue many times, with the consequent delay.

As you surely remember from the previous subsection, if you have a Hamiltonian, you can use it directly in a QAOA Runtime program in order to submit your problem to the queue just once. So, is it possible to obtain the Hamiltonian of our problem? It sure is! You can run the following code to further transform the QUBO problem into an equivalent Hamiltonian:

```
H1, offset = qubo.to_ising()

print("The Hamiltonian is", H1)
print("The constant term is", offset)
```

You can then use `H1` to solve the problem with the QAOA Runtime program and even recover the energy by adding back the `offset` term. But... that seems like a lot of work, doesn't it? What's more, you would need to deal with all those ugly slack variables that were introduced to transform the quadratic program into QUBO form. Surely, there has to be a simpler way.

Fortunately, the Qiskit developers are very thoughtful, and they have enabled us to use Qiskit Runtime directly with `MinimumEigenOptimizer`. To do that, though, you need something called the `QAOAClient`, which will take care of running everything smoothly with Runtime once you plug it into `MinimumEigenOptimizer`. Using it is as simple as selecting a device with enough qubits. We need at least 6, so we have selected `ibm_lagos`, which has 7; if you don't have access to a big enough device, you can always use `ibmq_qasm_simulator`, which supports up to 32. And once we have a device, we can just run the following instructions:

```
from qiskit_optimization.runtime import QAOAClient
from qiskit import IBMQ

provider = IBMQ.load_account()
qaoa_client = QAOAClient(provider=provider,
                    backend=provider.get_backend("ibm_oslo"), reps=1)

qaoa = MinimumEigenOptimizer(qaoa_client)
result = qaoa.solve(qp)
print(result)
```

This will yield the following output:

```
fval=-5.0, x=0.0, y=1.0, z=1.0, status=SUCCESS
```

And, of course, you can obtain further information about the execution, as we did in previous examples, by accessing and using the values of the variables `result.variables`, `result.samples`, and `result.min_eigen_solver_result`. Very convenient, right?

We have now learned how to work with QAOA in Qiskit and how to manage and solve our problems in many different ways. It is time for us to turn back to PennyLane and see what it can offer in order to solve our beloved QUBO problems.

5.3 Using QAOA with PennyLane

As we mentioned in *Chapter 2, The Tools of the Trade in Quantum Computing*, PennyLane is a quantum programming library focused mainly on quantum machine learning. As such, it doesn't include as many tools for quantum optimization algorithms – such as QAOA – as Qiskit does. However, it does provide some interesting features such as automatic differentiation – that is, analytical computation of gradients – that may make it an appealing alternative to Qiskit in some circumstances.

Let's begin by explaining how to declare and work with Hamiltonians in PennyLane. For that, we will use the `Hamiltonian` class. It provides a constructor that accepts a list of coefficients and a list of products of Pauli matrices. For instance, if you want to define $2Z_0Z_1 - Z_0Z_2 + 3.5Z_1$, you will pass `[2,-1,3.5]` as the first argument and `[PauliZ(0)@PauliZ(1),PauliZ(0)@PauliZ(2),PauliZ(1)]` as the second one. As we know from *Chapter 2, The Tools of the Trade in Quantum Computing*, `PauliZ` is the Z matrix in PennyLane. We are also using the `@` operator, which is PennyLane's symbol for the tensor product operation. Putting it all together, we get the following instructions:

```
import pennylane as qml
from pennylane import PauliZ
coefficients = [2,-1,3.5]

paulis = [PauliZ(0)@PauliZ(1),PauliZ(0)@PauliZ(2),PauliZ(1)]
```

```
H = qml.Hamiltonian(coefficients,paulis)
print(H)
```

The output when we execute that code will be the following:

```
(3.5) [Z1]
+ (-1) [Z0 Z2]
+ (2) [Z0 Z1]
```

As you can see, we have constructed exactly the Hamiltonian that we wanted. We can also obtain its matrix by using `print(qml.matrix(H))`, which would give us the following output:

```
[[4.5+0.j  0. +0.j  0. +0.j  0. +0.j  0. +0.j  0. +0.j  0. +0.j  0. +0.j]
 [0. +0.j  6.5+0.j  0. +0.j  0. +0.j  0. +0.j  0. +0.j  0. +0.j  0. +0.j]
 [0. +0.j  0. +0.j -6.5+0.j  0. +0.j  0. +0.j  0. +0.j  0. +0.j  0. +0.j]
 [0. +0.j  0. +0.j  0. +0.j -4.5+0.j  0. +0.j  0. +0.j  0. +0.j  0. +0.j]
 [0. +0.j  0. +0.j  0. +0.j  0. +0.j  2.5+0.j  0. +0.j  0. +0.j  0. +0.j]
 [0. +0.j  0. +0.j  0. +0.j  0. +0.j  0. +0.j  0.5+0.j  0. +0.j  0. +0.j]
 [0. +0.j  0. +0.j  0. +0.j  0. +0.j  0. +0.j  0. +0.j -0.5+0.j  0. +0.j]
 [0. +0.j  0. +0.j  0. +0.j  0. +0.j  0. +0.j  0. +0.j  0. +0.j -2.5+0.j]]
```

As expected, this is a diagonal matrix. We can visualize it in a more compact way by executing the following instructions, which will give us only the non-zero elements:

```
from pennylane.utils import sparse_hamiltonian

print(sparse_hamiltonian(H))
```

The result will be the following:

```
  (0, 0)	(4.5+0j)
  (1, 1)	(6.5+0j)
```

```
(2, 2)    (-6.5+0j)
(3, 3)    (-4.5+0j)
(4, 4)    (2.5+0j)
(5, 5)    (0.5+0j)
(6, 6)    (-0.5+0j)
(7, 7)    (-2.5+0j)
```

You can also define Hamiltonians in a more compact manner by specifying them in a mathematical expression like the following one:

```
H = 2*PauliZ(0)@PauliZ(1) - PauliZ(0)@PauliZ(2) +3.5*PauliZ(1)
```

If you print `H`, you will find that this definition is equivalent to the one that was introduced previously.

Exercise 5.7

Use PennyLane to define the $-3Z_0Z_1Z_2 + 2Z_1Z_2 - Z_2$ Hamiltonian in two different ways.

Now that we know how to define Hamiltonians, we can use them to create QAOA circuits with PennyLane. To this end, we will import the `qaoa` module, which will give us access to the `cost_layer` and `mixer_layer` functions. We will need a cost Hamiltonian – the one that encodes our optimization problem – to use with `cost_layer` and we will use $\sum_j X_j$ with `mixer_layer` (in the QAOA literature, our H_0 Hamiltonian is sometimes called the **mixer Hamiltonian**, hence the name of the function). With them, we can create a function that constructs the QAOA circuit and that computes the energy of the state prepared by the circuit with respect to H_1. This latter part is very easy to accomplish with PennyLane, because it provides the `expval` function, which computes exactly that, and it can be used instead of the types of measurements that we introduced in *Section 2.3.1*.

We can, thus, define a function that computes the energy of parameters with the following piece of code:

```
from pennylane import qaoa

H0 = qml.PauliX(0) + qml.PauliX(1)
H1 = 1.0*qml.PauliZ(0) @ qml.PauliZ(1)

wires = range(2)
dev = qml.device("default.qubit", wires=wires)

p = 2

@qml.qnode(dev)
def energy(angles):
    for w in wires:
        qml.Hadamard(wires=w)
    for i in range(p):
        qaoa.cost_layer(angles[2*i+1], H1)
        qaoa.mixer_layer(angles[2*i], H0)
    return qml.expval(H1)
```

There are several details that we need to explain here. First, we are working with a simple problem in which we want to find the ground state of Z_0Z_1. We have defined our H_0 Hamiltonian as X_0+X_1 with `H0 = qml.PauliX(0) + qml.PauliX(1)`. For H_1, we have used `1.0*qml.PauliZ(0) @ qml.PauliZ(1)` instead of just `qml.PauliZ(0) @ qml.PauliZ(1)`. If you do not include the `1.0` coefficient, the tensor product will not be converted to a `Hamiltonian` object, so you should be careful with that. Another important detail is that the `energy` function only receives as parameters the angles for the rotations in the QAOA circuit and we have declared `p` as a global variable. This is because we later want to optimize `energy` with respect to its parameters, and `p` is not something that we want to optimize, but a fixed value – in this case, we are setting it to 2.

Finally, notice that the exponentials for H_1 and H_0 receive their parameters from the `angles` list alternating between H_0 and H_1: first for the H_0 exponential (which is implemented by `mixer_layer`), then for the H_1 exponential (implemented by `cost_layer`), then again for the H_0 exponential, and so on. In the notation that we have been using throughout this chapter, if `angles` is `[1.0,2.0.3.0,4.0]`, then we would have $\beta_1 = 1$, $\gamma_1 = 2$, $\beta_2 = 3$, and $\gamma_2 = 4$. Now we are ready to run the optimization process. To do that, we can use the following code:

```
from pennylane import numpy as np
optimizer = qml.GradientDescentOptimizer()
steps = 20
angles = np.array([1,1,1,1], requires_grad=True)

for i in range(steps):
    angles = optimizer.step(energy, angles)

print("Optimal angles", angles)
```

We are using `GradientDescentOptimizer` as the classical minimizer. It uses the famous gradient descent algorithm — we will study this method in detail in *Part 3* of the book — by taking advantage of the fact that PennyLane implements automatic differentiation to compute all the required derivatives. That is why we use `requires_grad=True` when defining the initial angles, to inform PennyLane that these are parameters for which we will need to compute gradients. We run the process for 10 steps and... voilá! We obtain some (close to) optimal parameters. In this case, `[0.78178403 0.7203965 1.17250771 1.27995423]` was the answer found by the optimizer. The angles and energy that you find can be highly dependent on the initial parameters, so it is advisable to run your code with several different choices of initial angles.

In any case, we can now sample from the QAOA circuit with the parameters that we have found in order to obtain candidate solutions to our problem. We just need to modify slightly the `energy` function that we defined previously. We can do it, for instance, as follows:

```
@qml.qnode(dev)
def sample_solutions(angles):
    for w in wires:
        qml.Hadamard(wires=w)
    for i in range(p):
        qaoa.cost_layer(angles[2*i+1], H1)
        qaoa.mixer_layer(angles[2*i], H0)
    return qml.sample()
print(sample_solutions(angles, shots = 5))
```

The output when you run these instructions will be something like the following:

```
[[0 1]
 [0 1]
 [0 1]
 [1 0]
 [0 1]]
```

The five samples are, indeed, ground states of Z_0Z_1. Once more, we have been able to use QAOA to solve the problem, this time with PennyLane!

You surely have noticed that we have run our code on the `default.qubit` device, which is a simulator. Of course, you can replace it with a quantum device, as we learned to do in *Section 2.3.2*. However, this will mean that you will have to wait on the quantum computer execution queue every time the optimizer needs to evaluate the energy of some parameters.

Unfortunately, at the time of writing, PennyLane does not yet include an option to run QAOA programs using Qiskit Runtime. However, do not despair! As we will learn in *Chapter 7, VQE: Variational Quantum Solver*, there is a PennyLane implementation of

Runtime programs for some other algorithms. Hopefully, QAOA will receive the same treatment soon.

With this, we have now concluded our study of QAOA. In the next chapter, we will study a different method for finding solutions to optimization problems, and it will be based on one of the most famous of all quantum algorithms ever: Grover's algorithm.

Summary

In this chapter, you have learned about QAOA, one of the most popular quantum algorithms used to solve optimization problems with gate-based quantum computers. You now know that QAOA is derived as a discretization of quantum annealing and that it is implemented as a hybrid method that uses both a classical and a quantum computer to achieve its goal.

You also understand how to construct circuits for all the operations needed in the quantum part of the algorithm. In particular, you know how to use these circuits to estimate expectation values in an efficient way.

You have also mastered the tools that Qiskit provides in order to implement QAOA instances and to run them on both quantum simulators and quantum computers. You even know how to accelerate the process of running your code on quantum devices by using Qiskit Runtime. And, should you need to use QAOA with PennyLane, you also know how to do it with the help of some predefined utilities and PennyLane capabilities for automatic differentiation. This gives you the flexibility to solve optimization problems with QAOA in a number of different ways, depending on your needs and on the resources at your disposal.

Our next stop will be **Grover's Adaptive Search**, also known as **GAS**, a quite different quantum method that you can use to solve optimization problems, which we'll cover in the next chapter.

6

GAS: Grover Adaptive Search

If you do not expect the unexpected, you will not find it, for it is not to be reached by search or trail.

— Heraclitus

In this chapter, we are going to introduce another quantum method for solving combinatorial optimization problems. In this case, we are going to take Grover's algorithm — one of the most famous and celebrated quantum methods out there — as a starting point. Grover's algorithm is used to find elements that satisfy specific conditions in unsorted data structures. But, as we will soon see, it can be easily adapted to function minimization tasks — exactly what we need for our optimization problems! The resulting method is sometimes called **Grover Adaptive Search** or **GAS**.

It is important to note that GAS is essentially different from the kind of quantum algorithms that we have been studying so far in this part of the book. This method is not designed specifically for NISQ devices and would need fault-tolerant quantum computers to fully realize its potential. However, we have still decided to cover it because it is readily implemented in some quantum programming libraries — such as Qiskit — and it can be helpful in comparing and benchmarking other quantum optimization algorithms.

We will start the chapter by refreshing some details about Grover's algorithm, including the circuits that we need in order to implement it and the role that **oracles** play in it. Then, we will talk about the **Dürr-Høyer** method, which uses Grover's techniques to find the minimum of certain types of functions. After that, we will particularize the algorithm to QUBO problems and we will study how to implement the kind of oracle that they require.

With all those tools, we will have everything that we need in order to formulate and solve optimization problems with GAS, so we will turn to explain how to use Qiskit's implementation of the algorithm. We will study the different options that are available to run the method and we will test it on several different examples.

After reading this chapter, you will understand the theoretical foundations of Grover Adaptive Search, you will know how to implement efficient oracles for optimization problems and how to use them with GAS, and you will be able to run Qiskit's implementation of the algorithm to solve your own optimization problems.

The topics covered in this chapter are as follows:

- Grover's algorithm
- Quantum oracles for combinatorial optimization
- Using GAS with Qiskit

6.1 Grover's algorithm

In this section, we will cover the most important properties of Grover's algorithm. We will not cover all the theoretical details behind the procedure — for that, we recommend the book by Nielsen and Chuang [16] and, especially, the lecture notes by John Watrous [46] — but we need to at least get familiar with how the method operates, what oracles are and how they are used in the algorithm, and what kind of circuits are needed to implement it.

Let's start with the basics. Grover's algorithm is used for searching elements that satisfy certain conditions. More formally, the algorithm assumes that we have a collection of elements indexed by strings of n bits, and a Boolean function f that takes those binary strings and returns "true" (or 1) if the element indexed by the string satisfies the condition and "false" (or 0) otherwise. For instance, imagine that we are searching among 8 different elements and that the ones that satisfy the condition are indexed by the strings 010 and 100. Then, f will be the Boolean function such that $f(x) = 1$ if $x = 010$ or $x = 100$, and $f(x) = 0$ otherwise. To simplify the notation, from now on we will identify an element with the string x that is used to index it.

It is important to notice that, in this setting, we have no access to the inner workings of f. It acts like a black box. The only thing that we can do with the f function is call it on inputs and observe the outputs, thus checking whether the given input satisfies the condition that we are considering or not. Since we do not have any information about the indices of the elements that satisfy the condition, we cannot favour any position over any other. Thus, with a classical algorithm, if we are searching among N elements and only one of them satisfies the condition we are interested in, we will need to call f about $N/2$ times on average in order to find it. The element could be just anywhere! In fact, if we are extremely unlucky, we might need to use $N - 1$ calls (notice that we wouldn't need N calls: if we don't find the element after $N - 1$ different calls, we already know the remaining position to be the one where the element is located).

It may come as a big surprise, then, that with Grover's algorithm it is possible to find the hidden element with high probability (much more on this later in this section) by calling f around $\sqrt{N}$ times! This means that if we are searching among 1 000 000 elements, with a classical computer you would need to check f about 500 000 times on average, but calling f less than 1000 times would suffice in order to solve the problem with a quantum computer, at least with a high likelihood. What is more, the difference in the number of calls between the classical and the quantum methods grows bigger if N is higher.

How is this possible? It seems to defy all logic, but it rests on properties that we are already familiar with, such as superposition and entanglement. In fact, Grover's algorithm will query f with elements that are in superposition. But in order to understand this, we need to explore what quantum oracles are and how they can be used, so let's get to it!

6.1.1 Quantum oracles

We have mentioned that, in the setting of the search problem solved by Grover's algorithm, we are given a Boolean function f that we can use to determine whether an element is the one we are looking for or not. But what do we mean when we say that we are "given" this function?

In the classical case, this is more or less straightforward. If we were writing our code in Python, we could be given a function object that receives an n-bit string and returns `True` or `False`. Then, we could use that function in our own code to check the elements that we want to consider, without necessarily knowing how it is implemented.

But...what is the equivalent to that function definition when we are working with quantum circuits? The most natural assumption is that we are provided with a new quantum gate O_f that implements f and that we can use in our circuits whenever we need it. However, a quantum gate needs to be a unitary operation and, in particular, reversible, so we need to be a little bit careful in how we design it.

In the classical case, we had n inputs — the n bits of the string — and just one output. In the quantum case, we need at least n inputs — n qubits — but just one output would not work, because then it would be impossible to make the operation reversible, let alone unitary. In fact, as you surely remember, every quantum gate has the same number of inputs and outputs.

The usual approach, then, is to consider a quantum gate on $n + 1$ qubits. The first n of these qubits will serve as the input and the additional one will be used to store the output. More formally, on any input $|x\rangle|y\rangle$, where x is an n-bit string and y is a single bit, the output of the O_f gate will be $|x\rangle|y \oplus f(x)\rangle$, where $\oplus$ denotes addition modulo 2 (see *Appendix B, Basic Linear Algebra*, for a refresher on modular arithmetic). This defines the action of the gate on the computational basis states and then we can extend it to the rest of the quantum states by linearity, as usual.

This may look like an odd choice. The "natural" thing to do might seem to be requiring the output to be $|x\rangle|f(x)\rangle$, right? But that would not be reversible in general, because we would obtain the same output over the inputs $|x\rangle|0\rangle$ and $|x\rangle|1\rangle$. With our choice, though, the operation is reversible. If we applied O_f twice, we would obtain $|x\rangle|y \oplus f(x) \oplus f(x)\rangle$, which is equal to $|x\rangle|y\rangle$ because, when we are performing addition modulo 2, $f(x) \oplus f(x) = 0$ no matter the value of $f(x)$.

Exercise 6.1

Prove that O_f is not only reversible but also unitary and hence it deserves the name "quantum gate."

Usually, O_f is said to be a quantum oracle for f, because we can consult it to get the value of f on any input x without having to worry about its internal workings. In fact, if the input to O_f is $|x\rangle|0\rangle$, then the output is $|x\rangle|0 \oplus f(x)\rangle = |x\rangle|f(x)\rangle$ and we could hence recover $f(x)$ just by measuring the last qubit.

For any f, it is always possible to construct O_f by using just NOT and multi-controlled NOT gates – even if the resulting circuit is not the most efficient one in most cases. For instance, if f is a Boolean function on 3-bit strings such that f takes value 1 just on 101 and 011, then we can use the circuit depicted in *Figure 6.1*. Notice how we have used NOT gates before and after the multi-controlled gates to select those qubits that should be 0 in the input and to restore them to their original values.

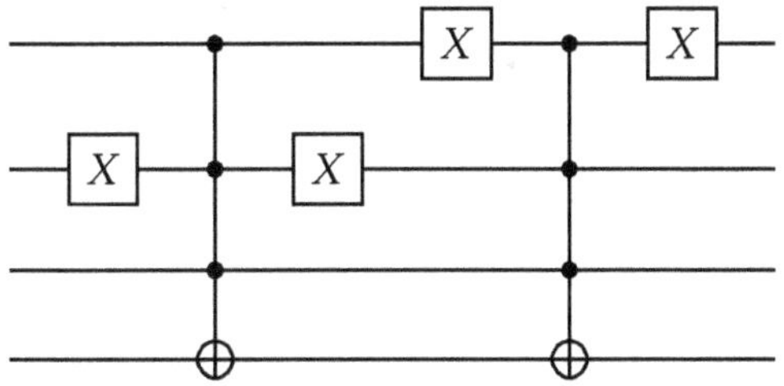

Figure 6.1: Oracle for the Boolean function f that takes value 1 on 101 and 011, and value 0 on the rest of the 3-bit strings

Exercise 6.2

Construct a circuit for O_f where f is a 4-bit Boolean function that takes value 1 on 0111, 1110, and 0101, and value 0 on any other input.

This settles how we are going to receive the Boolean function f that we can use to check whether a given element satisfies the conditions that we are interested in: the function will be given to us as a quantum oracle. Now it's time for us to show how we can use these quantum oracles in Grover's algorithm.

6.1.2 Grover's circuits

Let's say that we want to apply Grover's algorithm to a Boolean function f which receives binary strings of length n. In addition to the O_f oracle described in the previous section, the circuit used in Grover's algorithm involves two other blocks, as you can see in *Figure 6.2.*

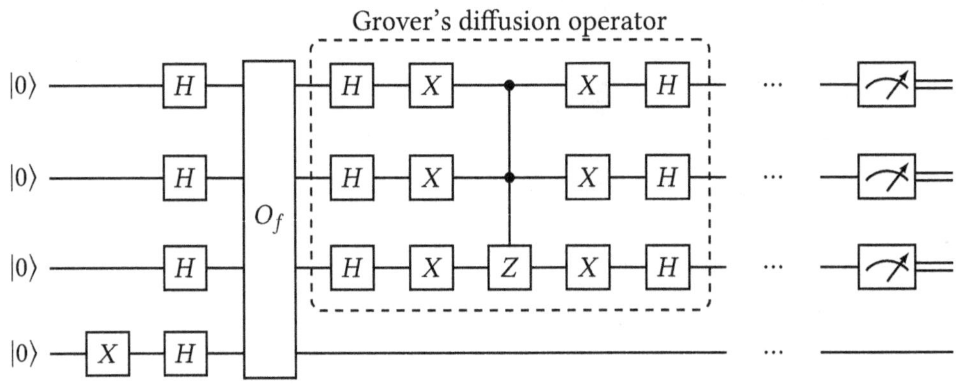

Figure 6.2: Circuit for Grover's algorithm in the case in which f receives strings of length 3 as input. The oracle O_f and Grover's diffusion operator are repeated, in that order, a number of times before the final measurements

The first block is composed of one-qubit gates that are applied to the initial state $|0\cdots0\rangle|0\rangle$, where the first register is of length n and the second one is of length 1. Thus, the state just before applying the oracle is

$$H^{\otimes n+1}\left|0\right\rangle^{\otimes n}\left|1\right\rangle = \left|+\right\rangle^{\otimes n}\left|-\right\rangle = \frac{1}{\sqrt{2^n}}\left(\left(\left|0\right\rangle + \left|1\right\rangle\right)\cdots\left(\left|0\right\rangle + \left|1\right\rangle\right)\right)\left|-\right\rangle = \frac{1}{\sqrt{2^n}}\sum_{x=0}^{2^n-1}\left|x\right\rangle\left|-\right\rangle,$$

because we apply the first X gate to $\left|0\right\rangle$ to obtain $\left|1\right\rangle$.

Notice that the first register of this state is a superposition of all basis states $\left|x\right\rangle$. This is exactly what we will use in order to evaluate f "in superposition" with our application of the O_f oracle. Indeed, by the definition of O_f, the state that we will have after the application of the oracle is

$$O_f\left(\frac{1}{\sqrt{2^n}}\sum_{x=0}^{2^n-1}\left|x\right\rangle\left|-\right\rangle\right) = O_f\left(\frac{1}{\sqrt{2^{n+1}}}\sum_{x=0}^{2^n-1}\left|x\right\rangle\left(\left|0\right\rangle - \left|1\right\rangle\right)\right) =$$
$$\frac{1}{\sqrt{2^{n+1}}}\sum_{x=0}^{2^n-1}O_f\left|x\right\rangle\left(\left|0\right\rangle - \left|1\right\rangle\right) = \frac{1}{\sqrt{2^{n+1}}}\sum_{x=0}^{2^n-1}\left|x\right\rangle\left(\left|0\oplus f(x)\right\rangle - \left|1\oplus f(x)\right\rangle\right),$$

where in the last two equalities he have used linearity together with the definition of O_f.

Let's focus on the $\left|0\oplus f(x)\right\rangle - \left|1\oplus f(x)\right\rangle$ term. If $f(x) = 0$, then it is just $\left|0\right\rangle - \left|1\right\rangle$. However, if $f(x) = 1$, we have

$$\left|0\oplus f(x)\right\rangle - \left|1\oplus f(x)\right\rangle = \left|0\oplus 1\right\rangle - \left|1\oplus 1\right\rangle = \left|1\right\rangle - \left|0\right\rangle = -\left(\left|0\right\rangle - \left|1\right\rangle\right),$$

because $1\oplus 1 = 0$. In both cases, we can write

$$\left|0\oplus f(x)\right\rangle - \left|1\oplus f(x)\right\rangle = (-1)^{f(x)}\left(\left|0\right\rangle - \left|1\right\rangle\right),$$

because $(-1)^0 = 1$ and $(-1)^1 = -1$.

Note how, thanks to these transformations, there is information about the value $f(x)$ coded into the amplitude of the state now. As you will soon see, this is a key ingredient of the algorithm.

If we take this to our expression for the state after the oracle application, we get

$$O_f\left(\frac{1}{\sqrt{2^n}}\sum_{x=0}^{2^n-1}|x\rangle|-\rangle\right)=\frac{1}{\sqrt{2^{n+1}}}\sum_{x=0}^{2^n-1}|x\rangle\left(|0\oplus f(x)\rangle-|1\oplus f(x)\rangle\right)=$$

$$\frac{1}{\sqrt{2^{n+1}}}\sum_{x=0}^{2^n-1}(-1)^{f(x)}|x\rangle\left(|0\rangle-|1\rangle\right)=\frac{1}{\sqrt{2^n}}\sum_{x=0}^{2^n-1}(-1)^{f(x)}|x\rangle\frac{1}{\sqrt{2}}\left(|0\rangle-|1\rangle\right)=$$

$$\frac{1}{\sqrt{2^n}}\sum_{x=0}^{2^n-1}(-1)^{f(x)}|x\rangle|-\rangle.$$

Notice how the application of O_f has introduced a relative phase in some of the states $|x\rangle$ of the superposition. This technique is called **phase kickback**, because we have only used the register in state $|-\rangle$ to create the phase but it ends up affecting the whole state. It is used in other famous quantum methods such as the Deutsch-Jozsa and Simon's algorithms (see the book by Yanofsky and Mannucci [47] for an excellent explanation of these methods).

As we have proved, the phase that goes with the basis state $|x\rangle$ depends only on $f(x)$ and it is 1 if $f(x)=0$ and -1 if $f(x)=1$. In this way, we say that we have **marked** those elements that satisfy the conditions that we are interested in, that is, those elements x such that $f(x)=1$. Remarkably, we have done this with just one call to O_f, exploiting the possibility of evaluating it in superposition. That is an exponential number of function evaluations with just one call! It sounds like magic, doesn't it?

However, although after applying O_f we have somehow separated the elements x that satisfy $f(x)=1$ from the rest, we do not seem to be closer to finding one of them. If we measure the state as it is, the probability of measuring an x such that $f(x)=1$ is the same as it was before applying O_f. The phase that we have introduced has an absolute value equal to 1 and, consequently, does not affect the measurement probability.

But, wait! There is more to Grover's algorithm. There is another circuit block that we apply after O_f: it's called **Grover's diffusion operator** and we will use it to increase the probability of measuring the marked states. Describing its inner workings in full detail would take us astray from our path — for that, we recommend checking out Dancing with

Qubits [7], by Robert Sutor, which offers a perfect explanation of its behaviour — but let's at least give a quick overview of what it does.

Grover's diffusion operator implements an operation called **inversion about the mean**. This may sound complicated, but in fact it is quite simple. First, the average value m of all the amplitudes of the states is computed. Then, every amplitude a is replaced with $2m - a$. After this transformation, the positive amplitudes will be a little bit smaller, but the negative ones will be a little bit bigger. This is why the technique used by Grover's algorithm is called **amplitude amplification**. Again, we recommend you checking Sutor's book [7] for a detailed description of how this operation works.

So, after this first application of Grover's diffusion operator, the amplitudes of the elements that we are interested in finding are a little bit larger. But, in general, this will still not be enough to guarantee a high probability of measuring one of them. For this reason, we will need to mark the elements again with O_f and then apply the diffusion operator once more. We will repeat this procedure, applying first O_f and then the diffusion operator, several times until the probability of measuring one of the states we are looking for is high enough (close to 1). And that is the moment when we can measure the whole state and observe the result to, hopefully, obtain one element that satisfies the conditions.

But how many times should we apply O_f followed by the diffusion operator? This is a crucial point in Grover's algorithm that we will study in more detail in the next subsection.

6.1.3 Probability of finding a marked element

As we have just seen, when using Grover's algorithm, we are repeatedly applying for a certain number of times the quantum oracle given to us followed by the diffusion operator. Of course, we would like the number of repetitions to be as small as possible — so that the algorithm runs faster — while guaranteeing a high probability of finding one of the marked elements. How can we go about this?

One possible approach in order to analyze the behaviour of Grover's algorithm could be studying the properties of the inversion about the mean operation that we mentioned in the previous subsection. However, there is a better way. It turns out that the combination of O_f and Grover's diffusion operator acts like a rotation in a two-dimensional space. We will not give the full details — check the lecture notes by John Watrous [46] for a very thorough and readable explanation — but, if we have n-bit strings and there is only one marked element x_1, it can be proved that the state that we reach after m applications of O_f followed by the diffusion operator is

$$\cos(2m+1)\theta \left|x_0\right\rangle + \sin(2m+1)\theta \left|x_1\right\rangle,$$

where

$$\left|x_0\right\rangle = \sum_{x\in\{0,1\}^n, x\neq x_1} \sqrt{\frac{1}{2^n-1}} \left|x\right\rangle$$

and $\theta \in (0, \pi/2)$ is such that

$$\cos\theta = \sqrt{\frac{2^n-1}{2^n}}, \qquad \sin\theta = \sqrt{\frac{1}{2^n}}.$$

Notice that $\left|x_0\right\rangle$ is just the uniform superposition of the states $\left|x\right\rangle$ such that $f(x) = 0$. Then, what we want to obtain is a state in which $\sin(2m+1)\theta$ is close to 1, because then we would have a high probability of finding x_1 when we measure. For that, ideally, we would like to have

$$(2m+1)\theta \approx \frac{\pi}{2},$$

because $\sin \pi/2 = 1$.

Solving for m, we obtain

$$m \approx \frac{\pi}{4\theta} - \frac{1}{2}.$$

What is more, we know that $\sin\theta = \sqrt{1/2^n}$, so, for a big enough n, we will have

$$\theta \approx \sqrt{\frac{1}{2^n}}$$

and then we can choose

$$m = \left\lfloor \frac{\pi}{4} \sqrt{2^n} \right\rfloor,$$

that is, the biggest integer that is less than or equal to $(\pi/4)\sqrt{2^n}$.

Notice that there are exactly 2^n elements but only one of them satisfies the conditions we are interested in. This means that, with a classical algorithm, if we can only use f to check if an element x is the one we are looking for — that is, to check if $f(x) = 1$ — then we would need about $2^n/2$ calls to f on average to find x. However, with Grover's algorithm, we only need about $\sqrt{2^n}$. That is a quadratic speedup!

Nevertheless, there is a subtlety here. In the classical setting, if we use f more times, the probability of finding the marked element increases. But with Grover's algorithm, if m is not selected wisely, we can overshoot and actually decrease the success probability instead of increasing it!

This sounds baffling. How is it possible that by searching more we find ourselves with less possibilities of finding the hidden element? The key is that, as we have shown, the probability of measuring x_1 is $(\sin(2m+1)\theta)^2$. This function is periodic and oscillates between 0 and 1, so after reaching values close to 1, it goes back down to 0.

Let's illustrate this with an example. In *Figure 6.3*, we consider the case $n = 4$ and we show how the probability of finding exactly one marked element changes as we vary the number of Grover iterations m, from 0 to 20. In this case, $\lfloor(\pi/4)\sqrt{2^n}\rfloor$ is 3 and, as you can see, the success probability with $m = 3$ is close to 1. However, for $m = 5$ the probability has decreased dramatically, and for $m = 6$ it is nearly 0.

This shows that we need to be very careful when selecting the number of iterations m in Grover's algorithm. For the case in which there is only one marked element, we have obtained a good choice for m. But what if there is more than one marked element? It turns out — check the lecture notes by John Watrous [46] — that if there are k marked elements,

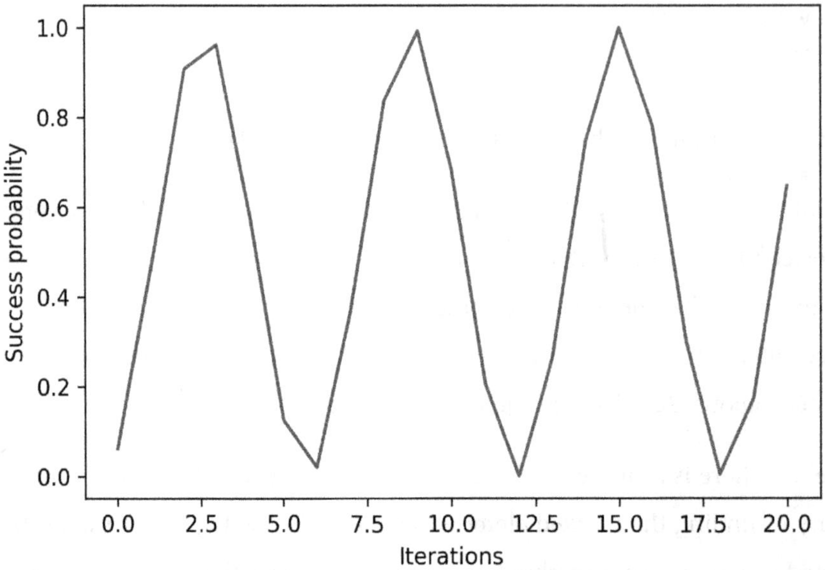

Figure 6.3: Probability of finding one marked element among 16 when using Grover's algorithm with a number of iterations that varies from 0 *to* 20

we can repeat our previous reasoning and show that a good value for m is

$$m = \left\lfloor \frac{\pi}{4}\sqrt{\frac{2^n}{k}} \right\rfloor,$$

provided that k is small compared to 2^n. If k is not small compared to 2^n, don't worry; then the probability of finding a marked element just by choosing at random is $k/2^n$, which will be sizeable, so you wouldn't even need a quantum computer in the first place.

This solves our problem if we know how many marked elements there are. But, in the most general case, we may lack that information. In that circumstance, we can apply the results of a very useful paper by Boyer, Brassard, Høyer, and Tapp [48]. They showed that by choosing m at random in a range that increases dynamically, we can still be guaranteed that will find a marked element with high probability while keeping the average number of iterations as $O(\sqrt{2^n})$ (see *Appendix C, Computational Complexity*, for a refresher on asymptotic notation).

In fact, they proved that the probability of finding a marked element with their method is at least $1/4$. This might seem unimpressive, but we can easily see how that is more than enough. Indeed, the probability of not finding a marked element is then no more than $3/4$. So, suppose that we repeat the process 1000 times. Then, the probability of failure is at most $(3/4)^{1000}$, which is extremely low. In fact, the chance of a meteorite hitting your quantum computer while running your circuits is much, much bigger than that!

So far in this section, we have covered all that we need to know in order to apply Grover's algorithm in search problems. However, our main goal is solving optimization problems. We explore the connection between both tasks in the next subsection.

6.1.4 Finding minima with Grover's algorithm

Optimization problems are obviously related to search problems. In fact, when solving an optimization problem, we are trying to find a value with a special property: it should be a minimum or maximum among all the possible values. This connection was exploited by Dürr and Høyer in a 1996 paper [49] in which they introduced a quantum algorithm, based on Grover's search, to find minima of functions. The main idea behind the algorithm is quite straightforward. Suppose we want to find a minimum of a function g that is computed over binary strings of length n. We select one such string x_0 at random and we compute $g(x_0)$. Now we apply Grover's algorithm with an oracle that, on input x, returns 1 if $g(x) < g(x_0)$ and 0 otherwise. If the element that we measure after applying Grover's search, call it x_1, really achieves a value that is lower than $g(x_0)$, we replace x_0 with it and repeat the process but now with an oracle that checks the condition $g(x) < g(x_1)$. If not, we keep using x_0. We repeat this process several times and we return the element with the lowest value among the ones that we have considered.

There are a couple of details that we need to flesh out here. The first one is how to construct the oracles. In general, of course, it will depend on the function g. For that reason, in the next section we will focus on circuits that we can use with the Dürr-Høyer algorithm to solve QUBO and HOBO problems.

On the other hand, we should take care of the number of iterations that we will use in each application of Grover's algorithm and, also, of the number of times that we need to repeat the procedure for selecting a new element and constructing a new oracle. The original paper by Dürr and Høyer gives all the details, but let's just mention that it uses the method proposed by Boyer, Brassard, Høyer, and Tapp [48] that we explained in the previous subsection, and it guarantees that a minimum will be found with a probability of at least $1/2$ with a number of calls to the oracle that is $O(\sqrt{2^n})$.

With this, we have now covered all the concepts that we need in order to apply this search method to solve QUBO and HOBO problems. We will devote the next section to explaining how to construct quantum oracles for these kinds of problems.

6.2 Quantum oracles for combinatorial optimization

As we have seen, the Dürr-Høyer algorithm can be used to find the minimum of a function g with high probability and with a quadratic speedup over brute force search. However, in order to use it, we need a quantum oracle that, given binary strings x and y, checks whether $g(x) < g(y)$.

In our case, we are interested in functions g that can appear in QUBO and HOBO problems. This means that g will be a polynomial with real coefficients and binary variables, and we could implement the quantum oracle with a straightforward approach: design a classical circuit for it using AND, OR, and NOT gates, and then simulate the classical gates with the Toffoli quantum gate, as we showed in *Section 1.5.2*.

However, in 2021, Gilliam, Woerner, and Gonciulea, introduced an improved way of implementing quantum oracles for QUBO and HOBO problems in a paper titled *Grover adaptive search for constrained polynomial binary optimization* [50].

In this section, we will study in detail the techniques that they proposed and how to use them to implement our quantum oracles. We will start by considering the case in which all the coefficients of the polynomial are integer numbers and, then, we will extend our study to the most general case when the coefficients are real numbers. But, before we get to that, we need to take a brief detour to talk about one of the most important subroutines in all of quantum computing: the **quantum Fourier transform**.

6.2.1 The quantum Fourier transform

The quantum Fourier transform (usually abbreviated as **QFT**) is, beyond any doubt, one of the most useful tools in quantum computing. It is an essential part of Shor's algorithm for integer factorization [6] and it is behind the speedups of other famous quantum algorithms such as HHL [14].

We will use the QFT to help us implement the arithmetical operations that we need to compute the values of the polynomial function of our QUBO and HOBO problems. We could, for instance, implement these operations in a basis representation. As an example, we might design a unitary transformation taking $|x\rangle|y\rangle|0\rangle$ to $|x\rangle|y\rangle|x+y\rangle$, where x and y are binary numbers and $x+y$ is their addition. However, this could involve a big number of one- and two-qubit gates.

Instead, we will use the approach proposed by Gilliam, Woerner, and Gonciulea in [50] and we will compute the arithmetical operations using the state amplitudes. We will explain in detail how to do that in the next subsections. But, before that, we will study how to use the QFT to recover information from the amplitudes of a quantum state.

The QFT on m qubits is defined as the unitary transformation that takes the basis states $|j\rangle$ to

$$\frac{1}{\sqrt{2^m}}\sum_{k=0}^{2^m-1} e^{\frac{2\pi ijk}{2^m}}|k\rangle,$$

where i is the imaginary unit. Its action is extended to the rest of the states by linearity.

We will not study the properties of the QFT in detail. For that, you can refer to *Dancing with Qubits*, by Robert Sutor [7]. However, we need to know that the QFT can be implemented with a number of one- and two-qubit gates that is quadratic in m. This is an exponential speedup over the best algorithm that we have for the analogous classical operation (the **discrete Fourier transform**).

For example, the circuit for the QFT on three qubits is shown in *Figure 6.4*. In it, the rightmost gate, which acts on the top and bottom qubits, is the SWAP gate. As we mentioned in *Section 1.4.3*, this gate swaps the states of two qubits and it can be implemented by means of CNOT gates. Moreover, this QFT circuit uses the **phase gate**, denoted by $P(\theta)$. This is a parametrized gate that depends on an angle θ and whose coordinate matrix is

$$\begin{pmatrix} 1 & 0 \\ 0 & e^{i\theta} \end{pmatrix}.$$

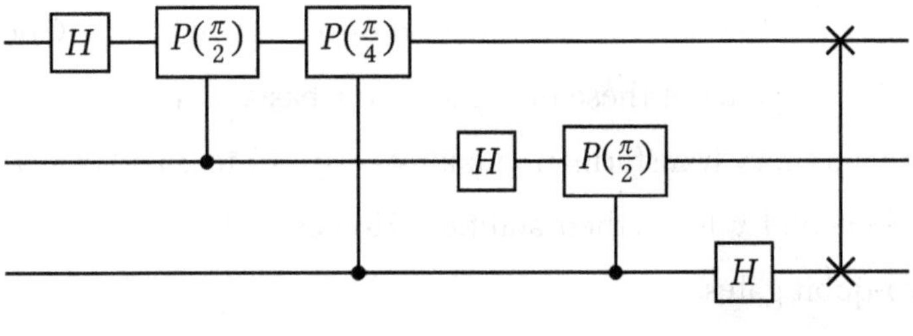

Figure 6.4: Circuit for the quantum Fourier transform on 3 qubits

Important note

The phase gate is very similar to the R_Z gate that we introduced in *Section 1.3.4*. In fact, when applied on its own to a qubit, $P(\theta)$ is equivalent to $R_Z(\theta)$ up to an unimportant global phase. However, in the QFT circuit, we are using a controlled version of the phase gate and the global phase becomes a relative one, which is not unimportant at all!

As we have seen, the QFT acts by introducing phases of the form $e^{2\pi ijk/2^m}$ when it is applied on basis states $|j\rangle$. Nevertheless, for the purposes of our computations, we are more interested in recovering the values j from those phases. For that, we will need the **inverse quantum Fourier transform**, usually denoted $\mathrm{QFT}^\dagger$. Of course, its action is the inverse of that of the QFT, meaning that it takes a state such as

$$\frac{1}{\sqrt{2^m}} \sum_{k=0}^{2^m-1} e^{\frac{2\pi ijk}{2^m}} |k\rangle$$

to the basis state $|j\rangle$.

The circuit for the inverse QFT can be obtained from that of the QFT by reading the circuit backwards and using the inverse of each gate we find. For example, the circuit for the inverse QFT on 3 qubits is shown in *Figure 6.5*. Notice that the inverse of $P(\theta)$ is $P(-\theta)$, while the H and SWAP gates are their own inverses.

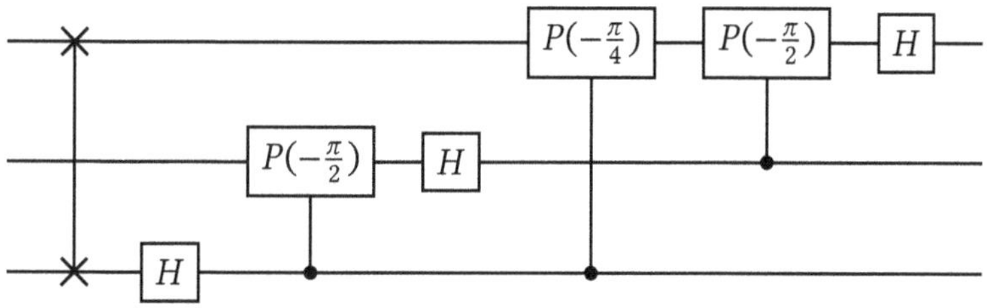

Figure 6.5: Circuit for the inverse quantum Fourier transform on 3 qubits

When designing a quantum oracle to minimize a function g, our goal will be to perform the computation in such a way that the $g(x)$ values appear as exponents in the amplitudes of our states so that we can later recover them by means of the inverse QFT. This may sound like a difficult endeavour, but as we will show in the following subsections, we already have all the tools that we need in order to succeed. We will start by showing how to encode integer values in exactly the way that we require.

6.2.2 Encoding and adding integer numbers

As will become apparent soon in this section, the most convenient way of working with integer numbers in the context of GAS oracles is using their **two's complement** representation. In it, we can encode numbers from -2^{m-1} to $2^{m-1} - 1$ by using m-bit strings.

Positive numbers are represented in the usual way for binary numbers, but a negative number x is represented by $2^m - x$.

For instance, if $m = 4$, we represent 3 by 0011, and −5 by 1011 (which is the binary representation of $11 = 16 - 5$). One advantage of this representation is that the most significant bit indicates the sign of the encoded number: positive numbers always start with 0, while negative numbers start with 1.

Another perk of two's complement representation is that, with it, we can compute additions involving both positive and negative numbers by simply performing regular binary addition and discarding the last carry-out, if it exists. For instance, if we add 0011 (which is 3) and 1011 (which is −5), we obtain 1110 which is, indeed, the encoding of −2 (because $14 = 16-2$). Similarly, if we add 0110 (which is 6) and 1100 (which is −4) we obtain 0010 (after discarding the last carry-out), which is 2, as expected. These facts about two's complement arithmetic will be very helpful in implementing our quantum oracle, as we show next.

Exercise 6.3

Using two's complement with 5 qubits, represent 10 and −7 and perform their addition.

As we have mentioned in the previous subsection, when computing $g(x)$ with an oracle, we are interested in obtaining the state

$$\frac{1}{\sqrt{2^m}} \sum_{k=0}^{2^m-1} e^{\frac{2\pi i g(x) k}{2^m}} |k\rangle$$

so that we can then apply the inverse QFT to get $|g(x)\rangle$. We will achieve this step by step.

Notice that $g(x)$ is always a sum of products of integer values. So, let's first deal with integer addition, and leave multiplication for the next subsection.

Following the notation of [51], we will call the state

$$\frac{1}{\sqrt{2^m}} \sum_{k=0}^{2^m-1} e^{\frac{2\pi i jk}{2^m}} |k\rangle$$

the **phase encoding** of j. Then, for our purposes, it is enough to be able to know how to prepare the phase encoding of 0 and to know how to add a given integer l to the phase encoding of any other integer. In that way, we can start from 0 and add all the terms in the polynomial expression of g one by one.

Preparing the phase encoding of 0 could not be easier. We just need to apply the Hadamard gate to each and every qubit that we are using to represent the integer values. In this way, we will obtain the state

$$\frac{1}{\sqrt{2^m}} \sum_{k=0}^{2^m-1} |k\rangle = \frac{1}{\sqrt{2^m}} \sum_{k=0}^{2^m-1} e^{\frac{2\pi i 0k}{2^m}} |k\rangle,$$

which is, indeed, the phase encoding of 0.

Suppose now that we have a state that phase-encodes j and we want to add l to it. We first assume that l is non-negative and, later, we will deal with negative numbers. To add l in phase encoding, we just need to apply the gates shown in *Figure 6.6.*

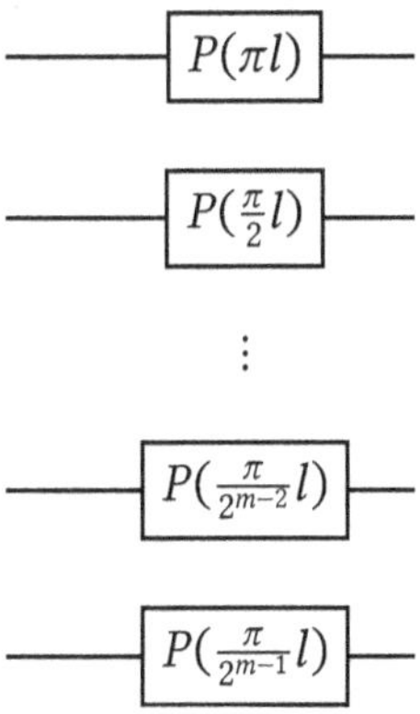

Figure 6.6: Circuit for adding l to a state in phase encoding when we have m qubits

Indeed, when we apply those gates to a basis state $|k\rangle$, we obtain $e^{2\pi ikl/2^m}|k\rangle$. To prove it, just notice how, if the h-th qubit of $|k\rangle$ is 1, the circuit of *Figure 6.6* adds a phase of value $e^{\pi il/2^h} = e^{2^{m-h}\pi il/2^m}$ (we start counting qubits from 0) and no phase otherwise. When we sum all these phases over the qubits of $|k\rangle$ that have value 1, we obtain exactly $e^{2\pi ilk/2^m}$. Thus, by linearity, when we apply the circuit to the phase encoding of j, we get

$$\frac{1}{\sqrt{2^m}}\sum_{k=0}^{2^m-1} e^{\frac{2\pi ijk}{2^m}} e^{\frac{2\pi ilk}{2^m}} |k\rangle = \frac{1}{\sqrt{2^m}}\sum_{k=0}^{2^m-1} e^{\frac{2\pi i(j+l)k}{2^m}} |k\rangle,$$

which is the phase encoding of $j + l$, as desired.

So, this works beautifully for non-negative numbers. But, what about negative ones? It turns out that, if l is negative, we can again use the very circuit in *Figure 6.6* — no further adjustments required. The key observation is that, for any integer $0 \leq h \leq m - 1$, it holds that

$$e^{\frac{\pi i(2^m+l)}{2^h}} = e^{\frac{\pi il}{2^h}} e^{\frac{\pi i2^m}{2^h}} = e^{\frac{\pi il}{2^h}} e^{\pi i2^{m-h}} = e^{\frac{\pi il}{2^h}},$$

because $m - h > 0$, making 2^{m-h} even and implying $e^{\pi i2^{m-h}} = 1$.

This means that if we plug in l or $2^m + l$ in the gates of *Figure 6.6*, we obtain exactly the same circuit. Thus, we can work with the two's complement representation of l instead of l and the results for the addition that we proved previously for non-negative integers will also hold for negative integers. The only concern could be that, when adding in two's complement a positive and a negative number, we get a carry-out (like, for instance, when we added 6 and −4 in a previous example). However, in that case, the carry-out will give us an even power of two and, again, the corresponding phase will be 1, leaving the result unchanged. Effectively, we are performing arithmetic modulo 2^m, so we are safe. Notice, nevertheless, that, if we add two positive or two negative integers and we get a carry-out, then we will get a wrong result — in this case, modular arithmetic turns against us!

Important note

You always need to use a number of qubits that is large enough to represent, in two's complement, any integer number that may arise from the computations. If you are working with a polynomial $g(x)$, you can simply add up the absolute value of all the coefficients in $g(x)$ to obtain a constant K. Then, you can choose any m such that $-2^{m-1} \leq -K \leq K \leq 2^{m-1} - 1$. If you want to be even more precise, you can select K as the maximum between the sum of all positive coefficients and the sum of the absolute value of all the negative coefficients.

As an example, in *Figure 6.7*, we present a circuit that prepares the phase representation of 0, adds 3 to it, and then adds -5 (or, equivalently, subtracts 5). Notice that some of the gates could be simplified. For instance, $P(3\pi)$ is just $P(\pi)$. We could also merge consecutive P gates into single gates by adding their angles together (for instance, $P(-5\frac{\pi}{2})P(3\frac{\pi}{2}) = P(-2\frac{\pi}{2}) = P(\pi)$). For the sake of clarity, throughout this section, we will keep the gates in their original form, without any simplification.

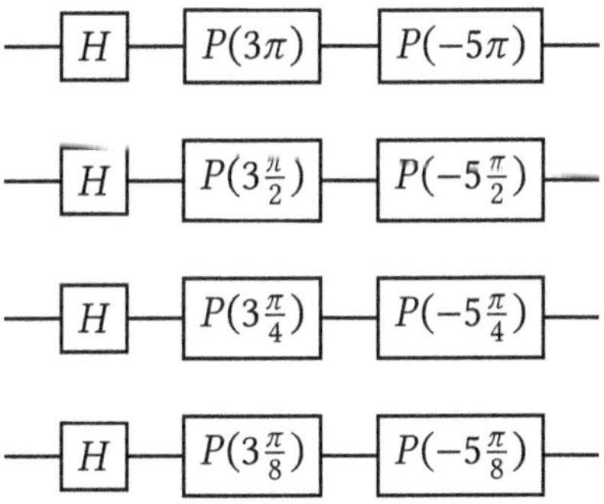

Figure 6.7: Circuit for preparing the phase representation of 0*, adding* 3 *to it and then subtracting* 5

Exercise 6.4

Derive a circuit that prepares the phase representation of 0, adds 6 to it and then subtracts 4. Use 4 qubits.

We now have the first ingredient that we need in order to compute the $g(x)$ polynomial: adding integers in phase encoding. In the next subsection, we will learn how to deal with the product of binary variables.

6.2.3 Computing the whole polynomial

You may be tempted to think that performing the multiplications that we need to compute our polynomial $g(x)$ will be much harder than performing the additions. But not quite! Let's look into this.

All the variables that we are considering are binary, and this means that, when we perform a multiplication such as x_0x_1, we always obtain either 0 or 1 as a result. Thus, if $g(x)$ is, for example, $3x_0x_1 - 2x_1x_2 + 1$, we will need to add 1 always (because it is the independent term and, as the name suggests, does not depend on the value of the variables), but we will only need to add 3 when both x_0 and x_1 are 1 and we will only need to subtract 2 when both x_1 and x_2 take value 1.

Does this sound familiar? Well, it should, because these computations that we have described correspond, precisely, to the application of controlled operations. Therefore, in order to calculate the contribution of a term such as $3x_0x_1$, we can use the circuit that we derived in the previous subsection to add 3 in phase encoding, but with each gate controlled by both x_0 and x_1. Notice that there is nothing special in using just two qubits as the controls, so we could also consider polynomials with terms such as $-2x_0x_2x_4$ or $5x_1x_2x_3x_5$.

To better illuminate these techniques, in *Figure 6.8*, we show a circuit that computes $3x_0x_1 - 2x_1x_2 + 1$. The first column of gates prepares the phase encoding of 0. The second one adds the independent term of the polynomial. The next one adds 3, but only if $x_0 = x_1 = 1$ (that is why all the gates are controlled by the $|x_0\rangle$ and $|x_1\rangle$ qubits). Similarly, the last column subtracts 2, but only when $x_1 = x_2 = 1$.

There are a couple of technical details to discuss about the circuit in *Figure 6.8*. First, we have adopted the usual convention of setting all the one-qubit gates that are controlled by the same qubits in a single column. In fact, we could consider them as a single multi-qubit

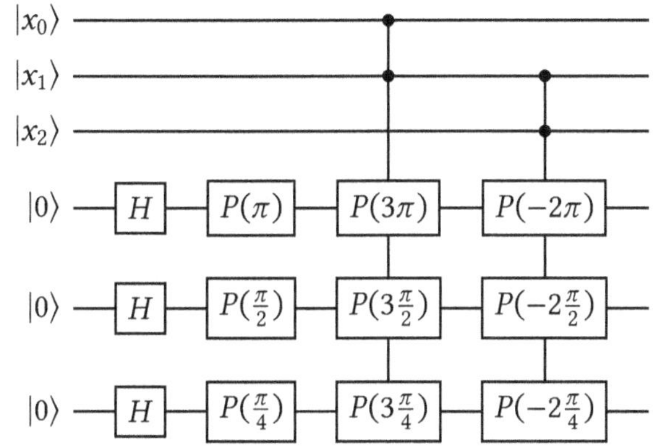

Figure 6.8: Circuit for computing $3x_0x_1 - 2x_1x_2 + 1$ *in phase encoding*

gate, but in some quantum computers you may need to separate them and apply them in sequence (in any case, this is something that the transpiler should take care of, don't worry). Also, notice that these gates are multi-controlled, but — using techniques like the ones described in *Section 4.3* of [16] — you can transform them into a combination of one- and two-qubit gates with Toffoli gates, which, in turn, can be decomposed into just one- and two qubit gates.

Exercise 6.5

Design a circuit for computing $x_1x_2 - 3x_0 + 2$ in phase encoding. Use multi-qubit and multi-controlled gates.

So now we know how to compute, in phase encoding, the values of polynomials on binary variables with integer coefficients. But what about the case in which the coefficients are real numbers? We have two options to deal with that situation. The first one is to approximate them by using fractions with the same denominator. For instance, if your coefficients are 0.25 and -1.17, you can represent them by $25/100$ and $-117/100$. Then, you can multiply the whole polynomial by 100 without changing the variable values at which the minimum is attained and work with 25 and -117, which are integers. The other option is to use the real numbers directly in the encoding. For instance, in the circuit of *Figure 6.6*, you would use l even if it is not an integer. In this case, you will work with a superposition of approximations of the real coefficient, with the better approximations having the larger amplitudes (see the discussion in [50] for all the details).

This completes our discussion on how to compute, in phase encoding, the value of any polynomial on binary variables. However, we are not quite done yet! In the next subsection, we will use our newly-acquired knowledge to finally implement the oracles that we need for the GAS algorithm.

6.2.4 Constructing the oracle

So far in this section we have covered a lot of ground. However, we should not forget what our final goal is: we want to implement an oracle that, given x and y, returns whether $g(x) < g(y)$ or not. This is what we need in order to use the **Dürr-Høyer** algorithm to find a minimum of g. In the previous subsection, we showed how to build a circuit that, given x, computes $g(x)$ in phase encoding. For the sake of simplicity, in the circuits that we will use in this subsection, we will denote the sequence of gates that implements $g(x)$, excluding the initial column of H gates, by just a big box with $g(x)$ inside. In a similar way, when we need to use the QFT or its inverse, we will use a box labeled QFT or $\text{QFT}^\dagger$.

Using this notation, an oracle to determine whether $g(x) < g(y)$ can be implemented by using the circuit depicted in *Figure 6.9*.

Let's explain bit by bit the elements of the circuit. First, notice that the upper qubits are reserved for the inputs x and y and, consequently, are registers of n qubits each. Next, we have m auxiliary qubits that we will use to compute the values of the polynomials (as we mentioned previously, you need to select m so that it is big enough to store all the intermediate results). Finally, the bottom qubit will store the result of checking whether $g(x) < g(y)$.

From what we have studied in this section and under the assumption that all the coefficients in g are integers, we know that the state just before the CNOT gate is $\left|x\right\rangle \left|y\right\rangle \left|g(x) - g(y)\right\rangle \left|0\right\rangle$.

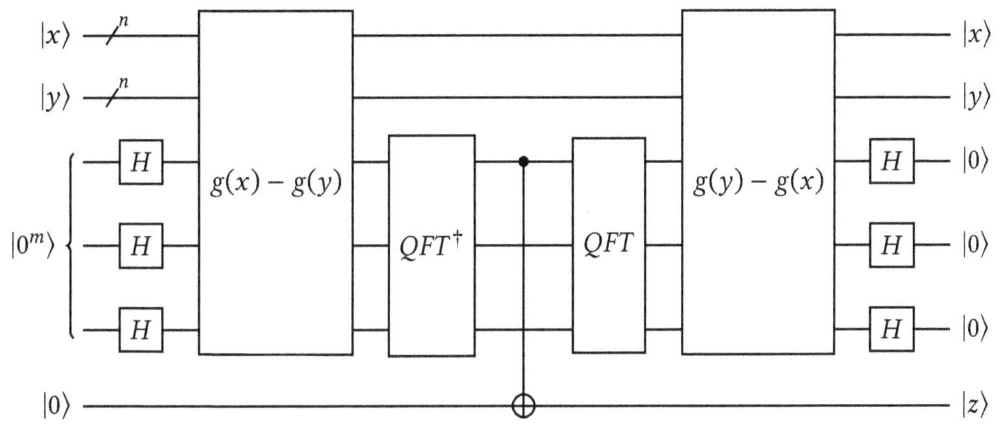

Figure 6.9: Oracle to determine whether $g(x) < g(y)$

Now, if $g(x) < g(y)$, then $g(x) - g(y) < 0$ and the most significant bit of $g(x) - g(y)$ will be 1, because we are working with two's complement representation. Thus, when we apply the CNOT gate, we will set the bottom qubit to $|1\rangle$ if $g(x) < g(y)$, and we will leave it in state $|0\rangle$ otherwise. This is the value that we will denote $|z\rangle$.

It would be natural to think that we could end the circuit after applying the CNOT gate. After all, we have already computed the result that we needed: z will be 1 if $g(x) < g(y)$ and it will be 0 otherwise. Nevertheless, we need to set the m auxiliary qubits back to $|0\rangle$. This is the value that is expected for the correct behaviour of the subsequent applications of the oracle (remember that we are using Grover's algorithm, so there will be several repetitions of the oracle circuit). What is more, we also need to set these qubits back to $|0\rangle$ to disentangle them from the rest of the qubits in the circuit. If they remain entangled, they may prevent the rest of the circuit from working correctly.

The process of setting the qubits back to $|0\rangle$ is known as **uncomputation** and it is a very important technique in many quantum algorithms. Since all quantum gates are reversible, we cannot just "erase" the content of some qubits (that would be extremely irreversible, because we would be forgetting the original values and it would be impossible to restore them). We need to perform the same computations that we carried out, but in reverse: hence the name "uncomputation." In our case, we use the QFT to go back to phase encoding and then we add $g(y)-g(x)$, which, of course, is the inverse of adding $g(x)-g(y)$. Consequently, after the $g(y) - g(x)$ gate, the auxiliary qubits contain the phase encoding of 0 and, when we apply the column of H gates, we obtain $|0\rangle$, as desired.

We have, finally, completed our construction of the oracle that we need for GAS. However, there are a couple of additional details that may be useful in practice. On the one hand, notice that, in each application of Grover's algorithm in GAS, the value of y is fixed (it is x_0, the best solution that we have found by then). Thus, we can simplify the design of the oracle in *Figure 6.9* by eliminating the qubits reserved for $|y\rangle$, computing $g(x_0)$ with a classical computer, and using $g(x) - g(x_0)$ and $g(x_0) - g(x)$ in the gates that compute the values of the polynomial.

On the other hand, using techniques similar to the ones that we have studied in this section, we can create oracles to check whether polynomial constraints are met or not. For instance, if one of the constraints in our problem is $3x_0 - 2x_0x_1 < 3$, we can easily adapt our oracle construction to check whether that condition is met. Thus, we do not always need to transform our optimization problems into a pure QUBO form, but we can keep (some of) the constraints and check them directly. This might be more convenient than working with penalty terms in some cases.

But enough of theoretical considerations for now. In the next section, we will explain how to use GAS in Qiskit in order to solve combinatorial optimization problems.

6.3 Using GAS with Qiskit

If you want to practice what you have learned in this chapter about Grover's search, the Dürr-Høyer algorithm, and the construction of oracles, you can try to implement your own version of GAS in Qiskit from scratch. It is not a difficult project and it can be very satisfactory. However, there is no need for that. In the Qiskit Optimization module, you can find a ready-to-use implementation of Grover Adaptive Search (we will be using **version 0.4.0** of the package). Let's see how to use it.

One additional advantage of working with Qiskit's GAS implementation is that it accepts the optimization problem format that we used with QAOA in *Section 5.2.2*. The simplest way of using it is by defining a QUBO problem like the one that we can create with the following piece of code:

```
from qiskit_optimization.problems import QuadraticProgram
qp = QuadraticProgram()
qp.binary_var('x')
qp.binary_var('y')

qp.minimize(linear = {'x':2,'y':2}, quadratic = {('x','y'):-3})

print(qp.export_as_lp_string())
```

The output of the execution is the following:

```
\ This file has been generated by DOcplex
\ ENCODING=ISO-8859-1
\Problem name: CPLEX

Minimize
 obj: 2 x + 2 y + [ - 6 x*y ]/2
Subject To

Bounds
 0 <= x <= 1
 0 <= y <= 1

Binaries
 x y
End
```

As you surely recognize, this is the type of problem that we have been extensively working with in the last few chapters. To solve it with GAS in Qiskit, we need to define a `GroverOptimizer` object as follows:

```
from qiskit_optimization.algorithms import GroverOptimizer
from qiskit import Aer
from qiskit.utils import algorithm_globals, QuantumInstance
seed = 1234
algorithm_globals.random_seed = seed
quantum_instance = QuantumInstance(Aer.get_backend("aer_simulator"),
    shots = 1024, seed_simulator = seed, seed_transpiler=seed)
grover_optimizer = GroverOptimizer(num_value_qubits = 3, num_iterations=2,
    quantum_instance=quantum_instance)
```

Notice that we have set seed values for reproducibility and we have created a quantum instance based on the Aer simulator. Of course, if you want to use a real quantum computer, you just need to create the quantum instance from one of the quantum devices, as we have seen in previous chapters. Then, we have defined a `GroverOptimizer` object that uses 3 qubits to represent the values of the polynomial (what we have denoted as m in the previous section) and that stops the execution if it has seen no improvement in 2 consecutive iterations (the `num_iterations` parameter). Notice that 3 qubits are enough to represent all the possible values of our polynomial in two's complement, but 2 qubits would be too few.

To use this `GroverOptimizer` object to solve our problem, we can run the following instructions:

```
results = grover_optimizer.solve(qp)
print(results)
```

This will give us the following output:

```
fval=0.0, x=0.0, y=0.0, status=SUCCESS
```

This is, indeed, the optimal solution to the problem, as you can check by trying the 4 possible options. That was easy, wasn't it?

Exercise 6.6

Write the code needed to use GAS in Qiskit to find the solution of the QUBO problem with binary variables x, y, and z and objective function $3x + 2y - 3z + 3xy$.

But what if you want to solve a more complicated problem? It turns out that the Grover Optimizer class also can work with problems with constraints. Imagine that we define a problem with the following instructions:

```
qp = QuadraticProgram()
qp.binary_var('x')
qp.binary_var('y')
qp.binary_var('z')
qp.minimize(linear = {'x':2}, quadratic = {('x','z'):1, ('z','y'):-2})
qp.linear_constraint(linear = {'x':2, 'y':-1, 'z':1},
    sense ="<=", rhs = 2)
print(qp.export_as_lp_string())
```

If we execute the code, we obtain the following output, which corresponds to a quadratic program with linear constraints:

```
\ This file has been generated by DOcplex
\ ENCODING=ISO-8859-1
\Problem name: CPLEX

Minimize
 obj: 2 x + [ 2 x*z - 4 y*z ]/2
Subject To
 c0: 2 x - y + z <= 2
```

```
Bounds
 0 <= x <= 1
 0 <= y <= 1
 0 <= z <= 1

Binaries
 x y z
End
```

We could create a `GroverOptimizer` object and directly use its `solve` method with `qp`. Then, the `GroverOptimizer` object will convert the constrained problem into a QUBO one and solve it. Easy peasy. However, there is a small problem: how can we know how many qubits we should use for the polynomial values? Since we don't know the penalty terms that will be introduced in the conversion, we don't know the coefficients of the polynomial. Of course, we could use a big enough value to be sure that there will be no problems, but that will make the execution slower, especially in the simulator. And if we use too few qubits, our results could be erroneous.

For that reason, we recommend converting the problem first into QUBO form and then solving it with GAS. In this way, we can more accurately determine the number of qubits that we need. For instance, for the problem that we have just defined, we can obtain the transformed QUBO problem with the following instructions:

```
from qiskit_optimization.converters import QuadraticProgramToQubo
qp_to_qubo = QuadraticProgramToQubo()
qubo = qp_to_qubo.convert(qp)
print(qubo.export_as_lp_string())
```

The output is the following:

```
\ This file has been generated by DOcplex
\ ENCODING=ISO-8859-1
```

```
\Problem name: CPLEX

Minimize
 obj: - 46 x + 24 y - 24 z - 24 c0@int_slack@0 - 48 c0@int_slack@1 + [ 48 x^2
      - 48 x*y + 50 x*z + 48 x*c0@int_slack@0 + 96 x*c0@int_slack@1 + 12 y^2
      - 28 y*z - 24 y*c0@int_slack@0 - 48 y*c0@int_slack@1 + 12 z^2
      + 24 z*c0@int_slack@0 + 48 z*c0@int_slack@1 + 12 c0@int_slack@0^2
      + 48 c0@int_slack@0*c0@int_slack@1 + 48 c0@int_slack@1^2 ]/2 + 24
Subject To

Bounds
 0 <= x <= 1
 0 <= y <= 1
 0 <= z <= 1
 0 <= c0@int_slack@0 <= 1
 0 <= c0@int_slack@1 <= 1

Binaries
 x y z c0@int_slack@0 c0@int_slack@1
End
```

As you can see, this is now a bona fide QUBO problem. Moreover, by inspecting the polynomial coefficients, we can notice that 10 qubits, for instance, are enough to store the polynomial values. Thus, we can solve the problem with the following piece of code:

```
grover_optimizer = GroverOptimizer(10,
    num_iterations=4, quantum_instance=quantum_instance)
results = grover_optimizer.solve(qubo)
print(results)
```

If we run it, we obtain the following, which is indeed the solution to the problem:

```
fval=-2.0, x=0.0, y=1.0, z=1.0, c0@int_slack@0=0.0, c0@int_slack@1=1.0,
status=SUCCESS
```

However, this involves the slack variables used in the transformation. If you don't want to see them, you can alternatively run GAS on the original problem, now that we know how many qubits to use:

```
grover_optimizer = GroverOptimizer(10, num_iterations=4,
    quantum_instance=quantum_instance)
results = grover_optimizer.solve(qp)
print(results)
```

In this case, the output is the following:

```
fval=-2.0, x=0.0, y=1.0, z=1.0, status=SUCCESS
```

This is exactly the same solution that we obtained with the transformed problem, but now without the slack variables.

This is all you need to know if you want to use GAS in Qiskit. In the next chapter, we will study the **Variational Quantum Eigensolver**, a generalization of QAOA that will allow us to solve many interesting optimization problems.

Summary

In this chapter, we have learned about Grover's search algorithm and how it can be adapted to find minima of functions with the Dürr-Høyer algorithm. We have also learned about quantum oracles and their role in these two methods.

After that, we learned how to perform arithmetic in phase encoding and how to retrieve the results by using the mighty Quantum Fourier Transform. We also studied how to use all these techniques to implement oracles that can be used in Grover's Adaptive Search to solve combinatorial optimization problems.

Finally, we also learned how to use GAS with Qiskit to obtain solutions of both QUBO problems and constrained quadratic programs.

Now, get ready for the next chapter: we will be studying the Variational Quantum Eigensolver and some of its most important applications!

7

VQE: Variational Quantum Eigensolver

From so simple a beginning endless forms most beautiful and most wonderful have been, and are being, evolved.

— Charles Darwin

In the previous chapters of this part of the book, we have studied how quantum algorithms can help us solve combinatorial optimization problems, but there are many other important types of optimization problems out there! This chapter will broaden the scope of our optimization methods to cover more general settings, including applications in fields such as chemistry and physics.

We will achieve this by studying the famous **Variational Quantum Eigensolver** (**VQE**) algorithm, which can be seen as a generalization of the Quantum Approximate Optimization Algorithm that we studied back in *Chapter 5, QAOA: Quantum Approximate Optimization Algorithm*. Actually, it would be more precise to say that we can see QAOA as a particular case of VQE; in fact, VQE was introduced earlier than QAOA in a now famous paper by Peruzzo et al. [52].

We shall begin by expanding our knowledge of Hamiltonians and by better understanding how to estimate their expectation values with quantum computers. That will allow us to define VQE in all its glory and to appreciate both the simplicity of its formulation and its wide applicability for finding the ground state of different types of Hamiltonians.

We will then show how to use VQE with both Qiskit and PennyLane using examples from the field of chemistry. We will also show how to study the influence of errors on the algorithm by running simulations of noisy quantum computers, and we will even discuss some techniques to mitigate the adverse effect of readout errors.

After reading this chapter, you will know both the theoretical foundations of VQE and how to use it in a wide variety of practical situations, on simulators and on actual quantum computers.

The topics that we will cover in this chapter are as follows:

- Hamiltonians, observables, and their expectation values
- Introducing the Variational Quantum Eigensolver
- Using VQE with Qiskit
- Using VQE with PennyLane

We have quite a lot to learn and, in fact, endless forms most beautiful to discover. So, let's not waste time and get started right away!

7.1 Hamiltonians, observables, and their expectation values

So far, we've found in Hamiltonians a way to encode combinatorial optimization problems. As you surely remember, in these optimization problems, we start with a function f that assigns real numbers to binary strings of a certain length n, and we seek to find a binary string x with minimum cost $f(x)$. In order to do that with quantum algorithms, we define a Hamiltonian H_f such that

$$\langle x| H_f |x\rangle = f(x)$$

holds for every binary string x of length n. Then, we can solve our original problem by finding a ground state of H_f (that is, a state $|\psi\rangle$ such that the expectation value $\langle\psi| H_f |\psi\rangle$ is minimum).

This was just a very quick summary of *Chapter 3, Working with Quadratic Unconstrained Binary Optimization Problems.* When you read that chapter, you may have noticed that the Hamiltonian associated to f has an additional, very remarkable property. We have mentioned this a couple of times already, but it is worth remembering that, for every computational basis state $|x\rangle$, it holds that

$$H_f |x\rangle = f(x) |x\rangle .$$

This means that each $|x\rangle$ is an eigenvector of H_f with associated eigenvalue $f(x)$ (if you do not remember what eigenvectors and eigenvalues are, check *Appendix B, Installing the Tools,* for all the relevant definitions and concepts). In fact, this is easy to see because we have always used Hamiltonians that are sums of tensor products of Z matrices, which are clearly diagonal. But tensor products of diagonal matrices are diagonal matrices themselves, and sums of diagonal matrices are still diagonal. Thus, since these Hamiltonians are diagonal, the computational basis states are their eigenvectors.

What is more, if we have a state $|\psi\rangle$, we can always write it as a linear combination of the computational basis states. In fact, it holds that

$$|\psi\rangle = \sum_x \alpha_x |x\rangle ,$$

where the sum is over all the computational basis states $|x\rangle$ and $\alpha_x = \langle x|\psi\rangle$. This is easy to check, because

$$\langle x|\psi\rangle = \langle x| \sum_y \alpha_y |y\rangle = \sum_y \alpha_y \langle x|y\rangle = \alpha_x.$$

The last identity follows from the fact that $\langle x|y\rangle$ is 1 if $x = y$ and 0 otherwise (remember that the computational basis is an orthonormal basis).

Then, the expectation value of H_f in the state $|\psi\rangle$ can be computed as

$$\begin{aligned}\langle\psi| H_f |\psi\rangle = \sum_y \alpha_y^* \langle y| H_f \sum_x \alpha_x |x\rangle = \sum_{x,y} \alpha_y^*\alpha_x \langle y| H_f |x\rangle = \sum_{x,y} \alpha_y^*\alpha_x f(x) \langle y|x\rangle \\ = \sum_x \alpha_x^*\alpha_x f(x) = \sum_x |\alpha_x|^2 f(x).\end{aligned}$$

Moreover, we know that $|\alpha_x|^2 = |\langle x|\psi\rangle|^2$ is the probability of obtaining $|x\rangle$ when measuring $|\psi\rangle$ in the computational basis; in this way, the expectation value matches the statistical expected value of the measurement. As you surely remember, this is exactly the fact that we used back in *Chapter 5, QAOA: Quantum Approximate Optimization Algorithm*, to estimate the value of the cost function when running QAOA circuits in a quantum computer.

These properties may seem dependent on the particular form of the Hamiltonians that we have been using. But, in fact, they are very general results, and we will use them extensively in our study of the VQE algorithm. But before we get to that, we will need to introduce the general notion of "observable", which is precisely the topic of the next subsection.

7.1.1 Observables

Up until this point, we have only considered measurements in the computational basis. This has worked well enough for our purposes, but, in doing so, we've ignored some details about how measurements are truly understood and described in quantum mechanics. We are now going to fill that gap.

We encourage you to go slowly through this section. Take your time and maybe prepare yourself a good cup of your favourite hot beverage. The ideas presented here may seem a little bit strange at first, but you will soon realize that they fit nicely with what we have been doing so far.

In quantum mechanics, any physical magnitude that you can measure — also known as a **(physical) observable** — is represented by a Hermitian operator. In case you don't remember, these are linear operators A that are equal to their adjoints (their conjugate transposes), that is, they satisfy $A^{\dagger} = A$.

To learn more...

You may remember how in *Chapter 3, Working with Quadratic Unconstrained Binary Optimization Problems*, we worked extensively with Hamiltonians. These, in general, are Hermitian operators that are, indeed, associated with an observable magnitude. That magnitude is none other than the energy of the system!

The nice thing about Hermitian operators is that, for them, one can always find an orthonormal basis of eigenvectors with real eigenvalues (please, check *Appendix B, Basic Linear Algebra*, if you need to review these notions). This means that there exist real numbers λ_j, $j = 1, \ldots, l$, all of them different, and states $\left|\lambda_j^k\right\rangle$, where $j = 1, \ldots, l$ and $k = 1, \ldots, r_j$, such that the states $\{\left|\lambda_j^k\right\rangle\}_{j,k}$ form an orthonormal basis and

$$A\left|\lambda_j^k\right\rangle = \lambda_j\left|\lambda_j^k\right\rangle,$$

for every $j = 1, \ldots, l$ and for every $k = 1, \ldots, r_j$.

Here, we are considering the possibility of having several eigenvectors $\left|\lambda_j^k\right\rangle$ associated with the same eigenvalue λ_j, hence the use of the superindices $k = 1, \ldots, r_j$, where r_j is the number of eigenvectors associated with the λ_j^k eigenvalue. If all the eigenvalues are different (a quite common case), then we will have $r_j = 1$ for every j and we can simply drop the k superindices.

What is the connection of these Hermitian operators with physical measurements? Let's consider an observable represented by a Hermitian operator A, and also an orthonormal basis of eigenvectors $\{\left|\lambda_j^k\right\rangle\}_{j,k}$ such that $A\left|\lambda_j^k\right\rangle = \lambda_j\left|\lambda_j^k\right\rangle$. This representation must be chosen to take the following into account:

- The possible outcomes of the measurement of the observable must be represented by the different eigenvalues λ_j

- The probability that a state $|\psi\rangle$ will, upon measurement, yield λ_j must be $\sum_k \left|\left\langle \lambda_j^k \middle| \psi \right\rangle\right|^2$

All of this is axiomatic. It is a fact of life that any physical observable can be represented by a Hermitian operator in such a way that those are requirements are satisfied. Moreover, it is a postulate of quantum mechanics that if the measurement returns the result associated to an eigenvalue λ_j, the state of the system will then become the normalized projection of $|\psi\rangle$ onto the space of eigenvectors with eigenvalue λ_j. This means that if we measure a state in a superposition such as

$$\sum_{j,k} \alpha_j^k \left|\lambda_j^k\right\rangle$$

and we obtain λ_j as the result, then the new state will be

$$\frac{\sum_k \alpha_j^k \left|\lambda_j^k\right\rangle}{\sqrt{\sum_k \left|\alpha_j^k\right|^2}}.$$

This is what we call the *collapse* of the original state and it is exactly the same phenomenon that we considered when studied measurements in the computational basis back in *Chapter 1, Foundations of Quantum Computing*.

The word "observable" is often used for both physical observables and for any Hermitian operators that represent them. Thus, we may refer to a Hermitian operator itself as an observable. To avoid confusions, we will usually not omit the "physical" adjective when referring to physical observables.

As a simple example, whenever we measure in the computational basis, we are indeed measuring some physical observable, and this physical observable can, of course, be represented by a Hermitian operator. This is, in a certain sense, the simplest observable in quantum computing and it is natural that it arises as a particular case of this, more general theory of quantum measurements.

The coordinated matrix of this measurement operator with respect to the computational basis could be the diagonal matrix

$$\begin{pmatrix} 0 & & & \\ & 1 & & \\ & & \ddots & \\ & & & 2^n - 1 \end{pmatrix}.$$

Exercise 7.1

Prove that, indeed, the previous matrix is the coordinate matrix on the computational basis of a Hermitian operator that represents a measurement in the computational basis.

When we measure a single qubit in the computational basis, the coordinate matrix with respect to the computational basis of the associated Hermitian operator could well be either of

$$N = \begin{pmatrix} 0 & 0 \\ 0 & 1 \end{pmatrix}, \qquad Z = \begin{pmatrix} 1 & 0 \\ 0 & -1 \end{pmatrix}.$$

Yes, that last matrix was the unmistakable Pauli Z matrix. Both of these operators represent the same observable; they only differ in the eigenvalues that they associate to the distinct possible outcomes. The first operator associates the eigenvalues 0 and 1 to the qubit's value being 0 and 1 respectively, while the second observable associates the eigenvalues 1 and −1 to these outcomes.

Important note

Measurements in quantum mechanics are represented by Hermitian operators, which we refer to as observables. One possible operator corresponding to measuring a qubit in the computational basis can be the Pauli Z matrix.

Now that we know what an observable is, we can study what its **expectation value** is and how it can be computed. The expectation value of any observable under a state $|\psi\rangle$ can be defined as

$$\langle A\rangle_\psi = \sum_{j,k} \left|\left\langle \lambda_j^k \middle| \psi \right\rangle\right|^2 \lambda_j,$$

which is a natural definition that agrees with the statistical expected value of the results obtained when we measure $|\psi\rangle$ according to A. As intuitive as this expression may be, we can further simplify it as follows:

$$\begin{aligned}\langle A\rangle_\psi &= \sum_{j,k} \left|\left\langle \lambda_j^k \middle| \psi \right\rangle\right|^2 \lambda_j = \sum_{j,k} \left\langle \psi \middle| \lambda_j^k \right\rangle \left\langle \lambda_j^k \middle| \psi \right\rangle \lambda_j = \sum_{j,k} \left\langle \lambda_j^k \middle| \psi \right\rangle \left\langle \psi \middle| \lambda_j^k \right\rangle \lambda_j \\ &= \sum_{j,k} \left\langle \lambda_j^k \middle| \psi \right\rangle \langle\psi| A \left|\lambda_j^k\right\rangle = \langle\psi| A \sum_{j,k} \left\langle \lambda_j^k \middle| \psi \right\rangle \left|\lambda_j^k\right\rangle = \langle\psi| A |\psi\rangle .\end{aligned}$$

Notice that we have used the fact that $A\left|\lambda_j^k\right\rangle = \lambda_j \left|\lambda_j^k\right\rangle$ and that $|\psi\rangle = \sum_{j,k} \left\langle \lambda_j^k \middle| \psi \right\rangle \left|\lambda_j^k\right\rangle$. This latter identity follows from the fact that $\{\left|\lambda_j^k\right\rangle\}_{j,k}$ is an orthonormal basis and, in fact, it can be proved in exactly the same way we did for the computational basis at the beginning of this section.

This expression for the expectation value agrees with our previous work in *Chapter 3, Working with Quadratic Unconstrained Binary Optimization Problems.*

Important note

The expectation value of any Hermitian operator (observable) A is given by

$$\langle A\rangle_\psi = \sum_{j,k} \left|\left\langle \lambda_j^k \middle| \psi \right\rangle\right|^2 \lambda_j = \langle\psi| A |\psi\rangle .$$

Notice that, from the very definition of the expectation value of an observable, we can easily derive the variational principle. This principle states, as you may recall from *Chapter 3, Working with Quadratic Unconstrained Binary Optimization Problems*, that the smallest expectation value of an observable A is always achieved at an eigenvector of that observable. To prove it, suppose that λ_0 is minimal among all the eigenvalues of A. Then, for any state ψ it holds that

$$\langle A\rangle_\psi = \sum_{j,k} \left|\left\langle \lambda_j^k \middle| \psi \right\rangle\right|^2 \lambda_j \geq \sum_{j,k} \left|\left\langle \lambda_j^k \middle| \psi \right\rangle\right|^2 \lambda_0 = \lambda_0,$$

where the last equality follows from the fact that $\sum_{j,k} \left|\left\langle \lambda_j^k \middle| \psi \right\rangle\right|^2 = 1$, since the sum of the probabilities of all the outcomes must add up to 1.

If we now take any eigenvector $\left|\lambda_0^k\right\rangle$ associated to λ_0, its expectation value will be

$$\left\langle \lambda_0^k \middle| A \middle| \lambda_0^k \right\rangle = \lambda_0 \left\langle \lambda_0^k \middle| \lambda_0^k \right\rangle = \lambda_0,$$

proving that the minimum expectation value is indeed achieved at an eigenvector of A. Obviously, if there were several orthogonal eigenvectors associated to λ_0, any normalized linear combination of them would also be a ground state of A.

In this subsection, we have studied the mathematical expression for the expectation of any observable. But we don't yet know how to estimate these expectation values with quantum computers. How could we do that? Just keep reading, because we will be exploring it in the next subsection.

7.1.2 Estimating the expectation values of observables

In the context of the VQE algorithm, we will need to estimate the expectation value of a general observable A. That is, we will no longer assume that A is diagonal, as we have done in all the previous chapters. For this reason, we will need to develop a new method for estimating the expectation value $\langle \psi | A | \psi \rangle$.

We know that, for a given state $|\psi\rangle$, the expectation value of A can be computed by

$$\sum_{j,k} \left|\left\langle \lambda_j^k \middle| \psi \right\rangle\right|^2 \lambda_j.$$

Thus, if we knew the eigenvalues λ_j and the eigenvectors $\{\left|\lambda_j^k\right\rangle\}_{j,k}$ of A, we could try to compute $\left|\left\langle\lambda_j^k\middle|\psi\right\rangle\right|^2$ and, hence, the expectation value of A. However, this is information that we usually don't know. In fact, the purpose of VQE is, precisely, finding certain eigenvalues and eigenvectors of a Hamiltonian! Moreover, the number of eigenvectors grows exponentially with the number of qubits of our system, so, even if we knew them, computing expectation values in this way might be very computationally expensive.

Thus, we need to take an indirect route. For this, we will use the fact that we can always express an observable A on n qubits as a linear combination of tensor products of Pauli matrices (see, for example, *Chapter 7* on the famous lecture notes by John Preskill [53]). Actually, A will be, in most cases, given to us in such a form, in the same way that the Hamiltonians of our combinatorial optimization problems were always expressed as sums of tensor products of Z matrices.

So, consider, for example, that we are given an observable

$$A = \frac{1}{2}Z \otimes I \otimes X - 3I \otimes Y \otimes Y + 2Z \otimes X \otimes Z.$$

Notice that, thanks to linearity,

$$\begin{aligned}
\langle\psi| A |\psi\rangle &= \langle\psi| \left(\frac{1}{2}Z \otimes I \otimes X - 3I \otimes Y \otimes Y + 2Z \otimes X \otimes Z\right) |\psi\rangle \\
&= \langle\psi| \left(\frac{1}{2}(Z \otimes I \otimes X)|\psi\rangle - 3(I \otimes Y \otimes Y)|\psi\rangle + 2(Z \otimes X \otimes Z)|\psi\rangle\right) \\
&= \frac{1}{2}\langle\psi|(Z \otimes I \otimes X)|\psi\rangle - 3\langle\psi|(I \otimes Y \otimes Y)|\psi\rangle + 2\langle\psi|(Z \otimes X \otimes Z)|\psi\rangle.
\end{aligned}$$

Then, in order to compute the expectation value of A, we can compute the expectation values of $Z \otimes I \otimes X$, $I \otimes Y \otimes Y$, and $Z \otimes X \otimes Z$ and combine their results. But wait a minute! Isn't that even more complicated? After all, we would need to compute three expectation values instead of just one, right?

The key observation here lies in the fact that, while we may not know the eigenvalues and eigenvectors of A in advance, we can very easily obtain those of $Z \otimes I \otimes X$ or any other tensor product of Pauli matrices. It is so easy, in fact, that you will now learn how to do it yourself in the following two exercises.

Exercise 7.2

Suppose that $\left|\lambda_j\right\rangle$ is an eigenvector of A_j with associated eigenvalue λ_j for $j = 1, \ldots, n$. Prove that $\left|\lambda_1\right\rangle \otimes \cdots \otimes \left|\lambda_n\right\rangle$ is an eigenvector of $A_1 \otimes \cdots \otimes A_n$ with associated eigenvalue $\lambda_1 \cdot \ldots \cdot \lambda_n$.

Exercise 7.3

Prove that:

1. The eigenvectors of Z are $|0\rangle$ (with associated eigenvalue 1) and $|1\rangle$ (with associated eigenvalue -1).
2. The eigenvectors of X are $|+\rangle$ (with associated eigenvalue 1) and $|-\rangle$ (with associated eigenvalue -1).
3. The eigenvectors of Y are $\left(1/\sqrt{2}\right)(|0\rangle + i|1\rangle)$ (with associated eigenvalue 1) and $\left(1/\sqrt{2}\right)(|0\rangle - i|1\rangle)$ (with associated eigenvalue -1).
4. Any non-null state is an eigenvector of I with associated eigenvalue 1.

Using the results in these exercises, we can readily deduce that $|0\rangle|+\rangle|0\rangle$, $|0\rangle|-\rangle|1\rangle$, $|1\rangle|+\rangle|1\rangle$, and $|1\rangle|-\rangle|0\rangle$ are eigenvectors of $Z \otimes X \otimes Z$ with eigenvalue 1 and that $|0\rangle|+\rangle|1\rangle$, $|0\rangle|-\rangle|0\rangle$, $|1\rangle|+\rangle|0\rangle$, and $|1\rangle|-\rangle|1\rangle$ are eigenvectors of $Z \otimes X \otimes Z$ with eigenvalue -1. All these states together form an orthonormal basis of eigenvectors of $Z \otimes X \otimes Z$, as you can easily check if you compute their inner products.

Exercise 7.4

Find orthonormal bases of eigenvectors for $Z \otimes I \otimes X$ and $I \otimes Y \otimes Y$. Compute their associated eigenvalues.

So, now we know how to obtain the eigenvalues and eigenvectors of any tensor product of Pauli matrices. How can we use this to estimate their expectation values? Remember that, given a Hermitian matrix A, we can compute $\langle\psi| A |\psi\rangle$ by

$$\sum_{j,k} \left|\left\langle \lambda_j^k \middle| \psi \right\rangle\right|^2 \lambda_j,$$

where the eigenvalues of A are λ_j and the associated eigenvectors are $\{\left|\lambda_j^k\right\rangle\}_{j,k}$. In our case, we only have two eigenvalues: 1 and -1. So, if we are able to estimate the values $\left|\left\langle \lambda_j^k \middle| \psi \right\rangle\right|^2$, we will have all the ingredients needed to "cook" our expectation values.

A priori, trying to get the values $\left|\left\langle \lambda_j^k \middle| \psi \right\rangle\right|^2$ out of a quantum computer can seem like a difficult task. For example, you may wonder whether it will be necessary to perform some weird fancy measurements on our quantum device in order to get these probabilities! Well, it turns out that we can easily estimate them on any quantum computer using ordinary measurements in the computational basis and a bunch of quantum gates. So, don't worry. If you've just bought yourself a flashy quantum computer, there's no need for a hardware upgrade just yet.

In any case, how can we actually estimate these $\left|\left\langle \lambda_j^k \middle| \psi \right\rangle\right|^2$ values with the tools that we have? Let's first work with an example.

Let's consider the observable $Z \otimes X \otimes Z$. We have previously in this section obtained its eigenvectors, so let's focus on one of them: $|0\rangle |+\rangle |0\rangle$. If we wanted to compute $|(\langle 0| \langle +| \langle 0|) |\psi\rangle|^2$, where $|\psi\rangle$ is a certain 3-qubit state, we could just notice that

$$|0\rangle |+\rangle |0\rangle = (I \otimes H \otimes I) |0\rangle |0\rangle |0\rangle$$

and hence

$$\begin{aligned}\langle 0| \langle +| \langle 0| = (|0\rangle |+\rangle |0\rangle)^\dagger = ((I \otimes H \otimes I) |0\rangle |0\rangle |0\rangle)^\dagger = \langle 0| \langle 0| \langle 0| (I \otimes H \otimes I)^\dagger \\ = \langle 0| \langle 0| \langle 0| (I \otimes H \otimes I),\end{aligned}$$

where we have used the fact that I and H are self-adjoint, and hence so is $I \otimes H \otimes I$. Keep in mind, however, that we will still write daggers throughout this example whenever we mean to consider the adjoint of $I \otimes H \otimes I$ — even if it still represents the same operator.

From this, it follows directly that

$$|(\langle 0| \langle +| \langle 0|) |\psi\rangle|^2 = \left|\langle 0| \langle 0| \langle 0| (I \otimes H \otimes I)^\dagger |\psi\rangle\right|^2.$$

But for any state $|\varphi\rangle$, we know that $|(\langle 0| \langle 0| \langle 0|) |\varphi\rangle|^2$ is the probability of obtaining $|0\rangle |0\rangle |0\rangle$ when measuring it in the computational basis. As a consequence, we can estimate the value $|(\langle 0| \langle +| \langle 0|) |\psi\rangle|^2$ by repeatedly preparing the state $(I \otimes H \otimes I) |\psi\rangle = (I \otimes H \otimes I)^\dagger |\psi\rangle$, measuring it in the computational basis, and then computing the relative frequency of $|0\rangle |0\rangle |0\rangle$.

And this is not the only eigenvector for which this works. It turns out that for each and every eigenvector $|\lambda_A\rangle$ of $Z \otimes X \otimes Z$, there is a unique state in the computational basis $|\lambda_C\rangle$ such that

$$|\lambda_A\rangle = (I \otimes H \otimes I) |\lambda_C\rangle.$$

Actually, the correspondence is bijective: for every state in the computational basis $|\lambda_C\rangle$, there is also a unique eigenvector $|\lambda_A\rangle$ of $Z \otimes X \otimes Z$ such that $|\lambda_C\rangle = (I \otimes H \otimes I)^\dagger |\lambda_A\rangle$, where we have used the fact that, for unitary operators, $U^\dagger = U^{-1}$. For example,

$$|1\rangle |-\rangle |1\rangle = (I \otimes H \otimes I) |1\rangle |1\rangle |1\rangle, \qquad |1\rangle |1\rangle |1\rangle = (I \otimes H \otimes I)^\dagger |1\rangle |-\rangle |1\rangle.$$

This is the reason why we call $I \otimes H \otimes I$ the **change of basis operator** between the computational basis and the basis of eigenvectors of $Z \otimes X \otimes Z$.

In this way, if we want to estimate the probabilities $\left|\left\langle \lambda_j^k \middle| \psi \right\rangle\right|^2$ when the states $\left|\lambda_j^k\right\rangle$ happen to be the eigenvectors of $Z \otimes X \otimes Z$, we just need to prepare $(I \otimes H \otimes I)^\dagger |\psi\rangle$ and measure it in the computational basis. Then, given any eigenvector $|\lambda_A\rangle$ of $Z \otimes X \otimes Z$, the probability $|\langle \lambda_A|\psi\rangle|^2$ can be estimated by the relative frequency of the measurement outcome associated to the eigenstate $\left|\lambda_C\right\rangle = (I \otimes H \otimes I)^\dagger \left|\lambda_A\right\rangle$ in the computational basis. That's because

$$\left\langle\lambda_C\right| \left((I \otimes H \otimes I)^\dagger \left|\psi\right\rangle\right) = \left\langle\lambda_A\right| \left((I \otimes H \otimes I)(I \otimes H \otimes I)^\dagger \left|\psi\right\rangle\right) = \left\langle\lambda_A|\psi\right\rangle,$$

where we have used the fact that, for any operator L and any states $\left|\alpha\right\rangle$ and $\left|\beta\right\rangle$, if $\left|\beta\right\rangle = L\left|\alpha\right\rangle$, then $\left\langle\beta\right| = \left\langle\alpha\right| L^\dagger$, and $L^{\dagger\dagger} = L$.

As a final note, in this example, when we set out to compute the probabilities $\left|\left\langle\lambda_A|\psi\right\rangle\right|^2$, we don't have to run executions for each of the probabilities individually: we can compute them all simultaneously. All we have to do is measure $(I \otimes H \otimes I)^\dagger \left|\psi\right\rangle$ in the computational basis a bunch of times and then retrieve the relative frequency of every outcome. This works because $(I \otimes H \otimes I)^\dagger$ transforms all the eigenvectors of A into the states of the computational basis. Then, the probability $\left|\left\langle\lambda_A|\psi\right\rangle\right|^2$ will be the relative frequency of the outcome, in the computational basis, associated to $(I \otimes H \otimes I)^\dagger \left|\lambda_A\right\rangle$. Of course, the higher the number of preparations and measurements, the more accurate our estimates will be.

To learn more...

Notice the similarity of this kind of procedure with the standard measurement in the computational basis. When we measure $\left|\psi\right\rangle$ in the computational basis, we have probability $\left|\left\langle x|\psi\right\rangle\right|^2$ of obtaining the outcome associated to $\left|x\right\rangle$. If we were measuring an observable that had all the $\left|\lambda_j^k\right\rangle$ as eigenvectors with a distinct eigenvalue for each of them – this is an observable that's able to distinguish all the eigenvectors in the basis – we would have probability $\left|\left\langle\lambda_j^k|\psi\right\rangle\right|^2$ of getting the outcome associated to $\left|\lambda_j^k\right\rangle$.

This is why we refer to the process of changing basis and, then, measuring in the computational basis, as performing a **measurement in the eigenvector basis** $\{\left|\lambda_j^k\right\rangle\}$ **of** A. It is exactly the same as if we had an observable that's able to measure and distinguish all the eigenvectors of A.

But wait, there's more! Our being able to change bases in this case is by no means a happy coincidence. It turns out that for every tensor product of Pauli matrices A, there is a simple change of basis matrix that defines a perfect correspondence between the states in the computational basis and the eigenvectors of A. Again, this can be readily verified, and we invite you to do it in the following two exercises.

Exercise 7.5

Since the computational basis is an eigenvector basis of Z, a change of basis operator of Z can be the identity I. Check that, in order to change from the computational basis to the basis of eigenvectors of the X, you can use the Hadamard matrix H, and that to change to the basis of eigenvectors of Y you can use SH.

Exercise 7.6

Prove that if U_1 and U_2 are the respective change of basis operators from the computational basis to the eigenvector basis of two observables A_1 and A_2, then $U_1 \otimes U_2$ is the change of basis operator from the computational basis to the eigenvector basis of $A_1 \otimes A_2$.

Putting everything together, we can easily deduce that, for instance, $I \otimes I \otimes H$ takes the eigenvectors of $Z \otimes I \otimes X$ to the computational basis and that $I \otimes (SH)^\dagger \otimes (SH)^\dagger$ takes the eigenvectors of $I \otimes Y \otimes Y$ to states in the computational basis as well.

Therefore, in order to estimate the expectation value $\langle\psi| (Z \otimes I \otimes X) |\psi\rangle$, we can use whatever circuit we need to prepare $|\psi\rangle$ followed by $(I \otimes I \otimes H)^\dagger = I \otimes I \otimes H$, then measure in the computational basis, and then get the probabilities as we have just discussed. In a similar way, to estimate $\langle\psi| (I \otimes Y \otimes Y) |\psi\rangle$, we will first prepare $|\psi\rangle$, then apply $I \otimes HS^\dagger \otimes HS^\dagger$ and, finally, measure in the computational basis.

Notice, by the way, how I and H are self-adjoint, so, when we took their adjoints, there was no observable (no pun intended) effect. That's not the case with SH, because $(SH)^\dagger = H^\dagger S^\dagger = HS^\dagger$.

To learn more...

For any Hermitian operator A, there is always a unitary transformation that takes any basis of eigenvectors of A to the computational basis, and vice versa. However, this transformation could very well be difficult to implement. In the case where A is a tensor product of Pauli matrices we have just proved that we can always obtain the transformation as the tensor product of very simple one-qubit operations.

Finally, after we have estimated the expectation value of every Pauli term in our observable (in our case, $Z \otimes I \otimes X$, $I \otimes Y \otimes Y$, and $Z \otimes X \otimes Z$) we can multiply them by the corresponding coefficients in the linear combination and add everything together to get the final result. And we are done!

Now you know how to estimate the expectation value of an observable by measuring in different bases. You can proudly say, as the famous internet meme goes, "All your base are belong to us." And, in fact, this was the last technical element that we needed in order to introduce VQE, something we will immediately do in the next section.

7.2 Introducing VQE

The goal of the **Variational Quantum Eigensolver** (**VQE**) is to find a ground state of a given Hamiltonian H_1. This Hamiltonian can describe, for instance, the energy of a certain physical or chemical process, and we will use some such examples in the following two sections, which will cover how to execute VQE with Qiskit and PennyLane. For the moment, however, we will keep everything abstract and focus on finding a state $|\psi\rangle$ such that $\langle\psi| H_1 |\psi\rangle$ is minimum. Note that in this section, we will be using H_1 to refer to the Hamiltonian so that it does not get confused with the Hadamard matrix that we will also be using in our computations.

To learn more...

VQE is by no means the only quantum algorithm that has been proposed to find the ground states of Hamiltonians. Some very promising options use a quantum subroutine known as **Quantum Phase Estimation** (**QPE**) (see, for instance, the excellent surveys by McArdle et al. [54] and by Cao et al. [55]). The main disadvantage of these approaches is that QPE uses the Quantum Fourier Transform that we studied in *Chapter 6, GAS: Grover Adaptative Search*, and, thus, requires quantum computers that are resilient to noise. An experimental demonstration of these limitations (and of the relative robustness of VQE) can be found, for instance, in the paper by O'Malley et al. [56]. For this reason, we will focus mainly on VQE and its applications, which seem to obtain better results with the NISQ computers that are available today.

The general structure of VQE is very similar to that of QAOA, which you surely remember from *Chapter 5, QAOA: Quantum Approximate Optimization Algorithm*: we prepare a parameterized quantum state, we measure it, we estimate its energy, and we change the parameters in order to minimize it; then, we repeat this process several times until some stopping criteria are met. The preparation and measurement of the state are done on the quantum computer, while the energy estimation and parameter minimization are handled by a classical computer.

The parametrized circuit, usually called **variational form** or **ansatz**, is usually chosen taking into account information from the problem domain. For instance, you could consider ansatzes that parametrize typical solutions to the kind of problem under study. We will show some examples of this in the last two sections of this same chapter. In any case, the ansatz is selected in advance and it is usually easy to implement on a quantum circuit.

Important note

In many applications, we can distinguish two parts in the creation of the parameterized state: the preparation of an initial state $\left|\psi_0\right\rangle$, that does not depend on any parameters, and then the variational form $V(\theta)$ itself, that obviously depends on θ. Thus, if we have $\left|\psi_0\right\rangle = U\left|0\right\rangle$, for some unitary transformation U implemented with some quantum gates, the ansatz gives us the state $V(\theta)U|0\rangle$. Notice, however, that we can always consider the whole operation $V(\theta)U$ as the ansatz and require the initial state to be $|0\rangle$. To simplify our notation, this is what we will usually do, although we will explicitly distinguish between initial state and ansatz in some practical examples that we will consider later in the chapter.

Algorithm 7.1 gives the pseudocode for VQE. Notice the similarities with Algorithm 5.1 from *Chapter 5, QAOA: Quantum Approximate Optimization Algorithm.*

Algorithm 7.1 (VQE).

Require: H_1 given as a linear combination of tensor products of Pauli matrices

Choose a variational form (ansatz) $V(\theta)$

Choose a starting set of values for θ

while the stopping criteria are not met **do**

Prepare the state $|\psi(\theta)\rangle = V(\theta)|0\rangle$ ▷ *This is done on the quantum computer!*

From the measurements of $|\psi(\theta)\rangle$ in different bases, estimate $\langle\psi(\theta)| H_1 |\psi(\theta)\rangle$

Update θ according to the minimization algorithm

Prepare the state $|\psi(\theta)\rangle = V(\theta)|0\rangle$ ▷ *This is done on the quantum computer!*

From the measurements of $|\psi(\theta)\rangle$ in different bases, estimate $\langle\psi(\theta)| H_1 |\psi(\theta)\rangle$

Let's remark on a couple of things about this pseudocode. Notice that we require that H_1 be given as a linear combination of tensor products of Pauli matrices; this is so that we can use the techniques that we introduced in the previous section to estimate $\langle\psi| H_1 |\psi\rangle$. Of course, the more terms we have in the linear combination, the bigger the number of bases in which we may need to perform measurements. Nevertheless, in some cases, we may group several measurements together. For example, if we have terms such as $I \otimes X \otimes I \otimes X$, $I \otimes I \otimes X \otimes X$, and $I \otimes I \otimes X \otimes X$, we can use $I \otimes H \otimes H \otimes H$ as our change of basis matrix (be careful! This H is the Hadamard matrix, not the Hamiltonian!) because it works for the three terms at the same time – keep in mind that any orthonormal basis is an eigenvector basis for I, not just $\{|0\rangle, |1\rangle\}$. Obviously, another hyperparameter that will impact the execution time of VQE is the number of times that we measure $|\psi\rangle$ in each basis. The higher this number, the more precise the estimation, but also the higher the time needed to estimate $\langle\psi| H_1 |\psi\rangle$.

Notice also that the pseudocode of Algorithm 7.1 concludes by estimating $\langle\psi(\theta)| H_1 |\psi(\theta)\rangle$ for the last state $|\psi(\theta)\rangle$ found by the minimization algorithm. This is a quite common use case, for instance, if we want to determine the ground state energy for a particular system. However, you are not restricted to just that. At the end of the VQE execution, you also know the θ_0 parameters that were used to build the ground state, and you could use them to reconstruct $|\psi(\theta_0)\rangle = V(\theta_0)|0\rangle$. This state could then be used for other purposes, such as being sent into another quantum algorithm.

In fact, in the next subsection, we are going to explore one of such uses: the computation of additional **eigenstates** (another name for our old friends, the eigenvectors) of Hamiltonians. You should be excited!

7.2.1 Getting excited with VQE

As we have just explained, VQE is used to search for a ground state of a given Hamiltonian H. However, with a small modification, we can also use it to find **excited states**: eigenstates with higher energies. Let's explain how to achieve this.

Suppose that you have been given a Hamiltonian H and you have used VQE to find a ground state $|\psi_0\rangle = V(\theta_0)|0\rangle$ with energy λ_0. Then, we may consider the modified Hamiltonian

$$H' = H + C|\psi_0\rangle\langle\psi_0|,$$

where C is a positive real number.

Before we move on to detailing why H' can enable us to find excited states, let's explain what the term $|\psi_0\rangle\langle\psi_0|$ in that expression means. First of all, notice that this term represents a square matrix: it is the product of a column vector ($|\psi_0\rangle$) and a row vector ($\langle\psi_0|$) of the same length. Moreover, it is a Hermitian matrix, because

$$(|\psi_0\rangle\langle\psi_0|)^\dagger = \langle\psi_0|^\dagger\,|\psi_0\rangle^\dagger = |\psi_0\rangle\langle\psi_0|\,.$$

Then, H' is the sum of two Hermitian matrices and is, therefore, also Hermitian. And what is its expectation value? If we have a generic quantum state $|\psi\rangle$, then

$$\langle\psi| H' |\psi\rangle = \langle\psi| H |\psi\rangle + C\langle\psi|\psi_0\rangle\langle\psi_0|\psi\rangle = \langle\psi| H |\psi\rangle + C|\langle\psi_0|\psi\rangle|^2.$$

That is, the expectation value of H' in a state $|\psi\rangle$ is the expectation value of H plus a non-negative value that quantifies the overlap of $|\psi\rangle$ and $|\psi_0\rangle$. Hence we have two extreme cases for $C|\langle\psi_0|\psi\rangle|^2$. If $|\psi\rangle = |\psi_0\rangle$, this term will be C. If $|\psi\rangle$ and $|\psi_0\rangle$ are orthogonal, the term will be 0.

Thus, if we make C big enough, $|\psi_0\rangle$ will no longer be a ground state of H'. Let's prove this is in a more formal way. To this end, let $\lambda_0 \le \lambda_1 \le \ldots \le \lambda_n$ be the eigenvalues of H associated to each eigenvector in an orthonormal eigenvector basis $\{|\lambda_j\rangle\}$ (since different eigenvectors may have the same eigenvalue, some of the energies may be repeated). As $|\psi_0\rangle$ is, by hypothesis, a ground state, we shall assume that $|\lambda_0\rangle = |\psi_0\rangle$. The states $\{|\lambda_j\rangle\}$ are also eigenvectors of H' because, on the one hand, if $j \neq 0$,

$$H'|\lambda_j\rangle = H|\lambda_j\rangle + C|\psi_0\rangle\langle\psi_0|\lambda_j\rangle = H|\lambda_j\rangle + C|\lambda_0\rangle\langle\lambda_0|\lambda_j\rangle = H|\lambda_j\rangle = \lambda_j|\lambda_j\rangle,$$

since $|\lambda_0\rangle$ and $|\lambda_j\rangle$ are orthogonal. On the other hand,

$$H'|\lambda_0\rangle = H|\lambda_0\rangle + C|\psi_0\rangle\langle\psi_0|\lambda_0\rangle = H|\lambda_0\rangle + C|\lambda_0\rangle\langle\lambda_0|\lambda_0\rangle = H|\lambda_0\rangle + C|\lambda_0\rangle = (\lambda_0 + C)|\lambda_0\rangle.$$

Thus, it follows that $\langle\lambda_j| H' |\lambda_j\rangle = \lambda_j$ when $j \neq 0$ and that $\langle\lambda_0| H' |\lambda_0\rangle = C + \lambda_0$. Hence, if $C > \lambda_1 - \lambda_0$, then $|\psi_0\rangle = |\lambda_0\rangle$ will no longer be a ground state of H', because the energy of $|\lambda_1\rangle$ will be lower than that of $|\psi_0\rangle$. Thanks to the variational principle, we know that the minimum energy is attained at an eigenvector of H', so $|\lambda_1\rangle$ must be a ground state of H'.

We can then use VQE to search for a ground state of H' and obtain the state $|\lambda_1\rangle$, as we intended.

To learn more...

Notice that it could be the case that $\lambda_1 = \lambda_0$. In that situation, $|\lambda_1\rangle$ would be another ground state of H. Otherwise, it will be the first excited state of H.

You may have also noticed that, even if the ground state is unique, the first excited eigenstate may not be so. This happens if and only if $|\lambda_2\rangle$ (and possibly other states in the basis) has the same energy as $|\lambda_1\rangle$, (that is, $\lambda_2 = \lambda_1$). In that case, any normalized linear combination of those eigenvectors will be a ground state of H'. Any of them will serve our purposes equally well.

Of course, once you obtain $|\lambda_1\rangle$, you can consider $H'' = H' + C' |\lambda_1\rangle \langle\lambda_1|$ and use VQE to search for $|\lambda_2\rangle$, and so on and so forth. Keep in mind that, in this process, we would have to pick the constants $C, C', \ldots$ properly — just to make sure that none of the eigenstates that we already know becomes a ground state again!

With this, our problem of finding eigenvectors of increasing energy is solved. Or is it?

There is just one little implementation detail that might be bothering you. In the previous section, we discussed how to estimate the expectation value of a Hamiltonian under the assumption that it was given as a sum of tensor products of Pauli matrices. However, the $|\psi_0\rangle \langle\psi_0|$ term is not of that form. In fact, we know $|\psi_0\rangle$ only as the result of applying VQE, so it is very likely that we will not know $|\psi_0\rangle$ explicitly; instead, we will have nothing more than some parameters θ_0 such that $V(\theta_0)|0\rangle = |\psi_0\rangle$. This, nonetheless, is enough to compute the expectation values that we need.

Let's step back a little bit and take a look at what we need to compute. At a given moment in the application of VQE, we have some parameters θ and we want to estimate the expectation value of $|\psi_0\rangle \langle\psi_0|$ with respect to $|\psi(\theta)\rangle = V(\theta)|0\rangle$. This quantity is

$$\langle\psi(\theta)|\psi_0\rangle \langle\psi_0|\psi(\theta)\rangle = |\langle\psi_0|\psi(\theta)\rangle|^2 = \left|\langle 0| V(\theta_0)^\dagger V(\theta)|0\rangle\right|^2.$$

But this is just the probability of obtaining $|0\rangle$ as the outcome of measuring $V(\theta_0)^\dagger V(\theta)|0\rangle$ in the computational basis! This is something that we can easily estimate because we can prepare $V(\theta_0)^\dagger V(\theta)|0\rangle$ by first applying our ansatz V, using θ as the parameters, to $|0\rangle$, and then applying the inverse of our ansatz, with parameters θ_0, to the resulting state. We will repeat this process several times, always measuring the resulting state $V(\theta_0)^\dagger V(\theta)|0\rangle$ in the computational basis and computing the relative frequency of the outcome $|0\rangle$. This is illustrated in *Figure 7.1.*

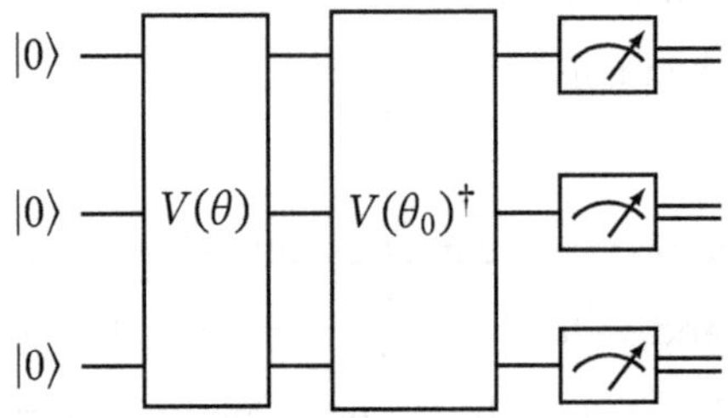

Figure 7.1: Circuit to compute $\left|\langle 0| V(\theta_0)^\dagger V(\theta)|0\rangle\right|^2$.

The only thing that may, at first, seem difficult with this method is to obtain the circuit for $V(\theta_0)^\dagger$. However, this is fairly easy. You just need to remember that every unitary gate is reversible. Thus, you can take the circuit for $V(\theta)$ and read the gates from right to left, reversing each one of them. As a simple example, if $\theta_0 = (a, b)$ and $V(\theta_0) = XR_Z(a)R_X(b)S$, then $V(\theta_0)^\dagger = S^\dagger R_X(-b)R_Z(-a)X$.

Do not forget about this technique for estimating $\left|\langle 0| V(\theta_0)^\dagger V(\theta)|0\rangle\right|^2$ because we will be using it again in *Chapter 9, Quantum Support Vector Machines*, in a completely different context.

This concludes our theoretical study of VQE. In the next section, we will learn how to use this algorithm with Qiskit.

7.3 Using VQE with Qiskit

In this section, we will show how we can use Qiskit to run VQE on both simulators and actual quantum hardware. To do that, we will use a problem taken from quantum chemistry: determining the energy of the H_2 or dihydrogen molecule. Our first subsection is devoted to defining this problem.

7.3.1 Defining a molecular problem in Qiskit

To illustrate how we can use VQE with Qiskit, we will consider a simple quantum chemistry problem. We will imagine that we have two atoms of hydrogen forming an H_2 molecule and that we want to compute its ground state and its energy. For that, we need to obtain the Hamiltonian of the system, which is a little bit different from the kind of Hamiltonian that we are used to. The Hamiltonians that we have considered so far are called **qubit Hamiltonians**, while the one that we need to describe the energy of the H_2 molecule is called a **fermionic Hamiltonian** – the name comes from the fact that it involves fermions, that is, particles such as electrons, protons, and neutrons.

We do not need to go into all the details of the computation of this type of Hamiltonian (if you are curious, you can refer to *Chapter 4* in the book by Sharkey and Chancé [57]), because all the necessary methods are provided in the Qiskit Nature package. What is more, no quantum computer is involved in the process; it is all computed and estimated classically.

To obtain the fermionic Hamiltonian for the dihydrogen molecule with Qiskit, we need to install the Qiskit Nature package as well as the pyscf library, which is used for the computational chemistry calculations (please, refer to *Appendix D, Installing the Tools*, for the installation procedure and note that we will be using version 0.4.5 of the package).

Then, we can execute the following instructions:

```
from qiskit_nature.drivers import Molecule
from qiskit_nature.drivers.second_quantization import \
    ElectronicStructureMoleculeDriver, ElectronicStructureDriverType
```

```
from qiskit_nature.problems.second_quantization import \
    ElectronicStructureProblem

mol = Molecule(geometry=[['H', [0., 0., -0.37]],
                         ['H', [0., 0., 0.37]]])
driver = ElectronicStructureMoleculeDriver(mol, basis='sto3g',
        driver_type=ElectronicStructureDriverType.PYSCF)
problem = ElectronicStructureProblem(driver)
secqop = problem.second_q_ops()
print(secqop[0])
```

Here, we are defining a molecule consisting of two hydrogen atoms located at coordinates $(0, 0, -0.37)$ and $(0, 0, 0.37)$ (measured in angstroms), which is close to an equilibrium state for this molecule. We are using the default parameters, such as, for instance, establishing that the molecule is not charged. Then, we define an electronic structure problem; that is, we ask the pyscf library, through the Qiskit interface, to compute the fermionic Hamiltonian that takes into account the different possible configurations for the electrons of the two hydrogen atoms. This is done with something called **second quantization** (hence the name `second_q_ops` for the method that we use).

When we run this piece of code, we obtain the following output:

```
Fermionic Operator
register length=4, number terms=36
  -1.2533097866459775 * ( +_0 -_0 )
+ -0.47506884877217725 * ( +_1 -_1 )
+ -1.2533097866459775 * ( +_2 -_2 )
+ -0.47506884877217725 * ( +_3 -_3 )
+ -0.3373779634072241 * ( +_0 +_0 -_0 -_0 )
+ -0.0 ...
```

This is a truncated view of the fermionic Hamiltonian, involving something called **creation** and **annihilation** operators that describe how electrons move from one orbital to another (more details can be found in *Chapter 4* of the book by Sharkey and Chancé [57]).

That is all very nice, but we can't yet use it on our shiny quantum computers. For that, we need to transform the fermionic Hamiltonian into a qubit Hamiltonian, involving Pauli gates. There are several ways to do this. One of the most popular ones is the **Jordan-Wigner** transformation (again, see the book by Sharkey and Chancé [57] for a thorough explanation), that we can use in Qiskit with the following instructions:

```
from qiskit_nature.converters.second_quantization import QubitConverter
from qiskit_nature.mappers.second_quantization import JordanWignerMapper

qconverter = QubitConverter(JordanWignerMapper())
qhamiltonian = qconverter.convert(secqop[0])
print("Qubit Hamiltonian")
print(qhamiltonian)
```

Upon running this code, we will obtain the following output:

```
Qubit Hamiltonian
-0.8121706072487122 * IIII
+ 0.17141282644776915 * IIIZ
- 0.22343153690813483 * IIZI
+ 0.17141282644776915 * IZII
- 0.22343153690813483 * ZIII
+ 0.12062523483390415 * IIZZ
+ 0.16868898170361205 * IZIZ
+ 0.04530261550379923 * YYYY
+ 0.04530261550379923 * XXYY
+ 0.04530261550379923 * YYXX
+ 0.04530261550379923 * XXXX
```

```
+ 0.16592785033770338 * ZIIZ
+ 0.16592785033770338 * IZZI
+ 0.1744128761226159 * ZIZI
+ 0.12062523483390415 * ZZII
```

Now we are back on known territory once again! This is, indeed, one of the Hamiltonians that we have come to know and love. In fact, this is a Hamiltonian on 4 qubits, involving tensor products of I, X, Y and Z gates, as the ones appearing in terms such as `0.17141282644776915 * IIIZ` or `0.04530261550379923 * XXYY`.

What is more important to us: this is the kind of Hamiltonian to which we can apply the VQE algorithm in order to obtain its ground state. And, without further ado, that is exactly what we will be doing in the next subsection.

7.3.2 Using VQE with Hamiltonians

Now that we have a qubit Hamiltonian describing our electronic problem, let's see how we can use VQE with Qiskit to find its ground state. Remember that, to use VQE, we first need to choose an ansatz. To start with, we will use something simple. We will select one of the variational forms provided by Qiskit: the `EfficientSU2` form. We can define it and draw its circuit for 4 qubits with the following instructions (remember that you need to install the pylatexenc library to draw with the `"mpl"` option; please, refer to *Appendix D, Installing the Tools*):

```
from qiskit.circuit.library import EfficientSU2

ansatz = EfficientSU2(num_qubits=4, reps=1, entanglement="linear",
    insert_barriers = True)
ansatz.decompose().draw("mpl")
```

Here, we have specified that we are using the variational form on 4 qubits, that we only use one repetition (that is, a single layer of CNOT gates) and that the entanglement that we want to use is linear: this means that each qubit is entangled with a CNOT gate to the following one. After running this piece of code, we will obtain the image depicted in *Figure 7.2*. As you can see, we are using R_Y and R_Z gates, together with entangling gates (CNOT gates, in this case). In total, we have 16 different tunable parameters, represented by $\theta[0], \ldots, \theta[15]$ in the figure. We will discuss more variational forms, similar to this one, in *Chapters 9* and *10*. For now, it is enough to notice that this is a circuit that we can easily implement in current quantum hardware (because it only involves simple one and two-qubit gates), but that allows us to create somewhat complicated quantum states, with entanglement among all the qubits.

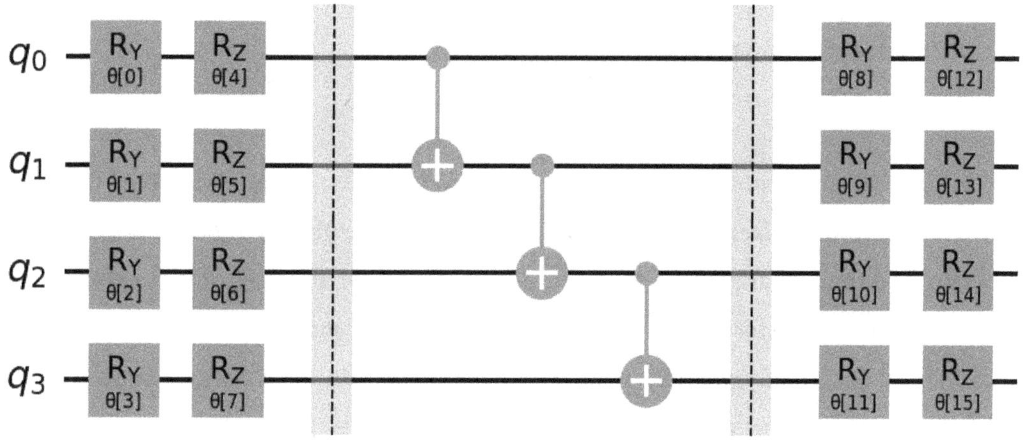

Figure 7.2: The EfficientSU2 variational form on 4 qubits.

Once we have selected our ansatz, we can define a VQE instance. In order to do that, we can use the following instructions:

```
from qiskit.algorithms import VQE
from qiskit import Aer
from qiskit.utils import QuantumInstance, algorithm_globals
import numpy as np
from qiskit.algorithms.optimizers import COBYLA

seed = 1234
np.random.seed(seed)
```

```
algorithm_globals.random_seed = seed

optimizer = COBYLA()

initial_point = np.random.random(ansatz.num_parameters)
quantum_instance = QuantumInstance(backend =
    Aer.get_backend('aer_simulator_statevector'))

vqe = VQE(ansatz=ansatz, optimizer=optimizer,
    initial_point=initial_point,
    quantum_instance=quantum_instance)
```

After the necessary imports, we set a seed for reproducibility. Then, we selected COBYLA as our classical optimizer; that is, the algorithm in charge of varying the parameters in order to find those that achieve the minimum energy. We also set some random initial values for our parameters and we declared a `QuantumInstance` that encapsulates a state vector simulator. Finally, we declared our VQE instance with the ansatz, optimizer, initial values, and quantum instance options.

Running the VQE is now very easy. We only need to execute the following instructions:

```
result = vqe.compute_minimum_eigenvalue(qhamiltonian)
print(result)
```

After a few seconds, we obtain the following output:

```
{   'aux_operator_eigenvalues': None,
    'cost_function_evals': 888,
    'eigenstate': array([ 1.55163279e-09+7.04522580e-10j,
        1.17994431e-06+6.29389934e-07j,
       -6.87287902e-05-1.19175176e-04j,  9.01607105e-09+1.75153048e-10j,
        3.17070261e-06-2.71251777e-05j, -9.23514532e-01-3.66685696e-01j,
```

```
        -6.50833666e-07-1.04178617e-06j, -6.40877389e-06-1.04499914e-05j,
        -1.33988128e-06+3.63309921e-07j,  1.08441415e-05+7.61755332e-08j,
         1.04578392e-01+4.15432635e-02j, -5.85921512e-06+4.47076415e-06j,
        -1.01179799e-09+1.85616927e-09j,  5.57085679e-05+5.29593190e-05j,
         1.47630244e-07+4.00357904e-08j,  1.51330159e-10+9.41869390e-10j]),
    'eigenvalue': (-1.8523881417094914+0j),
    'optimal_circuit': None,
    'optimal_parameters': {
        ParameterVectorElement(θ[7]): -0.10263498379273155,
        ParameterVectorElement(θ[6]): -0.13154223054201972,
        ParameterVectorElement(θ[8]): 3.1416468430294864,
        ParameterVectorElement(θ[13]): 0.6426987629297032,
        ParameterVectorElement(θ[9]): 2.4674114077579344e-05,
        ParameterVectorElement(θ[14]): -0.11387081297526412,
        ParameterVectorElement(θ[15]): 2.525254909939928,
        ParameterVectorElement(θ[12]): 1.8446272942674344,
        ParameterVectorElement(θ[11]): -0.0011789455587669483,
        ParameterVectorElement(θ[10]): 2.7179451047891577e-06,
        ParameterVectorElement(θ[3]): 3.1403232388683655,
        ParameterVectorElement(θ[1]): 9.061128731357842e-06,
        ParameterVectorElement(θ[2]): 3.141570826032646,
        ParameterVectorElement(θ[0]): -0.22553325325129397,
        ParameterVectorElement(θ[5]): 2.1513214842441912,
        ParameterVectorElement(θ[4]): 1.7045601611970793},
    'optimal_point': array([-2.25533253e-01,  9.06112873e-06,
        3.14157083e+00,  3.14032324e+00,
        1.70456016e+00,  2.15132148e+00, -1.31542231e-01, -1.02634984e-01,
        3.14164684e+00,  2.46741141e-05,  2.71794510e-06, -1.17894556e-03,
        1.84462729e+00,  6.42698763e-01, -1.13870813e-01,  2.52525491e+00]),
```

```
'optimal_value': -1.8523881417094914,
'optimizer_evals': None,
'optimizer_result': None,
'optimizer_time': 3.0011892318725586}
```

This may seem like a lot of information but, in fact, some data is presented several times in different ways and, all in all, the format is quite similar to what we are used to from our experience using QAOA in Qiskit back in *Chapter 5, QAOA: Quantum Approximate Optimization Algorithm*. As you can see, we have obtained the optimal values for the circuit parameters, the state that is generated with those parameters (the `'eigenstate'` field) and what we were looking for: the energy of that state, which happens to be about −1.8524 hartrees (the unit of energy commonly used in molecular orbital calculations). This means that...we have solved our problem! Or have we? How can we be sure that the value that we have obtained is correct?

In this case, the Hamiltonian that we are using is quite small (only 4 qubits), so we can check our result using a classical solver that finds the exact ground state. We will use `NumPyMinimumEigensolver`, just as we did with the combinatorial optimization problems that we considered back in *Chapter 5, QAOA: Quantum Approximate Optimization Algorithm*. For that, we can run the following piece of code:

```
from qiskit.algorithms.minimum_eigensolvers import \
    NumPyMinimumEigensolver
solver = NumPyMinimumEigensolver()
result = solver.compute_minimum_eigenvalue(qhamiltonian)
print(result)
```

The output of these instructions is the following:

```
{   'aux_operators_evaluated': None,
    'eigenstate': Statevector([-1.53666363e-17-4.93701060e-20j,
             -4.57234900e-16-4.65250782e-16j,
```

```
        1.25565337e-17-2.11612780e-17j,
        4.73690908e-16-1.33060132e-16j,
        1.52564317e-16-1.40021223e-16j,
       -6.67316913e-01-7.36221442e-01j,
       -1.62999711e-16-2.24584031e-16j,
       -8.42710421e-17+6.43081213e-17j,
       -7.98957973e-17-1.35250844e-17j,
        1.90408979e-16+3.25517112e-16j,
        7.55826341e-02+8.33870007e-02j,
       -3.56170534e-17+9.82948865e-17j,
       -4.51619835e-16+1.70721750e-16j,
        1.91645940e-17-1.45775129e-16j,
       -4.79331105e-17+5.57184037e-17j,
       -3.62080563e-17+4.86380668e-17j],
      dims=(2, 2, 2, 2)),
 'eigenvalue': -1.852388173569583}
```

This is certainly more concise than the VQE output, but the final energy is almost equal to the one we had obtained previously. Now we can really say it: we did it! We have successfully solved a molecular problem with VQE!

Of course, we can use VQE with any type of Hamiltonian, not just with the ones that come from quantum chemistry problems. We can even use it with Hamiltonians for combinatorial optimization problems, as we did back in *Chapter 5, QAOA: Quantum Approximate Optimization Algorithm*. With what we already know, this is easy... so easy that we entrust it to you as an exercise.

Exercise 7.7

Use Qiskit's VQE implementation to solve the Max-Cut problem for a graph of 5 vertices in which the connections are $(0, 1), (1, 2), (2, 3), (3, 4)$ and $(4, 0)$.

Once we know how to find the ground state of a Hamiltonian with VQE, why not be a little more ambitious? In the next subsection, we will also be looking for excited states!

7.3.3 Finding excited states with Qiskit

Back in *Section 7.2.1*, we learned how to use VQE iteratively to find not only the ground state of a Hamiltonian, but also states of higher energy that we call excited states. The algorithm that we studied is sometimes called **Variational Quantum Deflation** (this was the name used by Higgot, Wang, and Brierley when they introduced it [58]) or **VQD**, and it is implemented by Qiskit in the `VQD` class.

Using VQD in Qiskit is almost the same as using VQE. The only difference is that we need to specify how many eigenstates we want to obtain (of course, if we only request 1 this will be *exactly* like applying VQE). For instance, if we want to obtain two eigenstates (the ground state and the first excited state) in our molecular problem, we can use the following instructions:

```
from qiskit.algorithms import VQD
vqd = VQD(ansatz=ansatz,
    optimizer=optimizer,
    initial_point=initial_point,
    quantum_instance=quantum_instance,
    k = 2)
result = vqd.compute_eigenvalues(qhamiltonian)
print(result)
```

The `k` parameter is the one that we use to specify the number of eigenstates. Upon running these instructions, we obtain the following output (we have omitted part of it for brevity):

```
{   'aux_operator_eigenvalues': None,
    'cost_function_evals': array([ 888, 1000]),
    'eigenstates': ListOp([VectorStateFn(Statevector(
            [ 1.55163279e-09+7.04522580e-10j,
```

```
            1.17994431e-06+6.29389934e-07j,
            ...
            1.51330159e-10+9.41869390e-10j],
          dims=(2, 2, 2, 2)), coeff=1.0,
          is_measurement=False),
          VectorStateFn(Statevector(
          [-5.01605162e-02+4.38928908e-02j,
           -7.31117975e-01-3.69461649e-02j,
           -6.34876999e-03-5.19845422e-03j,
           ...
           -4.10301081e-02+2.77415065e-02j],
          dims=(2, 2, 2, 2)), coeff=1.0,
          is_measurement=False)], coeff=1.0,
          abelian=False),
 'eigenvalues': array([-1.85238814-1.11e-16j, -1.19536442+0.00e+00j]),
 'optimal_circuit': None,
 'optimal_parameters': [
 {    ParameterVectorElement(θ[0]): -0.22553325325129397,
      ParameterVectorElement(θ[1]): 9.061128731357842e-06,
      ...
      ParameterVectorElement(θ[15]): 2.525254909939928},
 {    ParameterVectorElement(θ[0]): 0.012174657752649348,
      ParameterVectorElement(θ[1]): -0.056812096977499754,
      ...
      ParameterVectorElement(θ[15]): 1.522408417522795}],
 'optimal_point':
 array([[-2.25533253e-01,  9.06112873e-06,  3.14157083e+00,
      3.14032324e+00,  1.70456016e+00,  2.15132148e+00,
      ...
```

```
        2.52525491e+00],
      [ 1.21746578e-02, -5.68120970e-02,  1.31641034e+00,
        4.59223490e-01,  7.25749716e-01,  9.54546607e-02,
        ...
        1.52240842e+00]]),
 'optimal_value': array([-1.85238814, -1.1952203 ]),
 'optimizer_evals': None,
 'optimizer_result': None,
 'optimizer_time': array([ 2.32541203, 53.26829457])}
```

As you can see, this output is structured like that of the VQE execution. However, in this case, we get two entries in each field, one for each of the eigenstates that we requested.

So far, we have learned how to use both VQE and VQD with Hamiltonians that may have come from any source. However, the use case of finding ground states of molecular Hamiltonians is so important that Qiskit provides special methods for dealing with them in particular. We will learn how in the next subsection.

7.3.4 Using VQE with molecular problems

In addition to using VQE to find ground states of any given Hamiltonian, we can use it directly with molecular problems that we define with the help of the Qiskit Nature utilities. For instance, we can use a VQE instance to solve the electronic problem that we defined in the previous subsection. To do that, we only need to run the following instructions:

```
from qiskit_nature.algorithms import GroundStateEigensolver

solver = GroundStateEigensolver(qconverter, vqe)
result = solver.solve(problem)
print(result)
```

As you can see, we have defined a `GroundStateEigensolver` object that we then use to solve our problem. This object, in turn, uses two objects that we had defined previously —

qconverter, which is used to transform the fermionic Hamiltonian into a qubit Hamiltonian, and the instance of VQE that we used two subsections ago. When we run these instructions, we get the following output:

```
=== GROUND STATE ENERGY ===

* Electronic ground state energy (Hartree): -1.852388141709
  - computed part:      -1.852388141709
~ Nuclear repulsion energy (Hartree): 0.715104339081
> Total ground state energy (Hartree): -1.137283802628

=== MEASURED OBSERVABLES ===

  0:  # Particles: 2.000 S: 0.000 S^2: 0.000 M: 0.000

=== DIPOLE MOMENTS ===

~ Nuclear dipole moment (a.u.): [0.0  0.0  0.0]

  0:
  * Electronic dipole moment (a.u.): [0.0  0.0  0.00001495]
    - computed part:      [0.0  0.0  0.00001495]
  > Dipole moment (a.u.): [0.0  0.0  -0.00001495]  Total: 0.00001495
                 (debye): [0.0  0.0  -0.000038]  Total: 0.000038
```

The information that we get in this case is at a higher level of abstraction than the one we obtained before. For instance, we get data about the number of particles, dipole moments, and so on (don't worry if you do not understand these concepts; they are meant to make sense to chemists and physicists that work with this kind of problem). However, the numerical result of the electronic ground state energy is the same that we obtained with our previous application of VQE. The difference is that now, we are providing the solver not only with the Hamiltonian, but with the whole problem, and it can use that information to reconstruct the meaning of the calculations in physical terms. For instance, we now get some bonus information such as the **total ground state energy**, which is the sum of the energy due to the electronic structure (the one that we had computed previously) and the energy due to nuclear repulsion.

This type of output is much more legible. That's why we will use this solver for the rest of this section.

As an additional example of how to use VQE to solve molecular problems, we are now going to consider a different, more elaborate ansatz. Earlier in this chapter, we mentioned how, when selecting the variational form and initial state to be used with VQE, it could prove useful to take into account information from the problem domain. This is the case of the **Unitary Coupled-Cluster Singles and Doubles** or **UCCSD** ansatz, which is widely used for molecular computations (see the survey by McArdle et al. [54] for more details).

In Qiskit, we can use the UCSSD ansatz with the following instructions:

```
from qiskit_nature.algorithms import VQEUCCFactory

vqeuccf = VQEUCCFactory(quantum_instance = quantum_instance)
```

The `VQEUCCFactory` class creates a whole VQE instance, with the UCSSD ansatz as the default variational form. Here, we are using the `quantum_instance` object that we had defined previously. We can visualize the circuit for the ansatz created by `VQEUCCFactory` with the following instruction:

```
vqeuccf.get_solver(problem, qconverter).ansatz.decompose().draw("mpl")
```

Notice that we are calling the `get_solver` method, to which we pass the `problem` object defined previously to provide the information about the Hamiltonian involved in the computation. Then, we access the ansatz circuit through the `ansatz` attribute and we proceed to draw it. Upon running this instruction, we obtain the circuit depicted in *Figure 7.3*. As you can see, the ansatz involves exponential functions of tensor products of Pauli matrices. There are also two X gates at the beginning of the circuit that set the initial state to which the variational form is later applied. In this case, the state is called the **Hartree-Fock** state, again a widely used option when solving molecular problems with quantum computers – and the default value with `VQEUCCFactory`.

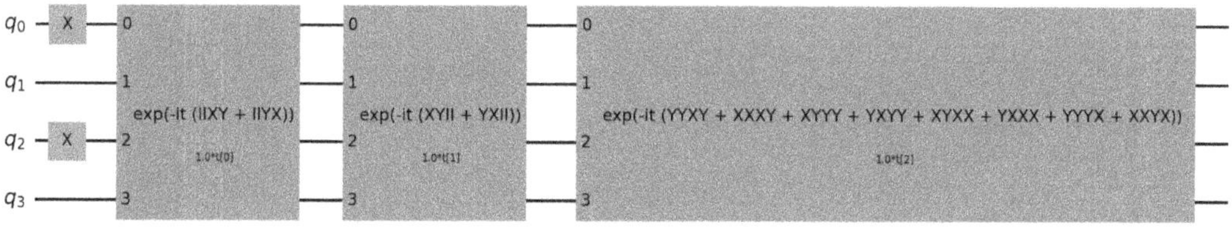

Figure 7.3: UCCSD ansatz for our problem

Now, we can easily use VQE to solve our problem with the selected ansatz by running the following piece of code:

```
solver = GroundStateEigensolver(qconverter, vqeuccf)
result = solver.solve(problem)
print(result)
```

This will give us the following output:

```
=== GROUND STATE ENERGY ===


* Electronic ground state energy (Hartree): -1.852388173513
  - computed part:      -1.852388173513
~ Nuclear repulsion energy (Hartree): 0.715104339081
> Total ground state energy (Hartree): -1.137283834432


=== MEASURED OBSERVABLES ===


  0:  # Particles: 2.000 S: 0.000 S^2: 0.000 M: 0.000


=== DIPOLE MOMENTS ===
```

```
~ Nuclear dipole moment (a.u.): [0.0  0.0  0.0]

  0:
  * Electronic dipole moment (a.u.): [0.0  0.0  -0.00000013]
    - computed part:      [0.0  0.0  -0.00000013]
  > Dipole moment (a.u.): [0.0  0.0  0.00000013]  Total: 0.00000013
                 (debye): [0.0  0.0  0.00000033]  Total: 0.00000033
```

This result is very similar to the one that we obtained with the `EfficientSU2` ansatz.

Exercise 7.8

Write code to use VQE with the UCCSD ansatz to compute the total ground state energy for two atoms of hydrogen that are at distances ranging from 0.2 to 2.0 angstroms, in steps of 0.01 angstroms. Plot the energy against the distance. This kind of plot is sometimes known as the **dissociation profile** of the molecule. *Hint:* when running VQE on a molecular problem, you can access the total ground state energy through the `total_energies` attribute of the result object.

Now that we know how to use VQE in different ways with simulators, we could try to run the algorithm on actual quantum computers. Nevertheless, before doing that, we will learn how to incorporate noise to our quantum simulator.

7.3.5 Simulations with noise

Going from a perfect, classical simulation of an algorithm to an execution on an actual quantum device can, sometimes, be too big a step. As we have mentioned in many occasions, current quantum computers suffer from the effect of different types of noise, including readout errors, imperfections in gate implementation, and decoherence, the loss of quantum properties of our states if the circuits are too deep.

For this reason, it is usually a good idea to perform a simulation of our algorithm under the effects of noise before going to the actual quantum device. In this way, we can study the performance of our algorithms in a controlled environment, and calibrate and adjust some of their parameters before running them on a quantum computer. For instance, if we observe that the results differ much from ideal simulation, we could decide to reduce the depth of our circuits by using a simpler ansatz.

There are several ways of conducting noisy simulations with Qiskit. Here, we will show how to use one that is both easy and very useful. We will create a simulator that mimics the behaviour of a real device, including the noise it is affected by. We can do this with the help of the `AerSimulator` class in the following way:

```
from qiskit.providers.aer import AerSimulator
from qiskit import IBMQ


provider = IBMQ.load_account()
backend = provider.get_backend('ibmq_manila')
quantum_instance = QuantumInstance(
    backend = AerSimulator.from_backend(backend),
    seed_simulator=seed, seed_transpiler = seed, shots = 1024)
```

Notice that we need to load an IBM account in order to access the calibration of a real device (`ibmq_manila`, in our example). This calibration is updated periodically to stay real to the state of the quantum computer and includes, among other things, information about readout errors, gate errors, and coherence times. Of course, this data will change from time to time, but we have decided to include seeds for our `QuantumInstance` object to make the result reproducible given the same calibration data. Notice that we are now specifying the number of shots, because we are not using state vector simulation anymore.

Now, we can run the VQE algorithm exactly as before:

```
vqe = VQE(
    ansatz=ansatz,
```

```
    optimizer=optimizer,
    initial_point=initial_point,
    quantum_instance=quantum_instance
)

solver = GroundStateEigensolver(qconverter, vqe)
result = solver.solve(problem)
print(result)
```

When we ran this code, we obtained the following output (your results may be different, depending on the device calibration):

```
=== GROUND STATE ENERGY ===

* Electronic ground state energy (Hartree): -1.763282965888
  - computed part:      -1.763282965888
~ Nuclear repulsion energy (Hartree): 0.715104339081
> Total ground state energy (Hartree): -1.048178626807

=== MEASURED OBSERVABLES ===

  0:  # Particles: 1.978 S: 0.080 S^2: 0.086 M: 0.001

=== DIPOLE MOMENTS ===

~ Nuclear dipole moment (a.u.): [0.0  0.0  0.0]

  0:
  * Electronic dipole moment (a.u.): [0.0  0.0  0.04634607]
    - computed part:      [0.0  0.0  0.04634607]
```

```
> Dipole moment (a.u.): [0.0  0.0  -0.04634607]  Total: 0.04634607
                (debye): [0.0  0.0  -0.11779994]  Total: 0.11779994
```

Observe how the effect of noise has affected the performance of the algorithm and degraded it, giving a result for the total ground state energy that is not that close to the correct one.

To learn more...

An alternative way of mimicking the behavior of real quantum computers is using objects of the `FakeProvider` class. The difference is that they use snapshots of past calibrations of the devices instead of the latest ones. You can find more details at `https://qiskit.org/documentation/apidoc/providers_fake_provider.html`.

Additionally, you can create custom noise models that include the different noise types implemented in the `qiskit_aer.noise` package. Check the documentation at `https://qiskit.org/documentation/apidoc/aer_noise.html` for further explanation.

A way to try of reducing the adverse effects of noise in our computations is using **readout error mitigation** methods. The idea behind the particular method that we are going to use is very simple. Imagine that we know that, when the state of our qubit is $|0\rangle$, there is a certain percentage of times it in which we obtain the incorrect value 1 when we measure it. Then, we can take into account this information to correct the measurement results that we have obtained.

In Qiskit, using readout error mitigation is very easy. We only need to create our Quantum Instance object in the following way:

```
from qiskit.utils.mitigation import CompleteMeasFitter

quantum_instance = QuantumInstance(
    backend = AerSimulator.from_backend(backend),
    measurement_error_mitigation_cls=CompleteMeasFitter,
    seed_simulator=seed, seed_transpiler = seed, shots = 1024)
```

Then, we can run VQE as usual, using this new `QuantumInstance` variable. In our case, that led to the following result (again, yours will likely differ because of the device calibration):

```
=== GROUND STATE ENERGY ===

* Electronic ground state energy (Hartree): -1.827326686753
  - computed part:      -1.827326686753
~ Nuclear repulsion energy (Hartree): 0.715104339081
> Total ground state energy (Hartree): -1.112222347671

=== MEASURED OBSERVABLES ===

  0:  # Particles: 1.991 S: -0.000 S^2: -0.000 M: -0.006

=== DIPOLE MOMENTS ===

~ Nuclear dipole moment (a.u.): [0.0  0.0  0.0]

  0:
  * Electronic dipole moment (a.u.): [0.0  0.0  -0.05906852]
    - computed part:      [0.0  0.0  -0.05906852]
  > Dipole moment (a.u.): [0.0  0.0  0.05906852]  Total: 0.05906852
                 (debye): [0.0  0.0  0.15013718]  Total: 0.15013718
```

As you can see, our current result — compared with our previous simulation with noise and no error mitigation — is now much closer to the real ground state energy (although you have surely noticed that there is still room for improvement).

In order to run this kind of error mitigation, we need to know the probability of measuring y when we actually have $|x\rangle$ for every pair of binary strings x and y. Of course, estimating these values is computationally very expensive, because the number of strings grows exponentially with the number of qubits. Alternatively, we could assume that the readout errors are local and estimate instead the probability of obtaining an incorrect result for each individual qubit only. In Qiskit, you choose to take this approach by replacing `CompleteMeasFitter` with `TensoredMeasFitter`. However, at the time of writing, not all backends support this possibility, so you'd better be careful if you decide to use it.

To learn more...

There is much more to say about trying to mitigate the effects of noise in quantum computations. Unfortunately, studying error mitigation further would make this chapter much, much longer (and it is already fairly long!). Should you be interested in this topic, we can recommend that you check the paper by Bravyi et al. [59] to learn more about measurement error mitigation and the papers by Temme et al. [60] and by Endo et al. [61] to learn more about how to mitigate errors in general, including the ones causes by imperfect gate implementation. You may also want to take a look at Mitiq, a very easy-to-use software package for error mitigation that is compatible with Qiskit and other quantum computing libraries [62].

The techniques that we have introduced to simulate noisy devices and to mitigate readout errors are not only applicable to the VQE algorithm. In fact, noisy simulation can be used when running any circuit, because we can just use a `backend` object created with the `AerSimulator.from_backend` function and a real quantum computer.

Moreover, readout error mitigation can be used with any algorithm that uses an object of the class `QuantumInstance` to run circuits. This includes QAOA and GAS, which we studied in *Chapters 5* and *6*, respectively, as well as the QSVMs, the QNNs and the QGANs that we will study in *Chapters 9*, *10*, *11*, and *12*.

But the possibilities don't end there. In fact, every `QuantumInstance` object provides an `execute` method that receives quantum circuits and executes them. So, you can create a `QuantumInstance` with a noisy backend and the `measurement_error_mitigation_cls` argument, and then invoke `execute` to obtain results with error mitigation.

Exercise 7.9

Create a noisy backend from a real quantum computer. Then, use it to run a simple two-qubit circuit consisting of a Hadamard gate on the first qubit, a CNOT gate with control on the first qubit and target in the second, and a final measurement of both qubits. Compare the results to those of ideal simulation. Then, create a `QuantumInstance` from your backend and using readout error mitigation. Run the circuit with it. Compare the results to what you obtained before.

Exercise 7.10

Run QAOA with a simple Hamiltonian on a noisy simulator with and without readout error mitigation. Compare the results.

Now that we know how to run simulations with noise, we are ready for the next big step: let's run VQE on actual quantum devices.

7.3.6 Running VQE on quantum computers

By now, you surely have guessed what we are going to say about running VQE on quantum devices. If you were thinking that we could just use a real backend when creating our `QuantumInstance` object, but that it would involve waiting multiple queues and that there must be a better way, you were completely spot on. In fact, we can use Runtime to send our VQE jobs to IBM's quantum computers, waiting only in one execution queue. The way in which we can use VQE with Runtime is very similar to what we showed in *Section 5.2.1* for QAOA. We can use the `VQEClient` as follows:

```
from qiskit_nature.runtime import VQEClient
backend = provider.get_backend('ibmq_manila')

vqe = VQEClient(
    ansatz=ansatz,
    provider=provider,
```

```
    backend=backend,
    shots=1024,
    initial_point = initial_point,
    measurement_error_mitigation=False
)

solver = GroundStateEigensolver(qconverter, vqe)
result = solver.solve(problem)
print(result)
```

This is completely analogous to how we run VQE on local simulators, but now we are sending the task to the real quantum device called `ibmq_manila`. Notice that we have specified the number of shots and that we have opted to use the default optimizer since we haven't provided a value for the optimizer argument. The default optimizer for this algorithm is SPSA.

The results that we obtained (after waiting some time in the queue) were the following:

```
=== GROUND STATE ENERGY ===

* Electronic ground state energy (Hartree): -1.745062049527
  - computed part:      -1.745062049527
~ Nuclear repulsion energy (Hartree): 0.715104339081
> Total ground state energy (Hartree): -1.029957710446

=== MEASURED OBSERVABLES ===

  0:  # Particles: 1.988 S: 0.131 S^2: 0.149 M: -0.005

=== DIPOLE MOMENTS ===
```

```
~ Nuclear dipole moment (a.u.): [0.0  0.0  0.0]

  0:
  * Electronic dipole moment (a.u.): [0.0  0.0  0.01726618]
    - computed part:      [0.0  0.0  0.01726618]
  > Dipole moment (a.u.): [0.0  0.0  -0.01726618]  Total: 0.01726618
                 (debye): [0.0  0.0  -0.04388625]  Total: 0.04388625
```

We can observe again the effect of noise in this execution. Of course, we can try to reduce it by setting `measurement_error_mitigation=True` and running the same code again. When we did that, we obtained the following output:

```
=== GROUND STATE ENERGY ===

* Electronic ground state energy (Hartree): -1.830922842008
  - computed part:      -1.830922842008
~ Nuclear repulsion energy (Hartree): 0.715104339081
> Total ground state energy (Hartree): -1.115818502927

=== MEASURED OBSERVABLES ===

  0:  # Particles: 2.020 S: 0.035 S^2: 0.036 M: 0.010

=== DIPOLE MOMENTS ===

~ Nuclear dipole moment (a.u.): [0.0  0.0  0.0]

  0:
  * Electronic dipole moment (a.u.): [0.0  0.0  -0.00999621]
    - computed part:      [0.0  0.0  -0.00999621]
```

```
> Dipole moment (a.u.): [0.0  0.0  0.00999621]  Total: 0.00999621
                (debye): [0.0  0.0  0.02540783]  Total: 0.02540783
```

That is a little bit better, right?

With this, we have covered everything we wanted to tell you about how to run VQE with Qiskit... or almost everything. In the next subsection, we will show you some new features that are being added to Qiskit and that can change the way in which algorithms such as VQE are used.

7.3.7 The shape of things to come: the future of Qiskit

Quantum computing software libraries are in constant evolution and Qiskit is no exception to this rule. Although everything that we have studied in this section is the main way of running VQE with the latest version of Qiskit (which is 0.39.2 at the time of writing this book), a new way of executing the algorithm is also being introduced and it will likely become the default one in the not-so-distant future.

This new way of doing things involves some modifications, the most important of which is replacing the use of `QuantumInstance` objects with `Estimator` variables. An `Estimator` is an object that is capable running a parametrized circuit to obtain a quantum state and then estimate (who would have guessed?) the expectation value of some observable on that state. Of course, this is exactly what we need in order to be able to run VQE, as you surely remember from *Section 7.2*.

Let's see an example of how this would work. The following code is a possible way of running VQE to solve the same molecular problem that we have been considering throughout this section with the new implementations that are being introduced in Qiskit:

```
from qiskit.algorithms.minimum_eigensolvers import VQE
from qiskit.primitives import Estimator

estimator= Estimator()
```

```
vqe = VQE(
    ansatz=ansatz,
    optimizer=optimizer,
    initial_point=initial_point,
    estimator=estimator
)

result = vqe.compute_minimum_eigenvalue(qhamiltonian)

print(result)
```

Notice that we are importing `VQE` from `qiskit.algorithms.minimum_eigensolvers`, not from `qiskit.algorithms` as before. Also notice how the `Estimator` object has replaced the `QuantumInstance` one that we used to use.

Running these instructions will give an output like the following one (shortened here for brevity):

```
{   'aux_operators_evaluated': None,
    'cost_function_evals': 1000,
    'eigenvalue': -1.8523881060316512,
    'optimal_circuit':
    <qiskit.circuit.library.n_local.efficient_su2.EfficientSU2
    object at 0x7f92367aac90>,
    'optimal_parameters': {
        ParameterVectorElement(θ[10]): -5.6469331359894016e-05,
        ParameterVectorElement(θ[7]): -0.07317113283182797,
        ...
        ParameterVectorElement(θ[15]): 2.5406547025358206},
    'optimal_point': array(
        [-2.25566150e-01, -3.48673819e-05, 3.14159358e+00,  3.13967948e+00,
```

```
        1.74766932e+00,  2.19381131e+00, -1.17362733e-01, -7.31711328e-02,
        3.14163959e+00, -5.24406909e-05, -5.64693314e-05, -1.86976530e-03,
        1.95315840e+00,  6.62795965e-01, -1.43666055e-01,  2.54065470e+00]),
    'optimal_value': -1.8523881060316512,
    'optimizer_evals': None,
    'optimizer_result':
    <qiskit.algorithms.optimizers.optimizer.OptimizerResult
    object at 0x7f9240af82d0>,
    'optimizer_time': 6.93215799331665}
```

This should sound familiar because it is the same kind of result that we obtained when using the current VQE implementation directly on a Hamiltonian.

As you can see, things are not going to change a lot in this new version. The main novelty is the use of `Estimator` objects. So, how do they work? Well, it depends. For instance, the one that we have imported from `qiskit.primitives` uses a state vector simulator to obtain a quantum state from a circuit. Then, it computes its expectation value by calling the `expectation_value` method, as we did back in *Section 3.2.2*. However, the `Estimator` class implemented in `qiskit_aer.primitives` uses the method that we explained in *Section 7.1.2* by appending additional gates to the parametrized circuit in order to perform measurements in different bases.

Unfortunately, at the time of writing this book, some of the features that we have covered in this section, such as noisy simulation and error mitigation, are still not completely supported by the new version of the algorithms. Moreover, some of the `Estimator` classes are not fully compatible with the new VQE implementation yet.

However, Qiskit changes rapidly, so maybe, in the future, you can fully reproduce our code with `Estimator` objects instead of `QuantumInstance` ones by the time you will be reading these lines. Time will tell!

Important note

The changes that we have described in this subsection are expected to also affect other algorithms implemented in Qiskit, such as VQD or QAOA. In the case of QAOA, instead of `Estimator` objects, you will need to use `Sampler` objects. As you can imagine, they will let you obtain samples from parametrized circuits, which can later be used by QAOA to estimate the value of the cost function.

And now, we promise, this is really all we wanted to tell you about running VQE with Qiskit. Our next stop is PennyLane.

7.4 Using VQE with PennyLane

In this section, we will illustrate how to run VQE with PennyLane. The problem that we will work with will be, again, finding the ground state of the dihydrogen molecule. This is a task we are already familiar with and, moreover, this will allow us to compare our results with those that we obtained with Qiskit in the previous section. So, without further ado, let's start by showing how to define the problem in PennyLane.

7.4.1 Defining a molecular problem in PennyLane

As with Qiskit, PennyLane provides methods to work with quantum chemistry problems. To study the H_2 molecule, we can use the following instructions:

```
import pennylane as qml
from pennylane import numpy as np

seed = 1234
np.random.seed(seed)

symbols = ["H", "H"]
coordinates = np.array([0.0, 0.0, -0.6991986158, 0.0, 0.0, 0.6991986158])
```

```
H, qubits = qml.qchem.molecular_hamiltonian(symbols, coordinates)
print("Qubit Hamiltonian: ")
print(H)
```

You may be thinking that there is something fishy here. When we defined this same molecule in Qiskit, we used coordinates `[0., 0., -0.37],[0., 0., 0.37]`, which seem different from the ones that we are using now. The explanation for this change is that, while Qiskit uses angstroms to measure atomic distances, PennyLane expects the values to be in atomic units. An angstrom is worth 1.8897259886 atomic units, hence the difference.

We can now obtain the qubit Hamiltonian that we need to use with VQE by running the following piece of code:

```
H, qubits = qml.qchem.molecular_hamiltonian(symbols, coordinates)
print("Qubit Hamiltonian: ")
print(H)
```

The output that we obtain is the following:

```
Qubit Hamiltonian:
  (-0.22343155727095726) [Z2]
+ (-0.22343155727095726) [Z3]
+ (-0.09706620778626623) [I0]
+ (0.17141283498167342) [Z1]
+ (0.1714128349816736) [Z0]
+ (0.12062523781179485) [Z0 Z2]
+ (0.12062523781179485) [Z1 Z3]
+ (0.16592785242008765) [Z0 Z3]
+ (0.16592785242008765) [Z1 Z2]
+ (0.16868898461469894) [Z0 Z1]
+ (0.17441287780052514) [Z2 Z3]
+ (-0.04530261460829278) [Y0 Y1 X2 X3]
```

```
+ (-0.04530261460829278) [X0 X1 Y2 Y3]
+ (0.04530261460829278) [Y0 X1 X2 Y3]
+ (0.04530261460829278) [X0 Y1 Y2 X3]
```

If you compare this Hamiltonian to the one that we obtained with Qiskit for the same problem you will notice that they are very different. But don't panic yet. While Qiskit gave us the Hamiltonian for the electronic structure of the molecule, PennyLane is accounting for the total energy, including nuclear repulsion. We will run the algorithm in a moment and, trust us, we will see how everything adds up.

7.4.2 Implementing and running VQE

Before using VQE, we need to decide what variational form we are going to use as the ansatz. To keep things simple, we will stick with the `EfficientSU2` that we used in the previous section.

This variational form is not included in PennyLane at the time of writing this book, but we can easily implement it with the following code:

```
nqubits = 4
def EfficientSU2(theta):
    for i in range(nqubits):
        qml.RY(theta[i], wires = i)
        qml.RZ(theta[i+nqubits], wires = i)
    for i in range(nqubits-1):
        qml.CNOT(wires = [i, i + 1])
    for i in range(nqubits):
        qml.RY(theta[i+2*nqubits], wires = i)
        qml.RZ(theta[i+3*nqubits], wires = i)
```

Notice that we have fixed the number of repetitions to 1, which was the case that we were using with Qiskit in the previous section.

Now that we have our variational form, we can use it to implement the VQE algorithm in PennyLane. To do that, we will define the energy function, which we compile as a quantum node because it needs to be evaluated on a device capable of running quantum circuits. We can do that with the following instructions:

```
dev = qml.device("lightning.qubit", wires=qubits)
@qml.qnode(dev)
def energy(param):
    EfficientSU2(param)
    return qml.expval(H)
```

Notice how we have used the `EfficientSU2` ansatz followed by an evaluation of the expectation value of our Hamiltonian (by using the `qml.expval` function that we introduced in *Chapter 5, QAOA: Quantum Approximate Optimization Algorithm*). Now, to execute VQE, we only need to select some initial values for the ansatz parameters and use a minimizer to find their optimal values. We can achieve that with the following piece of code:

```
from scipy.optimize import minimize

theta = np.array(np.random.random(4*nqubits), requires_grad=True)
result = minimize(energy, x0=theta)

print("Optimal parameters", result.x)
print("Energy", result.fun)
```

We have imported the `minimize` function from the `scipy.optimize` package (scipy is a powerful and very popular Python library for scientific computing). We have chosen at random some initial values for the variational form parameters. We have used `requires_grad=True` to allow the minimizer to compute gradients in order to optimize the parameters (we will have much more to say about this in *Part 3* of the book). Then, we have minimized the `energy` function using the default parameters of the `minimize` method. Notice how the `x0` argument is used to specify the initial values.

The result we obtain upon running this code is the following:

```
Optimal parameters
[ 2.25573385e-01  3.14158133e+00  1.91103424e-07 -1.88149577e-06
 -2.71613763e-03 -7.94107899e-01  4.52510610e-01  6.17686238e-01
  3.14158772e+00  6.28319382e+00  3.14158403e+00  3.14160984e+00
  2.21495304e-01  5.01302639e-01  6.51839333e-01  7.36625551e-02]
Energy -1.137283835001276
```

This includes the optimal values found by the optimizer (the x field) as well as the minimum energy. As you can check, this fits nicely with the results that we have obtained with Qiskit for the total molecular energy.

Now that we know how to run VQE on a simulator with PennyLane, we will turn to the task of executing the algorithm on actual quantum computers.

7.4.3 Running VQE on real quantum devices

You may remember that, back in *Section 5.3*, we mentioned that there is a PennyLane Runtime client that can be used to run VQE programs. This is exactly what we need now, so it is the perfect moment to learn how to use it.

In fact, using this Runtime implementation is very easy, because it is quite similar to the one we used with Qiskit. First, we need be sure that we have pennylane_qiskit installed and our IBM Quantum account enabled (see *Appendix D, Installing the Tools*, for directions). Then, we can run the following instructions:

```
from pennylane_qiskit import upload_vqe_runner, vqe_runner

program_id = upload_vqe_runner(hub="ibm-q", group="open", project="main")

job = vqe_runner(
    program_id=program_id,
    backend="ibm_oslo",
```

```
    hamiltonian=H,
    ansatz=EfficientSU2,
    x0=np.array(np.random.random(4*nqubits)),
    shots=1024,
    optimizer="SPSA",
    kwargs={"hub": "ibm-q", "group": "open", "project": "main"}
)
print(job.result())
```

The code is pretty much self-explanatory: we are just selecting the options for our VQE execution, including the device to run the circuits which, in this case, happens to be `ibm_oslo`. After waiting for the job to finish running, we will obtain an output similar the following:

```
     fun: -1.0125211856761642
 message: 'Optimization terminated successfully.'
    nfev: 300
     nit: 100
 success: True
       x: array([-0.02558326,  0.50137847,  1.49781722,  2.83016638,
       1.50688742, -0.00830098,  1.56006908, -0.01401641, -0.08208851,
       2.71490414, 1.39380584,  1.30662208,  1.5691855 ,  1.34979806,
       1.50345895, 0.39946571])
```

You may be wondering if we can also use error mitigation to try to improve our results. The answer is yes, of course. In fact, it is straightforward to set up, because we only need to include the additional parameter `use_measurement_mitigation = True` when creating the `vqe_runner` object. Running with this option will give you a result similar to the following one, which is closer to the real optimal value:

```
     fun: -1.0835711819668128
 message: 'Optimization terminated successfully.'
```

```
   nfev: 300
    nit: 100
success: True
      x: array([-0.06213913,  2.62825807,  2.85476345, -0.2260965,
      -0.07639407, -1.51018602,  1.73431192, -0.07301669, -0.16907148,
      2.60134032, 3.29831133, -0.2912491 ,  0.33893055,  1.90085806,
      1.7206114 , -1.49009082])
```

With this, we conclude our study of VQE and, in fact, we conclude the part of the book devoted to optimization problems. Starting with the next chapter, we will dive into the fascinating world of quantum machine learning. Hang tight and prepare for the ride!

Summary

In this chapter, we have studied Hamiltonians and observables in detail. In particular, we have learned how to derive mathematical expressions for their expectation values and how to estimate these quantities using quantum computers.

Then, we studied the VQE algorithm and how it can be used to find ground states of general Hamiltonians. We also described a modification of VQE called VQD that can also be used to compute excited states and not just states of minimum energy.

Then, we moved to practical matters and learned how to use Qiskit to run VQE and VQD. We illustrated this with a very interesting problem: that of finding the ground state of a simple molecule. We then introduced methods to simulate the behavior of quantum algorithms when there is noise and how to reduce the adverse effect of readout errors with some mitigation techniques. We also studied how to run VQE problems on actual quantum computers with IBM runtime.

After that, we also learned how to implement and run VQE on PennyLane, again solving a molecular structure problem. We even studied how to use Runtime from PennyLane to send VQE instances to real quantum computers.

After reading this chapter, you are now able to understand all the mathematical details behind the VQE algorithm. You also know how to run it on different types of problems with both Qiskit and PennyLane. Moreover, you now can run noisy simulations of all the algorithms that we have studied (and of any other quantum algorithm that you may learn in the future) as well as perform readout error mitigation on simulated and actual quantum devices.

In the next chapter, we will start studying the second big topic of the book: quantum machine learning. Prepare to learn how (quantum) machines learn!

Part 3

A Match Made in Heaven: Quantum Machine Learning

This part is devoted to studying different quantum machine learning models. You will learn how to implement and run Quantum Support Vector Machines and Quantum Neural Networks, and how to combine them with classical models to obtain interesting architectures, such as Quantum Generative Adversarial Networks.

This part includes the following chapters:

- *Chapter 8, What is Quantum Machine Learning?*
- *Chapter 9, Quantum Support Vector Machines*
- *Chapter 10, Quantum Neural Networks*
- *Chapter 11, The Best of Both Worlds: Hybrid Architectures*
- *Chapter 12, Quantum Generative Adversarial Networks*

8

What Is Quantum Machine Learning?

Tell me and I forget. Teach me and I remember. Involve me and I learn.

— Benjamin Franklin

We now begin our journey through **Quantum Machine Learning** (**QML**). In this chapter, we will set the foundation for the remainder of this part of the book. We will begin by reviewing some general notions from classical machine learning, and then we will introduce the basic ideas that underlie QML as a whole.

We'll cover the following topics in this chapter:

- The basics of machine learning
- Do you wanna train a model?
- Quantum-classical models

In this chapter, you will learn the basic principles behind general machine learning, and you will understand how to construct, train, and assess some simple classical models using industry-standard frameworks and tools. We will also present a general picture of the world of QML.

8.1 The basics of machine learning

Before discussing quantum machine learning, it may be a good idea to review some basic notions of Machine Learning (ML), in general. If you are familiar with the subject, feel free to skip this section. Please, keep in mind that the world of machine learning is extraordinarily vast; so much so that sometimes it is difficult to make general statements that can do justice to the overwhelming diversity of this field. For this reason, we will highlight those elements that will be more relevant for our purposes, while other aspects of machine learning — which are also of significant importance on their own — will be barely covered.

In addition to this, please keep in mind that this will be a very condensed and hands-on introduction to machine learning. If you'd like to dive deeper into this field, we can recommend some very good books, such as the one by Abu-Mostafa, Magdon-Ismail, and Lin [63], or the one by Aurélien Géron [64].

As mysterious as machine learning may seem, the ideas that underlie it are fairly simple. In broad terms, we could define the purpose of machine learning to be the design of algorithms that can make a "computational system" aware of patterns in data. These patterns can be truly anything. Maybe you want to design a system that can distinguish pictures of cats from pictures of rabbits. Maybe you would like to come up with a computational model that can transcribe verbal speech in English spoken with an Irish accent. Maybe you want to create a model able to generate realistic pictures of faces of people who do not exist, but that are indistinguishable from the real deal. The possibilities, as you surely know, are endless! What will be common to all these algorithms is that they will not be explicitly programmed to solve those tasks; instead, they will "learn" how to do it from data... hence the name "machine learning!"

The cats versus rabbits example is a particular case of an interesting type of model: a **classifier**. As the name suggests, a classifier is any kind of system that assigns, to every input, one label out of a finite set of possibilities. In many cases, there are just two of these labels, and they are commonly represented by 0 (read as **positive**) and 1 (**negative**); in physics applications, for instance, these labels are often read, respectively, as **signal** and **background**. In this situation, we say that the classifier is **binary**. Keep this in mind, for we will use this terminology in a few of the coming examples!

So now that we know where we are heading, we need to answer one basic question: how can we make machine learning a reality?

8.1.1 The ingredients for machine learning

In most machine learning setups, there are three basic ingredients:

- First of all, we need some sort of computational model that is "powerful enough" to tackle our problem. By this, we will often mean an algorithm that can be configured to solve the task at hand — at least to some level of accuracy.
- Then, if we want our model to capture patterns, we need to feed it some data so that it can do that. We will thus need data, preferably lots of it. The nature of this data will depend on the approach that we take, but, in most cases, we will need to transform it into numerical form. Most models expect data to be represented as vectors of real numbers called **attributes**, so this is what we will usually assume that we have.
- And, lastly, we need a **training procedure** that will allow us to optimize the configuration of our model to make it solve the task (or, at least, come close to solving it!). In ML jargon, we could say that we need to find a way to make our model **learn** in order to identify the patterns that hide behind the data in our problem.

That is a pretty solid — yet somewhat oversimplified — wish-list. Let's see how we can make more sense out of this.

The model

Let us first analyze that computational model that we have talked about. We said that it had to be "powerful enough," and this means that there should be a way to configure the model in such a way that it behaves as we intend it to.

At first sight, this requirement may seem suspicious: how can we possibly be sure that such a configuration exists? In most real-life problems, we can never be fully sure...but we can be certain to some degree! This certainty may come from experience or, desirably, also from some theoretical results that justify it. For instance, you may have heard of **neural networks**. We will discuss them shortly, but, for now, you should know that they are models that have been proven to be **universal function approximators**. That is, any function can be approximated up to any given accuracy, no matter its complexity, by a large-enough neural network. That makes neural networks natural good choices for many problems.

We will later discuss neural networks — and many other interesting models — in detail, but, to start with, we can consider a simplified version that, in fact, could be considered the grandparent of neural networks: the **perceptron**.

A perceptron is a computational model that takes N numerical inputs and returns a single bit as output. This model depends on a collection of weights w_i for $i = 1, \ldots, N$ and on a bias b, and it behaves as follows: for any input $x_1, \ldots, x_N$, if

$$\sum_{i=1}^{N} x_i w_i + b \geq 0,$$

then the model returns 1, and otherwise it returns 0.

This is a very simple computational model, but we could use it to set up a basic binary classifier by looking for some appropriate values for the weights and bias. That is, given a set of points on which we want the output to be 1 and another set of points on which the output should be 0, we can try to search for some values for the w_i and b coefficients that would make the perceptron return the desired outputs. In fact, in the dawn of the machine learning age, it was already proven that there is a simple learning algorithm that, under the condition that the problem data can be linearly separated, finds coefficients that can effectively classify the data.

There you have it, that could be your first baby machine learning model! Needless to say, a perceptron – at least on its own – is not a particularly powerful model, but it is, at least, a promising beginning!

Exercise 8.1

We can all agree that perceptrons are cute models. But just to get an idea of their limitations, prove that they cannot implement an XOR gate. That is, if you are given inputs $\{(0, 1), (1, 0)\}$ with desired output 1 and inputs $\{(0, 0), (1, 1)\}$ with desired output 0, there is no choice of perceptron weights and bias that works in this case.

The training procedure

Alright, so let's say that we have a model that we believe is just powerful enough to approach our problem (and not too powerful either...more on that later!). We will restrict ourselves to assuming that the configuration of our model depends on some numerical parameters θ; this would mean that we would be looking for some choice θ_0 of those parameters that will make our model work as well as possible. So, how do we find those parameters?

To learn more...

We will only discuss models whose behavior can be adjusted and defined solely by tweaking some numerical parameters, as in the case of the perceptron. Nevertheless, there also exist **non-parametric** models that don't behave in this manner. A popular example is the k-nearest neighbours algorithm; you can find some information in the references [64, Chapter 1].

To illustrate all of this, we will discuss how to train a parametric model to implement a binary classifier. That is, we aim to build a binary classifier on a certain domain D with some elements x that should be each classified as a certain y (where y can be either 0 or 1). For this, we will use a model M that depends on some parameters in a way that, for any choice θ of these parameters, it returns a label $M_\theta(x)$ for every element $x \in D$ in the dataset.

In this scenario, our goal is to look for a choice of parameters θ that can minimize the probability that any random input x be misclassified. To put it in slightly more formal terms, if y is the correct label to be assigned to an input x, we want to minimize $P(M_\theta(x) \neq y)$, that is, the probability of assigning an incorrect label to x. In this way, we have reduced the problem of training our model to the problem of finding some parameters θ that minimize $P(M_\theta(x) \neq y)$. This probability is known as the **generalization error** or the **true error**.

Now, if we had access to all the possible inputs in our domain D and we knew all their expected outputs, we would simply have to minimize the true error... and we would be done! Nevertheless, this is neither an interesting situation nor a common one.

If we had a problem in which we already knew all the possible inputs and their outputs... why should we bother with all this machine learning business? We could just implement an old-school algorithm! Indeed, the whole point of "learning" is being able to predict correct outputs for unseen data. Thus, when we resort to machine learning, we do so because we do not have full access to all the possible inputs and outputs in our domain — either because it is unfeasible or because such a domain might be infinite!

So now we are faced with a problem. We have a (potentially infinite) domain of data over which we have to minimize the true error, yet we only have access to a finite subset of it. But... how on earth can we compute the true error in order to minimize it? The answer is that, in general, we can't, because we would need complete information on how all the data and the labels of our problem are distributed, something that we usually don't have. Nevertheless, we still have access to a — presumably large — subset of data. Can we use it to our advantage? Yes, we surely can! The usual strategy is to divide the dataset that we have in two separate sets: a **training dataset** and a **test dataset**. The training set, usually much bigger than the test set, will be used to adjust the parameters of our model in an attempt to minimize the true error, while the test set will be used to estimate the true error itself.

Thus, what we can do is just take whichever training dataset T we are using, and – instead of minimizing the true error, to which we simply don't have access – we can try to minimize the **empirical error**: the probability of misclassifying an element within the training dataset. This empirical error would be computed as the proportion of misclassified elements in T:

$$R_{\text{emp}}(\theta) = \frac{1}{|T|} \sum_{(x,y)\in T} 1 - \delta(M_\theta(x), y),$$

where $|T|$ is the size of the training dataset and $\delta(a, b)$ is 1 if $a = b$ and 0 otherwise (this δ function is known as the Kronecker delta). We would do all of this, of course, hoping that the real error would take similar values to the empirical error. Naturally, this hope will have to be justified and rest on some evidence, and we will soon see how the test dataset can help us with that. In any case, if we want all this setup to work, we will need to use a large enough dataset.

Our goal then is to minimize the true error, and, so far, our only strategy is trying to achieve it by minimizing the empirical error. Nevertheless, in practice, we don't often work with these magnitudes directly. Instead, we take a more "general" approach: we seek to minimize the expected value of a **loss function**, which is defined for every choice of parameters θ and every possible input x and its desired output y. For instance, we could define the 0-1 loss function as

$$L_{01}(\theta; x, y) = 1 - \delta(M_\theta(x), y).$$

With this definition, it is trivial to see that the expected value, taken over the whole domain, of L_{01} is exactly the true error; this expected value is known as the **true risk**. In the same way, the expected value of this loss function over the training sample is the empirical error.

Important note

Keep in mind that the expected value of a loss function over a finite dataset will just be its average value.

So, in practice, our strategy for minimizing the true error will be minimizing the expected value of a suitable loss function over the training dataset. We will refer to this expected value as the **empirical risk**. For reasons that we will discuss later, we will usually consider loss functions different from L_{01}.

Assessing a trained model

We now have to address an important question. How can we be sure that – once we have trained a model – it will perform well on data outside the training dataset? For that, we cannot solely rely on $R_{\text{emp}}(\theta)$ because that average loss is computed on data that the classifier has already seen. That would be like testing a student only on problems that the teacher has already solved in class!

Thankfully, there's something that can save the day. Do you remember that test dataset we talked about before? This is its time to shine! In fact, we have kept this test set in a safe-deposit box to ensure that none of its examples were ever used in the training process. We can think of them as completely new problems that the student has never seen, so we can use them to assess their understanding of the subject. Thus, we can compute the average loss of M_{θ} on the examples of the test set – this is sometimes called the **test error**. Provided that they are representative of the classification problem as a whole and that the number of examples is big enough, we can be quite confident that the test error will be close to the true error of the model. This is just an application of some *central* theorems in statistics!

Now, if the test error is similar to the empirical risk (and if they are low enough), we are done. That's it! We have successfully trained a model. Nonetheless, as you can imagine, things can also go wrong. Terribly wrong.

What if the test error is much bigger than the empirical error, the one computed on the training set? This would be similar to having a student who knows how to repeat the solution to problems already solved by the teacher but who is unable to solve new problems. In our case, this would mean having a classifier that works beautifully on the training dataset but makes a lot of errors on any inputs outside of it.

This situation is called **overfitting**, and it is one of the biggest risks (no pun intended) in machine learning. It occurs whenever, somehow, our model has learned the particularities of the data it has seen but not the general patterns; that is, it has fitted the training data too well, hence the name "overfitting." This problem usually occurs when the training dataset is too small or when the model is too powerful. In the first case, there is simply not enough information to extract general patterns. That is why, in this chapter, we have insisted that the more data we have, the better. But what about the second case? Why can having a very powerful model end up being something bad?

An example can be very illustrative here. Let's say that we want to use machine learning to approximate some unknown real function. We haven't discussed how this setup would work, but the core ideas would be analogous to the ones we have seen (we would seek to minimize the expected value of a loss function, and so on). If we have a sample of 1000 points in the plane, we can always find a polynomial of degree 999 that fits the data perfectly, in the same way that we can always fit a line to just two points. However, if the points are just samples of $f(x) = x$ with some noise (which could result from some empirical sampling errors or some other reason), our polynomial will go out of its way to fit those points perfectly and will quickly deviate from the linear shape that it should have learned. In this way, being able to fit too much information can sometimes go against the goal of learning the general patterns of data. This is illustrated in *Figure 8.1*. In it, the degree of the "fitting" polynomial is so big that it can fit the training data perfectly, including its noise, but it misses the implicit linear pattern and performs very badly on test data.

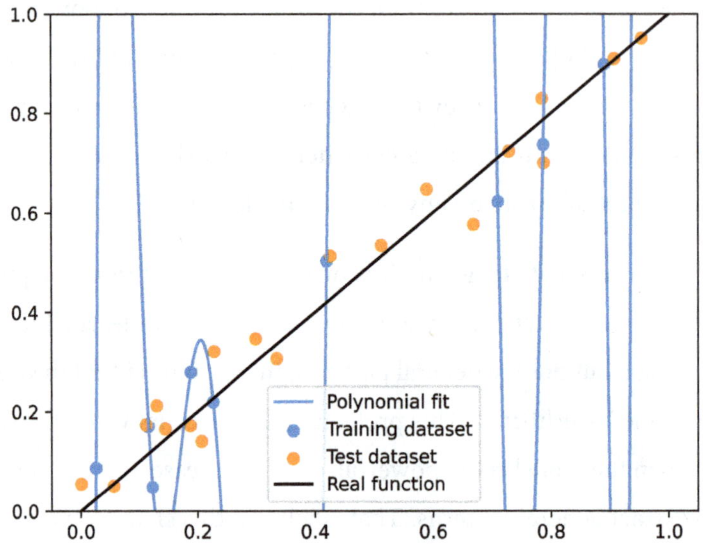

Figure 8.1: A simple example of overfitting that results from using too powerful a model.

Important note

Sometimes a machine learning model may only work properly on its training dataset. This phenomenon is known as **overfitting**. It usually occurs when the training dataset is too small or when the model is too powerful.

If you find yourself in a situation in which your model has overfitted the data, you can try obtaining more data — something that is not always possible — or somehow reducing the power of your model. For instance, with the neural networks that we will be studying later in the chapter, you can try reducing their size.

To learn more...

Another popular technique for avoiding overfitting is using **regularization**. Roughly speaking, regularization restricts the values that some of the parameters of your model can take, effectively making it less powerful and less prone to fit every single detail of the training data.

> To learn more about regularization techniques and their use in machine learning, we highly recommend checking the book by Aurélien Géron [64].

You may also want to know that your models can exhibit a type of problem that is the opposite of overfitting and has been aptly named **underfitting**. If your model is not expressive enough, you can find yourself with both a high error rate on the training set and a high error rate on the test set. For instance, if you are using a linear function to try to fit points that come from a quadratic polynomial and, thus, follow a parabolic shape, you will surely experience some form of underfitting. To fix this problem, use a more powerful model — or reduce regularization if you happen to be using it.

To summarize what we have discussed so far, remember that we want to obtain a model that has a low generalization error; that is, a model that works well even on data it has not been trained with. In order to achieve this, we consider a parametric model and look for those model parameters that minimize the error on the training set, because we cannot easily compute the true error. And to be sure that the model will behave well when confronted with new data, we compute the error on the test dataset as a way of assessing how representative the empirical risk is of the error on unseen data.

With this strategy, however, we may still be vulnerable to an additional problem. If we train a lot of different models, there is a risk that — just by pure chance! — one of them has great performance on the test dataset but not on the rest of the domain. In fact, this risk is higher the more models you train. Imagine that a thousand students take a test of 10 questions with 2 possible answers each. Even if they have not studied for the test and they answer completely at random, there is a very high probability that at least one of them will nail it. For this reason, you should never use the test dataset to select among your models, only to assess if their behavior is similar to the behavior they show during training.

This is definitely a problem because we usually want to train many different models and select the one we believe to be the best. What is more, many models have what are called **hyperparameters**. These are parameters that fix some property of the model, such as the size and number of layers in a neural network (more on that later), that cannot be optimized during training. Usually, we train many different models with different values of these hyperparameters, and then we select the best model from them.

This is where a third type of dataset comes into the equation: the **validation dataset**. This is an additional dataset that we could construct when splitting our global dataset; it should, of course, be fully independent of the training and test datasets.

What do we want the validation set for? Once we have trained our models with different choices of hyperparameters and configurations, we can compute the empirical risk on the validation set, and we may select the best one or maybe a handful of the best ones. Then, we could train those models again on the union of the training set and the validation set – to better extract all the information from our data – and then compute the error of the models on the test set, which we have held back until this very moment so that it remains a good estimator of the generalization error. In this way, we can select the best choice of hyperparameters or models while keeping the test dataset in pristine condition to be used in a final assessment process.

To learn more...

You may also want to know that, instead of using a fixed validation set, a popular way of selecting hyperparameters is to use ***k*-fold cross-validation**. With this technique, the training dataset is divided into k subsets or **folds** of equal size. The training of the model is repeated k times, each one with a different subset acting as a validation dataset and the rest used as the training dataset. The performance is computed over each validation set and averaged over the k repetitions. Of course, the estimation obtained with cross-validation is better than when using a fixed validation set, but the computational cost is much higher – k times higher, in fact! Software libraries such as scikit-learn – which we will be using in the next section of this chapter – provide implementations of cross-validation for hyperparameter selection. Take a look at the documentation of `GridSearchCV` – where CV stands for cross validation – if you want to see a concrete implementation of this technique.

Furthermore, sometimes training processes are iterative. In these cases, the **validation loss** (the average loss over the validation dataset) can be computed at the end of each iteration and compared against the training loss, just to know how the training is going – and to be able to stop it early should the model begin to overfit! It wouldn't be a good practice to use the test set for this purpose: the test set should only be used once all the training is complete and we just want some reassurance on the validity of our results.

To learn more...

All the informal notions that we are considering here can be formulated precisely using the language of probability theory. If you want to learn about the formal machinery behind machine learning, you should have a look at the book *Learning from data* [63] or at *Understanding machine learning* [65].

With all of this, we now have a good understanding of all the elements needed for machine learning to come alive. In the following section, we will try to make all of this more precise by studying some of the most common approaches that are taken when training ML models.

8.1.2 Types of machine learning

There are three big categories in which most, if not all, machine learning techniques can fit: **supervised learning**, **unsupervised learning**, and **reinforcement learning**. In this book, we will work mostly with supervised learning, but we will also consider some unsupervised learning techniques. Let's explain in a little bit more detail what each of these machine learning branches is about.

Supervised learning

The main goal of supervised learning is to learn to predict the values of a function on input data. These values can either be chosen from a finite set (the classification problems we have been talking about for the most part of this chapter) or be continuous values, such as the weight of a person or the value of some bonds a month from now. When the values we want to predict are continuous, we say that we are tackling a **regression** problem.

When we train a model using supervised learning, we need to work with a dataset that has both a large-enough collection of valid inputs and all the expected outputs that our model should return for these inputs. This is known as having a **labeled** dataset.

For example, if we were to train a cat-rabbit picture classifier using supervised learning, we would need to have a (large-enough) dataset with pictures of rabbits and cats, and we would also need to know, for each of those pictures, whether they are pictures of rabbits or pictures of cats.

With our labeled dataset, we would define a loss function that would depend on the inputs and the parameters of the model — so that we can compute the corresponding outputs — and the expected (correct) outputs. And, as we discussed before, then we would just apply an optimization algorithm to find a configuration of the model that could minimize the loss function on the training dataset — while ensuring that there is no overfitting.

We still have to discuss how that optimization algorithm is going to work, but that is for later!

Unsupervised learning

When we work with unsupervised learning, we have access to **unlabeled** datasets, in which there are no expected outputs. We let the algorithm learn on its own by trying to identify certain patterns. For instance, we may want to group similar data points together (this is known as **clustering**) or we may want to learn something about how the data is distributed.

In this latter case, our goal would be to train a **generative model** that we can use to create new data samples. An impressive example is the use of **Generative Adversarial Networks**, introduced by Ian Goodfellow and his collaborators in a highly influential paper [66] to create images that are similar — but completely different — to the ones used in the training phase. This is the kind of model that we will be working with in *Chapter 12, Quantum Generative Adversarial Networks,*...in a quantum form, of course!

Reinforcement learning

In reinforcement learning, the model — usually called the **agent** in this setting — interacts with an **environment**, trying to complete some task. This agent observes the **state** of the environment and takes some **actions** that in turn influence the state it observes. Depending on its performance, it receives "rewards" and "punishments"... and, of course, it wants to maximize the rewards while minimizing the punishments. To do that, it tries to learn a **policy** that determines what action to take for a given state of the environment.

For instance, the agent may be a robot and the environment a maze it needs to navigate. The state can consist of its position in the maze and the open paths it can follow, and its actions can be rotating in some direction and moving forward. The goal may be finding the exit to the maze in some predefined time, for which the robot will get a positive reward.

This kind of learning has been used extensively to train models designed to play games — AlphaGo, the computer program that in 2016 beat Go (human) grandmaster Lee Sedol in a five-games match, is a prominent example! To learn more about reinforcement learning, a good source is the book by Sutton and Barto [67].

Although there has been some interest in using quantum techniques in reinforcement learning (see, for instance [68]), this may very well be the machine learning branch in which quantum algorithms are less developed at the moment. For this reason, we will not cover this kind of learning in this book. Hopefully, in a few years there will be much more to tell about quantum reinforcement learning! Let's now try to make everything concrete by using supervised learning to implement a very simple classifier. For this, we will use TensorFlow.

8.2 Do you wanna train a model?

TensorFlow is a machine learning framework developed at Google, and it is very widely used. You should refer to *Appendix D, Installing the Tools*, for installation instructions. Keep in mind that we will be using version 2.9.1. We will use TensorFlow in some of our quantum machine learning models, so it is a good idea to become familiar with it early on.

To keep things simple, we will tackle an artificial problem. We are going to prepare a dataset of elements belonging to one of two possible categories, and we will try to use machine learning to construct a classifier that can distinguish to which category any given input belongs.

Before we do anything, let us quickly import NumPy and set a seed:

```
import numpy as np
seed = 128
np.random.seed(seed)
```

We will later use this same seed with TensorFlow. And now, let's generate the data!

Instead of generating a dataset by hand, we will use a function provided by the Python **scikit-learn** package (`sklearn`). This package is a very valuable resource for machine learning: not only does it include plenty of useful tools for everyday machine-learning-related tasks, but it also allows you to train and execute a wide collection of interesting models! We will use version 1.0.2 of `sklearn` and, as always, you should refer to *Appendix D, Installing the Tools*, for installation instructions.

In order to generate our dataset, we will use the `make_classification` function from `sklearn.datasets`. We will ask it to generate 2500 samples of a dataset with two features (variables). We will also ask for both features to be **informative** and not redundant; the variables would be redundant, for example, if one of them were just a multiple of the other. Lastly, we will ask for the proportions of the two categories in the dataset to be 20 % to 80 %. We can do this as follows:

```
from sklearn.datasets import make_classification

data, labels = make_classification(n_samples = 2500,
    n_features = 2, n_informative = 2, n_redundant = 0,
    weights = (0.2,0.8), class_sep = 0.5, random_state = seed)
```

The `class_sep` argument specifies how separable we want the two categories to be: the higher the value of this argument, the easier it is to distinguish them. Notice, also, that we have used the seed that we set earlier in order for the results to be repeatable.

You may now be wondering why we have specified that we want the two categories in the dataset to be in a proportion 20 % to 80 %, when it would be much more natural for the two categories to be balanced. Indeed, it is desirable for both categories to have the same number of representatives in a dataset...but life is difficult, and in many practical scenarios, that is not a possibility! So just think of this choice of ours as our own little way of feeling closer to real life.

Essentially, the `make_classification` function has returned an array `data` with the whole dataset (including all the elements from both categories, positive and negative), and an array `labels` such that the label of `data[i]` will be `labels[i]` (where 0 corresponds to positive and 1 to negative).

Just to get a feeling of what this dataset that we have created looks like, we can plot a simple histogram showing the distributions of the two features of our dataset:

```
import matplotlib.pyplot as plt


for i in range(2):
    plt.hist(data[:,i][labels == 1], bins=100, alpha=0.8, label = "Negative")
    plt.hist(data[:,i][labels == 0], bins=100, alpha=0.8, label = "Positive")
    plt.legend()
    plt.show()
```

Upon running this, we got the plots shown in *Figure 8.2*.

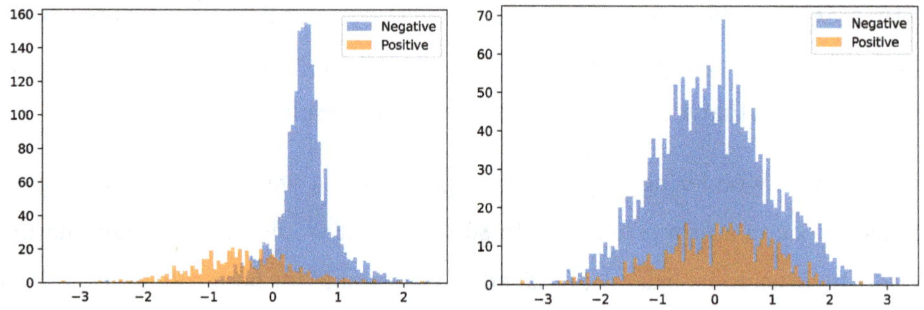

Figure 8.2: Histograms representing the distributions of the two features of our dataset.

Exercise 8.2

Visualizing the data you are working with through graphs can help you gain insights into how to approach the problem you have at hand. We have plotted our data using a histogram, which is usually a good choice. What other representations could we have used?

Our goal now is to use machine learning to come up with a system that can solve the classification problem that we have created. And the first step in doing so will be to pick a good model to tackle our problem!

8.2.1 Picking a model

Not long ago, we introduced the perceptron and we showed how, *on its own*, it wasn't the most powerful of models out there. We will now shed some light on why we emphasized "on its own," for we are about to introduce a very interesting model that can be thought of as being built by joining perceptrons together. Let's dive into **neural networks**!

You may remember how a perceptron took N numerical inputs x_i, used on a collection of N weights w_i and a bias b, and returned an output that depended on the value of

$$\sum_{i=1}^{N} w_i x_i + b.$$

Well, in this way, we can think of a neural network as being a collection of perceptrons – which we will, from now on, call **neurons** – organized in the following way:

- All the neurons are arranged into layers, and the output of the neurons in one layer is the input of the neurons in the next layer
- In addition to this, the "raw" linear output of every neuron will go through a (very possibly non-linear) **activation function** of our choice

That is the general idea, but let's now make it precise.

A neural network with N_0 inputs is defined from the following elements:

- An ordered sequence of **layers** ($l = 1, \ldots, L$), each with a fixed amount of **neurons** N_l.
- A bunch of **activation functions** h_{ln} for each neuron n in a layer l.
- A set of **biases** b_{ln} for every neuron, and, for every neuron n in a layer l, a set of N_{l-1} **weights** w_{kln} with $k = 1, \ldots, N_{l-1}$. These biases and weights are the adjustable parameters that we would need to tweak in order to get the model to behave as we want it to.

In *Figure 8.3*, we can see a graphical representation of a simple neural network.

These are the ingredients that we need to set up a neural network. So, how does it work, then? Easy! For any choice of activation functions h_{ln}, biases b_{ln} and weights w_{kln}, the neural network takes some numerical inputs a_{0n} and, from there on, these inputs are propagated through the layers of the neural network in the following way: the values a_{ln} of the neurons n in all layers l are determined according to the inductive formula

$$a_{ln} := h_{ln}\left(b_{ln} + \sum_{k=1}^{N_{l-1}} w_{kln} a_{l-1,k}\right).$$

With this procedure, we can assign a value to each neuron in the network. The values of the neurons in the last layer are the output of the model.

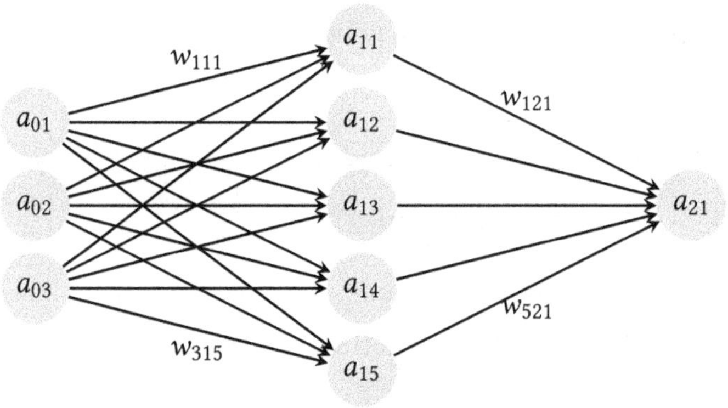

Figure 8.3: A simple neural network with two layers taking three inputs (a_{0n}). We have labeled some of the weights, but none of the biases or the activation functions

To be precise, what we have just described is known as an **artificial feed-forward dense neural network**. There are other possible architectures for neural networks, but this is the one that will be using for the most part of the rest of the book.

That is how you can define a neural network, but there is one element in the definition to which we have not paid much attention: the activation function. We have mentioned before that this can be any function of our choice, and we have seen what role it plays in the behavior of a neural network, but what are some reasonable choices for this function? Let's explore the most common ones:

- We may start off with a simple activation function, actually, the same one that we implicitly considered when we defined the perceptron. This is a **step function** given by

  $$h(x) = \begin{cases} 1, & x \geq 0 \\ 0, & x < 0. \end{cases}$$

  We could technically use this in a neural network, but...in truth...it would not be a very wise choice. It is not differentiable, not even continuous. And, as we will soon see, that usually makes any function a terrible candidate to be an activation function inside a neural network. In any case, it is an example of historical importance.

- Let's now consider a somewhat more sophisticated and interesting example: the **sigmoid** activation function. This function is smooth and continuous, and it outputs values between 0 and 1. This makes it an ideal candidate for the activation function in the final layer of, for example, a classifier. It is defined by

$$S(x) = \frac{e^x}{e^x + 1}.$$

  We have plotted it in *Figure 8.4a*.

- As beautiful as it may seem, when used in inner layers, the sigmoid function can easily lead to problems in the training process (see Aurelien's book for more on this [64]). In general, a better choice for inner layers is the **exponential linear unit** or **ELU** activation function, defined as

$$E(x) = \begin{cases} x, & x \geq 0 \\ e^x - 1, & x < 0. \end{cases}$$

  You can find its plot in *Figure 8.4b*.

- We will also discuss one last activation function: the **rectified linear unit** or **ReLU** function. In general, it yields worse results than the ELU function, but it is easier to compute and thus its use can speed up the training. It is defined as

$$R(x) = \max\{0, x\}.$$

  The plot can be found in *Figure 8.4c*.

Exercise 8.3

Check that the image of the sigmoid function S is $(0, 1)$. Prove that the ELU function E is smooth and that its image is $(-1, \infty)$. What is the image of the ReLU function? Is it smooth?

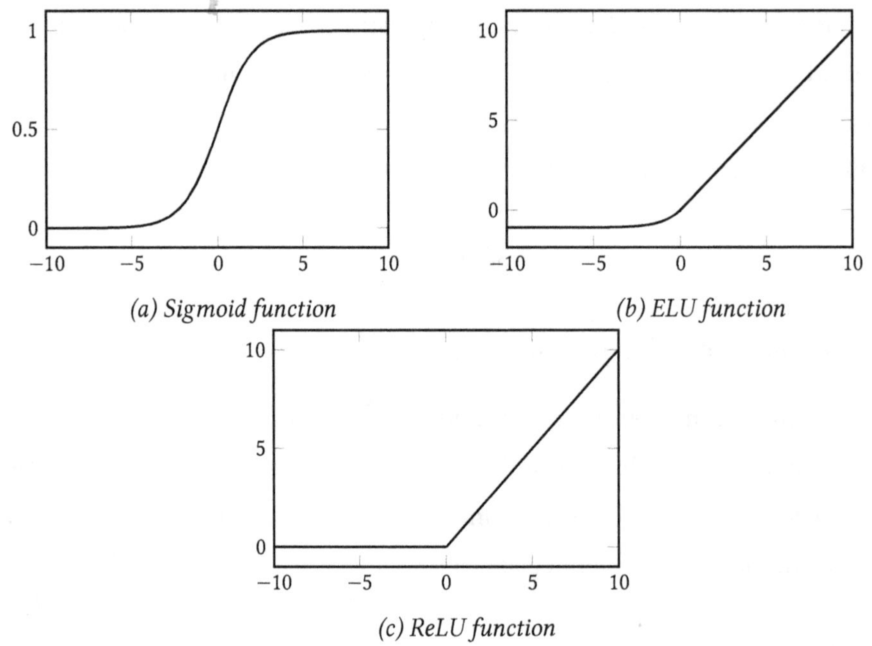

Figure 8.4: Some common activation functions in neural networks

As we mentioned at the beginning of the chapter, it has been proven that neural networks are universal function approximators [69], so they are interesting models to consider in any problem involving supervised machine learning. And thus, they will be the model we will use to build our classifier. We will consider a neural network with two inputs and some layers — we will later decide how many of them and how many neurons each will have. The final layer, of course, will have a single neuron, which will be the output. We will use ELU activation functions all throughout the network, except for the last layer; there, we will use a sigmoid activation function in order to get a normalized result. That way, we will get a continuous value between 0 and 1, and, as it is customary, we will define a threshold at $1/2$ to assign positive ($\geq 1/2$) or negative ($< 1/2$) to any given output.

Now that our model is ready, the next challenge that is waiting for us is finding a suitable loss function.

8.2.2 Understanding loss functions

When it comes to defining a loss function for supervised machine learning, with models that depend on some continuous parameters, we want to look for loss functions that are continuous and differentiable with respect to the trainable parameters of the model. The reason for this is the same reason why we want our activation functions to be differentiable, and it will become clear later on.

As we discussed earlier, the most natural loss function — and the one whose expected value we truly want to minimize — would be the 0-1 loss function, but this function would not have a continuous dependence on the parameters of the model: it would take "discrete jumps" as the classifier changes its behavior. Therefore, we need to look for alternative loss functions that are indeed continuous and differentiable while still measuring the loss in a manner that is reasonable and natural enough for classification problems.

Another somewhat naive yet much better choice would be to take the **mean squared error** as our loss function. For the purposes of our problem, we know that the neural network returns a continuous value between 0 and 1, and we know that — ideally — the closer this value is to 0 or 1, the more likely it corresponds to a negative or positive input respectively. In order to do the classification, we set a threshold at $1/2$ and get a discrete label, but, in order to compute the loss function, we should actually look at that continuous output! In this way, if we let $M_\theta(x)$ be the continuous value in $[0, 1]$ returned by the model for a given input x, and we let $y \in \{0, 1\}$ be its corresponding label, we could take our loss function to be

$$L(\theta; x, y) = (M_\theta(x) - y)^2 ,$$

where we have grouped in θ all the parameters (weights and biases) on which our neural network M depends.

Of course, in order to compute the training loss (the expected value over the training dataset), we would just take the average value over the training dataset, and analogously for the validation loss. This is usually called the **mean squared error** (**MSE**) because, well, it is the average of the error squared.

The MSE is a good loss function, but when it comes to binary classifiers, there is actually an even better candidate: the **binary cross-entropy**. It is computed as

$$H(\theta; x, y) = -y \log\left(M_\theta(x)\right) - (1 - y) \log\left(1 - M_\theta(x)\right).$$

Now, this may seem like a very complicated expression, but it is actually a very elegant and powerful loss function! For starters, if the output of the model is differentiable and continuous with respect to its trainable parameters, so is the loss (that is easy to check, just go back to Calculus 101). And that's not all. The following exercise may help you realize why the binary cross-entropy function is a great choice function for binary classifiers.

Exercise 8.4

Show that the output of the binary cross-entropy loss function $H(\theta; x, y)$ is 0 if $M_\theta(x) = y$ and that it diverges to ∞ as $M_\theta(x)$ approaches the opposite label to y (this is, as $M_\theta(x) \to 1$ if $y_i = 0$ and as $M(x) \to 0$ if $y = 1$).

And, with this, our shiny new loss function is ready to be used. However, there is one last element we still have to take care of, one that we have so far neglected and ignored. Yes, in the following section, we shall give optimization algorithms the attention and care that they deserve!

8.2.3 Gradient descent

You are now reading this book, probably in the comfort of your home, college library, or office. But life changes in the most unexpected ways, and, maybe, in a couple of weeks, you will find yourself at the top of a mountain, blindfolded (don't ask us why) and tasked with the mission of reaching the bottom of a nearby valley. If this happened, what would you do?

You don't have to be a survival expert to accomplish this task. It's true that — for undisclosed reasons — you are blindfolded, so you can't see where the valley is, but, hey, you can still move around, can't you? So, you could take some small steps in whichever direction you feel is leading you downwards with the highest slope. And you could just repeat that process several times and, eventually, you would reach the bottom of a valley.

Of course, as you descend, you will have to be careful with how big your steps are. Take steps that are too big, and you may go from the top of a mountain to the top of another one, skipping all the valleys in between (some medical doctors have suggested this might not be anatomically possible, but, well, you get what we mean). On the other hand, make your steps too small and it is going to take you forever to reach the valley. So, you will have to find a sweet spot!

Anyhow, how does this seemingly crazy thought experiment relate to machine learning? Let's see.

Gradient descent algorithms

We now have a powerful-enough model that depends on some parameters. Moreover, since we have made wise life choices, we also have a loss function L that depends continuously on and is differentiable with respect to these parameters (that is because we picked some smooth activation functions and the binary cross-entropy).

By doing this, we have effectively reduced our machine learning problem to the problem of minimizing a loss function, which is a differentiable function on some variables (the trainable parameters). And how do we do this? Using the Force…Sorry, we got carried away. We meant: using calculus!

The "getting to the valley" problem that we discussed before is — as you may have very well guessed by now — a simple analogy that will help us illustrate the **gradient descent method**. This method is just an algorithm that will allow us to minimize a differentiable function, and we can think of it as the mathematical equivalent of taking small steps in the steepest downward direction on a mountain. We should warn you that the remaining content of this subsection might be somewhat dense. Please, don't let technicalities overwhelm you. If this were a song, it'd be perfectly fine not to know its lyrics; all that would matter is for you to be familiar with its rhythm!

As you may remember from the sweet old days of undergraduate calculus, whenever you have a differentiable function $f : \mathbb{R}^N \longrightarrow \mathbb{R}$ (for those of you less familiar with mathematical notation, this is a fancy way of saying that f has N real-number inputs and returns a single real-number output), the direction in which it decreases more steeply at a point x is given by $-\nabla f(x)$, where $\nabla f(x)$ is the **gradient vector** at x, and is computed as

$$\nabla f(x) = \left(\left.\frac{\partial f}{\partial x_1}\right|_x, \ldots, \left.\frac{\partial f}{\partial x_n}\right|_x \right),$$

where $\partial/\partial x_i$ denotes the partial derivative operator with respect to a variable x_i. So, if we want to move towards a minimum at a given point, we will have to move in the direction of $-\nabla f(x)$, but by what amount?

The mathematical equivalent of the size of a step is going to be a parameter τ known as the **learning rate**. And, in this way, given a learning rate τ and an initial configuration θ_0 of the parameters of our model, we can try to find the parameters that minimize the loss function L by computing, iteratively, new parameters according to the rule

$$\theta_{k+1} = \theta_k - \tau \nabla L(\theta_k).$$

There are some algorithms that dynamically adjust this step size from an initial learning rate as the optimization progresses. One such algorithm is **Adam** (short for **Adaptive Moment Estimator**), which is one of the best gradient descents algorithms out there; it will actually be our go-to choice.

Important note

It is important to pick the learning rate wisely. If it is too small, the training will be very slow. If it is too large, you may find yourself taking huge strides that jump whole valleys, and the training may never be successful.

Of course, in order for gradient descent algorithms to work, you need to be able to compute the gradient of the loss function. There are several ways to do this; for example, you could always estimate gradients numerically. But, when working with certain models such as neural networks, you can employ a technique known as **backpropagation**, which enables the efficient computation of exact gradients. You may learn more about the technical details in Geron's exceptional book [64, Chapter 10].

To learn more...

The method of backpropagation has been one of the key developments leading to the great success of deep learning that we are experiencing today. Although this technique was already known in the 1960s, it was popularized for training neural networks by the work of Geoffrey Hinton and his collaborators. Hinton, together with Yoshua Bengio, Demis Hassabis, and Yann LeCun, received the 2022 Princess of Asturias Award for Technical and Scientific Research for outstanding work in the field of neural networks. You can learn a lot about the inception of backpropagation and about the history of neural networks research by reading the excellent *Architects of Intelligence*, in which Martin Ford interviews Bengio, Hassabis, Hinton, LeCun, and many other prominent figures in artificial intelligence [70]. By the way, Demis Hassabis is, in great part, responsible for the success of AlphaGo, one of the examples of reinforcement learning that we mentioned earlier in this chapter.

Mini-batch gradient descent

When the training dataset is large, computing the gradient of the loss function — as a function of the optimizable parameters of the model — can slow down the training significantly. In order to speed up the training, you can resort to the technique of **mini-batch gradient descent**. With this optimization method, the training dataset is split into batches of a fixed **batch size**. The gradient of the loss function is then computed on each of these batches, and the results are used to approximate the gradient of the global loss function: this is, the loss function on the whole training dataset. When we use this technique, we need to be careful with the batch size that we use: make it too small, the training will be very unstable; make it too large, the training will be too slow. As with the learning rate, it's all a matter of finding an equilibrium! However, in some cases, speed is of the essence, and we go to the extreme, using batches of just one input. This is called **stochastic gradient descent**. On the other hand, when the batch includes all the elements in the dataset, we say that we are using **batch gradient descent**.

Now we do have all that we need to train our first model. We have a dataset, we know what our model should look like, we have picked a loss function and we know how to optimize it. So let's make this work! For this, we will use TensorFlow and scikit-learn.

8.2.4 Getting in the (Tensor)Flow

We already have our dataset ready, and we could split it manually into training, validation, and test datasets, but there are already some good-quality machine learning packages with functions that help you do that. One of these packages is `sklearn`, which implements a `train_test_split` function. It splits a dataset into a training and test dataset (it doesn't return a validation dataset, but we can work our way around that). It does so by taking as arguments the dataset and the labels array; in addition, it has some optional arguments to specify whether the dataset should be shuffled and the proportions in which the dataset should be split. In order to get a training, validation, and test dataset with proportions 0.8, 0.1, and 0.1 respectively, we just need to use this function twice: once to get a training dataset (size 0.8) and a test dataset (size 0.2), and once more to split the test dataset in half, yielding a validation dataset and a test dataset of relative size 0.1 each.

Following convention, we will denote the datasets as variables `x` and the labels as variables `y`. In this way, we can run the following:

```
from sklearn.model_selection import train_test_split
# Split into a training and a test dataset.
x_tr, x_test, y_tr, y_test = train_test_split(
    data, labels, shuffle = True, train_size = 0.8)
# Split the test dataset to get a validation one.
x_val, x_test, y_val, y_test = train_test_split(
    x_test, y_test, shuffle = True, train_size = 0.5)
```

Notice how the function returns four arrays in the following order: the data for the training dataset, the data for the test dataset, the labels for the training dataset, and the labels for the test dataset. One important thing about the `train_test_split` function is that it can use **stratification**. If we had also provided the arguments `stratify = labels` and `stratify = y_test`, this would have meant that, when splitting the data into training and test examples, it would have kept the exact proportion of positive and negative classes from the original data (or at least as close to exact as possible). This can be important, especially if we are working with unbalanced datasets in which one class is much more abundant than the other. If we are not careful, we could end up with a dataset in which the minority class is non-existent.

Now that the data is perfectly prepared, it is time for us to focus on the model. For our problem, we are going to use a neural network with the following components:

- An input layer with two inputs
- Three intermediate (also known as **hidden**) layers with ELU activation functions and with 8, 16, and 8 neurons respectively
- An output layer with a single neuron that uses the sigmoid activation function

Let's now try to digest this specification a little bit. Because of the nature of the problem, we know that our model needs two inputs and one output, hence the sizes of the input and output layers. What is more, we want to get an output normalized between 0 and 1, so it makes sense to use the sigmoid activation function in the output layer. Now, we need to find a way to get from 2 neurons in the first layer to 1 neuron in the output layer. We could use hidden layers with 2 or 1 layers... but that wouldn't yield a very powerful neural network. Thus, we have progressively scaled the size of the neural network: first going from 2 to 8, then from 8 to 16, then down from 16 to 8, to finally reach the output layer with 1 neuron.

How do we define such a model in TensorFlow? Well, after doing the necessary imports and setting a seed (remember that it is an important part if we want this to be reproducible!), all it takes is to define what is known as a **Keras sequential model**.

The code is pretty self-explanatory:

```
import tensorflow as tf
tf.random.set_seed(seed)

model = tf.keras.Sequential([
    tf.keras.layers.Input(2),
    tf.keras.layers.Dense(8, activation = "elu"),
    tf.keras.layers.Dense(16, activation = "elu"),
    tf.keras.layers.Dense(8, activation = "elu"),
    tf.keras.layers.Dense(1, activation = "sigmoid"),
])
```

And that is how we can create our model, storing it as an object of the `Sequential` class.

To learn more...

Once you have defined a Keras `model`, like the sequential model that we have just considered, you can print a visual summary of it by running the instruction `print(model.summary())`. This summary lists all the layers of the model together with their shape, and also displays a count of all the model parameters.

Before we can train this model, we will need to **compile it**, associating it with an optimization algorithm and a loss function. This is done by calling the `compile` method and giving it the arguments `optimizer` and `loss`. In our case, we seek to use the Adam optimizer (just with its default parameters) and the binary cross entropy loss function. We can thus compile our model as follows:

```
opt = tf.keras.optimizers.Adam()
lossf = tf.keras.losses.BinaryCrossentropy()
model.compile(optimizer = opt, loss = lossf)
```

When we instantiate the Adam optimizer without providing any arguments, the learning rate is set, by default, to 10^{-3}. We may change this value — and we will very often do! — by setting a value for the optional argument `learning_rate`.

8.2.5 Training the model

Now we are ready to train our model. This will be done by calling the `fit` method. But before we do that, let's explore in some detail the most important arguments that we have to and can pass to this method:

- The first argument that `fit` admits is the dataset `x`. It should be an array containing the inputs that need to be passed to the model in order to train it. In our case, that would be `x_tr`.
- The second argument that we can send is the array of labels `y`. Of course, the dimensions of `x` and `y` need to match. In our case, we will set `y` to be `y_tr`.
- If you are using an optimizer that relies on gradient descent, you may want to resort to mini-batch gradient descent. For this purpose, you can give an integer value to the `batch_size` argument, which defaults to 32 (thus, by default, mini-batch gradient descent is used). If you do not want to use mini-batch gradient descent, you should set `batch_size` to `None`; that is what we will do.
- When we discussed gradient descent, we saw how these gradient descent algorithms are **iterative**: they work by computing a sequence of points that, in principle, should converge to a (local) minimum. But this raises the question of how many optimization cycles the algorithm should make — how many such points in the sequence it should compute. You may fix how many steps, also known as **epochs**, you want the optimization algorithm to take. This is done by setting a value for the `epochs` argument, which defaults to 1. In our case, we will use 8 epochs.
- If we want to use some validation data, as it is our case, we can pass it through the `validation_data` argument. The value of this argument should be a tuple with the

validation dataset in the first entry and the corresponding labels in the second one. Thus, in our case, we would set `validation_data` to `(x_val, y_val)`.

- You may have noticed that the whole process of extracting a training, validation, and test dataset can be somewhat tiresome. Well, it turns out that TensorFlow can help out here. In principle, we could just have given TensorFlow a dataset with both the training and validation data and told it in which proportions they should be split by setting a value in the `validation_split` argument. This value must be a float between 0 and 1 representing the proportion of the training dataset that should be used for validation.

  By doing this, we would save ourselves a "split", but we would still have to extract a test dataset on our own.

To learn more...

We have only covered some of the possibilities offered by TensorFlow — the ones that we will use most often. If you feel comfortable enough with the material that we have seen so far and want to explore TensorFlow in depth, you should check out the documentation (`https://www.tensorflow.org/api_docs/python/tf`).

The way we will then train our model will be the following:

```
history = model.fit(x_tr, y_tr,
    validation_data = (x_val, y_val), epochs = 8,
    batch_size = None)
```

And, upon executing this instruction on an interactive shell, we will get the following output:

```
Epoch 1/8
63/63 [==================] - 1s 3ms/step - loss: 0.6748
- val_loss: 0.4859
Epoch 2/8
63/63 [==================] - 0s 1ms/step - loss: 0.4144
```

```
- val_loss: 0.3095
Epoch 3/8
63/63 [====================] - 0s 1ms/step - loss: 0.3173
- val_loss: 0.2502
Epoch 4/8
63/63 [====================] - 0s 1ms/step - loss: 0.2908
- val_loss: 0.2315
Epoch 5/8
63/63 [====================] - 0s 1ms/step - loss: 0.2830
- val_loss: 0.2262
Epoch 6/8
63/63 [====================] - 0s 1ms/step - loss: 0.2793
- val_loss: 0.2221
Epoch 7/8
63/63 [====================] - 0s 1ms/step - loss: 0.2765
- val_loss: 0.2187
Epoch 8/8
63/63 [====================] - 0s 1ms/step - loss: 0.2744
- val_loss: 0.2185
```

When seeing this, the first thing we should do is comparing the training loss with the validation loss — just to stay away from overfitting! In our case, we see that these two are close enough and have evolved following similar decreasing trends during the training. That is indeed a good sign!

You may have noticed how we have saved the output of the `fit` method in an object that we have called `history` in which TensorFlow will store information about the training. For example, the training and validation losses at the end of each epoch is recorded in a dictionary that we could access as `history.history`.

Exercise 8.5

Plot on a single graph the evolution of the training and validation losses through the epochs, relying on the information contained in the `history` object.

In this case, we have manually set the number of epochs to 8, but this is not always the best strategy. Ideally, we would like to fix a maximum number of epochs that is reasonably large, but we would want the training to stop as soon as the loss is not improving. This is known as **early stopping**, and it can be easily used in TensorFlow.

In order to use early stopping in TensorFlow, we first need to create an `EarlyStopping` object in which we specify how we want early stopping to behave. Let's say that we want to train our model until, for three consecutive epochs, the validation loss doesn't decrease more than 0.001 after each epoch. To do this, we would have to invoke the following object:

```
early_stp = tf.keras.callbacks.EarlyStopping(
    monitor = "val_loss", patience = 3, min_delta = 0.001)
```

And then, when calling the `fit` method, we would just have to pass the optional argument `callbacks = [early_stp]`. It's as easy as that!

In any case, now we have trained our model. If we want our model to process any inputs, we can use the `predict` method, passing an array with any number of valid inputs. For example, in our case, if we wanted to get the output of the model on the test dataset, we could retrieve `model.predict(x_test)`. However, this will give us the continuous values returned by the model (which will range from 0 to 1), not a label! In order to get a discrete label (0 or 1), we need to set a threshold. Naturally, we will set it to 0.5. Thus, if we want to get the labels that our model would predict, we would have to run the following:

```
output = model.predict(x_test)
result = (output > 0.5).astype(float)
```

Of course, now we have to decide whether or not this training has been successful, so we should assess the performance of our model on the test dataset. In order to do this, we may simply compute the **accuracy** of our model on the test dataset, that is, we may compute the proportion of inputs in the test dataset that are correctly classified by our model.

In order to do this, we can use the `accuracy_score` function from `sklearn.metrics`:

```
from sklearn.metrics import accuracy_score
print(accuracy_score(result, y_test))
```

In our case, we got 89.2% accuracy. This seems like a pretty decent value, but we should always consider accuracy values in the context of each problem. For some tasks, 89.2% can indeed be marvelous, but for others it can be simply disappointing. Imagine, for instance, that you have a problem in which 99% of the examples belong to one class. Then, it is trivial to obtain at least 99% accuracy! You just need to classify all the inputs as belonging to the majority class. In the next few pages, we will introduce tools to take this kind of situation into account and better quantify classification performance.

Exercise 8.6

Re-train the model under the following conditions and compute the accuracy of the resulting model:

- Reducing the learning rate to 10^{-6}
- Reducing the learning rate to 10^{-6} and increasing the number of epochs to $1,000$
- Reducing the size of the training dataset to 20

In which cases is the resulting model less accurate? Why?

Does overfitting occur in any of these scenarios? How could you identify it?

So far, we have assessed the accuracy of our model just by measuring the proportion of elements that it would correctly classify by setting a threshold of 0.5. There are nevertheless other metrics of the performance of a binary classifier. We will study them in the next subsection!

8.2.6 Binary classifier performance

Whenever you have a binary classifier, any output can belong to one of the four categories depicted in the following table:

	Classified as positive	Classified as negative
Actual positive	True positive	False negative
Actual negative	False positive	True negative

The abbreviations TP, FN, FP, and TN are also used to denote the number of true positives, false negatives, false positives, and true negatives (respectively) produced by a classifier over a given dataset. These quantities are used very often. In fact, a common way of assessing the performance of a classifier is by looking at its **confusion matrix** (usually over the test dataset), which is nothing more than the matrix

$$\begin{pmatrix} \mathrm{TP} & \mathrm{FN} \\ \mathrm{FP} & \mathrm{TN} \end{pmatrix}.$$

To get started, we can now compute the confusion matrix for the binary classifier that we have just trained over the test dataset. For this, we can use the `confusion_matrix` function from `sklearn.metrics`, which requires two arguments: an array of predicted labels and an array of true labels:

```
from sklearn.metrics import confusion_matrix
confusion_matrix(y_true = y_test, y_pred = result)
```

Upon executing this piece of code, we get the following confusion matrix for our classifier:

$$\begin{pmatrix} 24 & 20 \\ 7 & 199 \end{pmatrix}.$$

This matrix shows that there are very few false positives compared to the number of true negatives, but almost as many false negatives as true positives. This means that our classifier does a very good job of picking up the negative class but it is not so good at identifying the positive one. In a moment, we will discuss how to quantify this more precisely.

To learn more...

Although we have focused just on binary classifiers, confusion matrices can also be defined for classification problems in which there are n classes. They have n rows and n columns, and the entry in row k column l represents the number of elements that actually belong to class k but that are labeled as class l by the system.

Additionally, if you fix one of the n classes as the positive one and consider the rest as negative, you can obtain TP, FP, TN, and FN for that particular class.

Confusion matrices are very informative, and the quantities in them can help us define several metrics of the performance of a binary classifier. For instance, the usual accuracy metric can be defined by

$$\mathrm{Acc} = \frac{\mathrm{TP} + \mathrm{TN}}{\mathrm{TP} + \mathrm{TN} + \mathrm{FP} + \mathrm{FN}}.$$

Other interesting metrics are the **positive predictive value** and the **sensitivity**, which are defined respectively as

$$P = \frac{\mathrm{TP}}{\mathrm{TP} + \mathrm{FP}}, \qquad S = \frac{\mathrm{TP}}{\mathrm{TP} + \mathrm{FN}}.$$

The positive predictive value is also known as the **precision** and the sensitivity is also known as the **recall** of the classifier.

There is a trade-off between P and S. Obtaining a perfect recall is trivial: you just need to classify every input as positive. But then, you will have a low precision. Similarly, it is easy to obtain very good values of precision: only classify an example as positive if you are extremely sure that it is positive. But then the recall will be very low.

For this reason, an interesting metric is the F_1 score, defined as the harmonic mean of P and S:

$$F_1 = \frac{2}{\frac{1}{P} + \frac{1}{S}} = \frac{2PS}{P + S}.$$

It is easy to see how this score can range from 0 (the score of the worst possible classifier) to 1 (the score of a perfect classifier). Moreover, a high F_1 score means that we are not favoring recall over precision or precision over recall.

If you are mathematically oriented, you may have realized that our expression for F_1 is actually undefined for $P = S = 0$, but we can trivially extend it by continuity to take the value $F_1 = 0$ there.

In order to compute these metrics, we may use the `classification_report` function from `sklearn.metrics`. In our case, we may run the following:

```
from sklearn.metrics import classification_report
print(classification_report(y_true = y_test, y_pred = result))
```

This yields the following output:

```
              precision    recall  f1-score   support

           0       0.77      0.55      0.64        44
           1       0.91      0.97      0.94       206

    accuracy                           0.89       250
   macro avg       0.84      0.76      0.79       250
weighted avg       0.89      0.89      0.88       250
```

And in this table, we can see all the metrics that we have mentioned. You can see that the scores are returned for both the case in which 0 is the positive class and for the case when 1 is positive instead (in our case, we have considered 0 to be positive, so we would look at the first row). By the way, the **support** of a class is meant to represent the number of elements in the class that can be found in the dataset. Also, the **macro average** of each metric is just the plain average of the values of the metric obtained by taking each class as positive. The weighted average is like the macro average, but weighted by the proportion of elements of each class in the dataset.

Let's say that we have a binary classifier that returns a continuous output between 0 and 1 before cutting through a threshold in order to assign a label. As we saw earlier, we could just measure the performance of our classifier by using a bunch of metrics. But if we want to get a broader perspective of how our classifier could work for any threshold, we can take another approach.

Using the entries of the confusion matrix over a dataset, we may define the **true positive rate** as the proportion

$$\mathrm{TPR} = \frac{\mathrm{TP}}{\mathrm{TP} + \mathrm{FN}},$$

that is, the proportion of examples from the positive class that are actually classified as positive. On the other hand, we can analogously define the **false positive rate** as the quotient

$$\mathrm{FPR} = \frac{\mathrm{FP}}{\mathrm{FP} + \mathrm{TN}}.$$

The **Receiver Operating Characteristic curve** or **ROC curve** of a classifier that returns continuous values is computed over a given dataset by plotting, for every possible choice of threshold, a point with a Y coordinate given by the corresponding TPR and an X coordinate with the FPR for that threshold. As the threshold increases from 0 to 1, this will give rise to a finite sequence of points. The curve is obtained by joining these through straight lines. Notice that we evaluate the performance of the classifier with different levels of "demand" for classifying an input as positive. When the threshold is high, it will be harder to classify something as positive; the FPR will be low — great! — but the TPR will probably be also low. On the other hand, for low values of the threshold, it will be easier for an input to be classified as positive: the TPR will be high — yay! — but that can also cause the false positives to go up.

Sounds familiar? This is the same kind of trade-off that we discussed when we defined precision and recall. The difference is that, in this case, we are taking into account the behavior of the classifier for every possible choice of threshold, giving us a global assessment. Plotting the ROC curve can be very informative because it can also help in selecting classification thresholds that are more suitable for our problem. For instance, if you are trying to detect whether a given patient has a certain serious illness, it may pay off to have some false positives – people that may need to undergo additional medical tests – at the cost of having very low false negatives. The ROC curve can help you there by identifying points at which the TPR is high and the FPR is acceptable.

In order to plot a ROC curve, we can use the `roc_curve` function from `sklearn.metrics`. It will return the X and Y coordinates of the points of the curve. In our particular case, we may run the following piece of code:

```
from sklearn.metrics import roc_curve
fpr, tpr, _ = roc_curve(y_test, output)
plt.plot(fpr, tpr)
plt.plot([0,1],[0,1],linestyle="--",color="black")
plt.xlabel("FPR"); plt.ylabel("TPR")
plt.show()
```

Notice how we have dropped part of the output of the `roc_curve` function; in particular, the return object that we ignore yields an array that includes the thresholds at which the classifier accuracy changes (you can refer to the documentation at `https://scikit-learn.org/stable/modules/generated/sklearn.metrics.roc_curve.html` for more information). The output that we got can be found in *Figure 8.5.* Notice that we have manually drawn a dashed line between $(0, 0)$ and $(1, 1)$. That is meant to represent the ROC curve that could be generated by a random classifier, one that assigns an input to a class with probability proportional to the size of that class, and it is an important visual aid. That's because any curves above that dashed line are ROC curves of classifiers that have some real classification power.

There are some interesting features in this ROC curve, so let's discuss it a little bit. To start with, notice that the points $(0, 0)$ and $(1, 1)$ always belong to the ROC curve of any classifier because they are achieved with the highest and lowest thresholds, respectively. In the first case, no input is assigned to the positive class, so we have neither TPs nor FPs.

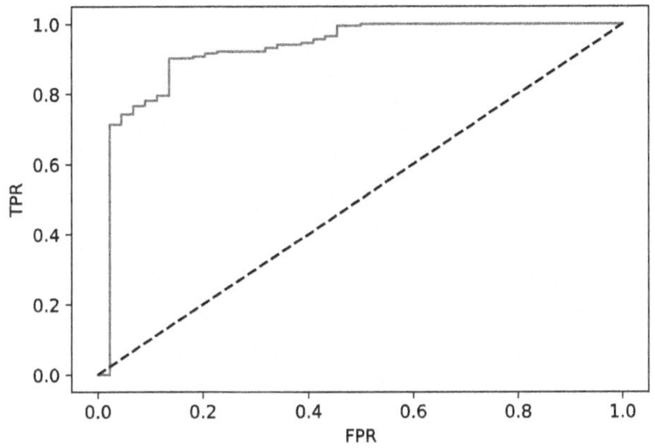

Figure 8.5: ROC curve (solid line) for the classifier that we have trained.

In the second one, all inputs are assigned to the positive class, so we have neither FNs nor TNs.

In addition to this, we can observe in our graph that, from $(0, 0)$, the ROC curve starts moving horizontally, increasing the FPR without increasing the TPR. This means that there are some examples in the test dataset that the model very confidently classifies as belonging to the positive class but that, in fact, are negative. This is undesirable, of course. We would like our ROC curve to go up — increasing the TPR — without moving to the right. And that is exactly what happens after that first hiccup. We observe a long segment in which the TPR goes up without any increase in the FPR. If we need our classifier to have high precision, we could select the threshold that achieves TPR of about 0.71 with FPR of only about 0.02. On the other hand, if we need high recall, we can select the point in the curve where the TPR is already 1 with a FPR of about 0.5. For a more balanced classifier, notice that there is a point in the ROC curve with TPR around 0.91 and FPR below 0.21.

Of course, the ideal classifier would have a ROC curve that goes all the way from $(0, 0)$ to $(1, 0)$. That would mean that there is a threshold for which all the positive examples are classified as positive, while no negative example is assigned to the positive class. That's just perfection! From there, the ROC curve would go straight to $(1, 1)$: we have already found all the positive examples so the TPR cannot increase, but by decreasing the threshold we will eventually increase the FPR from 0 to 1.

Obviously, that kind of perfect ROC curve is only achievable for extremely simple classification problems. However, we can still compare our actual model to that ideal classifier by computing the **area under the ROC curve**, often abbreviated as **AUC**. Since the ROC curve of the perfect classifier would have area equal to 1, we can consider that the closer the AUC of a classifier is to 1, the better its global performance is. In the same way, a random classifier would have an ROC curve that is a straight line from $(0, 0)$ to $(1, 1)$, so its AUC would be 0.5. Hence, classifiers whose AUC is higher than 0.5 have some actual classification power beyond just random guessing.

Having the coordinates of the points that define the ROC curve, we can easily get the AUC score using the `auc` function from `sklearn.metrics`:

```
from sklearn.metrics import auc
print(auc(fpr,tpr))
```

In our case, we get an AUC score of approximately 0.9271. Again, this seems like a great value, but let us stress once again that it all depended on the difficulty of the problem — and the one we have been considering is not particularly hard. Also, remember that the AUC is a global performance metric that takes into account every possible threshold of your classifier. At the end of the day, you need to commit to just one threshold value, and a high AUC might not mean much if, for your particular threshold choice, the accuracy, precision, and recall are not that great.

That was a lot of information! In any case, for most practical purposes, all that you will need to know is summarized in the following note.

Important note

Given a binary classifier with continuous output, we may compute its receiver operating characteristic curve (also known as the ROC curve) over a dataset. The higher the area under that curve, the higher the classifying power of the classifier.

We refer to the area under the ROC curve of a classifier as its AUC (short for "area under the curve"):

- An AUC of 1 corresponds to a perfect classifier
- An AUC of 0.5 would match that of a random classifier
- An AUC of 0 corresponds to a classifier that always returns the wrong output

By now, we should have a decent understanding of (classical) machine learning, and you may be wondering where does the "quantum" part begin? It begins now.

8.3 Quantum-classical models

In general terms, quantum machine learning refers to the application of machine learning techniques — only that quantum computing is involved at same stage of the process. Maybe you use a quantum computer in some part a model that you wish to train. Maybe you wish to use data generated by some quantum process. Maybe you use a quantum computer to process quantum-generated data. As you can imagine, the subject of quantum machine learning, as a whole, is broad enough to accommodate for a wide range of ideas and applications.

In an attempt to categorize it all a little bit, we can follow the useful classification shown in Schuld's and Petruccione's book [71] and divide quantum machine learning into four different flavors, which are depicted in *Figure 8.6*, according to the classical or quantum nature of the data and processing devices that are used:

- We could consider part of quantum machine learning all the quantum-inspired classical machine learning techniques; that is, all the classical machine learning methods that draw ideas from quantum computing. In this case, both the data and the computers are classical, but there is some quantum flavor involved in the process. This is represented as CC in the chart. Since there are no actual quantum computers involved in this approach, we will not study this kind of method.

- In addition, we can also consider part of quantum machine learning any classical machine learning algorithms that rely on quantum data; for our purposes, we can just think of it as data generated by quantum processes, or as the application of classical machine learning to quantum computing. This is the QC block in the chart. In this approach, machine learning is a tool rather than an end, so we will not be covering these techniques.

- The kind of machine learning that we will focus on in this book is the one represented by the CQ label in the chart: machine learning that relies on classical data and uses quantum computing in the model or the training.

- Lastly, there is also a very interesting QQ category. These techniques work on quantum data using quantum computing in the models themselves or in the training processes. Notice that – as opposed to CQ quantum machine learning – in this scenario, the quantum data need not be obtained from measurements: quantum states could be directly fed into a quantum model, for instance. This is an area of great promise (see, for instance, the recent paper by Huang et al. [72]), but the required technologies are still immature, so we will not be talking about this approach in much detail.

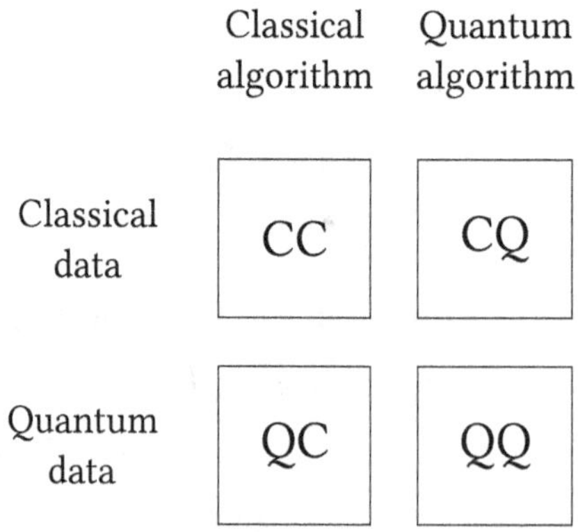

Figure 8.6: The four big families of quantum machine learning, categorized according to the nature of the models and data that they use

Our plan, then, is to focus on CQ quantum machine learning: machine learning on classical data that relies on quantum computing. Now, within this category, there is still a fairly broad range of possibilities. We could use quantum computing on the model and also in the optimization process. There are already many interesting proposals for how quantum computing could speed up traditional machine learning models, but these approaches cannot, in general, be used on our current quantum hardware. For this reason, we will not discuss them in this book — but if you are interested in learning more about them, we can recommend the excellent paper by Biamonte et al. [73].

Instead, we will devote ourselves, heart and soul, to the study of fully quantum-oriented models that can be run on NISQ devices. These models will be trained on classical data and, in general, we will use purely classical optimization techniques.

In the following chapters, we will study the following models:

- **Quantum support vector machines**. We will soon explore what support vector machines are and how they can be trained using classical machine learning. We will also see how their quantum version is just a particular case of a general support vector machine in which we use quantum computers to map data into a space of quantum states.
- **Quantum neural networks**. We will then explore a purely quantum model: quantum neural networks. This model runs fully on a quantum computer, and its behavior is inspired by classical neural networks.
- **Hybrid networks**. In the subsequent chapter, we will learn how to combine quantum neural networks with other classical models (most commonly, neural networks). We will refer to these models as hybrid networks.
- **Quantum generative adversarial networks**. Lastly, we will study generative adversarial networks and cover how the components of these models can be replaced by quantum circuits.

As in the rest of this book, our approach will be very hands-on and practical. If you wish to broaden your theoretical background on quantum machine learning, you can also have a look at the book by Maria Schuld and Francesco Petruccione [71].

Summary

In this chapter, we have explored some basic concepts and ideas that lie at the foundation of machine learning. And we haven't just explored them from a theoretical point of view: we have also seen them come to life.

We have learned what machine learning is all about, and we have discussed some of the most common approaches used to make it a reality. In particular, we have learned that many machine learning problems can be reduced to the minimization of a loss function through some optimization algorithm on a suitable model.

We have also studied in some depth classical neural networks, and we have used an industry-standard machine learning framework (TensorFlow) to train one.

Lastly, we have wrapped up this chapter by introducing what quantum machine learning is all about and having a sneak peek into the rest of the chapters of this part of the book.

9

Quantum Support Vector Machines

Artificial Intelligence is the new electricity

— Andrew Ng

In the previous chapter, we learned the basics of machine learning and we got a sneak peek into quantum machine learning. It is now time for us to work with our first family of quantum machine learning models: that of **Quantum Support Vector Machines** (often abbreviated as **QSVMs**). These are very popular models, and they are most naturally used in binary classification problems.

In this chapter, we shall learn what (classical) support vector machines are and how they are used, and we will use this knowledge as a foundation to understand quantum support vector machines. In addition, we will explore how to implement and train quantum support vector machines with Qiskit and PennyLane.

The contents of this chapter are the following:

- Support vector machines
- Going quantum
- Quantum support vector machines in PennyLane
- Quantum support vector machines in Qiskit

9.1 Support vector machines

QSVMs are actually particular cases of **Support Vector Machines** (abbreviated as **SVMs**). In this section, we will explore how these SVMs work and how they're used in machine learning. We will do so by first motivating the SVM formalism with some simple examples, and then building up from there: all the way up into how SVMs can be used to tackle complex classification problems with the kernel trick.

9.1.1 The simplest classifier you could think of

Let us forget about data for a moment and begin by considering a very naive problem. Let's say that we want to build a very simple classifier on the real line. In order to do this, all we have to do is split the real number line into two disjoint categories in such a way that any number belong to exactly one of these two categories. Thus, if we are given any input (a real number), our classifier will return the category to which it belongs.

What would be the easiest way in which you could do this? Odds are you would first pick a point a and divide the real number line into the set (category) of numbers smaller than a and the set of numbers larger than a. Then, of course, you would have to assign a to one of the two categories, so your categories would be either of the following:

- The set of real numbers x such that $x \leq a$ and the set of numbers x such that $x > a$
- The set of numbers x such that $x < a$ and the set of numbers x such that $x \geq a$

Either choice would be reasonable.

To learn more...

Actually, the choice as to in which category to include a is, to some extent, meaningless. At the end of the day, if you choose a real number at random, the probability that it be exactly a is zero. This fun fact is sponsored by probability and measure theory!

That was easy. Let's now say that we want to do the same with the real plane (the usual $\mathbb{R}^2$). In this case, a single point will not suffice to split it, but we could instead consider a good old line! This is exemplified in *Figure 9.1*. Any line can be used to perfectly split the real plane into two categories.

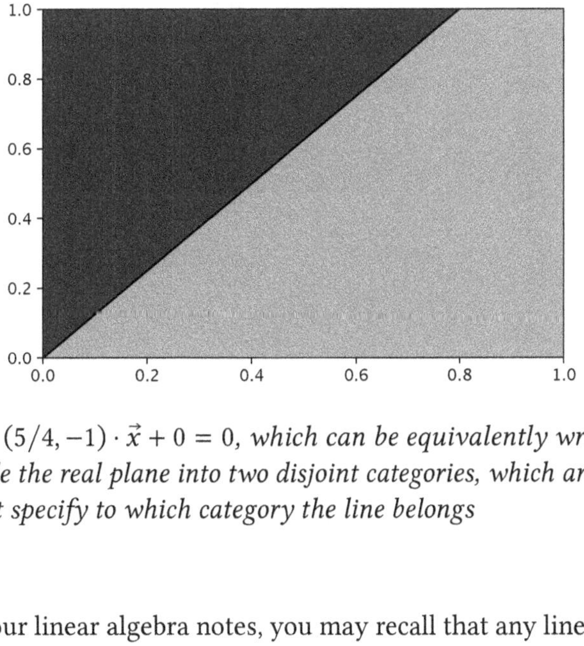

Figure 9.1: The line $(5/4, -1) \cdot \vec{x} + 0 = 0$, which can be equivalently written as $y = (5/4)x$, can be used to divide the real plane into two disjoint categories, which are colored differently. The picture does not specify to which category the line belongs

If you go back to your linear algebra notes, you may recall that any line in the plane can be characterized in terms of a vector $\vec{w} \in \mathbb{R}^2$ and a scalar $b \in \mathbb{R}$ as the set of points $\vec{x} = (x, y)$ such that $\vec{w} \cdot \vec{x} + b = 0$. Of course, we are using $\cdot$ to denote the scalar product (that is, $\vec{w} \cdot \vec{x} = w_1 x + w_2 y$, provided that $\vec{w} = (w_1, w_2)$). The vector $\vec{w}$ defines the **normal**, or perpendicular, direction to the line, and the constant b determines the intersection of the line with the X and Y axes.

When we worked on the one-dimensional case and used a point to split the real line, it was trivial to decide which category any input belonged to. In this case, it is slightly more complicated, but not too much. With some elementary geometry, you can check that any number $\vec{x}$ will be on one side or the other of the line defined by $\vec{w} \cdot \vec{x} + b = 0$ depending on the sign of the quantity $\vec{w} \cdot \vec{x} + b$. That is, if $\vec{w} \cdot \vec{x}_1 + b$ and $\vec{w} \cdot \vec{x}_2 + b$ have the same sign (both smaller than zero or both greater than zero), we will know that $\vec{x}_1$ and $\vec{x}_2$ will belong to the same category. Otherwise, we know they will not.

There is no reason for us to stop at two dimensions, so let's kick this up a notch and consider an n-dimensional Euclidean space $\mathbb{R}^n$. Just as we split $\mathbb{R}^2$ using a line, we could split $\mathbb{R}^n$ using…an $(n-1)$-dimensional hyperplane! For instance, we could split $\mathbb{R}^3$ using an ordinary plane.

These hyperplanes in $\mathbb{R}^n$ are defined by their normal vectors $\vec{w} \in \mathbb{R}^n$ and some constants $b \in \mathbb{R}$. In analogy to what we saw in $\mathbb{R}^2$, their points are the $\vec{x} \in \mathbb{R}^n$ that satisfy the equations of the form

$$\vec{w} \cdot \vec{x} + b = 0.$$

Moreover, we can determine to which side of the hyperplane a certain $\vec{x} \in \mathbb{R}^n$ is in terms of the sign of $\vec{w} \cdot \vec{x} + b$.

To learn more…

In case you are confused with all these equations and you are curious as to where they come from, let us quickly explain them. An (affine) hyperplane can be defined by a normal vector $\vec{w}$ and by a point $\vec{p}$ in the plane. Thus, a point $\vec{x}$ will belong to the hyperplane if and only if $\vec{x} = \vec{p} + \vec{v}$ for some vector $\vec{v}$ that is orthogonal to $\vec{w}$, that is, such that $\vec{w} \cdot \vec{v} = 0$. By combining these two expressions, we know that $\vec{x}$ will belong to the hyperplane if and only if $\vec{w} \cdot (\vec{x} - \vec{p}) = 0$, which can be rewritten as

$$\vec{w} \cdot \vec{x} + (-\vec{w} \cdot \vec{p}) = \vec{x} \cdot \vec{w} + b = 0,$$

where we have implicitly defined $b = -\vec{w} \cdot \vec{p}$.

Moreover, we have just seen how $\vec{w} \cdot \vec{x} + b$ is the scalar product of $\vec{x} - \vec{p}$ with $\vec{w}$, a fixed normal vector to the plane. This justifies why its sign determines on which side of the hyperplane $\vec{x}$ lies. Remember that, geometrically, the dot product of two vectors $\vec{u_1} \cdot \vec{u_2}$ is equal to $\|u\|_1 \cdot \|u\|_2 \cdot \cos\theta$, where θ denotes the smallest angle between them.

With what we have done so far, we have the tools required to construct (admittedly simple) binary classifiers on any Euclidean space. All it takes for us to do so is fixing a hyperplane!

Why is this important to us? It turns out that support vector machines do exactly what we have discussed so far.

Important note

A support vector machine takes inputs in an n-dimensional Euclidean space ($\mathbb{R}^n$) and classifies them according to which side of a hyperplane they are on. This hyperplane fully defines the behavior of the SVM. Of course, the adjustable parameters of an SVM are the ones that define the hyperplane: following our notation, the components of the normal vector $\vec{w}$ and the constant b.

In order to get the label of any point $\vec{x}$, all we have to do is look at the sign of $\vec{w} \cdot \vec{x} + b$.

As you may have suspected, vanilla SVMs, just on their own, are not the most powerful of binary classification models: they are intrinsically linear and they are not fit to capture sophisticated patterns. We will take care of this later in the chapter when we unleash the full potential of SVMs with "the kernel trick" (stay tuned!). In any case, for now, let us rejoice in the simplicity of our model and let's learn how to train it.

9.1.2 How to train support vector machines: the hard-margin case

Let's say that we have a binary classification problem, and we are given some training data consisting of datapoints in $\mathbb{R}^n$ together with their corresponding labels. Naturally, when we train an SVM for this problem, we want to look for the hyperplane that best separates the two categories in the training dataset. Now we have to make this intuitive idea precise.

Let the datapoints in our training dataset be $\vec{x}_j \in \mathbb{R}^n$ and their expected labels be $y_j = 1, -1$ (read as positive and negative, respectively). For now, we will assume that our data can be perfectly separated by a hyperplane. Later in the section, we will see what to do when this is not the case.

Notice that, under the assumption that there is at least one hyperplane separating our data, there will necessarily be an infinite number of such separating hyperplanes (see *Figure 9.2*). Will any of them be suitable for our goal of building a classifier? If we only cared about the training data, then yes, any of them would do the trick. In fact, this is exactly what the perceptron model that we discussed in *Chapter 8, What is Quantum Machine Learning?*, does: it just looks for a hyperplane separating the training data.

However, as you surely remember, when we train a classifier, we are interested in getting a low generalization error. In our case, one way of trying to achieve this is by looking for a separating hyperplane that can maximize the distance from itself to the training datapoints. And that is the way in which SVMs are actually trained. The rationale behind this is clear: we expect the new, unseen datapoints to follow a similar distribution to the one that we have seen in the training data. So it is very likely that new examples of one class will be closer to training examples of that same class. Therefore, if our separating hyperplane is too close to one of the training datapoints, we risk another datapoint of the same class crossing to the other side of the hyperplane and being misclassified. For instance, in *Figure 9.2*, the dashed line does separate the training datapoints, but it is certainly a much more risky choice than, for example, the continuous line.

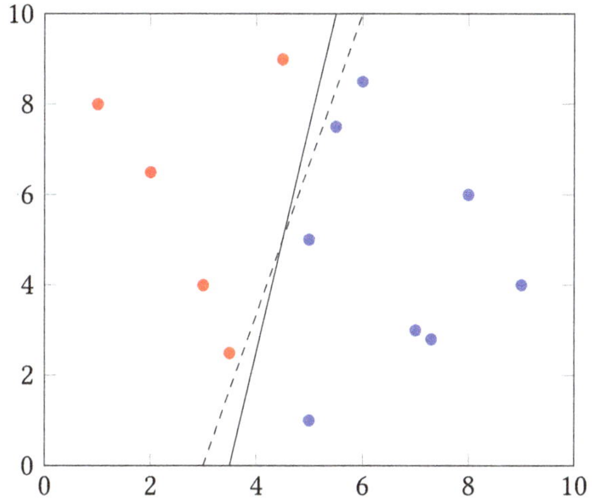

Figure 9.2: Both lines (hyperplanes) separate the two categories, but the continuous line is closer to the datapoints than the dashed line

The idea behind the training of an SVM is then clear: we seek to find not just any separating hyperplane, but one that is as far away from the training points as possible. This may seem difficult to achieve, but it can be posed as a rather straightforward optimization problem. Let's explain how to do it in a little bit more detail.

In a first approach, we could just consider the distance from a separating hyperplane H to all the points in the training dataset, and then try to find a way to tweak H in order to maximize that distance while making sure that H still separates the data properly. This is, however, not the best way to present the problem. Instead, we may notice how we can associate to each data point a unique hyperplane that is parallel to H and contains that datapoint. And, what is more, the parallel hyperplane that goes through the point that is closest to H will itself be a separating hyperplane — and so will be its reflection over H. This is illustrated in *Figure 9.3*.

This pair of hyperplanes — the parallel plane that goes through the closest point and its reflection — will be the two equidistant parallel hyperplanes, which are the furthest apart from each other while still separating the data. They are unique to H. The distance between

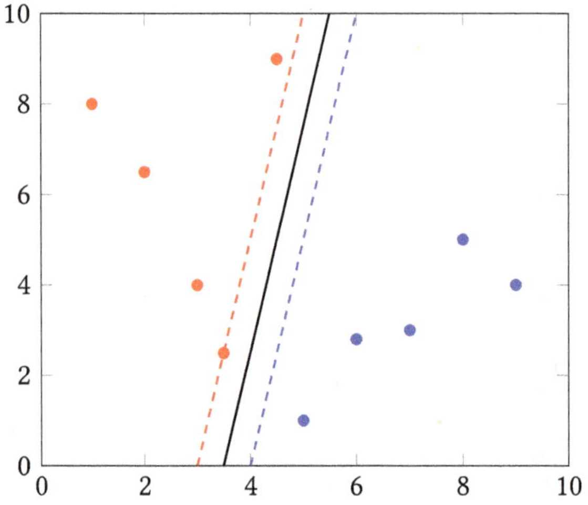

Figure 9.3: The continuous black line represents a separating hyperplane H. One of the dashed lines is the parallel hyperplane that goes through the closest point to H, and its reflection over H is the other dashed line

them is known as the **margin** and it is what we aim to maximize. This is illustrated in *Figure 9.4*.

We already know that any separating hyperplane H can be characterized by an equation of the form $\vec{w} \cdot \vec{x} + b = 0$. Moreover, any hyperplane that is parallel to H — in particular those that define the margin! — can be characterized as $\vec{w} \cdot \vec{x} + b = C$ for some constant C. And not only that, but their reflection over H will be itself characterized by the equation $\vec{w} \cdot \vec{x} + b = -C$. Hence, we know that, for some constant C, the hyperplanes that define the margin of H can be represented by the equations $\vec{w} \cdot \vec{x} + b = \pm C$.

Nevertheless, there is nothing preventing us here from dividing the whole expression by C. So, if we let $\tilde{w} = \vec{w}/C$ and $\tilde{b} = b/C$, we know that the hyperplane H will still be represented by $\tilde{w} \cdot \vec{x} + \tilde{b} = 0$, but the hyperplanes that define the margin will be characterized by

$$\tilde{w} \cdot \vec{x} + \tilde{b} = \pm 1,$$

which looks much more neat!

Let's summarize what we have. We want to find a hyperplane that, while separating the data properly, maximizes the distance to the points in the training dataset. We have seen how we can see this as the problem of finding a hyperplane that maximizes the margin: the distance between the two equidistant parallel hyperplanes that are the furthest away from each other while still separating the data. And we have just proven that, for any separating hyperplane, we can always find some values of $\vec{w}$ and b such that those hyperplanes that define the margin can be represented as

$$\vec{w} \cdot \vec{x} + b = \pm 1.$$

It can be shown that the distance between these two hyperplanes is $2/\|w\|$. Hence the problem of maximizing the margin can be equivalently stated as the problem of maximizing $2/\|w\|$ subject to the constraint that the planes $\vec{w} \cdot \vec{x} + b = \pm 1$ properly separate the data.

Exercise 9.1

Show that, as we claimed, the distance between the hyperplanes $\vec{w} \cdot \vec{x} + b = \pm 1$ is $2/\|w\|$.

Let's now consider an arbitrary element $\vec{p} \in \mathbb{R}^N$ and a hyperplane H characterized by $\vec{w} \cdot \vec{x} + b = 0$. When the value of $\vec{w} \cdot \vec{p} + b$ is zero, we know that $\vec{p}$ is in the hyperplane and, as this value drifts away from zero, the point gets further and further away from the hyperplane. If it increases and it is between 0 and 1, the point $\vec{p}$ is between the hyperplane H and the hyperplane $\vec{w} \cdot \vec{x} + b = 1$. When this value reaches 1, the point is in this latter hyperplane. And when the value becomes greater than 1, it moves beyond both hyperplanes. Analogously, if this value decreases and it is between 0 and -1, the point $\vec{p}$ is between the hyperplane H and $\vec{w} \cdot \vec{x} + b = -1$. When the value reaches -1, the point is in this last hyperplane. And when it is smaller than -1, it has moved beyond both H and $\vec{w} \cdot \vec{x} + b = -1$.

Since we are working under the assumption that there are no points inside the margin, the hyperplane $\vec{w} \cdot \vec{x} + b = 0$ will properly separate the data if, for all the positive entries, $\vec{w} \cdot \vec{x} + b \geq 1$, while all the negative ones will satisfy $\vec{w} \cdot \vec{x} + b \leq -1$. We can write this condition as

$$y_j\left(\vec{w}\cdot\vec{x}_j\right)\geq 1,$$

because we are considering $y_j = 1$ when the j-th example belongs to the positive class and $y_j = -1$ when it belongs to the negative one.

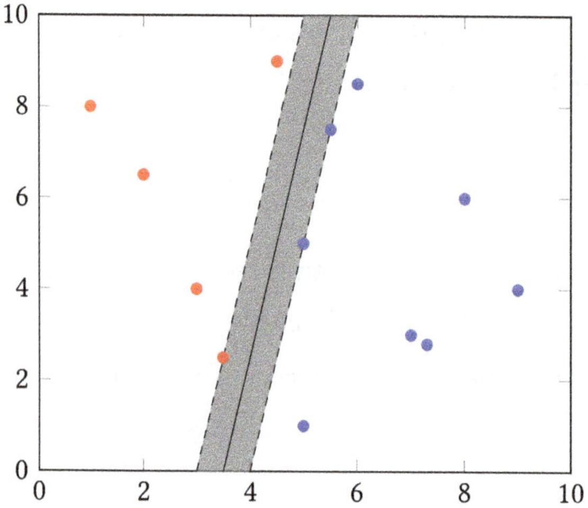

Figure 9.4: The hyperplane that could have been returned by an SVM is represented by a black continuous line, and the lines in dashed lines are the equidistant parallel hyperplanes that are the furthest apart from each other while still separating the data. The margin is thus half of the thickness of the colored region

For all this, the problem of finding the hyperplane that best separates the data can be posed as the following optimization problem:

$$\begin{aligned} \text{Minimize} \quad & \|w\| \\ \text{subject to} \quad & y_j(\vec{w}\cdot\vec{x}_j,+b)\geq 1, \end{aligned}$$

where, of course, each j defines an individual constraint. This formulation suffers from a small problem. The Euclidean norm is nice, visual, and geometric, but it has a square root. We personally have nothing against square roots — some of our best friends *are* square roots — but most optimization algorithms have some hard feelings against them. So just to make life easier for us, we may instead consider the following (equivalent) problem.

Important note

If the data in the training dataset can be separated by a hyperplane, the problem of training an SVM can be posed as the following optimization problem:

$$\begin{aligned} &\text{Minimize} \quad \frac{1}{2}\|w\|^2 \\ &\text{subject to} \quad y_j(\vec{w}\cdot\vec{x}_j + b) \geq 1. \end{aligned}$$

This is known as **hard-margin** training, because we are allowing no elements in the training dataset to be misclassified or even to be inside the margin.

That nice and innocent square will save us from so many troubles. Notice, by the way, that we've introduced a $1/2$ factor next to $\|w\|^2$. That's for reasons of technical convenience, but it isn't really important.

With hard-margin training, we need our training data to be perfectly separable by a hyperplane because, otherwise, we will not find any feasible solutions to the optimization problem that we have just defined. This scenario is, in most situations, too restrictive. Thankfully, we can take an alternative approach known as **soft-margin training**.

9.1.3 Soft-margin training

Soft-margin training is similar to hard-margin training. The only difference is that it also incorporates some adjustable **slack**, or "tolerance," parameters $\xi_j \geq 0$ that will add flexibility to the constraints. In this way, instead of considering the constraint $y_j(\vec{w}\cdot\vec{x}_j + b) \geq 1$, we will use

$$y_j(w \cdot x_j + b) \geq 1 - \xi_j.$$

Thus, when $\xi_j > 0$, we will allow $\vec{x}_j$ to be close to the hyperplane or even on the wrong side of the space (as separated by the hyperplane). What is more, the bigger the value of ξ_j, the further into the wrong side $\vec{x}_j$ will be.

Naturally, we would like these ξ_j to be as small as possible, so we need to include them in the cost function that we want to minimize. Taking all of this into account, the optimization problem that we shall consider in soft-margin training will be the following.

Important note

A support vector machine that may not be *necessarily* able to properly separate the training data with a hyperplane can be trained by solving the following optimization problem:

$$\begin{aligned} \text{Minimize} \quad & \frac{1}{2}\|w\|^2 + C\sum_j \xi_j \\ \text{subject to} \quad & y_j(\vec{w}\cdot\vec{x}_j + b) \geq 1 - \xi_j, \\ & \xi_j \geq 0. \end{aligned}$$

The value $C > 0$ is a hyperparameter that can be chosen at will. The bigger C is, the less tolerant we will be to training examples falling inside the margin or on the wrong side of the hyperplane.

This formulation is known as **soft-margin training** of an SVM.

Let us now try to digest this formulation. As expected, we also made the ξ_j contribute to our cost function, in such a way that their taking large values will be penalized. In addition, we've incorporated this C constant and said that it can be tweaked at will. As we mentioned before, in broad terms, the bigger it is, the more unwilling we will be to accept misclassified elements in the training dataset. Actually, if there is a hyperplane that can perfectly separate the data, setting C to a huge value would be equivalent to doing hard-margin training. At first, it might seem tempting to make C huge, but this would make our model more prone to overfitting. Perfect fits are not that good! Balancing the value of C is one of the many keys behind successful SVM training.

To learn more...

When we train an SVM, the actual loss function that we would like to minimize is

$$L(\vec{w}, b; \vec{x}, y) = \max\{0, 1 - y(\vec{w} \cdot \vec{x} + b)\},$$

which is called the **hinge loss**. In fact, our ξ_j variables are direct representatives of that loss. Minimizing the expected value of this loss function would be connected to minimizing the proportion of misclassified elements — which is what we want at the end of the day.

If, in our formulation, we didn't have the $\|w\|^2/2$ factor, that would be the training loss that we would be minimizing. We included this factor, however, because a small $\|w\|^2$ (that is, a large margin) makes SVM models more robust against overfitting.

We will conclude this analysis of soft-margin training by presenting an equivalent formulation of its optimization problem. This formulation is known as the **Lagrangian dual** of the optimization problem that we presented previously. We will not discuss why these two formulations are equivalent, but you can take our word for it — or you can check the wonderful explanation by Abu-Mostafa, Magdon-Ismail, and Lin [74].

Important note

The soft-margin training problem can be equivalently written in terms of some optimizable parameters α_j as follows:

$$\begin{aligned} \text{Maximize} \quad & \sum_j \alpha_j - \frac{1}{2} \sum_{j,k} y_j y_k \alpha_j \alpha_k \left(\vec{x}_j \cdot \vec{x}_k\right), \\ \text{subject to} \quad & 0 \leq \alpha_j \leq C, \\ & \sum_j \alpha_j y_j = 0. \end{aligned}$$

This formulation of the SVM soft-margin training problem is, most of the time, easier to solve in practice, and it is the one that we will be working with. Once we obtain the α_j

values, it is also possible to go back to the original formulation. In fact, from the α_j values, we can recover b and w. For instance, it holds that

$$\vec{w} = \sum_j \alpha_j y_j \vec{x}_j.$$

Notice that $\vec{w}$ only depends on the training points $\vec{x}_j$, for which $\alpha_j \neq 0$. These vectors are called **support vectors** and, as you can imagine, are the reason behind the name of the SVM model.

Furthermore, we can also recover b by finding some $\vec{x}_j$ that lies at the boundary of the margin and solving a simple equation — see [74] for all the details. Then, in order to classify a point x, we can just compute

$$\vec{w} \cdot \vec{x} + b = \sum_j \alpha_j y_j \left(\vec{x}_j \cdot \vec{x}\right) + b,$$

and decide whether $\vec{x}$ goes into the positive or negative class depending on whether the result is bigger than 0 or not.

We've now covered all we need to know about how to train a support vector machine. But, with our tools, we can only train these models to obtain linear separations between data, which is, well, not the most exciting of prospects. In the next section, we will overcome this limitation with a simple yet powerful trick.

9.1.4 The kernel trick

Vanilla SVMs can only be trained to find linear separations between data elements. For example, the data shown in *Figure 9.5a* cannot be separated effectively by any SVM, because there is no way to separate it linearly.

How do we overcome this? Using **the kernel trick**. This technique consists in mapping the data from its original space $\mathbb{R}^n$ to a higher dimensional space $\mathbb{R}^N$, all in the hope that, in that space, there may be a way to separate the data with a hyperplane. This higher dimensional space is known as a **feature space**, and we will refer to the function $\varphi : \mathbb{R}^n \longrightarrow \mathbb{R}^N$ – which takes the original data inputs into the feature space – as a **feature map**.

For instance, the data in *Figure 9.5a* is in the 1-dimensional real line, but we can map it to the 2-dimensional plane with the function

$$f(x) = (x, x^2).$$

As we can see in *Figure 9.5b*, upon doing this, there is a hyperplane that perfectly separates the two categories in our dataset.

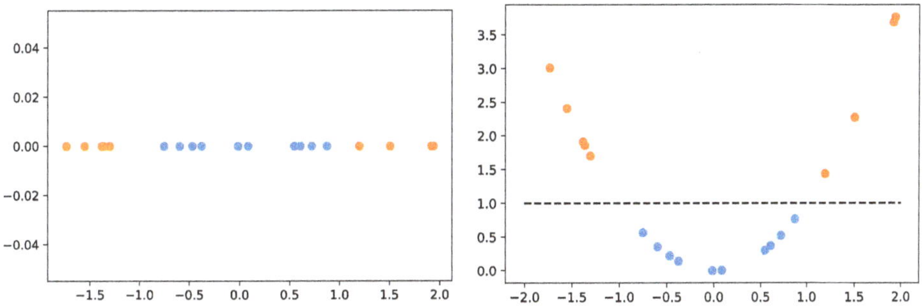

(a) Original data in the real line *(b) Data in the feature space*

Figure 9.5: The original data cannot be separated by a hyperplane, but – upon taking it to a higher-dimensional space with a feature map – it can. The separating hyperplane is represented by a dashed line

Looking at the dual form of the soft-margin SVM optimization problem, we can see how, in order to train an SVM – and to later classify new data – on a certain feature space with a feature map φ, all we need to "know" about the feature space is how to compute scalar products in it of elements returned by the feature map. This is because, during the whole training process, the only operation that depends on the $\vec{x}_j$ points is the inner product $\vec{x}_j \cdot \vec{x}_k$ – or the inner product $\vec{x}_j \cdot \vec{x}$ when classifying a new point $\vec{x}$. If instead of $\vec{x}_j$ we had $\varphi(\vec{x}_j)$, we would just need to know how to compute $\varphi(\vec{x}_j) \cdot \varphi(\vec{x}_k)$ – or $\varphi(\vec{x}_j) \cdot \varphi(\vec{x})$ to classify new data $\vec{x}$.

That is, it suffices to be able to compute the function

$$k(x, y) = \varphi(\vec{x}) \cdot \varphi(\vec{y}),$$

and that is the single and only computation that we need to perform in the feature space. This is a crucial fact. This function is a particular case of what are known as **kernel functions**. Broadly speaking, kernel functions are functions that *can be* represented as inner products in some space. Mercer's theorem (see [74]) gives a nice characterization of them in terms of certain properties such as being symmetric and some other conditions. In the cases that we will consider, these conditions are always going to be met, so we don't need to worry too much about them.

With this, we have a general understanding of how support vector machines are used in general, and in classical setups in particular. We now have all the necessary background to take the step to quantum. Get ready to explore quantum support vector machines.

9.2 Going quantum

As we have already mentioned, quantum support vector machines are particular cases of SVMs. To be more precise, they are particular cases of SVMs that rely on the kernel trick.

We have seen in the previous section how, with the kernel trick, we take our data to a feature space: a higher dimensional space in which, we hope, our data will be separable by a hyperplane with the right choice of feature map. This feature space is usually just the ordinary Euclidean space but, well, with a higher dimension. But we can consider other choices. How about...the space of quantum states?

9.2.1 The general idea behind quantum support vector machines

A QSVM works just like an ordinary SVM that relies on the kernel trick — with the only difference that it uses as feature space a certain space of quantum states.

As we discussed before, whenever we use the kernel trick, all we need from the feature space is a kernel function. That's the only ingredient involving the feature space that is necessary in order to be able to train a kernel-based SVM and make predictions with it. This idea inspired some works, such as the famous paper by Havlíček et al. [75], to try to use quantum circuits to compute kernels and, hopefully, obtain some advantage over classical computers by working in a sophisticated feature space.

Taking this into account, in order to train and then use a quantum support vector machine for classification, we will be able to do business as usual – doing everything fully classically – except for the computation of the kernel function. This function will have to rely on a quantum computer in order to do the following:

1. Take as input two vectors in the original space of data.
2. Map each of them to a quantum state through a **feature map**.
3. Compute the inner product of the quantum states and return it.

We will discuss how to implement these (quantum) feature maps in the next subsection, but, in essence, they are just circuits that are parametrized exclusively by the original (classical) data and thus prepare a quantum state that depends only on that data. For now, we will just take these feature maps as a given.

So, let's say that we have a feature map φ. This will be implemented by a circuit Φ that will depend on some classical data in the original space: for each input $\vec{x}$, we will have a circuit $\Phi(\vec{x})$ such that the output of the feature map will be the quantum state $\varphi(\vec{x}) = \Phi(\vec{x})\left|0\right\rangle$. With a feature map ready, we can then take our kernel function to be

$$k(\vec{a}, \vec{b}) = |\langle\varphi(a)|\varphi(b)\rangle|^2 = \left|\left\langle 0\right| \Phi^\dagger(a)\Phi(b)\left|0\right\rangle\right|^2.$$

And that is something that we can trivially get from a quantum computer! As you can easily check yourself, it is nothing more than the probability of measuring all zeros after preparing the state $\Phi^\dagger(\vec{a})\Phi(\vec{b})\left|0\right\rangle$. This follows from the fact that the computational basis is orthonormal.

In case you were wondering how to compute the circuit for $\Phi^\dagger$, notice that this is just the inverse of Φ, because quantum circuits are always represented by unitary operations. But Φ will be given by a series of quantum gates. So all you need to do is apply the gates in the circuit from right to left and invert each of them.

And that is how you implement a quantum kernel function. You take a feature map that will return a circuit $\Phi(\vec{x})$ for any input $\vec{x}$, you prepare the state $\Phi^\dagger(\vec{a})\Phi(\vec{b})\left|0\right\rangle$ for the pair of vectors on which you want to compute the kernel, and you return the probability of measuring zero on all the qubits.

In case you were concerned, by the way, all quantum kernels, as we have defined them, satisfy the conditions needed to qualify as kernel functions [76]. In fact, we'll now ask you to check one of those conditions!

Exercise 9.2

One of the conditions for a function k to be a kernel is that it be symmetric. Prove that, indeed, any quantum kernel is symmetric. ($k(\vec{a}, \vec{b}) = k(\vec{b}, \vec{a})$ for any inputs.)

Let's now study how to actually construct those feature maps.

9.2.2 Feature maps

A feature map, as we have said, is often defined by a parametrized circuit $\Phi(\vec{x})$ that depends on the original data and thus can be used to prepare a state that depends on it. In this section, we will study a few interesting feature maps that we will use throughout the rest of the book. They will also serve as examples that will allow us to better illustrate what feature maps actually are.

Angle encoding

We shall begin with a simple yet powerful feature map known as **angle encoding**. When used on an n-qubit circuit, this feature map can take up to n numerical inputs $x_1, \ldots, x_n$. The action of its circuit consists in the application of a rotation gate on each qubit j parametrized by the value x_j. In this feature map, we are using the x_j values as angles in the rotations, hence the name of the encoding.

In angle encoding, we are free to use any rotation gate of our choice. However, if we use R_Z gates and take $|0\rangle$ to be our initial state... the action of our feature map will have no effects whatsoever, as you can easily check from the definition of R_Z. That is why, when R_Z gates are used, it is customary to precede them by Hadamard gates acting on each qubit. All this is shown in *Figure 9.6*.

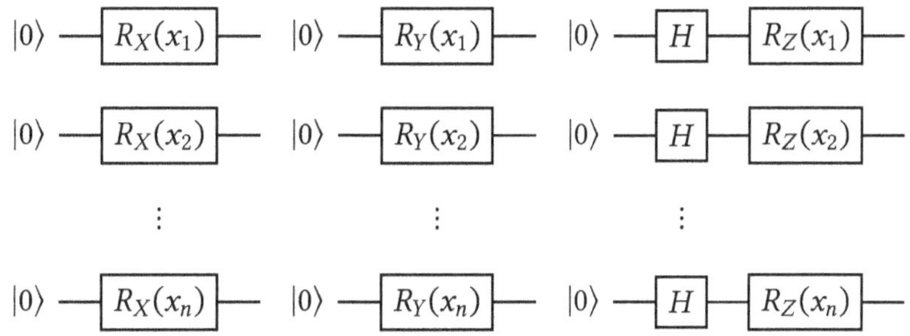

Figure 9.6: Angle encoding for an input $(x_1, \ldots, x_n)$ *using different rotation gates*

The variables that are fed to the angle encoding feature map should be normalized within a certain interval. If they are normalized between 0 and 4π, then the data will be mapped to a wider region of the feature space than if they were normalized between 0 and 1, for example. However, this would come at the cost of having the two extrema of the dataset identified under the action of the feature map. That's because 0 and 2π are exactly the same angle and, in our definition of rotation gates, we divided the input angle by 2.

The choice of normalization will thus be a trade-off between separating the extrema in the feature space and using the widest possible region in it.

Amplitude encoding

Angle encoding can take n inputs on n qubits. Does that seem good enough? Well, get ready for a big jump. The **amplitude encoding** feature map can take 2^n inputs when implemented on an n-qubit circuit. That is a lot, and it will enable us to effectively train QSVMs on datasets with a large number of variables. So, how does it work then?

If the amplitude encoding feature map is given an input $x_0, \dots, x_{2^n-1}$, it simply prepares the state

$$\left|\varphi(\vec{a})\right\rangle = \frac{1}{\sqrt{\sum_k x_k^2}} \sum_{k=0}^{2^n-1} x_k \left|k\right\rangle.$$

Notice how we've had to include a normalization factor to make sure that the output was, indeed, a quantum state. Remember from *Chapter 1, Foundations of Quantum Computing*, that all quantum states need to be normalized vectors! It's easy to see from the definition that amplitude encoding can work for any input except for the zero vector — for the zero vector, amplitude encoding is undefined. We can't divide by zero!

Implementing this feature map in terms of elementary quantum gates is by no means simple. If you want all the gory details, you can check the book by Schuld and Petruccione [71]. Luckily, it is built into most quantum computing frameworks.

By the way, when using amplitude encoding, there is an unavoidable loss of information if you decide to "push the feature map to its limit." In general, you won't be using all the 2^n parameters that it offers — you will only use some of them and fill the rest with zeros or any other value of your choice. But, if you use all the 2^n inputs to encode variables, there's a small issue: that the number of degrees of freedom of an n-qubit state is actually $2^n - 1$, not 2^n. This is, in any case, not a big deal. This loss of information can be ignored for sufficiently big values of n.

ZZ feature map

Lastly, we will present a known feature map that may bring you memories from *Chapter 5, QAOA: Quantum Approximate Optimization Algorithm*, where we implemented circuits for Hamiltonians with $Z_j Z_k$ terms. It's called the **ZZ feature map**. It is implemented by Qiskit and it can take n inputs $a_1, \dots, a_n$ on n qubits, just like angle embedding. Its parametrized circuit is constructed following these steps:

1. Apply a Hadamard gate on each qubit.
2. Apply, on each qubit j, a rotation $R_Z(2x_j)$.
3. For each pair of elements $\{j, k\} \subseteq \{1, \dots, n\}$ with $j < k$, do the following:

(a) Apply a CNOT gate targeting qubit k and controlled by qubit j.

(b) Apply, on qubit k, a rotation $R_Z\left(2(\pi - x_j)(\pi - x_k)\right)$.

(c) Repeat step 3a.

In *Figure 9.7* you can find a representation of the ZZ feature map on three qubits.

As with angle encoding, normalization plays a big role in the ZZ feature map. In order to guarantee a healthy balance between separating the extrema of the dataset and using as big a region a possible in the feature space, the variables could be normalized to $[0, 1]$ or $[0, 3]$, for example.

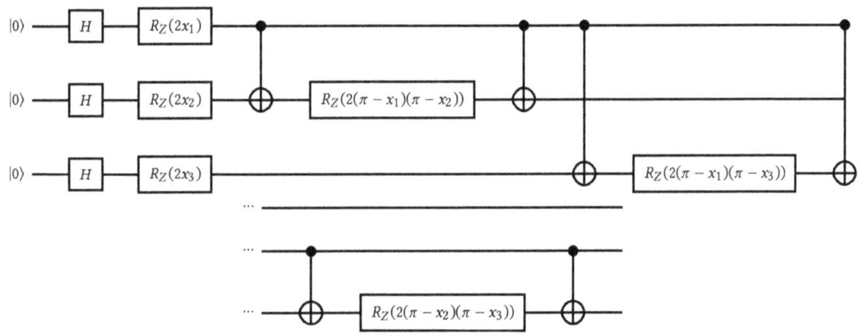

Figure 9.7: ZZ feature map on three qubits with inputs x_1, x_2, x_3

Of course, when designing a quantum feature map, your imagination is the only limit. The ones that we have presented here are some of the most popular ones — and the ones that you will find in frameworks such as PennyLane and Qiskit — but research on quantum feature maps and their properties is an active area. If you want to take a look at other possibilities, we can recommend the paper by Sim, Johnson, and Aspuru-Guzik [77].

But enough theory for now! Let's put into practice all that we have learned by implementing some QSVMs with both PennyLane and Qiskit.

9.3 Quantum support vector machines in PennyLane

It has been a long journey but, finally, we are ready to see QSVMs in action. In this section, we are going to train and run a bunch of QSVM models using PennyLane. Just to get started, let's import NumPy and set a seed so that our results are reproducible:

```
import numpy as np


seed = 1234
np.random.seed(seed)
```

9.3.1 Setting the scene for training a QSVM

Now, if we want to train QSVMs, we need some data to work with. In today's ever-changing job market, you should always keep your options open and, as promising as quantum machine learning may be, you may want to have a backup career plan. Well, we've got you covered. Have you ever dreamed of becoming a world-class sommelier? Today is your lucky day! (We are just kidding, of course, but we will use this wine theme to give some flavor to our example!)

We've already seen how the scikit-learn package offers lots of tools and resources for machine learning. It turns out that among them are a collection of pre-defined datasets on which to train ML models, and one of those datasets is a "wine recognition dataset" [78]. This is a labeled dataset with information about wines. In total, it has 13 numeric variables that describe the color intensity, alcohol concentration, and other fancy things whose meaning and significance we have no clue about. The labels correspond to the kind of wine. There are three possible labels, so, if we just ignore one, we are left with a beautiful dataset for a binary classification problem.

We can load the set with the `load_wine` function in `sklearn.datasets` as follows:

```
from sklearn.datasets import load_wine
x,y = load_wine(return_X_y = True)
```

We have set `return_X_y` to true so that we also get the labels.

You can find all the details about this dataset in its online documentation (`https://scikit-learn.org/stable/datasets/toy_dataset.html#wine-dataset` or `https://archive.ics.uci.edu/ml/datasets/Wine`, if you want to check the original source of the data). According to it, the 59 first elements in the dataset must belong to the first category (label 0) while the 71 subsequent ones have to belong to the second one (label 1). Thus, if we want to ignore the third category, we can just run the following piece of code:

```
x = x[:59+71]
y = y[:59+71]
```

And thus we have a labeled dataset with two categories. A perfect binary classification problem.

Before we proceed, however, a few disclaimers are in order. This wine recognition problem that we are going to work with is — from a machine learning point of view — very simple. You don't need very sophisticated models or a lot of computing power to tackle it. Thus, using a QSVM for this problem is overkill. It will work, yes, but that doesn't diminish the fact that we will be overdoing it. Quantum support vector machines can tackle complex problems, but we thought it would be better to keep things simple. You may call us overprotective, but we thought that using examples that could take two hours to run — or even two days! — might not be exactly ideal from a pedagogical perspective. We will also see how some examples yield better results than others. Unless we state otherwise, that won't be indicative of any general pattern. It will just mean that it so happens, some things work better than others for this particular problem. After all, from the few experiments that we will run, it wouldn't be sensible to draw hard conclusions!

With those remarks out of the way, let's attack our problem. We shall begin by splitting our dataset into a training dataset and a test dataset:

```
from sklearn.model_selection import train_test_split
x_tr, x_test, y_tr, y_test = train_test_split(x, y, train_size = 0.9)
```

We won't be making direct model comparisons, nor will we be using validation losses, so we will not use a validation dataset.

As we discussed previously, most feature maps expect our data to be normalized, and, regardless of that, normalizing your data is in general a good practice in machine learning. So that's what we shall do now! We will actually use the most simple of normalization techniques: scaling each of the variables linearly in such a way that the maximum absolute value taken by each variable be 1. This can be achieved with a `MaxAbsScaler` object from `sklearn.preprocessing` as follows:

```
from sklearn.preprocessing import MaxAbsScaler

scaler = MaxAbsScaler()
x_tr = scaler.fit_transform(x_tr)
```

And, with that, we know that — since all our variables were positive — all the values in our training dataset will be between 0 and 1. If there were negative values, our scaled variables would take values in $[-1, 1]$ instead. Notice that we have only normalized our training dataset. Normalizing the whole dataset *simultaneously* would be, in a way, cheating, because we could be polluting the training dataset with information from the test dataset. For instance, if we had an outlier in the test dataset with a very high value in some variable — a value never reached in the training dataset — this would be reflected in the normalization, and, thus, the independence of our test dataset could be compromised.

Now that the training dataset is normalized, we need to normalize the test dataset using the same proportions as the training dataset. In this way, the training dataset receives no information about the test dataset. This can be achieved with the following piece of code:

```
x_test = scaler.transform(x_test)
x_test = np.clip(x_test,0,1)
```

Notice how we have used the same `scaler` object as before, but we have called the `transform` method instead of `fit_transform`. In that way, the scaler uses the proportions that it saved before. In addition, we've run an instruction to "cut" the values in the test dataset at 0 and 1 – just in case there were some outliers and in order to comply with the normalization requirements of some of the feature maps that we will use.

9.3.2 PennyLane and scikit-learn go on their first date

We've said it countless times: QSVMs are like normal SVMs, but with a quantum kernel. So let's implement that kernel with PennyLane.

Our dataset has 13 variables. Using angle encoding or the ZZ feature map on the 13 variables would require us to use 13 qubits, which might not be feasible if we want our kernel to be simulated on some not especially powerful computers. Thus, we can resort to amplitude encoding using 4 qubits. As we mentioned before, this feature map can accept up to 16 inputs; we will fill the remaining ones with zeros – PennyLane will make that easy.

This is how we can implement our quantum kernel:

```
import pennylane as qml

nqubits = 4
dev = qml.device("lightning.qubit", wires = nqubits)

@qml.qnode(dev)
def kernel_circ(a, b):
    qml.AmplitudeEmbedding(
        a, wires=range(nqubits), pad_with=0, normalize=True)
    qml.adjoint(qml.AmplitudeEmbedding(
        b, wires=range(nqubits), pad_with=0, normalize=True))
    return qml.probs(wires = range(nqubits))
```

Now, there are a few things to digest here. We are first importing PennyLane, setting the number of qubits in a variable, and defining a device; nothing new there. And then comes the definition of the circuit of our kernel. In this definition, we are using `AmplitudeEmbedding`, which returns an operation equivalent to the amplitude encoding of its first argument. In our case, we use the arrays `a` and `b` for this first argument. They are the classical data that our kernel function takes as input. In addition to this, we also ask `AmplitudeEmbedding` to normalize each input vector for us, just as amplitude encoding needs us to do, and, since our arrays have 13 elements instead of the required 16, we set `pad_with = 0` to fill the remaining values with zeros. Also notice that we are using `qml.adjoint` to compute the adjoint (or inverse) of the amplitude encoding of `b`.

Lastly, we retrieve an array with the probabilities of measuring each possible state in the computational basis. The first element of this array (that is, the probability of getting a zero value in all the qubits) will be the output of our kernel.

Now we have our quantum kernel almost ready. If you'd like to check that the circuit works as expected, you can try it out on some elements from the training dataset. For instance, you could run `kernel_circ(x_tr[0], x_tr[1])`. If the two arguments are the same, keep in mind that you should always get 1 in the first entry of the returned array (which corresponds, as we have mentioned, to the output of the kernel).

Exercise 9.3

Prove that, indeed, any quantum kernel evaluated on two identical entries always needs to return the output 1.

Our next step will be using this quantum kernel in an SVM. Our good old scikit-learn has its own implementation, `SVC`, of support vector machines, and it works with custom kernels, so there we have it! In order to use a custom kernel, you are required to provide a `kernel` function accepting two arrays, `A` and `B`, and returning a matrix with entries (j, k) containing the kernel applied to `A[j]` and `B[k]`. Once the kernel is prepared, the SVM can be trained with the `fit` method. All of this is done in the following piece of code:

```
from sklearn.svm import SVC
def qkernel(A, B):
```

```
    return np.array([[kernel_circ(a, b)[0] for b in B] for a in A])


svm = SVC(kernel = qkernel).fit(x_tr, y_tr)
```

The training can take up to a few minutes depending on the performance of your computer. Once it is over, you can check the accuracy of your trained model with the following instructions:

```
from sklearn.metrics import accuracy_score


print(accuracy_score(svm.predict(x_test), y_test))
```

In our case, this gives an accuracy of 0.92, meaning that the SVM is capable of classifying most of the elements in the test dataset correctly.

This shows us how to train and run a quantum support vector in a fairly simple manner. But we can consider more sophisticated scenarios. Are you ready for that?

9.3.3 Reducing the dimensionality of a dataset

We have just seen how to use amplitude encoding to take full advantage of the 13 variables of our dataset while only using 4 qubits. In most cases, that is a good approach. But there are also some problems in which it may prove better to reduce the number of variables in the dataset — while trying to minimize the loss of information, of course — and thus be able to use other feature maps that could perhaps yield better results.

In this subsection, we are going to illustrate this approach. We shall try to reduce the number of variables in our dataset to 8 and, then, we will train a QSVM on the new, reduced variables using angle encoding.

If you want to reduce the dimensionality of a dataset while minimizing information loss, as we aim to do now, there are many options at your disposal. You may want to have a look at autoencoders, for instance. In any case, for the purposes of this section, we will consider a technique known as **principal component analysis**.

To learn more...

Before actually using principal component analysis, you may reasonably be curious about how this fancy-sounding technique works.

When you have a dataset with n variables, you essentially have a set of points in $\mathbb{R}^n$. With this set, you may consider what are known as the **principal directions**. The first principal direction is the direction of the line that best fits the data as measured by the mean squared error. The second principal direction is the direction of the line that best fits the data while being orthogonal to the first principal direction. This goes on in such a way that the k-th principal direction is that of the line that best fits the data while being orthogonal to the first, second, and all the way up to the $(k-1)$-th principal direction.

We thus may consider an orthonormal basis $\{v_1, \ldots, v_n\}$ of $\mathbb{R}^n$ in which v_j points in the direction of the j-th principal component. The vectors in this orthonormal basis will be of the form $v_j = (v_j^1, \ldots, v_j^n) \in \mathbb{R}^n$. Of course, the superscripts are not exponents! They are just superscripts.

When using principal component analysis, we simply compute the vectors of the aforementioned basis. And, then, we define the variables

$$\tilde{x}_j = v_j^1 x_1 + \cdots + v_j^n x_n.$$

And, lastly, in order to reduce the dimensionality of our dataset to m variables, we just keep the variables $\tilde{x}_1, \ldots, \tilde{x}_m$. This is all done under the assumption that the variables $\tilde{x}_j$ are, as we have defined them, sorted in decreasing order of relevance towards our problem.

So how do we use principal component analysis to reduce the number of variables in our dataset? Well, scikit-learn is here to save the day. It implements a `PCA` class that works in an analogous way to that of the `MaxAbsScaler` class that we used before.

This PCA class comes with a fit method that analyzes the data and figures out the best way to reduce its dimensionality using principal component analysis. Then, in addition, it comes with a transform method that can then transform any data in the way it learned to do when fit was invoked. Also, just like MaxAbsScaler, the PCA class has a fit_transform method that fits the data and transforms it simultaneously:

```
from sklearn.decomposition import PCA

pca = PCA(n_components = 8)

xs_tr = pca.fit_transform(x_tr)
xs_test = pca.transform(x_test)
```

And, with this, we have effectively reduced the number of variables in our dataset to 8. Notice, by the way, how we have used the fit_transform method on the training data and the transform method on the test data, all in order to preserve the independence of the test dataset.

We are now ready to implement and train a QSVM using angle encoding. For this, we may use the AngleEmbedding operator provided by PennyLane. The following piece of code defines the training; it is very similar to our previous kernel definition and, thus, pretty self-explanatory:

```
nqubits = 8
dev = qml.device("lightning.qubit", wires=nqubits)

@qml.qnode(dev)
def kernel_circ(a, b):
    qml.AngleEmbedding(a, wires=range(nqubits))
    qml.adjoint(qml.AngleEmbedding(b, wires=range(nqubits)))
    return qml.probs(wires = range(nqubits))
```

Once we have a kernel, we can train a QSVM as we did before, this time reusing the `qkernel` function, which will be using the new `kernel_circ` definition:

```
svm = SVC(kernel = qkernel).fit(xs_tr, y_tr)

print(accuracy_score(svm.predict(xs_test), y_test))
```

The returned accuracy on the test dataset is 1. Just a perfect classification in this case.

9.3.4 Implementing and using custom feature maps

PennyLane comes with a wide selection of built-in feature maps; you can find them all in the online documentation (`https://pennylane.readthedocs.io/en/stable/introduction/templates.html`). Nevertheless, you may want to define your own. In this subsection, we will train a QSVM on the reduced dataset using our own implementation of the ZZ feature map. Let's take feature maps into our own hands!

We can begin by implementing the feature map as a function with the following piece of code:

```
from itertools import combinations

def ZZFeatureMap(nqubits, data):

    # Number of variables that we will load:
    # could be smaller than the number of qubits.
    nload = min(len(data), nqubits)

    for i in range(nload):
        qml.Hadamard(i)
        qml.RZ(2.0 * data[i], wires = i)

    for pair in list(combinations(range(nload), 2)):
```

```
        q0 = pair[0]
        q1 = pair[1]

        qml.CZ(wires = [q0, q1])
        qml.RZ(2.0 * (np.pi - data[q0]) *
            (np.pi - data[q1]), wires = q1)
        qml.CZ(wires = [q0, q1])
```

In this implementation, we have used the `combinations` function from the `itertools` module. It takes two arguments: an array `arr` and an integer `l`. And it returns an array with all the sorted tuples of length `l` with elements from the array `arr`.

Notice how we have written the `ZZFeatureMap` function as we would write any circuit, taking advantage of all the flexibility that PennyLane gives us. Having defined this function for the ZZ feature map, we may use it on a kernel function and then train a QSVM just as we have done before:

```
nqubits = 4
dev = qml.device("lightning.qubit", wires = nqubits)

@qml.qnode(dev)
def kernel_circ(a, b):
    ZZFeatureMap(nqubits, a)
    qml.adjoint(ZZFeatureMap)(nqubits, b)
    return qml.probs(wires = range(nqubits))

svm = SVC(kernel=qkernel).fit(xs_tr, y_tr)
print(accuracy_score(svm.predict(xs_test), y_test))
```

In this case, the test accuracy is 0.77.

There's one detail to which you should pay attention here, and it is the fact that `qml.adjoint` is acting on the `ZZFeatureMap` function itself, not on its output! Remember that taking the adjoint of a circuit is the same as considering the inverse of that circuit.

That's all we had in store about QSVMs on PennyLane. Now it's time for us to see how things are done in Qiskit Land.

9.4 Quantum support vector machines in Qiskit

In the previous section, we mastered the use of QSVMs in PennyLane. You may want to review *subsection 9.3.1* and the beginning of *subsection 9.3.3*. That is where we prepare the dataset that we will be using here too. In addition to running the code in those subsections, you will have to do the following import:

```
from sklearn.metrics import accuracy_score
```

Now it's time for us to switch to Qiskit. In some ways, Qiskit can be easier to use than PennyLane — although this is probably a matter of taste. In addition, Qiskit will enable us to directly train and run our QSVM models using the real quantum computers available at IBM Quantum. Nevertheless, for now, let us begin with QSVMs on our beloved Qiskit Aer simulator.

9.4.1 QSVMs on Qiskit Aer

To get started, let us just import Qiskit:

```
from qiskit import *
```

When we defined a QSVM in PennyLane, we had to "manually" implement a kernel function in order to pass it to scikit-learn. This process is simplified in Qiskit, for all it takes to define a quantum kernel is to instantiate a `QuantumKernel` object. In the initializer, we are asked to provide a `backend` argument, which will be, of course, the backend object on which the quantum kernel will run. By default, the feature map that the quantum kernel will use is the ZZ feature map with two qubits, but we can use a different feature map by passing a value to the `feature_map` object. This value should be a parametrized circuit representing the feature map.

Defining parametrized circuits in Qiskit is actually fairly easy. If you want to use an individual parameter in a circuit, you can just import `Parameter` from `qiskit.circuit` and define a parameter object as `Parameter("label")` with any label of your choice. This object can then be used in quantum circuits. For example, we may define a circuit with an x-rotation parametrized by a value x as follows:

```
from qiskit.circuit import Parameter
parameter = Parameter("x")
qc = QuantumCircuit(1)
qc.rx(parameter, 0)
```

If you want to use an array of parameters in a circuit, you may define a `ParameterVector` object instead. It can also be imported from `qiskit.circuit` and, in addition to the mandatory label, it accepts an optional `length` argument setting the length of the array. By default, this length is set to zero. We may use these parameter vector objects as in the following example:

```
from qiskit.circuit import ParameterVector
parameter = ParameterVector("x", length = 2)
qc = QuantumCircuit(2)
qc.rx(parameter[0], 0)
qc.rx(parameter[1], 1)
```

Exercise 9.4

Define an `AngleEncodingX(n)` function that return the feature map for angle encoding using R_X rotations on `n` qubits.

Using parametrized circuits, we may define any feature map of our choice for its use in a quantum kernel; for instance, we could just send any of the `qc` objects that we have created in the previous pieces of code as the `feature_map` parameter in the `QuantumKernel` constructor. Nevertheless, Qiskit already comes with some pre-defined feature maps out of the box. For our case, we may generate a circuit for the ZZ feature map on eight qubits using the following piece of code:

```
from qiskit.circuit.library import ZZFeatureMap
zzfm = ZZFeatureMap(8)
```

As a matter of fact, this feature map can be further customized by providing additional arguments. We shall use them in the following chapter.

Once we have our feature map, we can trivially set up a quantum kernel reliant on the Aer simulator as follows:

```
from qiskit_machine_learning.kernels import QuantumKernel
from qiskit.providers.aer import AerSimulator
qkernel = QuantumKernel(feature_map = zzfm,
            quantum_instance = AerSimulator())
```

And that's all it takes! By the way, here we are using the Qiskit Machine Learning package. Please, refer to *Appendix D, Installing the Tools*, for installation instructions.

If we'd like to train a QSVM model using our freshly-created kernel, we can use Qiskit's own extension of the SVC class provided by scikit-learn. It's called `QSVC` and it can be imported from `quantum_machine_learning.algorithms`. It works just like the original `SVC` class, but it accepts a `quantum_kernel` argument to which we can pass `QuantumKernel` objects.

Thus, these are the instructions that we have to run in order to train a QSVM with our kernel:

```
from qiskit_machine_learning.algorithms import QSVC

qsvm = QSVC(quantum_kernel = qkernel)
qsvm.fit(xs_tr, y_tr)
```

As with PennyLane, this will take a few minutes to run. Notice, by the way, that we have used the reduced dataset (`xs_tr`), because we are using the ZZ feature map on 8 qubits.

Once the training is complete, we can get the accuracy on the test dataset as we have always done:

```
print(accuracy_score(qsvm.predict(xs_test), y_test))
```

In this case, the returned accuracy was 1.

That is all you need to know about how to run QSVMs on the Aer simulator. Now, let's get real.

9.4.2 QSVMs on IBM quantum computers

Training and using QSVMs on real hardware with Qiskit couldn't be easier. We will show how it can be done in this subsection.

Firstly, as we did back in *Chapter 2, The Tools of the Trade in Quantum Computing*, we will load our IBM Quantum account:

```
provider = IBMQ.load_account()
```

Naturally, for this to work, you should have saved your access token beforehand. At the time of writing, free accounts don't have access to any real quantum devices with eight qubits, but there are some with seven qubits. We can select the one that is the least busy with the following piece of code:

```
from qiskit.providers.ibmq import *

dev_list = provider.backends(
    filters = lambda x: x.configuration().n_qubits >= 7,
    simulator = False)
dev = least_busy(dev_list)
```

Of course, we will have to further reduce our data to seven variables, but we can do that very easily:

```
from sklearn.decomposition import PCA

pca = PCA(n_components = 7)

xss_tr = pca.fit_transform(x_tr)

xss_test = pca.transform(x_test)
```

And, with this, we have all the ingredients ready to train a QSVM on real hardware! We will have to follow the same steps as before — only this time using our real device as `quantum_instance` in the instantiation of our quantum kernel!

```
zzfm = ZZFeatureMap(7)
qkernel = QuantumKernel(feature_map = zzfm, quantum_instance = dev)
qsvm = QSVC(quantum_kernel = qkernel)
qsvm.fit(xss_tr, y_tr)
```

When you execute this code, all the circuit parameters are known in advance. For this reason, Qiskit will try to send as many circuits as possible at the same time. However, these jobs still have to wait in the queue. Depending on the number of points in your dataset and on your access privileges, this may take quite a long time to complete!

With this, we can bring our study of QSVMs in Qiskit to an end.

Summary

In this chapter, we first learned what support vector machines are, and how they can be trained to solve binary classification problems. We began by considering vanilla vector machines, and then we introduced the kernel trick — which opened up a world of possibilities! In particular, we saw how QSVMs are nothing more than a support vector machine with a quantum kernel.

From there on, we learned how quantum kernels actually work and how to implement them. We explored the essential role of feature maps, and discussed a few of the most well-known ones.

Finally, we learned how to implement, train, and use quantum support vector machines with PennyLane and Qiskit. In addition, we were able to very easily run QSVMs on real hardware thanks to Qiskit's interface to IBM Quantum.

And that pretty much covers how QSVMs can help you can identify wines – or solve any other classification task – like an expert, all while happily ignoring what the "alkalinity of ash" of a wine is. Who knows? Maybe these SVM models could open the door for you to enjoy a bohemian life of wine-tasting! No need to thank us.

In the next chapter, we will consider another family of quantum machine learning models: that of quantum neural networks. Things are about to get deep!

10

Quantum Neural Networks

The mind is not a vessel to be filled, but a fire to be kindled.

— Plutarch

In the previous chapter, we explored our first family of quantum machine learning models: quantum support vector machines. Now it is time for us to take one step further and consider yet another family of models, that of **Quantum Neural Networks** (**QNNs**).

In this chapter, you will learn how the notion of a quantum neural network can arise naturally from the ideas behind classical neural networks. Of course, you will also learn how quantum neural networks work and how they can be trained. Then, you will explore how quantum neural networks can actually be implemented, run, and trained using the two quantum frameworks that we have been working with so far: Qiskit and PennyLane.

These are the contents of this chapter:

- Building and training quantum neural networks
- Quantum neural networks in PennyLane

- Quantum neural networks in Qiskit: a commentary

Quantum support vector machines and quantum neural networks are probably the two most popular families of QML models, so, by the end of this chapter, you will already have a solid foundation in quantum machine learning.

To get started, let's understand how quantum neural networks work and how they can be effectively trained. Let's get to it!

10.1 Building and training a quantum neural network

Just like quantum support vector machines, quantum neural networks are what we called "CQ models" back in *Chapter 8, What is Quantum Machine Learning?,* – models with purely classical inputs and outputs that use quantum computing at some stage. However, unlike QSVMs, quantum neural networks are not a "particular case" of any classical model, although their behavior is inspired by that of classical neural networks. What is more, as we will soon see, quantum neural networks are "purely quantum" models, in the sense that their execution will only require classical computing for the preparation of circuits and the statistical analysis of measurements. Nevertheless, just like QSVMs, quantum neural networks will depend on classical parameters that will be optimized classically.

To learn more...

As you surely know by now, (quantum) machine learning is a vast field in which terms hardly ever have a unique meaning. The term "quantum neural network" can, in practice, be used to refer to any QML model that is inspired by the behavior of a classical neural network. Therefore, you should bear in mind that people may also use this name to refer to models different from the ones that we are considering to be quantum neural networks.

That should be enough of an introduction. Let's now get into the details. What actually are quantum neural networks and how do they relate to classical neural networks?

10.1.1 A journey from classical neural networks to quantum neural networks

If we do a small exercise of abstraction, we can think of the action of a classical neural network as consisting of the following stages:

1. **Data preparation**: This simply amounts to taking some (classical) input data and maybe carrying out some (simple) transformations on it. These may include normalizing or scaling the input data.

2. **Data processing**: Feeding the data through a sequence of layers that "transform" the data as it flows through them. The behavior of this processing depends on some optimizable parameters, which are adjusted in training.

3. **Data output**: Returning the output through a final layer.

Let's see how we can take this scheme and use it to define an analogous quantum model.

1. **Data preparation**: Quantum neural networks are given classical inputs (in the form of an array of numbers), but quantum computers don't work on classical data – they work on quantum states! So how can we take these classical inputs and embed them into the space of quantum states?

   That is a problem that we have already dealt with in *Section 9.2*. In order to encode the classical input of a QNN into a quantum state, we just have to use any feature map of our choice. As you know, we may also need to normalize or scale the data, of course.

   And that is how we actually "prepare the data" for a quantum neural network: feeding it into a feature map.

2. **Data processing**: At this point, we have successfully transformed our classical input into a "quantum input," in the form of a quantum state that encodes our classical data according to a certain feature map. Now, we need to figure out a way to process this input by drawing some inspiration from the processing in a classical neural network.

Trying to replicate the full, exact behavior of a classical neural network in a quantum neural network might prove not to be ideal given the state of current quantum hardware. Instead, we can look at the bigger picture.

In essence, the processing stage of a classical neural network consists in the application of some transformations that depend, exclusively, on some optimizable parameters. And that is an idea that we can very easily export to a quantum computer. We can simply define the "processing" stage of a quantum neural network as... the application of a circuit that depends on some optimizable parameters! In addition to this, as we will see later in this section, this circuit can be structured in layers in a way that somewhat reassembles the spirit of a classical neural network. This circuit will be said to be a **variational form** – they are just like the ones we studied back in *Chapter 7, VQE: Variational Quantum Eigensolver.*

3. **Data output**: Once we have a processed state, we need to return a classical output. And this shall be the result of some measurement operation; this operation can be whichever one suits our problem best!

   For instance, if we wanted to build a binary classifier with a quantum neural network, a natural choice for this measurement operation could be, for example, taking the expectation value of the first qubit when measured on the computational basis. Remember that the expectation value of a qubit simply corresponds to the probability of obtaining 1 upon measuring the qubit on the computational basis.

And those are all the ingredients that make up a quantum neural network.

As a matter of fact, feature maps and variational forms are both examples of **variational circuits**: quantum circuits that are controlled by some classical parameters. The only actual difference between a feature map and a variational form is their purpose: feature maps depend on the input data and are used to encode it, while variational forms depend on optimizable parameters and are used to transform a quantum input state.

This difference in purpose will materialize in the fact that we will often use different circuits for feature maps and variational forms. A good feature map need not be a good variational form, and vice versa.

You should keep in mind that – like all things QML – the terms "feature map" and "variational form" are not entirely universal, and different authors may refer to them with different expressions. For example, variational forms are commonly referred to as **ansatzs**, as we did back in *Chapter 7, VQE: Variational Quantum Eigensolver.*

Important note

A quantum neural network takes a classical input $\vec{x}$ and maps it to a quantum state through a feature map F. The resulting state then goes through a variational form V: a variational circuit dependent on some optimizable parameters $\vec{\theta}$. The output of the quantum neural network is the result of a measurement operation on the final state. All this can be seen, schematically, in the following figure:

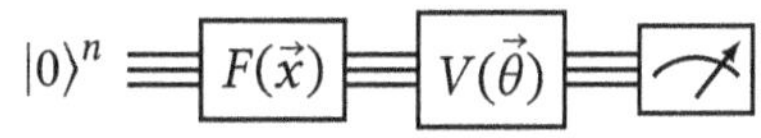

Thanks to our study of quantum support vector machines, we are already very familiar with feature maps, but we have yet to get acquainted with variational forms; that is what we will devote the next subsection to.

10.1.2 Variational forms

In principle, a variational form could be any variational circuit of your choice, but, in general, variational forms for QNNs follow a "layered structure," trying to mimic the spirit of classical neural networks. We can now make this idea precise.

If we wanted to define a variational form with k layers, we could consider k vectors of independent parameters $\vec{\theta}_1, \ldots, \vec{\theta}_k$. In order to define each layer j, we may take a variational circuit G_j dependent on the parameters $\vec{\theta}_j$. A common approach is to prepare variational forms by stacking these variational circuits consecutively and separating them by some

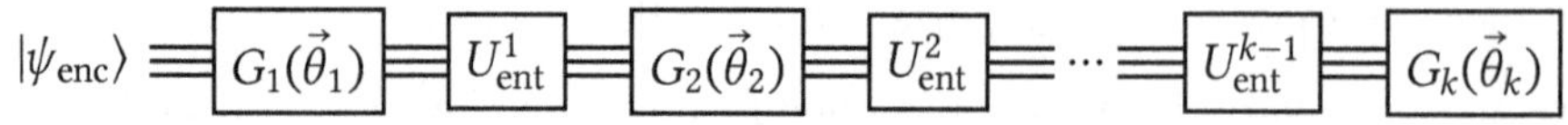

Figure 10.1: A variational form with k layers, each defined by a variational circuit G_j dependent on some parameters $\vec{\theta}_j$. The circuits U_{ent}^t are used to create entanglement, and the state $|\psi_{enc}\rangle$ denotes the output of the feature map

circuits U_{ent}^t, independent of any parameters, meant to create entanglement between the qubits. Just as in *Figure 10.1.*

We have now outlined one of the most common structures of variational forms, but variational forms are best illustrated by examples. There are lots of variational forms out there, and there is no way we could collect them all in this book — in truth, there would be no point either. For this reason, we will restrict ourselves to presenting just three variational forms, some of which we will use later in the book:

- **Two-local**: The **two-local variational form** with k repetitions on n qubits relies on $n \times (k+1)$ optimizable parameters, which we will denote as θ_{rj} with $r = 0, \ldots, k$ and $j = 1, \ldots n$. Its circuit is constructed as per the following procedure:

  procedure TwoLocal(n, k, θ)
      for all $r = 0, \ldots, k$ **do**
          ▷ *Add the r-th layer.* ◁
          for all $j = 1, \ldots, n$ **do**
              Apply a $R_Y(\theta_{rj})$ gate on qubit j.
          ▷ *Create entanglement between layers.* ◁
          if $r < k$ **then**
              for all $t = 1, \ldots, n-1$ **do**
                  Apply a CNOT gate with control on qubit t and target on qubit $t+1$.

  In *Figure 10.2* we have depicted the output of this procedure for $n = 4$ and $k = 3$. Sound familiar? The two-local variational form uses the same circuit as the angle encoding feature map for its layers, and then it relies on a cascade of controlled-NOT operations in order to create entanglement.

Notice, by the way, how the two-local variational form with k repetitions has $k + 1$ layers, not k. This tiny detail can sometimes be misleading.

The two-local variational form is very versatile, and it can be used with any measurement operation.

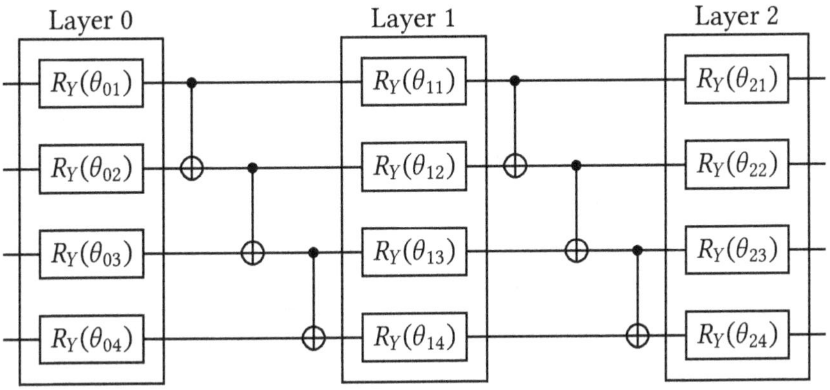

Figure 10.2: Two-local variational form on four qubits and two repetitions

- **Tree tensor**: The **tree tensor** variational form with $k + 1$ layers can be applied on $n - 2^k$ qubits. Each layer has half the number of parameters as the previous one, so the variational form relies on $2^k + 2^{k-1} + \cdots + 1$ optimizable parameters of the form

$$\theta_{rs}, \qquad r = 0, \ldots, k, \qquad s = 0, \ldots, 2^{k-r} - 1.$$

The procedure that defines is somewhat more opaque than that of the two-local variational form, and it reads as follows:

procedure TreeTensor(k, θ)

On each qubit j, apply a rotation $R_Y(\theta_{0j})$.

for all $r = 1, \ldots, k$ **do**

for all $s = 0, \ldots, 2^{k-r} - 1$ **do**

Apply a CNOT operation with target on qubit $1 + s2^r$ and controlled by qubit $1 + s2^r + 2^{r-1}$.

Apply a rotation $R_Y(\theta_{r,s})$ on qubit $1 + s2^r$.

An image is worth a thousand words, so, please, refer to *Figure 10.3* for a depiction of the output of this procedure for $k = 3$.

The tree tensor variational form fits best in quantum neural networks designed to work as binary classifiers. The most natural measurement operation that can be used in conjunction with it is the obtention of the expected value of the first qubit, as measured in the computational basis.

As a curiosity, the name of the tree tensor variational form comes from mathematical objects that are used for the simulation of physics systems and also in some machine learning models. See the survey paper by Román Orús for model details [79].

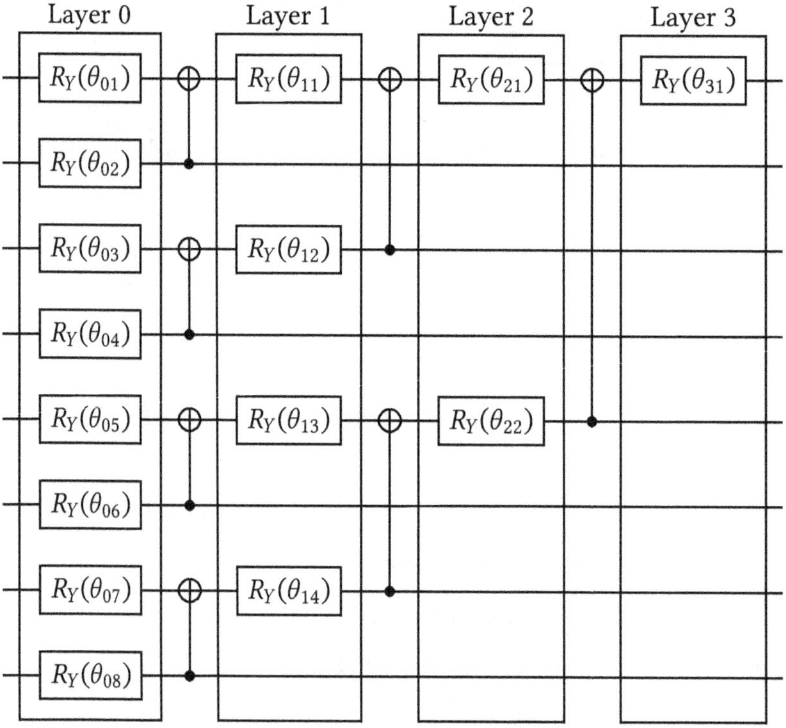

Figure 10.3: Tree tensor variational form on $8 = 2^3$ *qubits*

- **Strongly entangling layers**: The strongly entangling layers variational form acts on n qubits and can have any number k of layers. Each layer l is given a **range** r_l. In total, the variational form uses $3nk$ parameters of the form

$$\theta_{ljx}, \qquad l = 1, \ldots, k, \qquad j = 1, \ldots, n, \qquad x = 1, 2, 3.$$

The form is defined by the following algorithm:

procedure STRONGLYENTANGLINGLAYERS(n, k, r, θ)
  for all $l = 1, \ldots, k$ **do**
    for all $j = 1, \ldots, n$ **do**
      Apply a rotation $R_Z(\theta_{lj1})$ on qubit j.
      Apply a rotation $R_Y(\theta_{lj2})$ on qubit j.
      Apply a rotation $R_Z(\theta_{lj3})$ on qubit j.
    for all $j = 1, \ldots, n$ **do**
      Apply a CNOT operation controlled by qubit j and with target on qubit $[(j + r_l - 1) \bmod N] + 1$.

You may find a representation of a sample of this form in *Figure 10.4*.

As a final remark, our choice to use mostly Y rotations in the previous examples of variational forms is somewhat arbitrary. We could've also used X rotations, for example. The same goes for our choice to use controlled-X operations in the entanglement circuits. We could have used a different controlled operation, for instance. In addition to this, in the two-local variational form, there are more options for the distribution of gates in the entanglement circuit beyond the one that we have considered. Our entanglement circuit is said to have a "linear" arrangement of gates, but other possibilities are shown in *Figure 10.5*.

This is all we need to know, for now, about variational forms. Combined with our previous knowledge of feature maps, this ends our analysis of the elements of a quantum neural network…almost. We still have to dive deeper into that seemingly innocent measurement operation at the end of every quantum neural network.

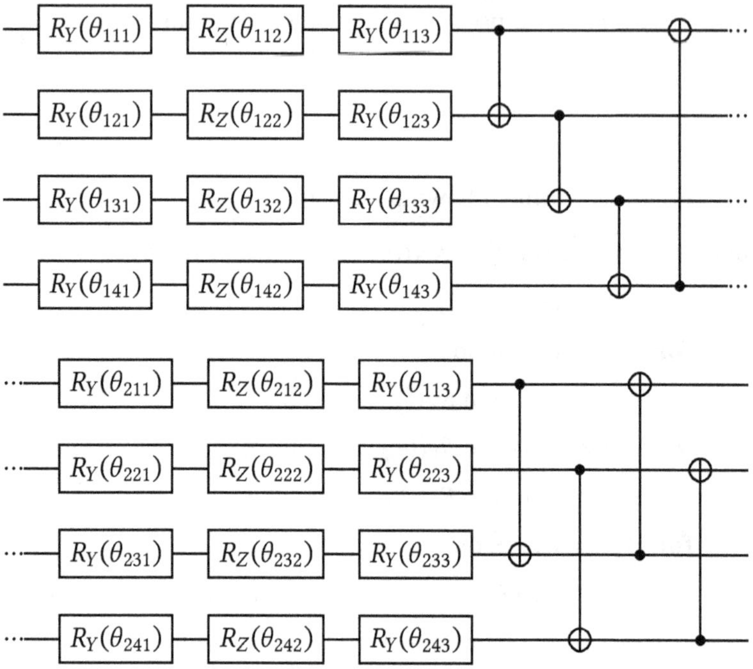

Figure 10.4: Strongly entangling layers form on four qubits and two layers with respective ranges 1 *and* 2

10.1.3 A word about measurements

As we saw back in *Chapter 7, VQE: Variational Quantum Eigensolver*, any physical observable can be represented by a Hermitian operator in such a way that all the possible outcomes of the measurement of the observable can be matched to the different eigenvalues of the operator. If you haven't done so already, please, have a look at *Section 7.1.1* if you are not familiar with this.

When we measure a single qubit in the computational basis, the coordinate matrix with respect to the computational basis of the associated Hermitian operator could well be either of

$$M = \begin{pmatrix} 1 & 0 \\ 0 & 0 \end{pmatrix}, \qquad Z = \begin{pmatrix} 1 & 0 \\ 0 & -1 \end{pmatrix}.$$

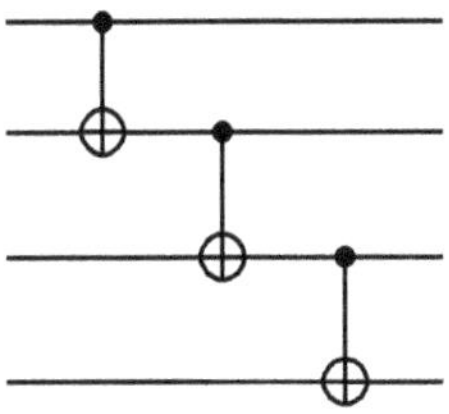

(a) Linear

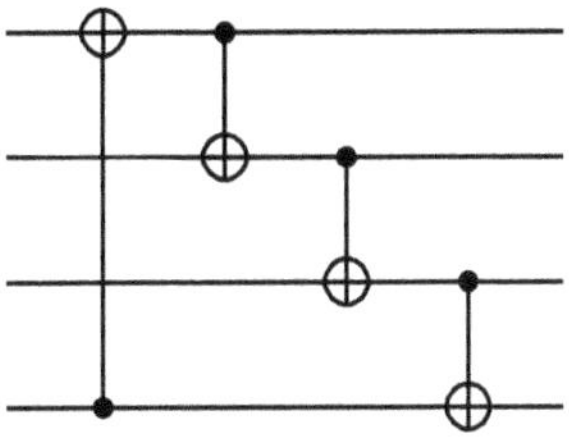

(b) Circular

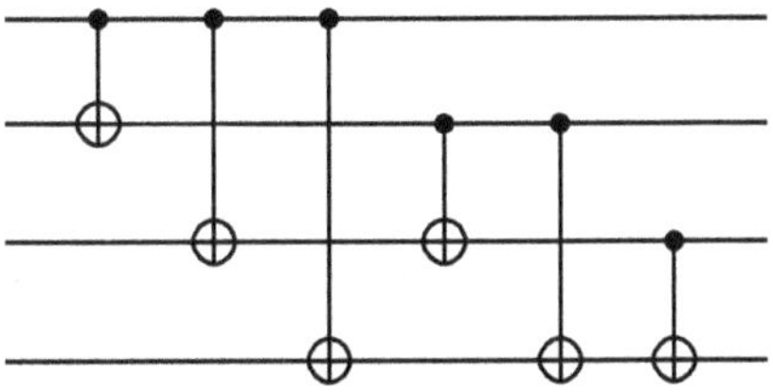

(c) Full

Figure 10.5: Different entanglement circuits

Both of these operators represent the measurement of a qubit, but they differ in the eigenvalues that they associate to the distinct outputs. The first operator associates the eigenvalues 1 and 0 to the qubit's value being 0 and 1 respectively, while the second observable associates the eigenvalues 1 and -1 to these outcomes.

Exercise 10.1

The purpose of this exercise is for you to get more familiar with Dirac notation. Show that the two previous Hermitian operators may be written, respectively, as

$$1\left|0\right\rangle\left\langle 0\right| + 0\left|1\right\rangle\left\langle 1\right| = \left|1\right\rangle\left\langle 1\right|, \quad \left|0\right\rangle\left\langle 0\right| - \left|1\right\rangle\left\langle 1\right|.$$

Hint: Remember that the product of a ket (column vector) and a bra (row vector) is a matrix. We saw an example of this back in *Section 7.2.1*.

As we will see later on in the chapter, frameworks such as PennyLane allow you to work with measurement operations defined by any Hermitian operator. This can give you a lot of flexibility when defining the measurement operation of a neural network. For instance, in an n-qubit circuit, you will be able to instruct PennyLane to compute the expectation value of the observable $M \otimes \cdots \otimes M$, which has as its coordinate representation in the computational basis the matrix

$$\begin{pmatrix} 0 & & & \\ & \ddots & & \\ & & 0 & \\ & & & 1 \end{pmatrix}_{2^n \times 2^n}.$$

Alternatively, you may want to consider the observable $Z \otimes \cdots \otimes Z$. It is easy to see how this observable will return $+1$ if an even number of qubits are measured as 0, and -1 otherwise. That's the reason why $Z \otimes \cdots \otimes Z$ is referred to as the **parity** observable.

Of course, you will also be able to take the measurement operation to be a good old expectation value on the first qubit. But, the point is, there's also a plethora of options available to you, should you want to explore them!

As we mentioned before, observables are the final building blocks of every quantum neural network architecture. Quantum neural networks accept an input, which usually consists of classical data being fed through a feature map. The resulting quantum state is then transformed by a variational form and, lastly, some (classical) numerical data is obtained through a measurement operation. In this way, we have a "black box" transforming some numerical inputs into outputs, that is, a model that — just like any other classical ML model — can be trained.

We have now defined what quantum neural networks are and learned how to construct them, at least in theory. That means we have a model. But this is quantum machine learning, so a model is not enough: we need to train it. And in order to do so, we will need, among other things, an optimization algorithm.

10.1.4 Gradient computation and the parameter shift rule

Although it is not the only option, the optimization algorithms that we shall use for quantum neural networks will be gradient descent algorithms; in particular, we will use the Adam optimizer. But, as we saw in *Chapter 8, What is Quantum Machine Learning?*, this algorithm needs to obtain the gradient of the expected value of a loss function in terms of the optimizable parameters.

Since our model uses a quantum circuit, the computation of these gradients is not entirely trivial. We shall now go briefly over the three main kinds of differentiation methods in which these gradient computations may be carried out:

- **Numerical approximation**: Of course, we have a method that always works. It may not always be the most efficient one, but it's always there. In order to compute gradients, we may just estimate them numerically. In order to do this, of course, we will have to run our quantum neural network plenty of times.

  Just to exemplify this a little bit, if we had a real-valued function taking n real inputs $f : \mathbb{R}^n \longrightarrow \mathbb{R}$, we could approximate its partial derivatives as

  $$\frac{\partial f}{\partial x_j} = \frac{f(x_1, \ldots, x_j + h, \ldots, x_n) - f(x_1, \ldots, x_n)}{h}$$

  for a sufficiently small value of h. That's, of course, the most naive way to numerically approximate a derivative, but hopefully it's enough to give you an intuition of how this works.

- **Automatic differentiation**: Given the current state of real quantum hardware, odds are that most of the quantum neural networks that you will train will run on simulators. As non-ideal as this may be, it comes with some advantages. Most notably, on simulated quantum neural networks, a classical computer may compute exact gradients using techniques similar to those employed on classical neural networks. If you are interested, the book Aurélien Géron [64, Chapter 10] and the one by Shai

Shalev-Shwartz and Shai Ben-David [65, §20.6] discuss these techniques for classical neural networks.

- **The parameter shift rule**: The standard automatic differentiation techniques can only be used on simulators. Fortunately, there is still another way to compute gradients when executing quantum neural networks on real hardware: using the **parameter shift rule**. As the name suggests, this technique enables us to compute gradients by using the same circuit in the quantum neural network, yet shifting the values of the optimizable parameters. The parameter shift rule can't always be applied, but it works on many common cases and can be used in conjunction with other techniques, such as numerical approximation.

  We won't get into the details of how this method works, but you may have a look at a research paper by Maria Schuld and others [80] for more information. For example, if you had a circuit consisting of a single rotation gate $R_X(\theta)$ and the measurement of its expectation value $E(\theta)$, you would be able to compute its derivative with respect to θ as

$$\nabla_{\theta}E(\theta) = \frac{1}{2}\left(E\left(\theta + \frac{\pi}{2}\right) - E\left(\theta - \frac{\pi}{2}\right)\right).$$

  This is similar to what happens with some trigonometric functions: for instance, you can express the derivative of the sine function in terms of shifted values of the same sine function.

  For our purposes, it will suffice to know that it exists and can be used. Of course, the parameter shift rule can also be used on simulators!

Important note

When quantum neural networks are run on simulators, gradients can be computed using automatic differentiation techniques analogous to those of classical machine learning. When they are run on either real hardware or simulators, these gradients can also be computed — at least on many cases — using the parameter shift rule.

Alternatively, numerical approximation is always an effective way to compute gradients.

As we have mentioned, all of these methods are already fully implemented in PennyLane, and we will try them all out in the following section.

To learn more...

Everything looks good and promising, but quantum neural networks also pose some challenges when it comes to training them. Most notably, they are known to be vulnerable to **barren plateaus**: situations in which the training gradients vanish and, thus, the training can no longer progress (see the paper by McClean et. al for further explanation [81]). It is also known that the kind of measurement operation used and the depth of the QNN play a role in how likely these barren plateaus are to be found. This is studied, for instance, in a paper by Cerezo and collaborators [82]. In any case, you should be vigilant when training your QNNs, and follow the literature for possible solutions should barren plateaus threaten the learning of your models.

We now have all the ingredients necessary to construct and train quantum neural networks. But before we get to do that in practice, we will discuss a few techniques and tips that will help you get the most of our brand new quantum machine learning models.

10.1.5 Practical usage of quantum neural networks

The following are a collection of ideas that you should keep in mind when designing QNN models and training them. You can think of it as a summary of the previous sections, with a few highlights from *Chapter 8, What is Quantum Machine Learning?*:

- **Make wise choices**: When you set out to design a QNN, you have three important decisions to make: you have to pick a feature map, a variational form, and a measurement operation. Be intentional about these choices and consider the problem and the data that you are working with. Your decisions can influence how likely you are to find barren plateaus, for instance. A good recommendation is to check the literature for similar problems and to build up from there.

- **Size matters**: When you use a well-designed variational form, in general, the power of the resulting quantum neural network will be directly related to the number of optimizable parameters it has. Use too many parameters, and you may have a model that overfits. Use very few, and your model may end up underfitting.
- **Optimize optimization**: For most problems, the Adam optimizer can be your go-to choice for training a quantum neural network. Remember that, as we discussed in *Chapter 8, What is Quantum Machine Learning?*, you will have to pick a learning rate and a batch size when using Adam.

  A smaller learning rate will make the algorithm more accurate, but also slower. Analogously, a higher batch size should make the optimization more effective, to the detriment of execution time.
- **Feed your QNN properly**: The data that is fed to a quantum neural network should be normalized according to the requirements of the feature map in use. In addition, depending on the dimensions of the input data, you may want to rely on dimensionality reduction techniques.

  Of course, the more data you have, the better. Nonetheless, one additional fact that you may want to take into account is that, under some conditions, quantum neural networks have been shown to need fewer data samples than classical neural networks in order to be successfully trained [83].

To learn more...

If you want to further boost the power of your quantum neural networks, you may want to consider the **data reuploading** technique [84]. In a vanilla QNN, you have a feature map F dependent on some input data $\vec{x}$, which is then followed by a variational form V dependent on some optimizable parameters $\vec{\theta}_0$. Data reuploading simply consists in repeating this scheme – any number of times you want – before performing the measurement operation of the QNN. The feature maps use the same input data in each repetition, but each instance of the variational form takes its own, independent, optimizable parameters.

This is represented in the following diagram, which shows data reuploading with k repetitions:

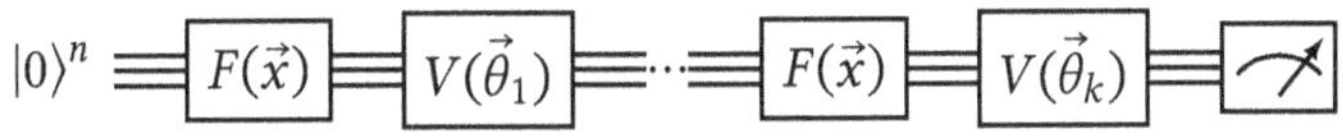

This has been shown, both in practice and in theory [85], to offer some advantages over the simpler, standard approach at the cost of increasing the depth of the circuits that are used. In any case, it is good to have it in mind when implementing your own QNNs.

This concludes our theoretical discussion of quantum neural networks. Now it's time for us to get our hands dirty with the actual implementation of all the fancy artifacts and techniques that we have discussed. In this regard, we will focus mostly on PennyLane. Let's begin!

10.2 Quantum neural networks in PennyLane

We are now ready to implement and train our first quantum neural network with PennyLane. The PennyLane framework is great for many applications, but it shines the most when it comes to the implementation of quantum neural network models. This is all due to its flexibility and good integration with classical machine learning frameworks. We, in particular, are going to be using PennyLane in conjunction with TensorFlow to train a QNN-based binary classifier. All that effort that we invested in *Chapter 8, What is Quantum Machine Learning?*, is finally going to pay off!

Important note

Remember that we are using **version 2.9.1** of the TensorFlow package and **version 0.26** of PennyLane.

Let's begin by importing PennyLane, NumPy, and TensorFlow and setting some seeds for these packages, just to make sure that our results are reproducible. We can achieve this with the following piece of code:

```
import pennylane as qml
import numpy as np
import tensorflow as tf


seed = 4321
np.random.seed(seed)
tf.random.set_seed(seed)
```

Keep in mind that you may still get slightly different results from ours if you are using different package versions. However, the results you obtain will be fully reproducible in your own machine.

Before we get to our problem, there's one last detail that we need to sort out. PennyLane works with doubles while TensorFlow uses ordinary floats. This isn't always an issue, but it's a good idea to ask TensorFlow to work with doubles just as PennyLane does. We can accomplish this as follows:

```
tf.keras.backend.set_floatx('float64')
```

With this out of the way, let's meet our problem.

10.2.1 Preparing data for a QNN

As we have already mentioned, we are going to train a QNN model to implement a binary classifier. Our recurrent use of binary classifiers is no coincidence, for binary classifiers are perhaps the simplest machine learning models to train. Later in the book, however, we will explore more exciting use cases and architectures.

For our example problem, we are going to use one of the toy datasets provided by the scikit-learn package: the "Breast cancer Wisconsin dataset" [78]. This dataset has a total of 569 samples with 30 numerical variables each. These variables describe features that can be used to characterize whether a breast mass is benign or malignant. The label of each sample can be either 0 or 1, corresponding to malignant or benign, respectively. You may find the documentation of this dataset online at `https://scikit-learn.org/stable/datasets/toy_dataset.html#breast-cancer-dataset` (the original documentation of the dataset can also be found at `https://archive.ics.uci.edu/ml/datasets/breast+cancer+wisconsin+(diagnostic)`).

We can get this dataset by calling the `load_breast_cancer` function from `sklearn.datasets`, setting the optional argument `return_X_y` to true in order to retrieve the labels in addition to the samples. For that, we can use the following instructions:

```
from sklearn.datasets import load_breast_cancer

x,y = load_breast_cancer(return_X_y = True)
```

When we trained QSVMs, since we were not going to make any comparisons between models, a training and test dataset sufficed. In our case, however, we are going to train our models with early stopping on the validation loss. This means – in case you don't remember – that we will be keeping track of the validation loss and we will halt the training as soon as it doesn't improve – according to some criteria that we will define. What is more, we will keep the model configuration that best minimized the validation loss. Using the test dataset for this purpose wouldn't be good practice, for then the test dataset would have played a role in the training and it would not give a good estimate of the true error; that's why we will need a separate validation dataset.

We can split our dataset into a training, validation, and test dataset as follows:

```
from sklearn.model_selection import train_test_split

x_tr, x_test, y_tr, y_test = train_test_split(
    x, y, train_size = 0.8)
x_val, x_test, y_val, y_test = train_test_split(
    x_test, y_test, train_size = 0.5)
```

All the variables in the dataset are non-zero, but they are not normalized. In order to use them with any of our feature maps, we shall normalize the training data between 0 and 1 using `MaxAbsScaler` as follows:

```
from sklearn.preprocessing import MaxAbsScaler

scaler = MaxAbsScaler()
x_tr = scaler.fit_transform(x_tr)
```

And we then normalize the test and validation datasets in the same proportions as the training dataset:

```
x_test = scaler.transform(x_test)
x_val = scaler.transform(x_val)

# Restrict all the values to be between 0 and 1.
x_test = np.clip(x_test, 0, 1)
x_val = np.clip(x_val, 0, 1)
```

Just as we did when we trained a QSVM in the previous chapter!

So far, we have simply done some fairly standard data preprocessing, without having to think too much about the actual architecture of our future quantum neural network. But that changes now. We have a problem to address: our dataset has 30 variables, and that can be a pretty large number for current quantum hardware. Since we don't have access to quantum computers with 30 qubits, we may consider the following choices:

- Use the amplitude encoding feature map on 5 qubits, which can accommodate up to $2^5 = 32$ variables
- Use any of the other feature maps that we have used, but in conjunction with a dimensionality reduction technique

We will go for the latter choice. You can try the other possibility on your own: it's fairly straightforward if you use the `qml.AmplitudeEmbedding` template that we studied back in *Chapter 9, Quantum Support Vector Machines.*

Exercise 10.2

As you follow along this section, try to implement a QNN using all the original variables with amplitude encoding on five qubits.
Keep in mind that, when feeding the data to the `qml.AmplitudeEmbedding` object through the features argument, instead of using the `inputs` variable, you should use `[a for a in inputs]`. This is needed because of some internal type conversions that PennyLane needs to perform.

Training a quantum neural network on a simulator is a very computationally-intensive task. We don't want anyone's computer to crash, so, just to make sure everyone can run this example smoothly, we will restrict ourselves to using 4-qubit circuits. Thus, we will use a dimensionality reduction technique to shrink the number of variables to 4, and then set up a QNN with a feature map that will take the resulting 4 input variables.

As we did in the previous chapter, we will use principal component analysis in order to reduce the number of variables in our dataset to 4:

```
from sklearn.decomposition import PCA
pca = PCA(n_components = 4)

xs_tr = pca.fit_transform(x_tr)
xs_test = pca.transform(x_test)
xs_val = pca.transform(x_val)
```

Now that we have our data fully ready, we need to choose how our quantum neural network is going to work. This is exactly the focus of the next subsection.

10.2.2 Building the network

For our case, we will choose the ZZ feature map and the two-local variational form. Neither is built into PennyLane, so we have to provide our own implementation of these variational circuits. PennyLane includes, however, a version of the two-local form with circular entanglement (`qml.BasicEntanglerLayers`), in case you want to use it in your QNNs. To implement the circuits that we need, we can just use the pseudocode that we provided in *Section 10.1.2* and do something like the following:

```
from itertools import combinations

def ZZFeatureMap(nqubits, data):

    # Number of variables that we will load:
    # could be smaller than the number of qubits.
    nload = min(len(data), nqubits)

    for i in range(nload):
        qml.Hadamard(i)
        qml.RZ(2.0 * data[i], wires = i)

    for pair in list(combinations(range(nload), 2)):
        q0 = pair[0]
        q1 = pair[1]

        qml.CZ(wires = [q0, q1])
        qml.RZ(2.0 * (np.pi - data[q0]) *
            (np.pi - data[q1]), wires = q1)
```

```
        qml.CZ(wires = [q0, q1])

def TwoLocal(nqubits, theta, reps = 1):

    for r in range(reps):
        for i in range(nqubits):
            qml.RY(theta[r * nqubits + i], wires = i)
        for i in range(nqubits - 1):
            qml.CNOT(wires = [i, i + 1])

    for i in range(nqubits):
        qml.RY(theta[reps * nqubits + i], wires = i)
```

Remember that we already implemented the ZZ feature map in PennyLane in the previous chapter.

In this chapter, we have talked about observables, and how these are represented by Hermitian operators in quantum mechanics. PennyLane allows us to work directly with these Hermitian representations.

Remember how every circuit in PennyLane returns the result of some measurement operation? For instance, you may use **return** `qml.probs(wires = [0])` at the end of the definition of a circuit in order to get the probabilities of every possible measurement outcome on the computational basis. Well, it turns out that PennyLane offers a few more possibilities. For instance, given any Hermitian matrix A (encoded as a numpy array `A`), we may retrieve the expectation value of A on an array of wires `w` at the end of a circuit simply by calling **return** `qml.expval(A, wires = w)`. Of course, the dimensions of A must be compatible with the length of `w`. This is useful in our case, for in order to get the expectation value on the first qubit, we will just have to compute the expectation value of the Hermitian

$$M = \begin{pmatrix} 1 & 0 \\ 0 & 0 \end{pmatrix}.$$

The matrix M can be constructed as follows:

```
state_0 = [[1], [0]]
M = state_0 * np.conj(state_0).T
```

In this construction, we have used the fact that $M = |0\rangle\langle 0|$, as we discussed in an exercise earlier in this chapter. This will give us, as output, a value between 0 and 1, which is perfect to construct a classifier: as usual, we will assign class 1 to every data instance with a value of 0.5 or higher, and class 0 to all the rest.

Now we have all the pieces gathered in order to implement our quantum neural network. We are going to construct it as a quantum node with two arguments: `inputs` and `theta`. The first argument is mandatory: in order for PennyLane to be able to train a quantum neural network with TensorFlow, its first argument must accept an array with all the inputs to the network, and the name of this argument must be `inputs`. After this argument, we may add as many as we want. These can correspond to any parameters of the circuit, and, of course, they need to include the optimizable parameters in the variational form.

Thus, we may implement our quantum neural network as follows:

```
nqubits = 4
dev = qml.device("default.qubit", wires=nqubits)

def qnn_circuit(inputs, theta):
    ZZFeatureMap(nqubits, inputs)
    TwoLocal(nqubits = nqubits, theta = theta, reps = 1)
    return qml.expval(qml.Hermitian(M, wires = [0]))

qnn = qml.QNode(qnn_circuit, dev, interface="tf")
```

To keep things simple, we have chosen to use just one repetition of the variational form. If your dataset is more complex, you may want to increase this number in order to have more trainable parameters.

Notice, by the way, how we have added the argument `interface = "tf"` to the quantum node initializer. This is so that the quantum node will work with tensors (TensorFlow's data object) in lieu of with arrays, just to allow PennyLane to communicate smoothly with TensorFlow. Had we used the `@qml.qnode` decorator, we would've had to include the argument in its call.

This defines the quantum node that implements our quantum neural network. Now we need to figure out a way to train it, and, for that purpose, we will rely on TensorFlow. We'll do exactly that in the next subsection.

10.2.3 Using TensorFlow with PennyLane

In *Chapter 8, What is Quantum Machine Learning?*, we already learned how TensorFlow can be used to train a classical neural network. Well, thanks to PennyLane's great interoperability, we will now be able to train our quantum neural network with TensorFlow almost as if it were a classical one.

To learn more...

PennyLane can also be integrated with other classical machine learning frameworks such as PyTorch. In addition, it provides its own tools to train models based on the NumPy package, but these are more limited.

Remember how we could construct TensorFlow models using Keras layers and joining them in sequential models? Look at this:

```
weights = {"theta": 8}
qlayer = qml.qnn.KerasLayer(qnn, weights, output_dim=1)
```

That is how you can create a Keras layer containing our quantum neural network – just as if it were any other layer in a classical model! In order to do this, we've had to call `qml.qnn.KerasLayer`, and we've had to pass a few things to it. First, of course, we've sent the quantum node with the neural network. Then, a dictionary is indexed by the names of all the node arguments that take the optimizable parameters, and specifies, for each of these arguments, the number of parameters that they take. Since we only have one such argument, `theta`, and it should contain 8 optimizable parameters (that is, it will be an array of length 8), we have sent in `{"theta: 8}`. Lastly, we've had to specify the dimension of the output of the quantum node; since it only returns a numerical expectation value, this dimension is 1.

Once we have a quantum layer, we can create a Keras model easily:

```
model = tf.keras.models.Sequential([qlayer])
```

The ability to integrate quantum nodes into neural networks with such a level of flexibility will enable us to easily construct more complex model architectures in the following chapter.

Having our model ready, we now have to pick an optimizer and a loss function, and then we can compile the model just like any classical model. In our case, we will use the binary cross entropy loss (because we are training a binary classifier, after all) and we will rely on the Adam optimizer with a learning rate of 0.005. For the remaining parameters of the optimizer, we will trust the default values. Our code is, then, as follows:

```
opt = tf.keras.optimizers.Adam(learning_rate = 0.005)
model.compile(opt, loss=tf.keras.losses.BinaryCrossentropy())
```

In addition to this, we will use early stopping on the validation loss with a patience of two epochs by using the following instructions:

```
earlystop = tf.keras.callbacks.EarlyStopping(
    monitor = "val_loss", patience = 2, verbose = 1,
    restore_best_weights = True)
```

And we are now ready to send the final instruction to get our model trained.

To learn more...

You may remember that, at some point in this chapter, we discussed the different ways in which gradients involving quantum neural networks could be computed. And you might wonder why we haven't had to deal with that in order to get our model trained.

It turns out that PennyLane already picks the best differentiation method for us in order to compute gradients. Each quantum node can use certain differentiation methods — for instance, nodes with devices that act as interfaces to real hardware can't use automatic differentiation methods, but nodes with simulators can, and most do.

Later in this section, we will discuss in detail all the differentiation methods that can be used in PennyLane.

To train our model, we just have to call the `fit` method. Since we will be using early stopping, we will be generous with the number of epochs and set it to 50. Also, we will fix a batch size of 20. For that, we can use the following piece of code:

```
history = model.fit(xs_tr, y_tr, epochs = 50, shuffle = True,
    validation_data = (xs_val, y_val),
    batch_size = 20,
    callbacks = [earlystop])
```

The output that you will get upon running this instruction will be similar to the following:

```
Epoch 1/50
23/23 [====] - 22s 944ms/step - loss: 0.8069 - val_loss: 0.7639
Epoch 2/50
23/23 [====] - 21s 932ms/step - loss: 0.7485 - val_loss: 0.7174
Epoch 3/50
23/23 [====] - 21s 930ms/step - loss: 0.7022 - val_loss: 0.6819
```

```
Epoch 4/50
23/23 [====] - 22s 957ms/step - loss: 0.6685 - val_loss: 0.6554
Epoch 5/50
23/23 [====] - 21s 925ms/step - loss: 0.6433 - val_loss: 0.6362
Epoch 6/50
23/23 [====] - 21s 915ms/step - loss: 0.6249 - val_loss: 0.6232
Epoch 7/50
23/23 [====] - 21s 916ms/step - loss: 0.6122 - val_loss: 0.6141
Epoch 8/50
23/23 [====] - 21s 931ms/step - loss: 0.6029 - val_loss: 0.6081
Epoch 9/50
23/23 [====] - 21s 931ms/step - loss: 0.5961 - val_loss: 0.6052
Epoch 10/50
23/23 [====] - 22s 951ms/step - loss: 0.5918 - val_loss: 0.6027
Epoch 11/50
23/23 [====] - 22s 948ms/step - loss: 0.5889 - val_loss: 0.6007
Epoch 12/50
23/23 [====] - 22s 964ms/step - loss: 0.5865 - val_loss: 0.5997
Epoch 13/50
23/23 [====] - 21s 926ms/step - loss: 0.5855 - val_loss: 0.5998
Epoch 14/50
23/23 [====] - 22s 956ms/step - loss: 0.5841 - val_loss: 0.5993
Epoch 15/50
23/23 [====] - 22s 958ms/step - loss: 0.5835 - val_loss: 0.5994
Epoch 16/50
23/23 [====] - 22s 946ms/step - loss: 0.5831 - val_loss: 0.5997
Epoch 16: early stopping
Restoring model weights from the end of the best epoch: 14.
```

To learn more...

If you followed all that we've done so far without having asked TensorFlow to work with doubles, everything would work just fine – although you would get slightly different results. Nonetheless, if you try to fit a model using the Lightning simulator, you do need to ask TensorFlow to use doubles.

Note that we have manually shrunk the progress bar so that the output could fit within the width of the page. Also, keep in mind that the execution time may vary from device to device, but, in total, the training shouldn't take more than 20 minutes on an average computer.

Just by looking at the raw output, we can already see that the model is indeed learning, because there is a very significant drop in both the training and validation losses as the training progresses. It could be argued that there might be a tiny amount of overfitting, because the drop in the training loss is slightly greater than that in the validation loss. In any case, let's wait until we have a look at the accuracies before coming to any final conclusions.

In this case, the training has only run for 16 epochs, so it's easy to get insights from the output returned by TensorFlow. Nonetheless, in the real world, training processes can go on for up to very large numbers of epochs, and, needless to say, in those situations the console output isn't particularly informative. In general, it's always a good practice to plot both the training and validation losses against the number of epochs, just to get a better insight into the performance of the training process. We can do this with the following instructions:

```
import matplotlib.pyplot as plt


def plot_losses(history):
    tr_loss = history.history["loss"]
    val_loss = history.history["val_loss"]
```

```
    epochs = np.array(range(len(tr_loss))) + 1
    plt.plot(epochs, tr_loss, label = "Training loss")
    plt.plot(epochs, val_loss, label = "Validation loss")
    plt.xlabel("Epoch")
    plt.legend()
    plt.show()

plot_losses(history)
```

We've decided to define a function just so that we can reuse it in future training processes. The resulting plot is shown in *Figure 10.6*.

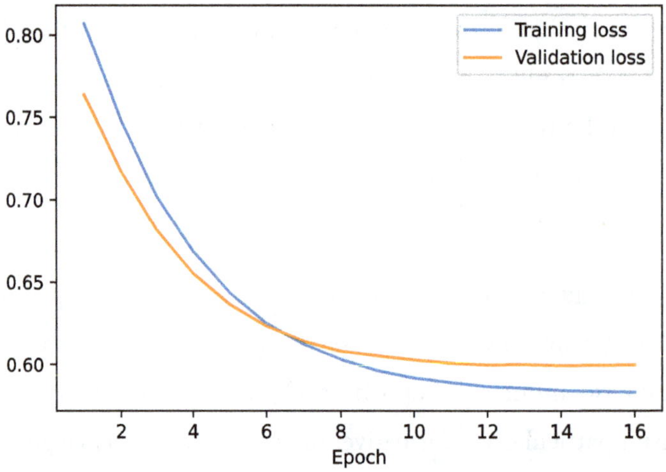

Figure 10.6: Training and validation loss functions for every epoch

And now it's time for our final test. Let's check the accuracy of our model on all our datasets to see if its performance is acceptable. This can be done with the following piece of code:

```
from sklearn.metrics import accuracy_score

tr_acc = accuracy_score(model.predict(xs_tr) >= 0.5, y_tr)
```

```
val_acc = accuracy_score(model.predict(xs_val) >= 0.5, y_val)
test_acc = accuracy_score(model.predict(xs_test) >= 0.5, y_test)

print("Train accuracy:", tr_acc)
print("Validation accuracy:", val_acc)
print("Test accuracy:", test_acc)
```

Upon running this, we get a training accuracy of 71%, a validation accuracy of 72%, and a test accuracy of 72%. These results don't reflect any kind of overfitting.

Instead of implementing your own variational forms, you may prefer to use one of PennyLane's built-in circuits. For instance, you could use the `StronglyEntanglingLayers` class. You should keep in mind, however, that the resulting variational form – as opposed to our own implementation of two-local – won't take a one-dimensional array of inputs, but a three dimensional one! In particular, this form on n qubits with l layers will take as input a three-dimensional array of size $n \times l \times 3$. Remember how, in this variational form, we need 3 arguments for the rotation gates, and there are n such gates in each of the l layers (you can take another look at *Figure 10.4*).

If you are ever in doubt, you may call the `StronglyEntanglingLayers.shape` function specifying the number of layers and the number of qubits in the respective arguments `n_layers` and `n_wires`. This will return a three-tuple with the shape that the variational form expects.

For example, we could redefine our previous QNN to use this variational form as follows:

```
nqubits = 4
dev = qml.device("default.qubit", wires=nqubits)

nreps = 2
weights_dim = qml.StronglyEntanglingLayers.shape(
    n_layers = nreps, n_wires = nqubits)
nweights = 3 * nreps * nqubits
```

```
def qnn_circuit_strong(inputs, theta):

    ZZFeatureMap(nqubits, inputs)
    theta1 = tf.reshape(theta, weights_dim)
    qml.StronglyEntanglingLayers(weights = theta1,
                                 wires = range(nqubits))

    return qml.expval(qml.Hermitian(M, wires = [0]))

qnn_strong = qml.QNode(qnn_circuit_strong, dev)

weights_strong = {"theta": nweights}
```

In this piece of code, we have stored in `nreps` the number of repetitions that we want in each instance of the variational form, in `weights_dim` the dimensions of the input that the variational form expects, and in `nweights` the number of inputs that each instance of the variational form will take. The rest is pretty self-explanatory. Inside the circuit, we've had to reshape the `theta` array of parameters to make it fit into the shape that the variational form expects; in order to do this, we've used the `tf.reshape` function, which can reshape TensorFlow's tensors while preserving all their metadata. The `weights_strong` dictionary that we defined at the end is the one that we would send to TensorFlow when constructing the Keras layer.

We've already learned how you can train a quantum neural network using PennyLane and TensorFlow. We shall now discuss a few technical details in depth before bringing this section to an end.

10.2.4 Gradient computation in PennyLane

As we have already mentioned, when you train a model with PennyLane, the framework itself figures out the best way in which to compute gradients. Different quantum nodes may be compatible with different methods of differentiation based on a variety of factors, most notably the kind of device they use.

To learn more...

For an up-to-date reference of the differentiation methods that the `default.qubit` simulator supports, you may check the online documentation at `https://docs.pennylane.ai/en/stable/introduction/interfaces.html#supported-configurations`.

You will see that the compatibility of a quantum node with a differentiation method not only depends on the device itself but also on the return type of the node and the machine learning interface (in our case, the interface was TensorFlow).

These are the differentiation methods that can be used in PennyLane:

- **Backpropagation**: Just the good old backpropagation method that is used in classical neural networks. Of course, this differentiation method only works on simulators that are compatible with automatic differentiation, because that is what is needed in order to analytically compute the gradients.

  The name of this method in PennyLane is `"backprop"`.

- **Adjoint differentiation**: This is a more efficient version of backpropagation that relies on some of the nice computational "oddities" of quantum computing, such as the fact that all the quantum circuits are implemented by unitary matrices, which are trivially invertible. Like backpropagation, this method only works on the simulators that are compatible with automatic differentiation, but it is more restrictive.

  The name of this method in PennyLane is `"adjoint"`.

- **Finite differences**: Ever took a numerical analysis course at college? Then this will sound familiar. This method implements the old-school way of computing a numerical approximation of a gradient that we discussed in the previous section. It works on almost every quantum node.

The name of this method in PennyLane is `"finite-diff"`.

- **Parameter shift rule**: PennyLane fully implements the parameter-shift rule that we introduced previously. It works on most quantum nodes.

  The name of this method in PennyLane is `"parameter-shift"`.

- **Device gradient computation**: Some devices provide their own way of computing gradients. The name of the corresponding differentiation method is `"device"`.

There are a couple of things that deserve clarification; the first of them is how a simulator could not be compatible with automatic differentiation. Oversimplifying a little bit, most simulators work by computing the evolution of the quantum state of a circuit and returning an output that is differentiable with respect to the parameters. The operations required to do all of this are themselves differentiable, and hence it's possible to use automatic differentiation on quantum nodes that use that simulator. But simulators may work differently. For instance, a simulator could return individual shots in a way that "breaks" the differentiability of the computation.

Another thing that may have caught your attention is that the finite difference method can be used on "most" quantum nodes, but not on all of them. That's because some quantum nodes may return outputs that don't make it possible for the finite differences method to work with them. For instance, if a node returns an array of samples, the differentiability is broken. If instead, it returned an expectation value — even if it were just an empirical approximation obtained from a collection of samples — then a gradient would exist and the finite differences method could be used to compute it.

Exercise 10.3

List all the PennyLane differentiation methods that can be used on quantum hardware and all the differentiation methods that can be used on simulators.

The way in which you can ask PennyLane to use a specific differentiation method — let's say one named `"method"` — is by passing the optional argument `diff_method = "method"` to the quantum node decorator or initializer. That is, if you use the QNode decorator, you should write

```
@qml.qnode(device, interface = "tf", diff_method = "method")
def qnn():
    # Circuit goes here.
```

Alternatively, if you decided to assemble a circuit `circuit` and a device `device` into a quantum node directly, you should call the following:

```
qnn = qml.QNode(circuit, device, interface = "tf",
                diff_method = "method")
```

By default, `diff_method` is set to `"best"`, which, as we've said before, lets PennyLane choose on our behalf the best differentiation method.

In our particular case, PennyLane has been using the backpropagation differentiation method all this time — without us even noticing!

To learn more...

If you want to know which differentiation method PennyLane uses by default on a device `dev` and on a certain interface `inter` (in our case, `"tensorflow"`), you can just call the following function:

```
qml.QNode.best_method_str(dev, inter)
```

Our quantum node is compatible with all the differentiation methods except with device differentiation, because `default.qubit` doesn't implement its own special way of computing gradients. Thus, just to better understand the differences in performance, we can try out all the differentiation methods and see how they behave.

To learn more...

You may remember that, when using the Lightning simulator, we do need to ask TensorFlow to use doubles all across the Keras model instead of floats — it's not an option, but a necessity. The same happens when we use differentiation methods other than backpropagation with `default.qubit`.

Let's begin with adjoint differentiation. In order to retrain our model with this differentiation method, we will rerun all our previous code, but changing the quantum node definition to the following:

```
qnn = qml.QNode(qnn_circuit, dev,
    interface="tf", diff_method="adjoint")
```

Reasonably enough, instead of rerunning all your code, you may want to add the execution of alternative differentiation methods as part of it — particularly if you are keeping your code in a notebook. If you want to do so while ensuring that the training is done in identical conditions (the same environment and seeds), these are the lines that you would have to run:

```
method = "adjoint" # Set it to whatever you want!

tf.random.set_seed(seed)

qnn = qml.QNode(qnn_circuit, dev, interface="tf",
                diff_method = method)
qlayer = qml.qnn.KerasLayer(qnn, weights, output_dim=1)
model = tf.keras.models.Sequential([qlayer])
opt = tf.keras.optimizers.Adam(learning_rate = 0.005)
model.compile(opt, loss=tf.keras.losses.BinaryCrossentropy())
history = model.fit(xs_tr, y_tr, epochs = 50, shuffle = True,
    validation_data = (xs_val, y_val),
    batch_size = 20,
    callbacks = [earlystop])
```

Upon running this, you will get the exact same training behavior that we got with backpropagation — the same evolution of the training and validation losses and, of course, the same accuracies. Where there is a noticeable change, however, is in training time. In our case, training with backpropagation took a rough average of 21 seconds per epoch. Using adjoint differentiation, in contrast, the training took, on average, 10 seconds per epoch. That's a big gain!

Actually, if you wanted to further reduce the training time, you should try the Lightning simulator with the adjoint method. Depending on the hardware configuration of your computer, it can yield very significant boosts in performance.

Let's now train our model with the two remaining differentiation methods, which are the hardware-compatible ones: the parameter-shift rule and finite differences. In order to do that, we will just have to rerun our code changing the value of the differentiation method in the quantum node definition. In order to avoid redundancy, we won't rewrite everything here — we trust these small changes to you!

When retraining with these two models, these are the results we obtained:

- Using the parameter shift rule yielded the very same results as the other differentiation methods. Regarding training time, each epoch took, on average, 14 seconds to complete. That's better than the 21 seconds that we got with backpropagation, but not as good as the 10 seconds that the adjoint method gave us.
- When using finite differences differentiation, we got, once again, the same results that the other methods yielded. On average, each epoch took 10 seconds to complete, which matches the training time of adjoint differentiation.

Keep in mind that this comparison holds for the particular model that we have considered. The results may vary as the complexity of the models increases and, in particular, hardware-compatible methods may perform more poorly on simulators when training complex QNN architectures.

And that's probably all you need to know about the differentiation methods that are available in PennyLane. Let's now have a look at what Qiskit has to offer in terms of quantum neural networks.

10.3 Quantum neural networks in Qiskit: a commentary

In the previous section, we had a chance to explore in great depth the implementation and training of quantum neural networks in PennyLane. We won't do an analogous discussion for Qiskit in such a level of detail, but we will at least give you a few ideas about how to get started should you ever need to use Qiskit in order to work with quantum neural networks.

PennyLane provides a very homogeneous and flexible experience. No matter if you're training a simple binary classifier or a complex hybrid architecture like the ones we will study in the following chapter, it's all done in the same way.

Qiskit, by contrast, provides a more "structural" approach. It gives you a suite of classes that can be used to train different kinds of neural networks and that allow you to define your networks in different ways. It's difficult to judge whether this is a better or worse approach; in the end, it's just a matter of taste. On the one hand, training basic models in Qiskit might be simpler than training them in PennyLane because of the ease of use of some of these purpose-built classes. On the other hand, having different ways of accomplishing the same thing — one could argue — might generate some unnecessary complexity.

The classes provided by Qiskit for the implementation of quantum neural networks can be imported from `qiskit_machine_learning.neural_networks` (please, refer to *Appendix D, Installing the Tools*, for installation instructions). These are some of them:

- **Two-layer QNN**: The `TwoLayerQNN` class can be used to implement a quantum neural network with a single feature map, a variational form, and an observable. It works for any vanilla quantum neural network.
- **Circuit QNN**: The `CircuitQNN` class allows you to implement a quantum neural network from a parametrized circuit. The final state of the circuit will be measured on

the computational basis, and each measurement result can be mapped to an integer label through an interpreter function. This can be useful, for instance, if you want to build a classifier.

By the way, in Qiskit lingo, variational forms are called **ansatzs**. As you surely remember, this is also the name that was used in the context of the VQE algorithm that we studied in *Chapter 7, VQE: Variational Quantum Eigensolver.*

If, when designing a neural network in Qiskit, you want to use the ZZ feature map or the two-local variational form, there's no need for you to re-implement them; they are bundled with Qiskit. You can get them as follows:

```
from qiskit.circuit.library import ZZFeatureMap, TwoLocal
nqubits = 3 # We'll do it for three qubits.
zzfm = ZZFeatureMap(nqubits, reps = 1)
twol = TwoLocal(nqubits, 'ry', 'cx', 'linear', reps = 1)
# Change rep(etition)s above to suit your needs.
```

In the call to the ZZ feature map class, we have set the number of repetitions to 1 – any other number would yield a feature map with that number of repetitions of the ZZ feature map scheme. In the call to the two-local class, we have also specified – in addition to the repetitions – the rotation gates, the controlled gates, and the entanglement layout that we want to use.

For the sake of example, we can define a `TwoLayer` quantum neural network on three qubits with the ZZ feature map and two-local variational form that we have just instantiated. We can do this as follows:

```
from qiskit_machine_learning.neural_networks import TwoLayerQNN
from qiskit.providers.aer import AerSimulator

qnn = TwoLayerQNN(nqubits, feature_map = zzfm, ansatz = twol,
                  quantum_instance = AerSimulator(method="statevector"))
```

Since we haven't specified an observable, the resulting QNN will return the expectation value of the $Z \otimes Z \otimes Z$ observable measured after feeding the execution of the network's circuit.

We can simulate analytically the network that we have just created on some random inputs and optimizable parameters as follows:

```
qnn.forward(np.random.rand(qnn.num_inputs),
            np.random.rand(qnn.num_weights))
```

The first argument is an array with some (random) classical inputs while the second argument is an array with (random) values for the optimizable parameters. Notice how we've used the `qnum_inputs` and `num_weights` properties of the quantum neural network.

All the neural network classes that we have presented are subclasses of a `NeuralNetwork` class. For example, should you want to train a neural network as a classifier, you could rely on Qiskit's `NeuralNetworkClassifier` class. This class can be initialized with a `NeuralNetwork` object and specifying a loss function and an optimizer among other things.

In addition to this, there is a subclass of `NeuralNetworkClassifier` that can be used to readily create a trainable neural network classifier directly, providing a feature map, a variational form, an optimizer, a loss, and so on.

This subclass is called `VQC` (short for Variational Quantum Classifier) and it can also be imported from the Qiskit module `qiskit_machine_learning.algorithms.classifiers`.

If you wanted to create a neural network classifier object from our previous `qnn` object using the default parameters provided by Qiskit, you could run the following instructions:

```
from qiskit_machine_learning.algorithms.classifiers import \
    NeuralNetworkClassifier

classifier = NeuralNetworkClassifier(qnn)
```

By default, the classifier will use the squared error loss function and rely on the SLSQP optimizer [86].

Then, if you had some training data `data_train` with labels `labels_train`, you could train your newly-created classifier by calling the `fit` method as follows:

```
classifier.fit(data_train, labels_train)
```

If you then wanted to compute the outcomes of the trained classifier on some data `data_test`, you could use the `predict` method like so:

```
outcomes = classifier.predict(data_test)
```

Alternatively, if you wanted to compute the accuracy score of the trained model on some test dataset (`data_test` and `labels_test`), you could run the following instruction:

```
acc = classifier.score(data_test, labels_test)
```

Nevertheless, you shouldn't care too much about the `NeuralNetworkClassifier` and `VQC` classes because, as it turns out, there is an alternative – and, in our opinion, better – way to train QNNs in Qiskit. We will discuss it in the following chapter, and it will involve an interface with an existing machine learning framework, PyTorch. What is more, being able to work with this interface will allow us to explore Qiskit's "Torch Runtime": a Qiskit utility that will enable us to more efficiently train QNNs on IBM's real quantum hardware. This is the same technique that we used in *Chapter 5, QAOA: Quantum Approximate Optimization Algorithm*, to run QAOA executions on quantum hardware. Exciting, isn't it? Bear with us until the end of the next chapter.

Summary

This has been a long journey, hasn't it? In this chapter, we first introduced quantum neural networks as quantum analogs of classical neural networks. We have seen how the training of a quantum neural network is very similar to that of a classical one, and we've also explored the differentiation methods that make this possible.

With the theory out of the way, we got our keyboards ready to do some work. We learned how to implement and train a quantum neural network using PennyLane, and we also discussed some technicalities about this framework, such as details about the differentiation methods that it provides.

PennyLane comes with some wonderful simulators, but — as we already mentioned in *Chapter 2, The Tools of the Trade in Quantum Computing* — it's also integrated with quantum hardware platforms such as Amazon Braket and IBM Quantum. Thus, your ability to train quantum neural networks on actual quantum computers is at your fingertips!

We concluded the chapter with a short overview of how to work with quantum neural networks in Qiskit.

By now, you have a solid understanding of quantum neural networks. Combined with your previous knowledge of quantum support vector machines, this gives you a fairly solid foundation in quantum machine learning. In the following chapter — which will be very practically-oriented — we will explore more complex model architectures based on quantum neural networks.

11

The Best of Both Worlds: Hybrid Architectures

Unity makes strength.

— English aphorism

By now, we have a solid understanding of both classical and quantum neural networks. In this chapter, we will leverage this knowledge to explore an interesting kind of model: hybrid architectures of quantum neural networks.

In this chapter, we will discuss what these models are and how they can be useful, and we will also learn how to implement and train them with PennyLane and Qiskit. The whole chapter is going to be very hands-on, and we will also take the time to fill in some gaps regarding the actual practice of training models in real-world scenarios. In addition to this — and just to spice things up a bit — we will go beyond our usual binary classifiers and also consider other kinds of problems.

We'll cover the following topics in this chapter:

- The what and why of hybrid architectures
- Hybrid architectures in PennyLane (with a brief overview of best practices for training models in real-world scenarios and an introduction to multi-class classification problems)
- Hybrid architectures in Qiskit (with an introduction to PyTorch)

This is going to be a very exciting chapter. Let's begin by giving meaning to these hybrid architectures.

11.1 The what and why of hybrid architectures

Up until now, we've used the adjective "hybrid" to describe algorithms that rely on both classical and quantum processing; algorithms such as QAOA or VQE fit in this category, as well as the training of QSVMs and QNNs. When we talk about **hybrid architectures** or **hybrid models**, however, we refer to something more specific: we speak about models that combine classical models with other quantum-based models by joining them together and training them as a single unit. Of course, the training of hybrid models will itself be a hybrid algorithm. We know that the terminology might be confusing, but what can we do? Hybrid is too versatile a word to give it up.

In particular, we will combine quantum neural networks with classical neural networks, for they are the two models that fit more naturally together. The way we will go about doing this will be by taking a usual classical neural network and plugging in a quantum neural network as one of its layers. In this way, the "quantum layer" will take as input the outputs of the previous layer (or the inputs to the model, if there's no layer before it) and will feed its output to the next layer (should there be any). The output of the quantum neural network will be a numerical array of length k; thus, in the eyes of the next layer, the quantum layer will behave as if it were a classical layer with k neurons.

These hybrid architectures combining classical and quantum neural networks are said to be, to the surprise of no one, **hybrid quantum neural networks**.

Important note

In summary, a hybrid QNN is a classical neural network in which one or more of its layers have been replaced by quantum layers. These are quantum neural networks that get inputs from the outputs of the previous layer and feed their outputs to the next one. Of course, if there's no next layer, the output of the quantum layer will be the output of the network. Analogously, if there's no previous layer, the input to the quantum network will be the model's input.

As we've already hinted, a hybrid neural network is trained as a single unit: the training process involves the optimization of both the parameters of the classical layers and those of the quantum neural networks inside the quantum layers.

To make the whole definition of hybrid QNNs more clear, let us consider a simple example of how one such network may be constructed:

1. The hybrid QNN must begin taking some classical inputs. Let's say it takes 16.
2. We may then feed the input into a usual classical layer with 8 neurons and use the sigmoid activation function.
3. Then, we will add a quantum layer. This quantum layer will have to accept 8 inputs from the previous layer. For example, we could use a QNN with three qubits using amplitude encoding. The output of this quantum layer could be, for instance, the expectation values of the first and second qubits, both measured on the computational basis. In this case, this quantum layer that we have added will return two numeric values.
4. Finally, we may add a classical layer with a single neuron that uses the sigmoid activation function. This layer will take inputs from the quantum layer, so it will accept two inputs. It will essentially treat the quantum layer as if it were a classical layer with two neurons.

And that's how you can build yourself a simple hybrid QNN — at least in theory! But the question is... why would we want to do such a thing? What are these hybrid models good for? Let's illustrate it with a typical example.

In the previous chapter, we learned how to use a QNN to tackle a (binary) classification task. But, due to the limitations of current quantum hardware and simulators, we were forced to apply some dimensionality reduction techniques on our data before we could use it. That's a situation where hybrid QNNs may prove useful: why not combine, in a single model, classical dimensionality reduction carried out by a classical neural network with classification performed by a quantum neural network?

In this way, instead of first reducing the dimensionality of our data and then classifying it with a quantum neural network, we could consider a hybrid QNN with

- a bunch of classical layers that would reduce the dimensionality of our data,
- joined to a quantum layer that would be in charge of making the classification.

Of course, since the whole network would be trained as a single unit, there would be no way to truly tell whether the classical part of the network is only doing dimensionality reduction and the quantum part is only doing classification. Most likely, both parts will work on both tasks to some degree.

Before proceeding any further, a few disclaimers are in order. First and foremost: quantum layers are not any sort of magical tool that will surely lead to great improvements in the performance of a classical neural network. Actually, if used unwisely, quantum layers could very easily have a negative impact on your model! The key takeaway is that you shouldn't blindly use a quantum layer solely as a replacement for a classical layer in a network. Be intentional. If you are going to include a quantum layer in your model, think about the role it's going to play in it.

In addition, when working with hybrid QNNs, you should watch out for how you are joining classical and quantum layers together. For instance, if you have a quantum layer using a feature map that requires its inputs to be normalized, maybe using an ELU activation function in the previous layer isn't the best of ideas, because it is in no way bounded. On the other hand, in that case, a sigmoid activation function could be a great fit for the previous layer.

In the use case that we discussed a few paragraphs ago (combining classical data reduction with quantum classification), we can witness the "intentionality" that we've just talked about. We do know that, in principle, a neural network can do a good job at reducing data dimensionality; in case you didn't know, it's a known fact that, using something called **autoencoders** [64, Chapter 17], one can train an **encoder** network that can reduce the dimensionality of a dataset. And we know that a quantum neural network can do a good job at classifying data coming from a dimensionality reduction technique (just have a look at the previous chapter!). So there must be some choice of parameters such that the combined hybrid model will successfully accomplish both tasks. Hence, with the right training, our hybrid model should be able to perform at least as well as it would if a classical encoder and a quantum classifier were trained separately. And the important bit is the "at least," because when training the classical encoder and the quantum classifier together we can join their powers!

And that's the heuristic justification behind this interesting application of hybrid neural networks. Actually, this is the use case that we will devote this chapter to. However, this is by no means the only application of hybrid models!

To learn more...

Hybrid architectures can also be used in regression problems, as we will later see in an exercise. In fact, this is a very interesting application, for Skolit et al. [68] have shown that adding a final layer with trainable parameters that transform the output of a quantum neural network can be very beneficial in certain reinforcement learning problems.

Now we promised that this chapter would be very hands-on, and we are going to honor that. That should have been enough of a theoretical introduction, so let's gear up! Get ready to train a bunch of hybrid QNNs to classify some data.

11.2 Hybrid architectures in PennyLane

In this section, we are going to use PennyLane to implement and train a couple of hybrid QNNs in order to solve some classification problems. Firstly, we will tackle a binary classification problem, just to better understand how hybrid QNNs work in a familiar setting. Then, we will take one step further and do the same for a multi-class classification problem.

Before we get to the problems, though, let us set things up.

11.2.1 Setting things up

As on previous occasions, we shall begin by importing NumPy and TensorFlow and setting a seed for both packages — all to ensure the reproducibility of our results:

```
import numpy as np
import tensorflow as tf

seed = 1234
np.random.seed(seed)
tf.random.set_seed(seed)
```

Now we can also import some useful functions from scikit-learn. We've already used them extensively — they need no introduction!

```
from sklearn.metrics import accuracy_score
from sklearn.model_selection import train_test_split
```

In this chapter, we will generate our own datasets to have more flexibility. In order to create them, we will rely on the `make_classification` function in the scikit-learn package. Remember that we introduced it in *Chapter 8, What Is Quantum Machine Learning?*:

```
from sklearn.datasets import make_classification
```

Also, in this section, we will use the Lightning simulator with adjoint differentiation in order to get a good performance. Thus, we need to change the default datatype used by Keras models:

```
tf.keras.backend.set_floatx('float64')
```

We can now import PennyLane and define the hermitian matrix M that we used in the previous chapter. Recall that it corresponds to the observable that assigns the eigenvalue 1 to $|0\rangle$ and the eigenvalue 0 to $|1\rangle$; that is, $M = |0\rangle\langle 0|$.

```
import pennylane as qml

state_0 = [[1], [0]]
M = state_0 * np.conj(state_0).T
```

Lastly, we may import Matplotlib and reuse the function that we defined in the previous chapter for plotting training and validation losses:

```
import matplotlib.pyplot as plt

def plot_losses(history):
    tr_loss = history.history["loss"]
    val_loss = history.history["val_loss"]
    epochs = np.array(range(len(tr_loss))) + 1
    plt.plot(epochs, tr_loss, label = "Training loss")
    plt.plot(epochs, val_loss, label = "Validation loss")
    plt.xlabel("Epoch")
    plt.legend()
    plt.show()
```

And that's all we need to get started. Let's go for our first problem.

11.2.2 A binary classification problem

We are now ready to build our first hybrid QNN and train it to solve a binary classification task. Of course, the first thing we need is data and, as we discussed in the previous section, we shall generate it using the `make_classification` function. Using a hybrid QNN that will "combine classical encoding with quantum classification" can make sense if, for instance, we have a large number of variables (features) in our dataset, so we will generate a dataset with 20 variables – that might already be quite large for current quantum hardware! Just to make sure that we have enough data, we will generate 1000 samples. This is how we can do it:

```
x, y = make_classification(n_samples = 1000, n_features = 20)
```

By default, the `make_classification` functions generate datasets with two possible classes. Just what we wanted!

As usual, we will have to split this dataset into some training, validation, and test datasets:

```
x_tr, x_test, y_tr, y_test = train_test_split(
    x, y, train_size = 0.8)
x_val, x_test, y_val, y_test = train_test_split(
    x_test, y_test, train_size = 0.5)
```

With our data ready, we need to think about the model that we will use. Let's begin by constructing the quantum layer (the QNN) that we will include at the end of the network.

For this problem, we will use the two-local variational form that we introduced in the previous chapter (see *Figure 10.2*). As you surely remember, we can implement it in PennyLane as follows:

```
def TwoLocal(nqubits, theta, reps = 1):

    for r in range(reps):
        for i in range(nqubits):
            qml.RY(theta[r * nqubits + i], wires = i)
```

```
        for i in range(nqubits - 1):
            qml.CNOT(wires = [i, i + 1])

    for i in range(nqubits):
        qml.RY(theta[reps * nqubits + i], wires = i)
```

We will take the quantum layer to be a simple QNN on four qubits using angle embedding as a feature map followed by the two-local variational form that we have just implemented. The measurement operation in the QNN will be the computation of the expectation value of M on the first qubit; that's a sensible choice for binary classifiers in general, because it returns a value between 0 and 1. The QNN can be defined as follows:

```
nqubits = 4
dev = qml.device("lightning.qubit", wires = nqubits)

@qml.qnode(dev, interface="tf", diff_method = "adjoint")
def qnn(inputs, theta):
    qml.AngleEmbedding(inputs, range(nqubits))
    TwoLocal(nqubits, theta, reps = 2)
    return qml.expval(qml.Hermitian(M, wires = [0]))

weights = {"theta": 12}
```

Notice how we have already declared the weights dictionary that we will have to send to the TensorFlow interface in order to create the quantum layer. In it, we've specified that our variational form uses $4 \cdot (2 + 1) = 12$ weights.

We will define our hybrid QNN to have an input layer with 20 inputs in order to match the dimension of our data. This will be followed by a classical layer, which will be immediately followed by the quantum neural network (the quantum layer). Since our QNN accepts 4 inputs, the classical layer will have 4 neurons itself. Moreover, for the QNN to work optimally, we need the data to be normalized, so the classical layer will use a sigmoid activation function. We can define this model in Keras as follows:

```
model = tf.keras.models.Sequential([
    tf.keras.layers.Input(20),
    tf.keras.layers.Dense(4, activation = "sigmoid"),
    qml.qnn.KerasLayer(qnn, weights, output_dim=1)
])
```

To learn more...

When defining the Keras model, you may be tempted to store the quantum layer in a variable and then use it in the model definition, as follows:

```
qlayer = qml.qnn.KerasLayer(qnn, weights, output_dim=1)
model = tf.keras.models.Sequential([
    tf.keras.layers.Input(20),
    tf.keras.layers.Dense(4, activation = "sigmoid"),
    qlayer
])
```

This code will work and, a priori, there's nothing wrong with it. However, if you decide to reset or modify your model, you will also have to rerun the first line, with the definition of `qlayer`, if you want to re-initialize the optimizable parameters (weights) in the quantum neural network!

Having the model ready, we can also define our usual early stopping callback:

```
earlystop = tf.keras.callbacks.EarlyStopping(
    monitor="val_loss", patience=2, verbose=1,
    restore_best_weights=True)
```

We've set the patience to 2 epochs in order to speed up the training; having a higher patience may easily lead to better results!

And now, all it takes for us to train our model is to — just as we've always done on TensorFlow — pick an optimizer, compile our model with the binary cross entropy loss function, and call the `fit` method with the appropriate arguments:

```
opt = tf.keras.optimizers.Adam(learning_rate = 0.005)
model.compile(opt, loss=tf.keras.losses.BinaryCrossentropy())

history = model.fit(x_tr, y_tr, epochs = 50, shuffle = True,
    validation_data = (x_val, y_val),
    batch_size = 10,
    callbacks = [earlystop])
```

Et voilà! In just a matter of minutes, your flashy hybrid model will have finished training. Take a moment to reflect on how easy this was. You have been able to train a hybrid QNN with full ease, just as if it were a simple QNN. With PennyLane, quantum machine learning is a piece of cake.

To check how well the training went, we can plot the training and validation losses with our custom function:

```
plot_losses(history)
```

The generated plot can be found in *Figure 11.1*. Those losses look really good; there don't seem to be signs of overfitting and the model appears to be learning. In any case, let's compute the test accuracy. We may also compute the training and validation accuracies, just for reference:

```
tr_acc = accuracy_score(model.predict(x_tr) >= 0.5, y_tr)
val_acc = accuracy_score(model.predict(x_val) >= 0.5, y_val)
test_acc = accuracy_score(model.predict(x_test) >= 0.5, y_test)
print("Train accuracy:", tr_acc)
print("Validation accuracy:", val_acc)
print("Test accuracy:", test_acc)
```

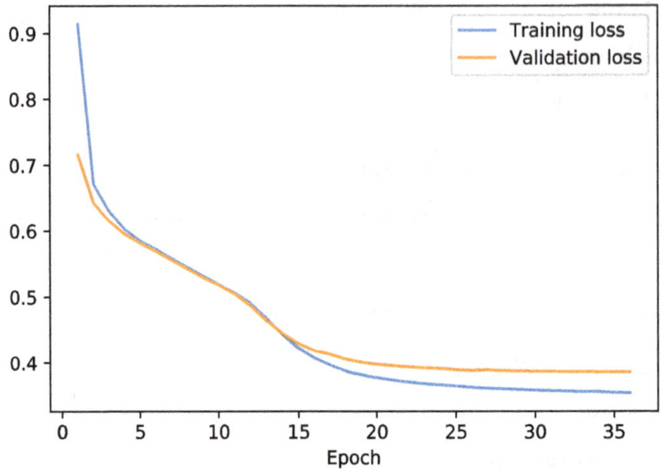

Figure 11.1: Evolution of the training and validation loss functions in the training of a hybrid QNN binary classifier

When running the preceding code, we can see how our model has a training accuracy of 95%, a validation accuracy of 90%, and a test accuracy of 96%.

That's a very satisfactory result. We have just trained our first hybrid QNN binary classifier, and we've seen how it can be effectively used to solve classification tasks.

> **Exercise 11.1**
>
> Try to solve this problem using two additional (dense) classical layers, with 16 and 8 neurons each. Compare the results.

Now, we said that this chapter was going to be hands-on and we truly meant it. So far, we have just trained models and gotten them right in one shot, but that's something that rarely happens in practice. That's why we've put together a small subsection on how to optimize models in real-world conditions.

11.2.3 Training models in the real world

Whether you believe it or not, we care for you, our dear reader. All this time, behind each and every model that we've trained, we've invested hours of meticulous parameter selection and model preparation — all to make sure that the results we give you are good enough, if not optimal.

When you set out to train models on your own, you will soon find out that things don't always work as well as you expected. For each well-performing model, there will be tens or even hundreds of discarded ones. And that's something you need to prepare yourself for.

At the early stages of a machine learning project in general — and a quantum machine learning project in particular — you should address two main following questions:

- **How will you log all your results?** When you train lots of models, you need to find a way to log their performances together with their architectures and the parameters used in their training. That way, you can easily identify what works and what doesn't, and you can avoid repeating the same mistakes.
- **How will you explore variations of your models?** Keeping a separate script for every model can be manageable when you are not training many models, but this isn't a solution for large-scale projects. Oftentimes, you want to try a wide range of configurations and see which one works best. And automation can truly make your life easier in this regard.

We leave the first question to you. In truth, there's no universal way to address it — it all depends on the problem at hand and on the training strategy that you take. However, in regard to the second question, we do have something to offer.

When training a model, choosing good hyperparameters — such as a good batch size or learning rate — is not an easy task, but it is a crucial one. Should you use a smaller or a larger learning rate? How many layers should you use? Of what type? Decisions, decisions, decisions! The number of possibilities grows exponentially, so it is impossible to explore every one of them. But, in machine learning, finding a good configuration can be the difference between success and failure. How can we do this systematically and (kind of) effortlessly?

There are quite a few packages and utilities out there that can help you automate the search for optimal training parameters. One of the most popular ones is the Optuna package, which we are about to demonstrate. Please refer to *Appendix D, Installing the Tools*, for installation instructions.

To learn more...

The process of automatically searching for optimal training parameters in a machine learning problem fits into what is known as **automated machine learning**, usually abbreviated as **AutoML**. This refers to the use of automation in order to solve machine learning problems. Having machines in charge of training other machines!

Once you've installed Optuna, you can import it as follows:

```
import optuna
```

We are going to use Optuna to find the best possible learning rate between the values 0.001 and 0.1. In order to do this, we need to define a function (which we shall call `objective`) with a single argument (`trial`). The objective function should use the training parameters that we want to optimize — in a manner that we will soon make precise — and it should return whichever metric we want to optimize. For instance, in our case, we would like to maximize the validation accuracy, so the objective function should train a model and return the validation accuracy.

The `trial` argument of the `objective` function is meant to represent an object of the `Trial` class that can be found in the `optuna.trial` module. We will use this object to define, within the objective function itself, the training parameters that we want to optimize, while also specifying their constraints: whether we want them to be integers or floats, the ranges within which we want our values to be, and so on.

For our case, this is the objective function that we would have to define:

```
def objective(trial):
    # Define the learning rate as an optimizable parameter.
    lrate = trial.suggest_float("learning_rate", 0.001, 0.1)
```

```
    # Define the optimizer with the learning rate.
    opt = tf.keras.optimizers.Adam(learning_rate = lrate)

    # Prepare and compile the model.
    model = tf.keras.models.Sequential([
        tf.keras.layers.Input(20),
        tf.keras.layers.Dense(4, activation = "sigmoid"),
        qml.qnn.KerasLayer(qnn, weights, output_dim=1)
    ])
    model.compile(opt, loss=tf.keras.losses.BinaryCrossentropy())

    # Train it!
    history = model.fit(x_tr, y_tr, epochs = 50, shuffle = True,
        validation_data = (x_val, y_val),
        batch_size = 10,
        callbacks = [earlystop],
        verbose = 0 # We want TensorFlow to be quiet.
    )

    # Return the validation accuracy.
    return accuracy_score(model.predict(x_val) >= 0.5, y_val)
```

Notice how we have defined the learning rate as an optimizable parameter by calling the `trial.suggest_float("learning_rate", 0.001, 0.1)` method. In general, if you want to optimize a parameter named `"parameter"`, the following applies:

- If the data type of the parameter is a float and the parameter is bounded between `m` and `M`, you should call the `suggest_float("parameter", m, M)` method. If you only want your parameter to take discrete values between `m` and `M` separated by a step s, you can send the optional argument `step = s`, which defaults to `None` (by default, the parameter will take continuous values).

- If the data type of the parameter is an integer bounded between `m` and `M`, you should call `suggest_int("parameter", m, M)`. Also, if the values of the parameter should be separated by a step `s` from `m` to `M`, you can send in `step = s`.
- If your parameter takes values out of a list `values` of possible values, you should call `suggest_categorical("parameter", values)`. For instance, if we wanted to try out different activation functions on a layer of a neural network, we could use something like the following:

```
activation = trial.suggest_categorical(
    "activation_function", ["sigmoid", "elu", "relu"]).
```

Of course, a single objective function can have as many optimizable parameters as desired. They would just be defined with separate invocations of the methods that we've just outlined.

So that's how you can create an objective function and specify the parameters that you want to optimize within it. Now, how do we optimize them? The first step is to create a `Study` object with the `create_study` function, just as follows:

```
from optuna.samplers import TPESampler

study = optuna.create_study(direction='maximize',
    sampler=TPESampler(seed = seed))
```

Here we have specified that we want to create a study in order to maximize some objective function and using `TPESampler` with a seed. By default, Optuna will try to minimize objective functions – that's why we had to send in that argument. The sampler that we've passed is just the object that, during the optimization process, is going to look for values to try. The one we've selected is the default one, but we have passed it manually so that we could give it a seed and get reproducible results. There are many other samplers.

Most notably, `GridSampler` allows you to try all the combinations of parameters out of a pre-defined "search space." For instance, we could use the following sampler:

```
values = {"learning_rate": [0.001, 0.003, 0.005, 0.008, 0.01]}
sampler = optuna.samplers.GridSampler(values)
```

This would make Optuna try out the values 0.001, 0.003, 0.005, 0.008, and 0.01 — and no others.

If you want to learn more about how these samplers work, you may have a look at their online documentation (`https://optuna.readthedocs.io/en/stable/reference/samplers/index.html`).

With the `Study` object ready, all we have to do is call the `optimize` method specifying the objective function and the number of trials that we will let Optuna run:

```
study.optimize(objective, n_trials=6)
```

Upon running this (it can take a while), you will get an output similar to the following:

```
Trial 0 finished with value: 0.9 and parameters:
    {'learning_rate': 0.01996042558751034}.
    Best is trial 0 with value: 0.9.

Trial 1 finished with value: 0.9 and parameters:
    {'learning_rate': 0.06258876833294336}.
    Best is trial 0 with value: 0.9.

Trial 2 finished with value: 0.9 and parameters:
    {'learning_rate': 0.04433504616170433}.
    Best is trial 0 with value: 0.9.

Trial 3 finished with value: 0.91 and parameters:
    {'learning_rate': 0.07875049978766316}.
```

```
    Best is trial 3 with value: 0.91.

Trial 4 finished with value: 0.92 and parameters:
    {'learning_rate': 0.07821760500376156}.
    Best is trial 4 with value: 0.92.

Trial 5 finished with value: 0.9 and parameters:
    {'learning_rate': 0.02798666792298152}.
    Best is trial 4 with value: 0.92.
```

With the parameter variations that we have considered, we haven't seen any significant differences in performance. But, still, at least we've learned how to use Optuna!

Exercise 11.2

Use Optuna to simultaneously optimize the learning rate and the batch size of the model.

As a final remark, notice how, in the objective function, we have used the validation accuracy and not the test accuracy. The test dataset, remember, should only be used once we've already picked our best model. Otherwise, its independence is compromised. For instance, if we had saved the models following each Optuna trial, now it would make sense for us to compute the test accuracy on the trial 4 model in order to make sure that we have a low generalization error.

Exercise 11.3

Optuna can be used on any framework, not just TensorFlow – it can be used to optimize any parameters that you want for any purpose! All you have to do is build a suitable objective function. To further illustrate this, use Optuna to find the minimum of the function $f(x) = (x - 3)^2$.

To learn more...

In these few pages, we haven't been able to cover all there is to know about Optuna. If you would like to learn more, you should have a look at its online documentation. You can find it at `https://optuna.readthedocs.io/en/stable/index.html`.

That was a short overview of how to train (quantum) machine learning models in real-world scenarios. In the following subsection, we will leave our comfort zone and use PennyLane to solve a new problem for us: a multi-class classification task.

11.2.4 A multi-class classification problem

This is going to be an exciting subsection, for we are about to consider a new kind of problem on which to apply our QML knowledge. Nonetheless, every long journey begins with a first step, and ours shall be to reset the seeds of NumPy and TensorFlow, just to make reproducibility easier:

```
np.random.seed(seed)
tf.random.set_seed(seed)
```

We are about to consider a multi-class classification problem and, of course, the first thing we need is data. Our good old `make_classification` function can help us here, for we can give it the optional argument `n_classes = 3` in order for it to generate a dataset with 3 distinct classes, which will be labeled as 0, 1, and 2. However, there's a catch. Increasing the number of classes means that, as per the function's requirements, we will also have to tweak some of the default parameters; a valid configuration can be reached by setting the argument `n_clusters_per_class` to 1. Thus, we can generate our dataset for ternary classification as follows:

```
x, y = make_classification(n_samples = 1000, n_features = 20,
    n_classes = 3, n_clusters_per_class = 1)
```

Now that we have data, it's time for us to think about the model. We are approaching a new kind of problem, so we need to go back to the basics. For now, let's forget about the hybrid component of the network, and let's try to think about how we could design a QNN capable of solving a ternary classification problem.

A general perspective on multi-class classification tasks

In this regard, it is useful to look at how this kind of problem is handled with classical neural networks. We know that, when solving binary classification problems, we consider neural networks having a single neuron in the final layer with a bounded activation function; in this way, we assign a label depending on whether the output is closer to 0 or 1. Such an approach might not be as effective, in general, when having multiple classes.

When working with k-class classification problems, neural networks are usually designed to have k neurons in their final layer — again, with bounded activation functions that make the values lie between 0 and 1. And how is a label assigned from the output of these neurons? Easy. Each neuron is associated to a label, so we just assign the label of the neuron that has the highest output. Heuristically, you may think of each of these k neurons in the final layer as light bulbs — whose brightness is determined by their output — indicating how likely it is that the input will belong to a certain category. All we do in the end is assigning the category of the light bulb that shines the most!

Porting this idea to quantum neural networks is easy. Instead of taking the expectation value of the observable M on the first qubit, we return an array of values with the expectation values of the M observable on the first k qubits — assigning to each qubit a label. It couldn't be easier.

To learn more...

There are other ways to build classifiers in problems with multiple classes. For instance, two popular approaches are the **one-versus-all** and **one-versus-one** methods. They involve training multiple binary classifiers and combining their results. We invite you to have a look at chapter 3 of Geron's book if you are curious [64].

That solves the problem of designing a QNN that can handle our task, but we still have an issue left: we don't yet have a suitable loss function for this kind of problem. In binary classification, we could rely on the binary cross-entropy function, but it doesn't work for problems with multiple categories. Luckily for us, there's a loss function that generalizes the binary cross entropy. Please, let us introduce you to the **categorical cross-entropy** loss.

Let us consider an arbitrary neural network N that, for any choice of parameters θ and any input x, returns an array $N_\theta(x)$ with k entries, all of them between 0 and 1. The categorical cross-entropy loss function depends on the parameters of the neural network θ, the inputs x, and the targets y, but there is an important subtlety: the loss function expects the targets y to be in **one-hot form**. This means that y shouldn't be a number representing a label $(0, 1, \ldots, k-1)$. Instead, it should be a vector (an array) with k entries that are all set to 0 except for the entry in the position of the label, which should be set to 1. Thus, instead of having $y = 0$, we would have $y = (1, 0, \ldots, 0)$, or, instead of having $y = k - 1$, we would have $y = (0, \ldots, 0, 1)$. Under these assumptions, the categorical cross-entropy is defined as follows:

$$H(\theta; x, y) = -\sum_{j=1}^{k} y_j \log\big(N_\theta(x)_j\big).$$

Of course, we have used the subindex j in y and $N_\theta(x)$ to denote their j-th entries. Notice how, in this definition, we have implicitly assumed that the first neuron in the final layer is associated to the label 0, the second neuron is associated to 1, and so on.

Exercise 11.4

Prove that the binary cross-entropy loss is a particular case of the categorical cross-entropy loss for $k = 2$.

Of course, the categorical cross-entropy function is a reasonable loss function for multi-class classification, and it shares some nice properties with the binary cross-entropy loss function. For instance, it is zero if a classifier gets an output completely right (it assigns 1 to the correct output and 0 to the rest), but it diverges if a classifier assigns 1 to a wrong output and 0 to the rest.

So far, we already know how to implement our QNN and we have a loss function, so we just have to finalize the details of our architecture. Regarding the quantum layer, we already know which observable we are going to use, so that's not a problem. For the feature map, we will rely on angular encoding and, for the variational form, we shall use the two-local variational form. To keep things somewhat efficient, we will take our QNN to have four qubits, and we will leave the rest of the hybrid architecture just as it was in the previous subsection.

That's enough abstract thinking for now; let's get to the code. And be prepared, because things are about to get hot.

Implementing a QNN for a ternary classification problem

According to our plan, the first thing that we need to do is encode our array of targets `y` in one-hot form.

Exercise 11.5

There is a variation of the categorical cross entropy loss that doesn't require the targets to be in one-hot form. It is the **sparse categorical cross entropy loss**. Try to replicate what follows using this loss function and the unencoded targets. You may access it as `tf.keras.losses.SparseCategoricalCrossentropy`.

We could implement our own one-hot encoder, but there's no need to. The scikit-learn package — once again to our rescue! — already implements a `OneHotEncoder` class, which you can import from `sklearn.preprocessing`. You can work with this class just as you would with other familiar scikit-learn classes, such as `MaxAbsScaler`.

In order to one-hot-encode an array of targets, you would need a `OneHotEncoder` object and you would just have to pass the array to the `fit_transform` method. But with a catch: the array should be a column vector! Our array of targets `y` is one-dimensional, so we will have to reshape it before we can feed it to the `fit_transform` method. Thus, this is how we can encode our array of targets in one-hot form:

```
from sklearn.preprocessing import OneHotEncoder
hot = OneHotEncoder(sparse = False)
y_hot = hot.fit_transform(y.reshape(-1,1))
```

Notice how we have added the argument `sparse = False`. This Boolean value, which defaults to `True`, determines whether or not the encoder should return sparse matrices. Sparse matrices are datatypes that can be very memory-efficient when storing matrices with lots of zeros, such as one-hot encoded arrays. Essentially, instead of logging the value of each entry in a matrix, a sparse matrix only keeps track of the non-zero entries in it. When working with very large matrices, it can save a ton of memory, but, sadly, using sparse matrices would lead to problems in the training, so we need our one-hot encoder to give us an ordinary array.

To learn more...

The neat thing about the `OneHotEncoder` class is that, once we have encoded an array of targets with representatives from each class using `fit_transform`, we can use the `transform` method on any array of targets. In our case, the `hot` object will remember that there are 3 classes in our dataset, and hence `hot.transform` will encode any targets correctly: even if it's given an input with nothing other than zeros, it will still encode them as arrays of length 3.

We have to do nothing more to our data, so we can now split it into some training, validation, and test datasets:

```
x_tr, x_test, y_tr, y_test = train_test_split(
    x, y_hot, train_size = 0.8)
x_val, x_test, y_val, y_test = train_test_split(
    x_test, y_test, train_size = 0.5)
```

And we can now implement the QNN that will constitute the quantum layer of our model. In truth, there's nothing particularly special about this quantum neural network other than the fact that it will return an array of values rather than a single one. We can define it, according to our previous specification, as follows:

```
nqubits = 4
dev = qml.device("lightning.qubit", wires = nqubits)

@qml.qnode(dev, interface="tf", diff_method = "adjoint")
def qnn(inputs, theta):
    qml.AngleEmbedding(inputs, range(nqubits))
    TwoLocal(nqubits, theta, reps = 2)
    return [qml.expval(qml.Hermitian(M, wires = [0])),
            qml.expval(qml.Hermitian(M, wires = [1])),
            qml.expval(qml.Hermitian(M, wires = [2]))]

weights = {"theta": 12}
```

The code is pretty self-explanatory. Notice that, as usual, we have taken the chance to define the weights dictionary that we will use in the definition of the quantum Keras layer. In this case, we will be using 12 weights, exactly as in the case of our model in *Subsection 11.2.2*, because we are using the same variational form and the same number of qubits and repetitions.

With our QNN ready, we can define the Keras model for our hybrid QNN. This is just analogous to what we did in the previous subsection, with a few important differences — don't copy-paste so fast! First of all, in this case, we need to set the output dimension of the quantum layer to three, not one. And, much more importantly, we need to add an extra activation function on the QNN output.

The categorical cross entropy loss function expects probability distributions. In principle, it assumes that the output of the j-th neuron is the probability that the input belong to category j. Thus, the data that the model outputs should be normalized: it should add up to 1. Nevertheless, a priori, there's no way for us to guarantee that the QNN will return some normalized outputs with our current setup. In order to ensure this, we may use the **softmax** activation function, which is defined as

$$\sigma(x_1, \dots, x_n) = \frac{1}{\sum_{j=1}^{n} e^{x_j}}(e^{x_1}, \dots, e^{x_n}).$$

It's easy to check that σ is a vector with components bounded by 0 and 1 which add up to 1 and, hence, is a probability distribution.

In addition to these modifications, we will add an extra classical layer with 8 neurons:

```
model = tf.keras.models.Sequential([
    tf.keras.layers.Input(20),
    tf.keras.layers.Dense(8, activation = "elu"),
    tf.keras.layers.Dense(4, activation = "sigmoid"),
    qml.qnn.KerasLayer(qnn, weights, output_dim = 3),
    tf.keras.layers.Activation(activation = "softmax")
])
```

And we can now compile our model with the Adam optimizer and the categorical cross-entropy loss before training it with the `fit` method; nothing particularly exciting here. As a fun fact, if you were forgetful enough to tell TensorFlow to use the binary cross-entropy loss instead of the categorical cross-entropy one, it would still use the categorical cross-entropy loss (don't look at us; we don't say it from experience, right?). This is a rather nice and thoughtful feature from the guys behind TensorFlow.

```
opt = tf.keras.optimizers.Adam(learning_rate = 0.001)
model.compile(opt, loss=tf.keras.losses.CategoricalCrossentropy())

history = model.fit(x_tr, y_tr, epochs = 50, shuffle = True,
    validation_data = (x_val, y_val),
```

```
    batch_size = 10,
    callbacks = [earlystop])
```

After a few minutes of training, we may get a plot of the evolution of the training and validation losses with the following instruction:

```
plot_losses(history)
```

The resulting plot can be found in *Figure 11.2*, which shows the evolution of both losses.

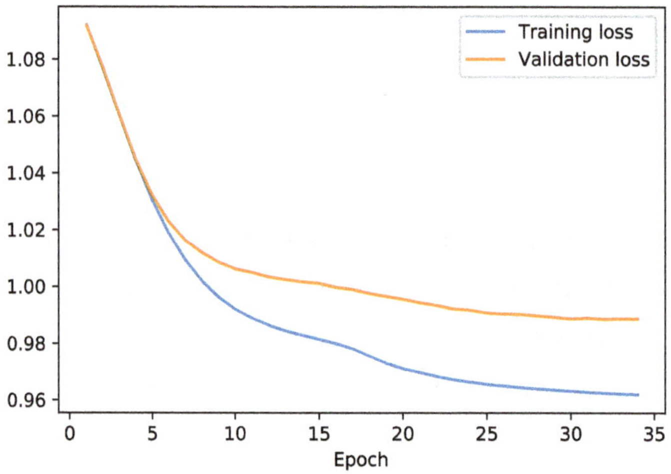

Figure 11.2: Evolution of the training and validation loss functions in the training of a hybrid QNN multi-class classifier

We may now compute the training, validation, and test accuracies of our freshly-trained models, but, in order to do so, the `accuracy_score` function needs the predicted and actual labels to be represented by numbers, not encoded in one-hot form as arrays. Hence, we need to undo the one-hot encoding. For this purpose, we can just use the `argmax` method, which returns the entry of the maximum value in an array, and it can be given an optional `axis` argument for it to be applied only in one axis. Thus, we may compute the accuracy scores as follows:

```
tr_acc = accuracy_score(
    model.predict(x_tr).argmax(axis = 1),
```

```
    y_tr.argmax(axis = 1))
val_acc = accuracy_score(
    model.predict(x_val).argmax(axis = 1),
    y_val.argmax(axis = 1))
test_acc = accuracy_score(
    model.predict(x_test).argmax(axis = 1),
    y_test.argmax(axis = 1))
print("Train accuracy:", tr_acc)
print("Validation accuracy:", val_acc)
print("Test accuracy:", test_acc)
```

This returns a training accuracy of 67%, a validation accuracy of 53%, and a test accuracy of 60%. Notice that the low accuracy on the validation dataset – compared to that of the training dataset – seems to indicate an overfitting problem. This might be fixed, for example, by using a larger training dataset; of course, this would lead to longer training times.

Exercise 11.6

Just to further leave our "classifier comfort zone," try to implement a hybrid model able to do regression. This model should be trained on some data with inputs x and target values y for which there is a continuous function $f(x)$ such that $f(x) \simeq y$ (you can create such a dataset, for instance, with the `make_regression` method from scikit-learn). The model should try to learn the function f for all the points in the dataset.

You may design this model using some classical layers, followed by a quantum layer like the ones that we have considered, and a final classical layer with no activation functions and just one neuron. You should train it with the mean squared error loss.

That concludes our study of hybrid architectures in PennyLane. It's time for us to get to Qiskit, and that's going to be a very different adventure!

11.3 Hybrid architectures in Qiskit

In the previous section, we discussed how hybrid QNNs could be implemented and trained using PennyLane in conjunction with TensorFlow, an ML framework that we already know how to use. We will devote this section to studying how to work with these hybrid architectures in Qiskit, and in this mission we will need to face a new challenge.

For better or for worse, Qiskit doesn't have a built-in TensorFlow interface at the time of writing. It only has native support for a different ML framework: PyTorch. So, if we want to get those hybrid NNs working on Qiskit, we better learn a thing or two about PyTorch. As daunting as this task may seem, it won't be such a hassle and it will greatly pay off in the future — and, yes, the future is our next chapter on QGANs.

Important note

We will be using **version 1.13** of the PyTorch package. If you are using a different version, things may be slightly different!

What's so special about PyTorch to be worth our time beyond this short section? Come and see.

11.3.1 Nice to meet you, PyTorch!

So far, we have worked with TensorFlow. In our experience, this framework provides a very easy and streamlined experience for the implementation and training of all sorts of network-based models. However, there's a small catch behind all that ease of use. In this book, we haven't been using "pure TensorFlow," but we have been relying heavily on Keras. In spite of being fully integrated into TensorFlow, Keras is a component that creates some additional layers of abstraction in order to simplify the handling of neural-network models in TensorFlow. All this time, Keras has been taking care of lots of things for us behind the scenes.

At the time of writing, there are two very popular ML frameworks out there: TensorFlow and PyTorch. The former we already know, the latter we soon will. PyTorch, unlike TensorFlow, doesn't come with its own Keras (although there are some third-party packages that provide similar functionalities). In PyTorch, we will have to take care of many details ourselves. And that's great. Granted, learning how to use PyTorch will require a tiny bit more effort on our part, but PyTorch will offer us a level of flexibility that TensorFlow's Keras simply can't. Let's get started then.

We will be using version 1.13 of the PyTorch package. Please refer to *Appendix D, Installing the Tools,* for instructions on how to install it.

As usual, we shall begin by importing NumPy and a few utilities from scikit-learn. We will also set a seed for NumPy:

```
import numpy as np
from sklearn.metrics import accuracy_score
from sklearn.model_selection import train_test_split
from sklearn.datasets import make_classification

seed = 1234
np.random.seed(seed)
```

With those imports out of the way, we can get to our main dish. This is how you can import PyTorch and give it a seed to ensure reproducibility:

```
import torch
torch.manual_seed(seed)
```

Most functionality related to the implementation of models is in the `torch.nn` module, and most activation functions can be found in the `torch.nn.functional` module, so let's import these as well:

```
import torch.nn as nn
import torch.nn.functional as F
```

Those are all the imports that we need for now.

Setting up a model in PyTorch

In order to understand how the PyTorch package works, we will implement and train a simple binary classifier as a (classical) neural network. This neural network will take 16 inputs and return a unique output between 0 and 1. As usual, the two possible labels will be 0 and 1 and the output label will be decided based on whether the network output is closer to 0 or 1.

Let's see how we can implement this neural network classifier. In PyTorch, model architectures are defined as subclasses of the `nn.Module` class, and individual models are objects of these subclasses. When defining subclasses of `nn.Module`, you should implement an initializer that first calls the parent's initializer and then prepares all the variables of the model architecture; for instance, all the network layers should be initialized here. In addition, you need to provide a `forward` method that defines the behavior of the network: this method should take any input to the network as an argument and return its output.

Our desired neural network could be implemented as follows (don't worry, we will discuss this piece of code right away):

```
class TorchClassifier(nn.Module):

    def __init__(self):

        # Initialize super class.
        super(TorchClassifier, self).__init__()

        # Declare the layers that we will use.
        self.layer1 = nn.Linear(16, 8)
        self.layer2 = nn.Linear(8, 4)
        self.layer3 = nn.Linear(4, 2)
        self.layer4 = nn.Linear(2, 1)

    # Define the transformation of an input.
```

```
    def forward(self, x):
        x = F.elu(self.layer1(x))
        x = F.elu(self.layer2(x))
        x = F.elu(self.layer3(x))
        x = torch.sigmoid(self.layer4(x))

        return x
```

There are a few things to digest in this implementation. Let us first look at the initializer. As expected, we are defining a subclass of `nn.Module` and we are first calling the parent's initializer; so far, so good. Then we are defining what seem to be the layers of the neural network, and here is where some confusion may arise. Our first issue arises from terminology: "linear layers" are PyTorch's equivalent of Keras' "dense" layers – not a big deal. But then we have a deeper issue. Back in our Keras days, we defined the layers of a neural network by specifying the number of neurons they had and their activation function. But here there's no trace of activation functions and the layers take what seem to be two-dimensional arguments. What's going on?

In a neural network, you have a bunch of neurons that are arranged into arrays, and these arrays are connected by some "linear wiring" between them. In addition, each array of neurons has a (usually non-linear) activation function. In Keras, layers were associated to these arrays of neurons themselves (with their activation functions) and to the "linear wiring" before them. In PyTorch, on the other hand, when we speak of layers, we only refer to the linear wiring between these arrays of neurons. Hence, `nn.Linear(16, 8)` is nothing more than the linear wiring – with its weights and biases – between an array of 16 neurons and an array of 8 neurons. This will make more sense when we look at the `forward` method.

The `forward` method defines what happens to any input that gets into the network. In its implementation, we can see how any input, which will be a PyTorch tensor of length 16, goes through the first layer. This first layer is the "linear wiring" between an array of 16 neurons and an array of 8 neurons; it has its own weights w_{jk} and biases b_k and, for any input $(x_1, \ldots, x_{16})$, it returns a vector $(\hat{x}_1, \ldots, \hat{x}_8)$ with

$$\hat{x}_k = \sum_{j=1}^{16} w_{jk} x_j + b_k.$$

Then, each entry in the resulting tensor goes through the ELU activation function. The rest of the code is self-explanatory and simply defines a neural network that matches our specifications.

To learn more...

Layers in PyTorch define their own weights and biases. If you wish to remove the biases – setting them to zero for all eternity – you may do so by sending the optional argument `bias = False` when initializing a layer.

Now that we have our model architecture defined, we can instantiate it into an individual model by initializing an object of the `TorchClassifier` class. A nice thing about PyTorch models, by the way, is that they can be printed; their output gives you an overview of the different model components. Let's create our model object and see this in action:

```
model = TorchClassifier()
print(model)
```

Upon running this, we get the following output from the print instruction:

```
TorchClassifier(
  (layer1): Linear(in_features=16, out_features=8, bias=True)
  (layer2): Linear(in_features=8, out_features=4, bias=True)
  (layer3): Linear(in_features=4, out_features=2, bias=True)
  (layer4): Linear(in_features=2, out_features=1, bias=True)
)
```

This is somewhat analogous to the model summaries that we could print in Keras.

By default, the weights and biases of models are random, so our newly-created `model` should already be ready to be used. Let's try it out! The `torch.rand` function can create a random tensor of any specified size. We will use it to feed our model some random data and see if it works:

```
model(torch.rand(16))
```

This is the output that we get:

```
tensor([0.4240], grad_fn=<SigmoidBackward0>)
```

And there we have it! As expected, our model returns a value between 0 and 1. By the way, notice one little thing in the output: right next to the tensor value, there is a `grad_fn` value that somehow remembers that this output was last obtained from the application of a sigmoid function. Interesting, isn't it? Well, you may remember that TensorFlow used its own tensor datatype, and PyTorch has its own tensors too. The cool thing about them is that every PyTorch tensor keeps track of how it was computed in order to enable gradient computation through backpropagation. We will further discuss this later on in this subsection.

In any case, now that our network is all set up, let us generate some data and split it into some training, validation, and test datasets:

```
x, y = make_classification(n_samples = 1000, n_features = 16)


x_tr, x_test, y_tr, y_test = train_test_split(
    x, y, train_size = 0.8)
x_val, x_test, y_val, y_test = train_test_split(
    x_test, y_test, train_size = 0.5)
```

Training a model in PyTorch

In principle, we could work with this raw data just as we did in TensorFlow — perhaps converting it to PyTorch tensors, but still. However, we know that PyTorch will require us to take care of many things ourselves; one of which will be splitting our data into batches should we want to. Doing that ourselves could be tedious to say the least. Thankfully, PyTorch comes with some tools that can assist us in the process, so we better give them a shot.

The best way to deal with datasets in PyTorch is by storing data in subclasses of a `Dataset` class, which can be found in the `torch.utils.data` module. Any subclasses of `Dataset` should implement an initializer, a `__getitem__` method (to access data items by indexing), and a `__len__` method (returning the number of items in the dataset). For our purposes, we can create a subclass in order to create datasets from our NumPy arrays:

```
from torch.utils.data import Dataset


class NumpyDataset(Dataset):
    def __init__(self, x, y):


        if (x.shape[0] != y.shape[0]):
            raise Exception("Incompatible arrays")


        y = y.reshape(-1,1)


        self.x = torch.from_numpy(x).to(torch.float)
        self.y = torch.from_numpy(y).to(torch.float)


    def __getitem__(self, i):
        return self.x[i], self.y[i]


    def __len__(self):
        return self.y.shape[0]
```

Notice how we have added some size-checking to ensure that the data array and the labels vector have matching dimensions, and how we have reshaped the array of targets — that's in order to avoid problems with the loss functions, which expect them to be column vectors. With this class set up, we may create dataset objects for the training, validation and test datasets as follows:

```
tr_data = NumpyDataset(x_tr, y_tr)
val_data = NumpyDataset(x_val, y_val)
test_data = NumpyDataset(x_test, y_test)
```

Just to check whether our implementation was successful, let us try to access the first element in `tr_data` and get the length of the training dataset:

```
print(tr_data[0])
print("Length:", len(tr_data))
```

This is the output returned by these instructions:

```
(tensor([ 1.4791,  1.4646,  0.0430,  0.0409, -0.3792, -0.5357,
          0.9736, -1.3697, -1.2596,  1.5159, -0.9276,  0.6868,
          0.5138,  0.4751,  1.0193, -1.7873]),
tensor([0.]))

Length: 800
```

We can see how, indeed, it gave us a tuple with a tensor of length 16 and its corresponding label. Also, a call to the **`len`** function did return the correct number of items in our dataset. Now, you may reasonably wonder why we should bother with all this mess of creating dataset classes. There are a couple of reasons. For one, this allows us to have our data organized and structured in a more orderly manner. What is more, using dataset objects, we can create data loaders. The `DataLoader` class can be imported from `torch.utils.data` and its objects allow us to easily iterate through batches of data. An example may help clarify this.

Let's say that we want to iterate over the training dataset in batches of 2. All we would have to do is to create a data loader with the `tr_data` dataset specifying the batch size and the fact that we would like it to shuffle the data. Then, we could create an iterator object out of the data loader with the **iter** function and iterate over all the batches. This is shown in the following piece of code:

```
from torch.utils.data import DataLoader
tr_loader = iter(DataLoader(
    tr_data, batch_size = 2, shuffle = True))
print(next(tr_loader))
```

You may recall from Python 101 that calling `next(tr_loader)` for the first time would be equivalent to running a `for x in tr_loader` loop and extracting the value of `x` in the first iteration. This is the output that we get:

```
[tensor([[-1.2835, -0.4155,  0.4518,  0.6778, -1.3869, -0.4262, -0.1016,
          1.4012, -0.9625,  1.0038,  0.3946,  0.1961, -0.7455,  0.4267,
         -0.8352,  0.9295],
        [-1.4578, -0.4947, -1.1755, -0.4800, -0.3247,  0.7821, -0.0078,
         -0.5397, -1.0385, -1.3466,  0.4591,  0.5761,  0.2188, -0.1447,
          0.3534,  0.5055]]),
 tensor([[0.],
        [0.]])]
```

And there you have it! In each iteration of the data loader, we get an array with the training data in the batch and its corresponding array of targets. All is shuffled and taken care of by PyTorch automatically. Neat, isn't it? That can and will save us a good deal of effort.

We must say that, in truth, you could technically use data loaders without going through the whole process of defining datasets — just sending in the numpy arrays. But it wouldn't be the most "PyTorchy" of practices. Anyhow, this settles our preparation of datasets.

In the training process, we will use, as always, the binary cross-entropy loss. We can save its function in a variable as follows:

```
get_loss = F.binary_cross_entropy
```

Thus, the `get_loss` function will take a tensor of values between 0 and 1 and a matching tensor of labels, and will use them to compute the binary cross entropy loss. To see if it works as expected, we may compute a simple loss:

```
print(get_loss(torch.tensor([1.]), torch.tensor([1.])))
```

Since the only value in the tensor matches the expected value, we should get a loss of 0 and, indeed, this instruction returns `tensor(0.)`.

We are already preparing ourselves for the training. In our case, since our dataset has 1000 elements, it could make sense to use a batch size of 100, so let us prepare the training data loader to that effect:

```
tr_loader = DataLoader(tr_data, batch_size = 100, shuffle = True)
```

As usual, we will rely on the Adam optimizer for the training. The optimizer is implemented as a class in the `torch.optim` module, and, in order to use it, we need to specify which parameters it is going to optimize; in our case, that will be the parameters in our model, which we can retrieve with the `parameters` method. In addition, we can further configure the optimizer by passing optional arguments for the learning rate, among other adjustable parameters. We will use a learning rate of 0.005 and trust the default values of the remaining parameters. Thus, we can define our optimizer as follows:

```
opt = torch.optim.Adam(model.parameters(), lr = 0.005)
```

Now we have all the ingredients ready and we can finally get to the training itself. In Keras, this would've been as easy as calling a method with a bunch of parameters, but here we have to work the training out ourselves! We will begin by defining a function that will perform one full training epoch. It will be the following:

```
def run_epoch(opt, tr_loader):
    # Iterate through the batches.
    for data in iter(tr_loader):
        x, y = data # Get the data in the batch.
        opt.zero_grad() # Reset the gradients.
        # Compute gradients.
        loss = get_loss(model(x), y)
        loss.backward()
        opt.step() # Update the weights.
    return get_loss(model(tr_data.x), tr_data.y)
```

The code is pretty much self-explanatory, but a few details deserve clarification. We have used two new methods: `backward` and `step`. Oversimplifying a bit, the `backward` method on `loss` computes the gradient of the loss by tracing back how it was computed and saving the partial derivatives in the optimizable parameters of the model on which the loss depends. This is the famous backpropagation technique that we talked about in *Chapter 8, What Is Quantum Machine Learning?*. Then, `opt.step()` prompts the optimizer to update the optimizable parameters using the derivatives that `loss.backward()` computed.

To learn more...

If you are curious about how differentiation works with the `backward` method on PyTorch tensors, we can run a quick example to illustrate. We may define two variables, `a` and `b`, taking the values 2 and 3 respectively as follows:

```
a = torch.tensor([2.], requires_grad = True)
b = torch.tensor([3.], requires_grad = True)
```

Notice how we set `requires_grad = True` to tell PyTorch that these are variables it should keep track of. We may then define the function $f(a,b) = a^2 + b$ and compute its gradient as follows:

```
f = a**2 + b
```

```
f.backward()
```

We know that $\partial f/\partial a = (\partial/\partial a)a^2 + b = 2a$, which in our case is equal to $2a = 2 \cdot 2 = 4$. When we run the `backward` method, PyTorch has already computed this partial derivative for us, and we can access it by calling `a.grad`, which, as expected, returns `tensor([4.])`. Analogously, $\partial f/\partial b = 1$, and, as expected, `b.grad` returns `tensor([1.])`.

In principle, we could train our model by calling `run_epoch` manually as many times as we wanted, but why suffer like that when we can leave Python in charge?

Let us define a training loop in which, at each iteration, we will run an epoch and log the training and validation loss obtained over the whole dataset. Instead of fixing a specific number of epochs, we will keep iterating until the validation loss increases — this will be our own version of the early stopping callback that we used in TensorFlow. The following piece of code gets the job done:

```
tr_losses = []
val_losses = []
while (len(val_losses) < 2 or val_losses[-1] < val_losses[-2]):
    print("EPOCH", len(tr_losses) + 1, end = " ")
    tr_losses.append(float(run_epoch(opt, tr_loader)))
    # ^^ Remember that run_epoch returns the training loss.
    val_losses.append(float(
        get_loss(model(val_data.x), val_data.y)))
    print("| Train loss:", round(tr_losses[-1], 4), end = " ")
    print("| Valid loss:", round(val_losses[-1], 4))
```

Notice how, when logging the losses in `tr_losses`, we have converted the PyTorch tensors to floats. This is the output that we get after executing this loop:

```
EPOCH 1 | Train loss: 0.6727 | Valid loss: 0.6527
EPOCH 2 | Train loss: 0.638 | Valid loss: 0.6315
EPOCH 3 | Train loss: 0.5861 | Valid loss: 0.5929
EPOCH 4 | Train loss: 0.5129 | Valid loss: 0.5277
EPOCH 5 | Train loss: 0.4244 | Valid loss: 0.4428
EPOCH 6 | Train loss: 0.3382 | Valid loss: 0.3633
EPOCH 7 | Train loss: 0.2673 | Valid loss: 0.3024
EPOCH 8 | Train loss: 0.2198 | Valid loss: 0.2734
EPOCH 9 | Train loss: 0.1938 | Valid loss: 0.2622
EPOCH 10 | Train loss: 0.1819 | Valid loss: 0.2616
EPOCH 11 | Train loss: 0.1769 | Valid loss: 0.2687
```

An image is worth a thousand words, so, just to get a visual overview of the performance of our training, let us recycle the `plot_losses` function that we had for TensorFlow and run it:

```
import matplotlib.pyplot as plt
def plot_losses(tr_loss, val_loss):
    epochs = np.array(range(len(tr_loss))) + 1
    plt.plot(epochs, tr_loss, label = "Training loss")
    plt.plot(epochs, val_loss, label = "Validation loss")
    plt.xlabel("Epoch")
    plt.legend()
    plt.show()

plot_losses(tr_losses, val_losses)
```

The resulting plot can be found in *Figure 11.3*. The plot does show some signs of overfitting, but likely not something to be concerned about; in any case, let's wait until we get the accuracy over the test dataset.

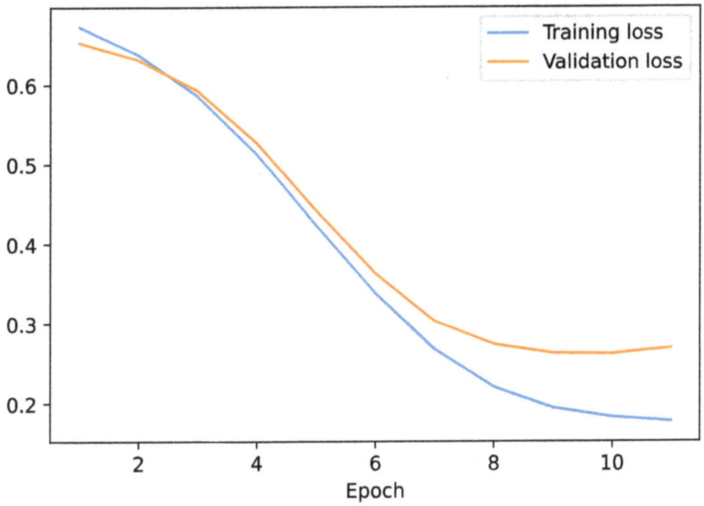

Figure 11.3: Evolution of the training and validation losses over the training of a classical binary classifier with PyTorch

In order to get the accuracy of our classifier on the training, validation, and test datasets, we can run the following instructions:

```
train_acc = accuracy_score(
    (model(tr_data.x) >= 0.5).to(float), tr_data.y)
val_acc = accuracy_score(
    (model(val_data.x) >= 0.5).to(float), val_data.y)
test_acc = accuracy_score(
    (model(test_data.x) >= 0.5).to(float), test_data.y)
print("Training accuracy:", train_acc)
print("Validation accuracy:", val_acc)
print("Test accuracy:", test_acc)
```

This returns a training accuracy of 94%, a validation accuracy of 92%, and a test accuracy of 96%.

We have just concluded our not-that-short introduction to PyTorch. Let's go quantum!

11.3.2 Building a hybrid binary classifier with Qiskit

In this subsection, we will implement our first hybrid QNN with Qiskit. The process will be fairly straightforward, and we will be able to rely on a good deal of the code that we already have. To get started, let us import the Qiskit package and the ZZ feature map and two-local variational form that come bundled with it:

```
from qiskit import *
from qiskit.circuit.library import ZZFeatureMap, TwoLocal
```

With a QNN, it will be advisable to use smaller datasets in order for the training time to be reasonable on our simulators. We can prepare them, along with the corresponding dataset and data loader objects, as follows:

```
x, y = make_classification(n_samples = 500, n_features = 16)
x_tr, x_test, y_tr, y_test = train_test_split(x, y, train_size = 0.8)
x_val, x_test, y_val, y_test = train_test_split(x_test, y_test, train_size = 0.5)

tr_data = NumpyDataset(x_tr, y_tr)
val_data = NumpyDataset(x_val, y_val)
test_data = NumpyDataset(x_test, y_test)

tr_loader = DataLoader(tr_data, batch_size = 20, shuffle = True)
```

Our quantum layer will be a simple 4-qubit QNN with one instance of the ZZ feature map and the two-local variational form. Thus, the components that we will use in our QNN circuit will be the following:

```
zzfm = ZZFeatureMap(2)
twolocal = TwoLocal(2, ['ry','rz'], 'cz', 'linear', reps = 1)
```

Here, we have instantiated the two-local form as in *Chapter 10, Quantum Neural Networks*.

Also, just as we did in the previous chapter, we could use the `TwoLayerQNN` class in order to generate our quantum neural network according to our specifications. We may import it as follows:

```
from qiskit_machine_learning.neural_networks import TwoLayerQNN
```

We are now ready to define our model architecture with PyTorch. Its structure will be analogous to that of a classical architecture. The only difference is that we will have to define a quantum neural network object in the initializer, and we will have to rely on the `TorchConnector` in order to use the QNN in the `forward` method. This `TorchConnector` is analogous to the `qml.qnn.KerasLayer` that we used in PennyLane — only that it's for Qiskit and PyTorch! This is how we may then define our hybrid network and instantiate a model:

```
from qiskit_machine_learning.connectors import TorchConnector
from qiskit.providers.aer import AerSimulator

class HybridQNN(nn.Module):

    def __init__(self):

        # Initialize super class.
        super(HybridQNN, self).__init__()

        # Declare the layers that we will use.
        qnn = TwoLayerQNN(2, zzfm, twolocal, input_gradients = True,
            quantum_instance = AerSimulator(method="statevector"))
        self.layer1 = nn.Linear(16, 2)
        self.qnn = TorchConnector(qnn)
        self.final_layer = nn.Linear(1,1)
```

```
    def forward(self, x):
        x = torch.sigmoid(self.layer1(x))
        x = self.qnn(x)
        x = torch.sigmoid(self.final_layer(x))
        return x

model = HybridQNN()
```

Notice how we've passed the optional argument `input_gradients = True` to the `TwoLayer` initializer; that is required for the PyTorch interface to work properly. Apart from that, the construction of the quantum neural network was fully analogous to what we did in *Chapter 10, Quantum Neural Networks.* A detail that perhaps deserves an explanation is the reason why we have included a final classical layer after the quantum one. This is because our QNN will return values between −1 and 1, not between 0 and 1; by including this final layer followed by the classical sigmoid activation function, we can ensure that the output of our network will be bounded between 0 and 1, as we expect.

Now all we have left to do before we can start the training is prepare the optimizer, and send the model parameters to it:

```
opt = torch.optim.Adam(model.parameters(), lr = 0.005)
```

And we can simply reuse the `run_epoch` function to complete the training, just as we did in the previous subsection:

```
tr_losses = []
val_losses = []

while (len(val_losses) < 2 or val_losses[-1] < val_losses[-2]):
    print("EPOCH", len(tr_losses) + 1, end = " ")
    tr_losses.append(float(run_epoch(opt, tr_loader)))
    val_losses.append(float(get_loss(model(val_data.x), val_data.y)))
```

```
    print("| Train loss:", round(tr_losses[-1], 4), end = " ")
    print("| Valid loss:", round(val_losses[-1], 4))
```

This is the output that the execution will yield:

```
EPOCH 1 | Train loss: 0.6908 | Valid loss: 0.696
EPOCH 2 | Train loss: 0.6872 | Valid loss: 0.691
EPOCH 3 | Train loss: 0.6756 | Valid loss: 0.6811
EPOCH 4 | Train loss: 0.6388 | Valid loss: 0.6455
EPOCH 5 | Train loss: 0.5661 | Valid loss: 0.5837
EPOCH 6 | Train loss: 0.5099 | Valid loss: 0.5424
EPOCH 7 | Train loss: 0.4692 | Valid loss: 0.5201
EPOCH 8 | Train loss: 0.4425 | Valid loss: 0.5014
EPOCH 9 | Train loss: 0.4204 | Valid loss: 0.4947
EPOCH 10 | Train loss: 0.4019 | Valid loss: 0.4923
EPOCH 11 | Train loss: 0.3862 | Valid loss: 0.4774
EPOCH 12 | Train loss: 0.3716 | Valid loss: 0.4668
EPOCH 13 | Train loss: 0.3575 | Valid loss: 0.451
EPOCH 14 | Train loss: 0.3446 | Valid loss: 0.4349
EPOCH 15 | Train loss: 0.3332 | Valid loss: 0.4323
EPOCH 16 | Train loss: 0.3229 | Valid loss: 0.4259
EPOCH 17 | Train loss: 0.3141 | Valid loss: 0.4253
EPOCH 18 | Train loss: 0.3055 | Valid loss: 0.422
EPOCH 19 | Train loss: 0.2997 | Valid loss: 0.4152
EPOCH 20 | Train loss: 0.2954 | Valid loss: 0.4211
```

As before, we can get a plot of the loss evolution as follows:

```
plot_losses(tr_losses, val_losses)
```

This returns the plot shown in *Figure 11.4*. There does seem to be some overfitting, which could likely be fixed by giving more data to the classifier.

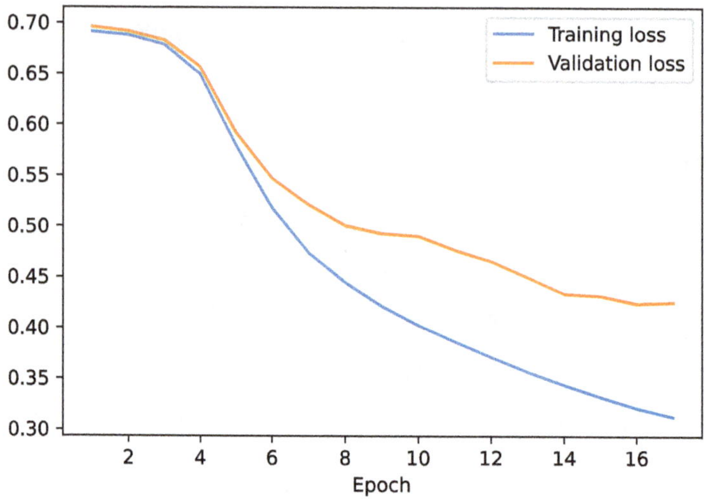

Figure 11.4: Evolution of the training and validation losses over the training of a hybrid binary classifier with PyTorch

In any case, let's compute the training, validation, and test accuracies to get a better insight into the performance of the classifier. We may do that by executing the following instructions:

```
tr_acc = accuracy_score(
    (model(tr_data.x) >= 0.5).to(float), tr_data.y)
val_acc = accuracy_score(
    (model(val_data.x) >= 0.5).to(float), val_data.y)
test_acc = accuracy_score(
    (model(test_data.x) >= 0.5).to(float), test_data.y)
print("Training accuracy:", tr_acc)
print("Validation accuracy:", val_acc)
print("Test accuracy:", test_acc)
```

Upon running this, we get a training accuracy of 92%, a validation accuracy of 86%, and a test accuracy of 74%. This confirms our suspicions regarding the existence of overfitting. As in other cases, should we want to fix this, we could try training the model with additional data, for instance.

Of course, all that we've learned about how to train hybrid QNNs with PyTorch and Qiskit also works for ordinary QNNs. If you want to train a simple Qiskit QNN using PyTorch, you've just learned how to do it; all it will take is defining a model with no classical layers.

This concludes our study of hybrid neural networks in Qiskit. But we still have one thing left before bringing this section to an end.

One of the advantages of Qiskit is its tight integration with IBM's quantum hardware. Nevertheless, as was the case in our study of quantum optimization, queueing times make the training of any QNN model on real hardware unfeasible through the usual interfaces to IBM's hardware – that is, just using a real hardware backend, as we discussed in *Chapter 2, The Tools of the Trade in Quantum Computing*. Thankfully, there's a better way.

11.3.3 Training Qiskit QNNs with Runtime

Using Qiskit's Runtime service, as we did in *Chapters 5* and *7*, we can effectively train any QNN model defined in PyTorch through a Qiskit Torch connector on any of the devices and simulators provided by IBM Quantum. All it takes is waiting on a single queue, and the whole training process is executed as a unit – with all the executions on quantum hardware included. The folks at IBM refer to this use case of Qiskit Runtime as "Torch Runtime."

That is very convenient. However, we must warn you that, at the time of writing, the queuing times to run these Torch Runtime programs can be somewhat long: around the order of a few hours. Also, you should keep in mind that – again, at the time of writing – this service enables you to train QNNs defined on PyTorch, but not hybrid QNNs! That is, your PyTorch model should not have any classical layers whatsoever.

We will train a simple QNN model on a real device. As usual, we should firstly load our IBMQ account and pick a device. We will pick the least busy device among all the real devices with at least four qubits:

```
from qiskit.providers.ibmq import *

provider = IBMQ.load_account()
dev_list = provider.backends(
    filters = lambda x: x.configuration().n_qubits >= 4,
                        simulator = False)

dev = least_busy(dev_list)
```

We may define a simple QNN model with the PyTorch connector as follows:

```
class QiskitQNN(nn.Module):

    def __init__(self):

        super(QiskitQNN, self).__init__()

        qnn = TwoLayerQNN(2, zzfm, twolocal, input_gradients = True)
        self.qnn = TorchConnector(qnn)

    def forward(self, x):
        x = self.qnn(x)
        return x

model = QiskitQNN()
```

Then, we may generate some data on which to train this model using the `make_classification` function:

```
x, y = make_classification(n_samples = 100, n_features = 2,
    n_clusters_per_class = 1, n_informative = 1, n_redundant = 1)
x_tr, x_test, y_tr, y_test = train_test_split(x, y, train_size = 0.8)
```

```
x_val, x_test, y_val, y_test = train_test_split(x_test, y_test,
    train_size = 0.5)
tr_data = NumpyDataset(x_tr, y_tr)
val_data = NumpyDataset(x_val, y_val)
test_data = NumpyDataset(x_test, y_test)
```

Notice how we have adjusted some of the parameters of the `make_classification` function in order to comply with its requirements (check its documentation at `https://scikit-learn.org/stable/modules/generated/sklearn.datasets.make_classification.html` for more details).

Our model should return values between 0 and 1, but the observable that we have chosen for our circuit – the default one, the parity observable (check *Chapter 10, Quantum Neural Networks*, for reference) – returns two possible values: 1 or -1, not 0 and 1. Thus we need to update the targets mapping $0 \mapsto -1$ and $1 \mapsto 1$. This can be done with the following instructions:

```
tr_data.y = 2 * (tr_data.y - 1/2)
val_data.y = 2 * (val_data.y - 1/2)
test_data.y = 2 * (test_data.y - 1/2)
```

Let us now set up some data loaders for the training, validation, and test data:

```
tr_loader = DataLoader(tr_data, batch_size = 20, shuffle = True)
val_loader = DataLoader(val_data)
test_loader = DataLoader(test_data)
```

And the only ingredients that we have left to define are the optimizer and the loss function. We can still rely on Adam as an optimizer, but the binary cross entropy loss will no longer work since our labels are now -1 and 1 instead of 0 and 1; thus, we will use the mean squared error loss instead:

```
get_loss = F.mse_loss
```

```
opt = torch.optim.Adam(model.parameters(), lr = 0.005)
```

In order to be able to use our model with Torch Runtime, we will have to define a Torch Runtime Client, `client`, specifying a few self-explanatory parameters. This is done as follows:

```
from qiskit_machine_learning.runtime import TorchRuntimeClient


client = TorchRuntimeClient(provider = provider, backend = dev,
    model = model, optimizer = opt, loss_func = get_loss,
    epochs = 5)
```

We have set the number of epochs to 5 in order to get some quick results, but feel free to increase it.

And now this is the instruction that we need to execute if we want to train our model:

```
result = client.fit(train_loader = tr_loader, val_loader = val_loader)
```

This will likely take a while because of the queue time required to run a Torch Runtime program. Sit back and relax. Eventually, your model will be trained. Once that happens, you can get information about the training from the `result` object, whose type is `TorchRuntimeResult`. In particular, the attributes `train_history` and `val_history` will show you the evolution of the training and validation losses throughout the training process.

If you'd like to get the model's prediction on some data — for instance, the test dataset — all you have to do is send a data loader object with the data to the `predict` method. And this is how you can get your predictions:

```
pred = client.predict(test_loader).prediction
```

Don't expect to get great results! The model that we have defined is not very powerful and we only trained for a few epochs. As if that were not enough, when you run on real hardware, there's always the issue of having to deal with noise. Of course, you could use error mitigation as we did back in *Chapter 7, VQE: Variational Quantum Eigensolver*, by setting `measurement_error_mitigation = True` in the `TorchRuntimeClient` instantiation.

11.3.4 A glimpse into the future

The way in which we have worked with Torch Runtime is supported by IBM at the time of writing, but change is the only constant in Qiskit land.

In the future, Torch Runtime will no longer be supported and, instead, it will be necessary to use a different interface in order to train quantum neural networks with Qiskit Runtime. This interface — which, at the time of writing, is still in active development — will rely on the `Sampler` and `Estimator` objects that we mentioned in *Section 7.3.7*. In this subsection, we will present to you a simple example that will showcase how to work with this new interface.

The following piece of code can be used to train a simple variational quantum classifier (a `VQC` object) using the "new" Qiskit Runtime on the `ibmq_lima` device:

```
from qiskit_ibm_runtime import QiskitRuntimeService,Session,Sampler,Options
from qiskit_machine_learning.algorithms.classifiers import VQC

# channel = "ibmq_quantum" gives us access to IBM's quantum computers.
service = QiskitRuntimeService(channel = "ibm_quantum", token = "TOKEN")

with Session(service = service, backend = "ibmq_lima"):
    sampler = Sampler()
    vqc = VQC(sampler = sampler, num_qubits = 2)
    vqc.fit(x_tr, y_tr)
```

Please note that you need to install the `qiskit_ibm_runtime` package (refer to *Appendix D, Installing the Tools*, for instructions) and replace `"TOKEN"` with your actual IBM Quantum token.

As a matter of fact, when you send a program through this new Qiskit Runtime interface, you will likely see a fairly big collection of jobs on your IBM Quantum dashboard. Don't worry, Runtime is working just fine. All those jobs correspond to different calls to the quantum computer, but they are all executed without the need to wait in the queue after each and every job execution.

And that's all we wanted to share with you about the Torch Runtime utility. Let's wrap up this chapter.

Summary

This has been a long and intense chapter. We began by learning what hybrid neural networks actually are and in which use cases they can be useful. We then explored how to implement and train these hybrid networks in PennyLane and, along the way, we discussed a few good practices that apply to any machine learning project. In addition, we left our comfort zone and considered a new kind of QML problem: the training of multi-class classifiers.

Once we finished our study of PennyLane, we dived into Qiskit, and a big surprise was waiting for us there. Since Qiskit relied on an interface with the PyTorch ML package for the implementation of hybrid QNNs, we invested a good deal of effort in learning how to use PyTorch. In the process, we saw how PyTorch provided us with a level of flexibility that we simply couldn't get using TensorFlow and Keras. At the point where we had a solid understanding of the PyTorch package, we got to work with Qiskit and its PyTorch connector and we trained a hybrid QNN with them.

Lastly, we concluded the chapter by fulfilling a promise we made in *Chapter 10, Quantum Neural Networks*, and we discussed how to train quantum neural networks on IBM's quantum hardware using Torch Runtime.

12

Quantum Generative Adversarial Networks

Fake it 'till you make it

— Someone, somewhere

So far, we have only dealt with quantum machine learning models in the context of supervised learning. In this final chapter of our QML journey, we will discuss the wonders and mysteries of a QML model that will lead us into the domain of unsupervised learning. We will discuss quantum versions of the famous **Generative Adversarial Networks** (often abbreviated as **GANs**) that are called **Quantum Generative Adversarial Networks**, **quantum GANs**, or **QGANs**.

In this chapter, you will learn what classical and quantum GANs are, what they are useful for, and how they can be used. We will begin from the basics, exploring the intuitive ideas that lead to the concept of a GAN. Then, we will get into some of the details and discuss QGANs. In particular, we will talk about the different types of QGANs out there and their (possible) advantages. You will also learn how to work with them using PennyLane (with its TensorFlow interface) and Qiskit.

We'll cover the following topics in this chapter:

- GANs and their quantum counterparts
- Quantum GANs in PennyLane
- Quantum GANs in Qiskit

Excited about this last chapter? Let's begin by understanding what these GANs are all about.

12.1 GANs and their quantum counterparts

Quantum GANs are **generative models** that can be trained in a perfectly unsupervised manner. By the fact that they are generative models we mean that quantum GANs will be useful for generating data that can mimic a training dataset; for instance, if you had a large dataset with pictures of people, a good generative model would be able to generate new pictures of people that would be indiscernible from those coming from the original distribution. The fact that QGANs can be trained in an unsupervised fashion simply means that our datasets will not have to be labeled; we won't have to tell the generator whether its output is good or bad, the model will figure that out on its own. How exactly? Stay tuned!

That's the big picture of GANs, but, before we can explore all their details, there's something we need to talk about. Let's talk about how to counterfeit money.

12.1.1 A seemingly unrelated story about money

Of course, all of us reading these lines are law-abiding citizens — no need to call the police just now — but, for the purposes of intellectual illustration, let's put ourselves in the place of the bad guys for one day. In the process of counterfeiting money, there are two main actors involved:

- We, the bad guys who create (generate) counterfeit money, trying to make it as close to the real thing as possible
- Some authority, usually a central bank, which is in charge of designing tools and techniques to discern real money from counterfeit money

This is shown in *Figure 12.1*. By the way, we have drawn the fake dollar ourselves. Graphic design is our passion.

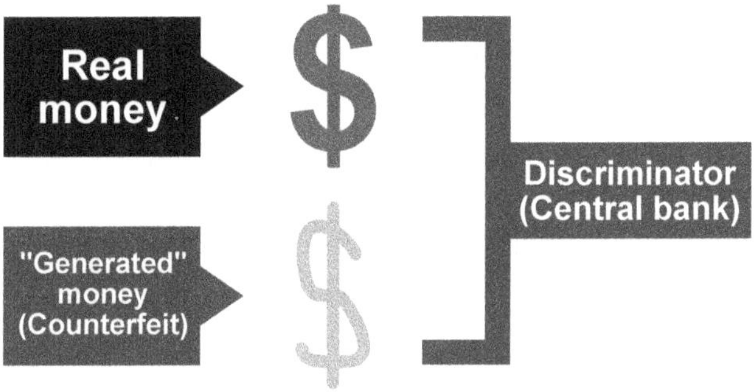

Figure 12.1: Schematic representation of the agents involved in the generation of counterfeit money

Now that we are all set, we can imagine what our counterfeiting career could look like. Since none of us has any experience in this field, our first attempts at faking banknotes would be extremely disastrous: any of our generated banknotes would be very easily identified as fake by the central bank. However, that would just be the beginning of the story. Along the process — and assuming we didn't get arrested — we could always try to study how the central bank is discerning real notes from counterfeit ones and use it to our advantage by trying to fool its detection mechanisms. Naturally, however, that would only be a temporary solution, for it wouldn't take long for the central bank to notice our improved fake notes and design better detection systems, which would take us back to the drawing board, starting the process all over again.

Banknotes have a finite amount of defining features, so, after a large enough number of iterations of this process, at some point, we would likely end up producing banknotes that would be identical to the real ones. And, thus, a beautiful equilibrium would be reached in which the central bank would no longer be able to detect our fake notes. Sadly for us, this adventure would most surely end with the central bank changing the notes completely and sending us before a judge. But let's ignore those tiny details!

Important note

Just in case it wasn't obvious, we are joking when we talk about imagining ourselves doing counterfeiting. Counterfeiting money is, as you hopefully know, a serious criminal offense that we, of course, don't encourage or endorse in any way. Please, don't do illegal stuff. The editorial team thought — with good reason! — that this was worth a disclaimer; so here it is!

Now, you may wonder why we have discussed this. Well, because, as it turns out, the process of training a GAN is just like that of counterfeiting money — minus the risk of ending up in prison. Let's see how it works!

12.1.2 What actually is a GAN?

GANs were introduced in 2014 in a very influential paper [66] by Goodfellow et al. As we mentioned in the introduction, a GAN is a machine learning model that can be trained to generate data closely reassembling the patterns and properties of a given dataset. In order to accomplish this, a GAN has two main components:

- A "generative" neural network (generator), which will be nothing more than a neural network taking arbitrary seeds as input and returning outputs that match the datatype of the elements in the original dataset. The goal of this neural network will be, by the end of the training, to generate new data that be indistinguishable from the data in the original dataset.
- A discriminator neural network, which will be a binary-classifier neural network taking as input the original data in the dataset and the output of the generative network. This discriminator network will be tasked with trying to discern the generated data from the original data.

These components are depicted in *Figure 12.2*. By the way, we have drawn the fake tree ourselves. Graphic design is our passion.

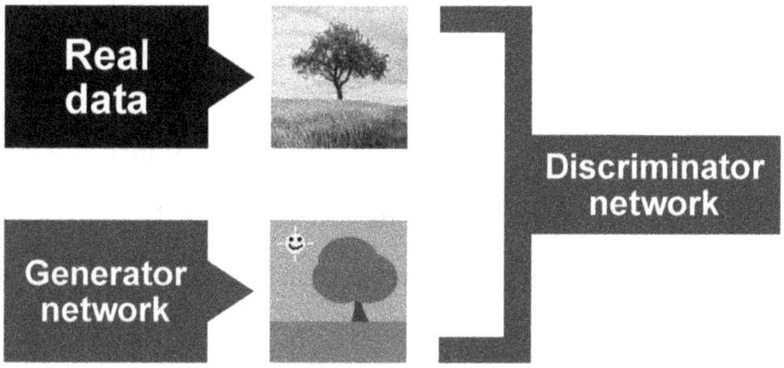

Figure 12.2: Schematic representation of the agents involved in a generative adversarial network

To learn more...

GANs have been very successfully used in practical generative tasks. For instance, StyleGANs are GANs introduced by NVIDIA researchers [87] that are able to generate extremely realistic human faces. Their code is open source (you can find it at `https://github.com/NVlabs/stylegan`) and they power the mesmerizing website "This Person Does Not Exist" (`https://www.thispersondoesnotexist.com/`).

This description settles the question of what a GAN is, but now we need to understand how these GANs are actually trained. In essence, this is how the whole training process works:

1. You initialize the generator and the discriminator to some random configuration.

2. You train the discriminator to discern the real data from the output of the generator. At this initial stage, this should be a very easy task for the discriminator.

3. You then train the generator to fool the discriminator: you train it in a way that the discriminator — as trained in the previous step — will classify as many of the generated outputs as real. Once trained, you use it to generate a bunch of fake data.

4. And here is where the fun begins. You re-train the discriminator on the new generated dataset, and then you re-train the generator to fool the new discriminator. And you repeat this process in as many iterations as you want. Ideally, in each iteration, it

will be harder for the discriminator to tell the generated data from the real data. And, eventually, an equilibrium will be reached in which the generated data will be indiscernible from the original data. Just like in our previous counterfeiting adventure – and with no legal troubles on the horizon!

This process is exemplified schematically in *Figure 12.3*, where we present a schematic illustration of the training process of a GAN meant to generate pictures of cute cats. When the GAN is initialized, the generator just produces random noise. After subsequent training iterations, the output of the generator will more closely resemble the images in the original dataset – which, in this example, should be a dataset with pictures of cats. By the way, we have drawn the fake cats ourselves. Have we mentioned that graphic design is our passion?

We should highlight that this scheme is very oversimplified. In truth, you usually don't "fully" train the discriminator and the generator alternately, but you optimize them in an alternate fashion. For example, if you were using gradient descent with a given batch size, then, on each epoch and on each batch, you would optimize the weights of the discriminator in a single optimizer step, and then you would do the same for the weights of the generator.

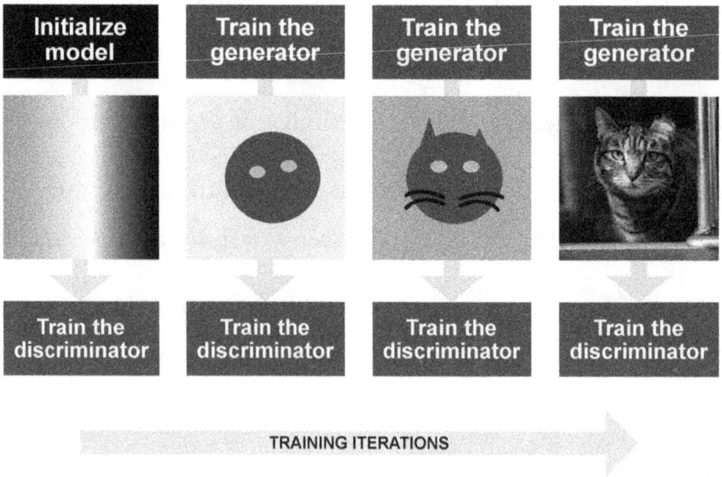

Figure 12.3: Schematic illustration of the training process of a GAN meant to generate pictures of cute cats

With this description of the training process, we can now make sense of the term GAN. These models are "generative" because they are aimed at generating data. They are "networks" because, well, they use neural networks. And they are "adversarial" because the whole training process consists in a competition between a generator network and a discriminator network. These networks engage in a fierce competition in which we, their programmers and creators, shall be the only true winners.

To learn more...

All this time, we have been talking about how GANs use neural networks in both the discriminator and the generator. However, these neural networks are not always like the ones we have discussed in this book.

The neural networks that we have studied are known as "dense" neural networks. In these networks, all the layers are dense, which means that neurons in subsequent layers are fully connected. However, when neural networks are designed to handle images – whether it be generating them, classifying them, or manipulating them – a different kind of layer is often employed: convolutional layers. We won't get into the details of how these layers work (check Chapter 14 in Gerón's book [64] for a thorough explanation), but you should at least know that they exist.

GANs are often used in image generation tasks, so, should you ever decide to study classical GANs, be aware that you will surely have to deal with these layers at some point. And, yes, there are quantum versions of convolutional layers and convolutional networks [88], [89] that, sadly, we do not have the time to cover in this book.

There are a few details that we should highlight about the training process of a GAN. The first and most important one is the fact that at no point in the training is the generator network "exposed" to or fed the original data. The only way the generator network can learn about the data it has to replicate is through the discriminator. In this way, instead of us having to tell the generator network what its output should look like, the discriminator takes up our role as teachers and enables us to train the whole network in a fully unsupervised manner.

Another issue to which we should pay attention is that GANs, like any other machine learning model, are vulnerable to problems in training. For instance, how could we have any guarantee that the generated outputs are not just slightly distorted copies of the original data, rather than new data elements that match the patterns in the original dataset? For instance, in the cat GAN that we considered in *Figure 12.3*, how could we have guarantees that the generated images are new cat pictures rather than, say, blurred copies of our original images that have lost any resemblance to cats but that were nevertheless able to fool the discriminator network? This could happen, for example, if our discriminator weren't powerful enough compared to the generator.

To learn more...

The training of a GAN can also fail if the resulting GAN is unable to generate all the possible variations (or **modes**) of data that can be found in the dataset. For instance, in the example that we have been considering, we would find this problem if our GAN were only able to generate pictures of a small selection of cats, maybe even only one! This occurrence is known as **mode collapse**. To try to avoid it, several modified GANs have been proposed, including **Wasserstein GANs (WGANs)** [90], which derive their loss function from a distance called the Wasserstein metric.

In the models that we considered in previous chapters, there was always a simple, straightforward way to effectively assess their performance — namely evaluating loss functions on test datasets. When working with GANs, things can be more subtle. In general, you should always take a look at the generated data and check if the results are satisfactory.

Important note

A GAN consists of two neural networks: a generator and a discriminator. They compete against each other in an iterative training process. The discriminator is tasked with discerning a dataset of real data from the output of the generator network, while the generator network is tasked with generating data that the discriminator will mistakenly identify as real.

Just to conclude this overview of classical GANs, let's discuss a few technicalities about the training of the generator and discriminator networks.

12.1.3 Some technicalities about GANs

We have already mentioned how the generator and discriminator networks are ordinary neural network models — even if they may be different from the ones that we've discussed so far — that are constantly re-trained in an iterative process. We will now briefly talk about how this training is carried out.

Let X be a set of real data and let S be a set of "seeds" that we give to the generator. In the case of the discriminator neural network, we are just training a binary classifier and, as is standard, this classifier will return an output bounded between 0 and 1. Without loss of generality, we will assume that values closer to 1 are meant to represent inputs from the real dataset while values closer to 0 are labeled as generated inputs — that's an arbitrary choice; it could perfectly be the other way around.

As with any other binary classifier, the most natural loss function to use will be the binary cross-entropy loss, and hence this classifier will be trained as it would in supervised learning: assigning the "true label" 1 to any input from the real dataset and the "true label" 0 to any generated input. In this way, if we let G and D denote the actions of the generator and the discriminator, the discriminator training loss, L_D, would be computed as

$$L_D = -\frac{1}{|X| + |S|}\left(\sum_{x \in X} \log D(x) + \sum_{s \in S} \log\left(1 - D(G(s))\right)\right),$$

where we are using $|X|$ and $|S|$ to denote the sizes of the sets X and S, respectively. The job of the discriminator would be to minimize this loss.

Now, what about the generator network? What could be a good choice for the loss function that we would like to minimize in its training process? Our goal when training the generator is to fool the discriminator trying to get it to classify our generated data as real data. Hence, the goal in the training of the generator is to maximize the loss function of the discriminator, that is, to minimize

$$-L_D = \frac{1}{|X| + |S|}\left(\sum_{x \in X} \log D(x) + \sum_{s \in S} \log\left(1 - D(G(s))\right)\right).$$

Nevertheless, the contribution of the first term in the sum is necessarily constant in the generator training since it does not depend on the generator in any way. Thus, equivalently, we may consider the goal of the generator training to be the minimization of the generator loss function

$$L'_G = \frac{1}{|S|}\sum_{s \in S} \log\left(1 - D(G(s)\right).$$

That is how things are in theory. However, in practice, it has been shown [66] that it is usually more stable to take the goal of the generator training to be the minimization of the loss

$$L_G = -\frac{1}{|S|}\sum_{s \in S} \log\left(D(G(s)\right).$$

The crucial thing here is that, with both definitions, if these generator loss functions decrease while training the generator, it will be more likely for our generated data to be (mistakenly) classified as real data by our classifier. That will mean, in turn, that our data should be gradually getting more and more similar to the data in the original dataset.

It has also been shown that, in the optimal equilibrium between the generator and the discriminator, the discriminator assigns values $D(x)$ and $G(D(s))$ equal to $1/2$ (because it cannot distinguish between real and generated data), and, hence, when $L_D = L_G = -\log 1/2 = \log 2 \approx 0.6931$. You can find the proof (with slightly different but equivalent loss functions) in the original GANs paper [66].

To learn more...

It can be shown that the optimal configuration of a GAN is a Nash equilibrium of an adversarial game between the generator and the discriminator (see, for instance, the helpful tutorial given by Goodfellow at NIPS [91]). In this equilibrium, the configuration of the GAN is a (local) minimizer of both the generator and discriminator losses.

That should be enough of an introduction to classical GANs. Let's now see what quantum GANs are and what they have to offer.

12.1.4 Quantum GANs

What is a quantum GAN? It's just a GAN, with its competing discriminator and generator, where a part of the model is implemented by a quantum model (usually some form of a quantum neural network), and it is trained just like a classical GAN. In other words, training a quantum GAN is just like counterfeiting money — but you don't risk going to prison and you get to play with quantum stuff.

To learn more...

By the way, did you know that there are proposals of quantum money that cannot be counterfeited at all? The original idea was proposed by Stephen Wiesner [92] and it became the inspiration for unbreakable quantum cryptographical protocols such as the famous BB84 proposed by Bennett and Brassard [4].

In truth, that is as close to a precise definition as we can get, because the range of models that can fit into the category of QGAN is vast. Depending on the kind of problem that you want to tackle, you may want to use quantum GANs with completely different architectures which, still, will share the same core elements of a competing discriminator and generator. The examples that we will consider in the following sections will help us exemplify this.

Broadly speaking, any quantum GAN could fit into one of the following categories:

- **Uses quantum data and both the generator and the discriminator are quantum:** This quantum data will just be some quantum states, and the generator and discriminator will be implemented by quantum circuits.

  This situation allows for a very special QGAN architecture, with a fully quantum model. Since we are dealing with quantum data (states), and all the components of the GAN are quantum circuits, they can be perfectly joined together without having to resort to feature maps or measurement operations in the middle of the model.

  Later in the chapter, we will study an example of this purely quantum architecture on PennyLane.

- **Uses quantum data and a quantum generator with a classical discriminator:** If the discriminator is classical, the architecture of our QGANs will be more similar to that of classical GANs. The generator will produce quantum states but, ultimately, they will be transformed into classical data by some measurement operation in order to feed them into the classifier. Of course, the original quantum data will also have to be measured.

- **Uses classical data with a quantum generator or discriminator:** This is the scenario in which QGANs can best match their classical counterparts. The use of QGANs in these cases essentially mounts up to replacing the generator or the discriminator (or both) with a quantum model with classical inputs and outputs. In the case of a quantum discriminator, for example, we would have to use a feature map to load classical data into a quantum state.

  Because the availability of classical data is much bigger than that of quantum data, this is the type of architecture that has been studied more widely by the quantum computing community.

  Later in the chapter, we will consider a QGAN with classical data and a classical classifier, but with a quantum generator. That will be in our Qiskit section.

To learn more...

In the literature, there are many different proposals of quantum versions of GANs. Some of the earliest ones include works by Lloyd and Weedbrook [93], by Dallaire-Demers and Killoran [94], and by Zoufal, Lucchi, and Woerner [95].

Exercise 12.1

In this book, we have discussed four different QML models: quantum support vector machines, quantum neural networks, hybrid QNNs and quantum GANs. Decide which of these models would be suitable for the following tasks:

1. Distinguishing cat pictures from dog pictures.
2. Generating pictures of dogs.
3. Deciding whether a financial transaction is fraudulent based on its metadata.
4. Assessing the risk of heart failure from a patients' medical records and data from an electrocardiogram.
5. Creating a dataset of random images of electrocardiograms in order to train future doctors.

We should give you a word of caution. GANs aren't the easiest models to train. As we mentioned previously, when you train a GAN, you don't have a single and straightforward loss function that can measure how successful your training is. Training a GAN is not a simple optimization problem, but a more intricate process. Using quantum models, of course, only makes matters more difficult, and training quantum GANs can be... complicated.

We will now consider a couple of interesting QGAN examples, in both PennyLane and Qiskit. Naturally, since we've picked them, our quantum GANs will learn smoothly. But you have been warned: quantum GANs are usually wild creatures.

12.2 Quantum GANs in PennyLane

In this section, we are going to train a purely quantum GAN that will learn a one-qubit state. In our previous counterfeiting example, we imagined ourselves as behaving like a

GAN in order to replicate some training data (a banknote) to produce fake banknotes that, ideally, would get closer and closer to the real thing in each iteration. In this case, our training data will be a one-qubit state, characterized by some amplitudes, and the job of our QGAN will be to replicate that state without the generator having direct access to it. Our dataset, then, will consist of multiple copies of a one-qubit state, and our goal will be to train a generator able to prepare that state (or something very close to it).

To learn more...

Notice that this setting does not violate the no-cloning theorem that we proved in *Section 1.4.5*. We will have multiple copies of the same quantum state and we will perform operations on them, including measuring them (and, hence, collapsing their states). From that, we will learn some properties of the state that we will use to reproduce it with the generator. But we won't be having a unitary operation (a quantum gate) that creates an additional, independent copy of a given state. In fact, we will destroy the original copies in the process!

What we will be doing here is more similar to **quantum state tomography** (see, for instance, the review by Altepeter, James, and Kwiat [96]), which can be defined as the process of applying quantum operations and measurements to multiple copies of a state and, from the results, learning to reconstruct the original state.

For this example, we will use the PyTorch machine learning package. Please, have a look at *Subsection 11.3.1* if you haven't already.

The reason behind our choice to use PyTorch is simple. As much as we have used TensorFlow so far, we only know how to use it at a basic level, relying heavily on the Keras interface. On the other hand, we have studied PyTorch extensively in the previous chapter, which makes it a better tool for us when it comes to dealing with more complex architectures. In other words, this choice isn't grounded in any technical superiority of any package over the other, but solely on what's most practical given the content that we've covered in this book. In fact, virtually any model that can be built and trained on PyTorch can also be dealt with on TensorFlow and vice versa.

With those preliminaries aside, let's get to our model.

12.2.1 Preparing a QGAN model

The purely quantum GAN that we seek to implement and train will run on a device with two qubits, and it will be made up of the following components:

- A quantum circuit that will be able to prepare the one-qubit state $|\psi_1\rangle$ that we want our QGAN to learn. This circuit will run on the first qubit of the device. We should regard it as a black box, the inner working of which is fully opaque to our model.

  The state $|\psi_1\rangle$, which we will refer to as the "true state," is the quantum training data that we will use in our QGAN. This circuit will just provide us with a way of accessing the training data: as many copies of the $|\psi_1\rangle$ state as we may need in the training process. This emulates, for instance, a physics experiment that produces some quantum state that we want to learn.

- A quantum generator, which will also run on the first qubit of the device and which aims to prepare a state similar to $|\psi_1\rangle$ on the first qubit. The quantum generator will be implemented by a variational form dependent on some trainable parameters.

- A quantum discriminator, which will run on the first and second qubits of the device. Its "input" will be the state on the first qubit, which can either be the state we want our QGAN to learn or the state prepared by the generator. Of course, the job of the discriminator will be to try to distinguish these two states. We implement it with two qubits (instead of just one) to be sure that it has enough discriminative power.

  Since this discriminator already takes a quantum input, it only needs to consist of a variational form followed by a measurement operation – there will be no need to use feature maps, as we had to do when working with classical data. As usual, we will place the measurement operation on the first qubit.

All the components that we have just described are depicted in *Figure 12.4*.

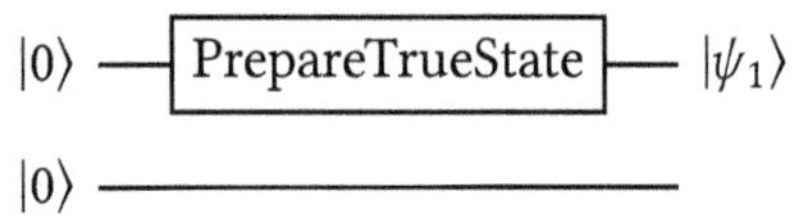

(a) Circuit preparing the state $|\psi_1\rangle$ that we want our QGAN to learn.

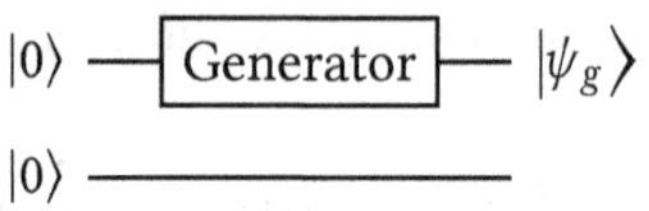

(b) Generator circuit that outputs a state $|\psi_g\rangle$. We aim for $|\psi_g\rangle$ to be similar to $|\psi_1\rangle$.

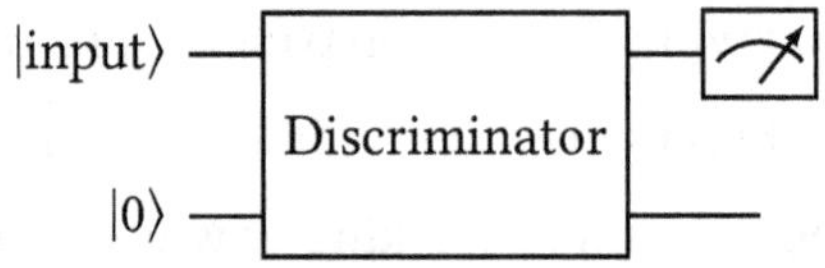

(c) Discriminator circuit, tasked with deciding whether the state $|input\rangle$ is the state $|\psi_1\rangle$ or the output of the generator.

Figure 12.4: Components of the quantum GAN that we will train to generate $|\psi_1\rangle$

Now that we have a sense of where we are heading, let's get ready to write some code. First of all, we will do our usual imports and set some seeds to ensure the reproducibility of our results:

```
import pennylane as qml
import numpy as np

import torch
import torch.nn as nn

seed = 1234
np.random.seed(seed)
torch.manual_seed(seed)
```

We will construct the state $|\psi_1\rangle$ using the universal one-qubit gate $U_3(\varphi, \theta, \delta)$ since, as we learned back in *Chapter 1, Foundations of Quantum Computing*, it allows us to create any one-qubit state. In particular, we will feed it the values $\varphi = \pi/3$, $\theta = \pi/4$, and $\delta = \pi/5$:

```
phi = np.pi / 3
theta = np.pi / 4
delta = np.pi / 5
```

With these values set, we can define a function that will construct the circuit that will prepare $|\psi_1\rangle$:

```
def PrepareTrueState():
    qml.U3(theta, phi, delta, wires = 0)
```

Notice that we have defined this as a function and not as a quantum node. That's because, for the purposes of the training, we are not interested in running any of the components of the quantum GAN individually. We will instead have to run them in composition. For instance, we will have to run this circuit that we've just defined composed with the discriminator.

Now that we have a circuit that can prepare $|\psi_1\rangle$, it's time for us to think about the two core components of our QGAN: the generator and the discriminator. Specifically, we will have to find some suitable variational forms for them.

For the generator, we will simply use a parametrized U3 gate, whereas, for the discriminator, we will use a variation of the two-local variational form. These can be implemented as follows:

```
def GeneratorVF(weights):
    qml.U3(weights[0], weights[1], weights[2], wires = 0)


def DiscriminatorVF(nqubits, weights, reps = 1):
    par = 0 # Index for parameters.
    for rep in range(reps):
        for q in range(nqubits):
            qml.RX(weights[par], wires = q)
            par += 1
```

```
            qml.RY(weights[par], wires = q)
            par += 1
            qml.RZ(weights[par], wires = q)
            par += 1
        for i in range(nqubits - 1):
            qml.CNOT(wires = [i, i + 1])

    for q in range(nqubits):
        qml.RX(weights[par], wires = q)
        par += 1
        qml.RY(weights[par], wires = q)
        par += 1
        qml.RZ(weights[par], wires = q)
        par += 1
```

You can see a graphical representation of the discriminator variational form in *Figure 12.5*; its implementation is mostly analogous to that of the two-local variational form with just a few small differences. On a very minor note, we have renamed the vector of optimizable parameters to `weights` (instead of `theta`) to avoid any sort of confusion with the angle `theta` that defines $|\psi_1\rangle$.

Taking advantage of these newly-defined variational forms, we will define the circuits of the generator and the discriminator as follows:

```
def Generator(weights):
    GeneratorVF(weights)


def Discriminator(weights):
    DiscriminatorVF(2, weights, reps = 3)
```

We are now ready to define the quantum nodes that we will use in the training. In the classifier, we shall take the measurement operation to be the computation of the expectation

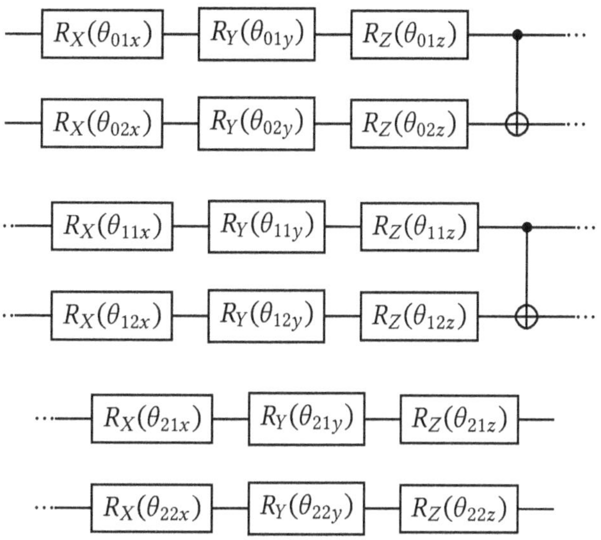

Figure 12.5: Discriminator variational form on two qubits and two repetitions

value of $M = |0\rangle\langle 0|$ on the first qubit. For this purpose, we may construct the matrix M as follows:

```
state_0 = [[1], [0]]
M = state_0 * np.conj(state_0).T
```

And we can now define two quantum nodes: one concatenating the generation of the state $|\psi_1\rangle$ with the discriminator, and one concatenating the generator with the discriminator. We can achieve this with the following piece of code:

```
dev = qml.device('default.qubit', wires = 2)


@qml.qnode(dev, interface="torch", diff_method = "backprop")
def true_discriminator(weights_dis):
    PrepareTrueState()
    Discriminator(weights_dis)
    return qml.expval(qml.Hermitian(M, wires = [0]))
```

```
@qml.qnode(dev, interface="torch", diff_method = "backprop")
def generator_discriminator(weights_gen, weights_dis):
    Generator(weights_gen)
    Discriminator(weights_dis)
    return qml.expval(qml.Hermitian(M, wires = [0]))
```

The measurement operation is the computation of the expectation value of M on the first qubit in both nodes; since this operation is the output of the discriminator, these measurement operations need be identical. Notice, by the way, that, since the discriminator works on the two qubits of our device, we could have also used the expectation value of M on the second qubit.

The training process

Now we have fully set up our model, and we have defined all the nodes that we will use in its training. But there's something essential that we haven't yet defined: the loss functions of the discriminator and the generator.

As we discussed before, a reasonable choice for the loss function of the discriminator of a GAN is the binary cross-entropy. In our case, our discriminator only has to classify two data points: the true state $|\psi_1\rangle$ with intended label 1, and the generated state $|\psi_g\rangle$ with intended label 0. Therefore, if we let D denote the action of the discriminator under a certain configuration, the binary cross-entropy loss would be

$$L_D = -\frac{1}{2}\left(\log\left(1 - D(|\psi_g\rangle)\right) + \log\left(D(|\psi_1\rangle)\right)\right).$$

This loss function can be implemented with our previously-defined nodes as follows:

```
def discriminator_loss(weights_gen, weights_dis):

    # Outcome of the discriminator with a generated state.
```

```
    out_gen = generator_discriminator(weights_gen, weights_dis)

    # Outcome of the discriminator with the true state.
    out_true = true_discriminator(weights_dis)

    return -(torch.log(1 - out_gen) + torch.log(out_true))/2
```

Now, what about the loss of the generator? We already know that the goal of the generator is to fool the discriminator into misclassifying the generated state as real. Moreover, we have also mentioned a reasonable generator loss function is

$$L_G = -\log\left(D(|\psi_g\rangle)\right).$$

This would be the binary cross-entropy loss of the discriminator if it were tasked with classifying the generated state as the true state.

We can easily implement this loss as follows:

```
def generator_loss(weights_gen, weights_dis):
    out_gen = generator_discriminator(weights_gen, weights_dis)
    return -torch.log(out_gen)
```

And that defines all our losses. Let's now prepare ourselves for the training process. First and foremost, let's initialize the weights of the generator and the discriminator to a tensor with random values:

```
weights_gen = torch.rand(3, requires_grad = True)
weights_dis = torch.rand((3 + 1) * 2 * 3, requires_grad = True)
```

The dimensions of these arrays are justified from the fact that the generator uses 3 weights and the variational form of the discriminator has 3 + 1 groups of parametrized gates, with 3 parameters being used on each of the 2 qubits on which the form acts. Also, remember that we need to set `requires_grad = True` in order for PyTorch to be able to compute gradients on these weights later on.

Now we can define the optimizers that we will use in the training. For this problem, we will rely on the stochastic gradient descent algorithm, which is a more simple version of the Adam optimizer that we used in previous chapters (see *Section 8.2.3* for a refresher). When invoking the optimizers, we have to provide an array or dictionary with the parameters that we want our optimizer to look after. Back when we defined PyTorch models as subclasses of `nn.Module`, we could just get this with the `parameters` method, but in this case, we will create the list ourselves. This can be done as follows:

```
optg = torch.optim.SGD([weights_gen], lr = 0.5)
optd = torch.optim.SGD([weights_dis], lr = 0.5)
```

In this call to the optimizers, we have set their learning rate to 0.5.

And those are all the ingredients needed to train our model. We can execute the following piece of code in order to do so:

```
dis_losses = [] # Discriminator losses.
gen_losses = [] # Generator losses.
log_weights = [] # Generator weights.

ncycles = 150 # Number of training cycles.

for i in range(ncycles):
    # Train the discriminator.
    optd.zero_grad()
    lossd = discriminator_loss(weights_gen.detach(), weights_dis)
    lossd.backward()
    optd.step()

    # Train the generator.
```

```
    optg.zero_grad()
    lossg = generator_loss(weights_gen, weights_dis.detach())
    lossg.backward()
    optg.step()

    # Log losses and weights.
    lossd = float(lossd)
    lossg = float(lossg)
    dis_losses.append(lossd)
    gen_losses.append(lossg)
    log_weights.append(weights_gen.detach().clone().numpy())

    # Print the losses every fifteen cycles.
    if (np.mod((i+1), 15) == 0):
        print("Epoch", i+1, end= " ")
        print("| Discriminator loss:", round(lossd, 4), end = " ")
        print("| Generator loss:", round(lossg, 4))
```

There is quite a lot to digest here. In the first few lines of code, we are simply defining some arrays in which we will store data as the training progresses. The arrays `dis_losses` and `gen_losses` will save the discriminator and generator losses in each training cycle, and the array `log_weights` will store the generator weights obtained at the end of each training cycle. We will later use this information in order to assess the effectiveness of the training.

We have fixed the training to run for 150 optimization cycles. In each of them, we will optimize the values of the discriminator, then optimize those of the generator, and, finally, log all the results. Let's go through it step by step:

1. When we optimize the discriminator, we reset its optimizer (`optd`) and then compute the discriminator loss function and store it in `lossd`. Observe that, when we send the generator weights, we pass them through the `detach` method. This method removes

the need to compute gradients for these weights. The discriminator optimizer is not going to touch those weights either way, so this will save us some computation time. Once we have the loss, we just compute its gradients with the `backward` method and run a step of the discriminator optimizer.

2. The optimization of the generator is fully analogous. We simply use the generator optimizer `optg` on the gradients obtained from the generator loss `lossg`. Of course, we detach the discriminator weights in the call to the generator loss function instead of the generator weights.

3. Finally, we log the values of the losses. For this purpose, we simply store the values of the losses that we computed in the training cycle. These will probably be different from the ones at the end of the cycle, but they will still be informative enough.

   After this, we store the generator weights. Please observe the call to the `clone` method. This call ensures that we are getting a copy of the weights and not a reference to the weights tensor. If we didn't call this method, all the weight arrays in `log_weights` would reference the same tensor and their values would all be the same and would change (simultaneously) as the training progresses!

   Finally, we print some information about the training. Since we are going to execute this loop for 150 training cycles and the training will be fast, we shall only print information every 15 cycles.

Notice how, instead of fully training the discriminator and the generator in an alternating fashion, we are optimizing them in an alternating fashion in every training cycle.

The output that we get upon running the preceding code is the following:

```
Epoch 15 | Discriminator loss: 0.6701 | Generator loss: 0.7065
Epoch 30 | Discriminator loss: 0.6987 | Generator loss: 0.6791
Epoch 45 | Discriminator loss: 0.6931 | Generator loss: 0.6992
Epoch 60 | Discriminator loss: 0.6931 | Generator loss: 0.6924
Epoch 75 | Discriminator loss: 0.6932 | Generator loss: 0.6927
```

```
Epoch 90 | Discriminator loss: 0.6931 | Generator loss: 0.6934
Epoch 105 | Discriminator loss: 0.6931 | Generator loss: 0.6931
Epoch 120 | Discriminator loss: 0.6931 | Generator loss: 0.6931
Epoch 135 | Discriminator loss: 0.6931 | Generator loss: 0.6932
Epoch 150 | Discriminator loss: 0.6931 | Generator loss: 0.6931
```

Just by looking at this raw output, we can see that there is a chance that our training may have been successful: the discriminator loss and the generator loss are both approaching $-\log 1/2$, just as they should do at the optimal point. This is a good sign!

In order to have a better insight on the evolution of these losses, we may use the `gen_losses` and `dis_losses` array in order to plot their evolution. This can be done as follows:

```
import matplotlib.pyplot as plt
epochs = np.array(range(len(gen_losses))) + 1
plt.plot(epochs, gen_losses, label = "Generator loss")
plt.plot(epochs, dis_losses, label = "Discriminator loss")
plt.xlabel("Epoch")
plt.legend()
```

The resulting graph can be found in *Figure 12.6* and, indeed, we can see a nice trend from which to draw some optimism.

But now comes the moment of truth. Let's see if, indeed, our model has learned as we wanted it to. We mentioned in the previous section that, when training generative adversarial networks, the best criteria for determining whether a training process was successful or not depends on the problem at hand. In our case, our training will be successful if the state returned by the generator is close to $|\psi_1\rangle$.

Now, how do we determine the state vector of a qubit? It turns out that the state of a qubit is fully characterized (up to an unimportant global phase, as we saw back in *Section 1.3.4*) by its Bloch sphere coordinates. And now that we've come across these coordinates, let's learn

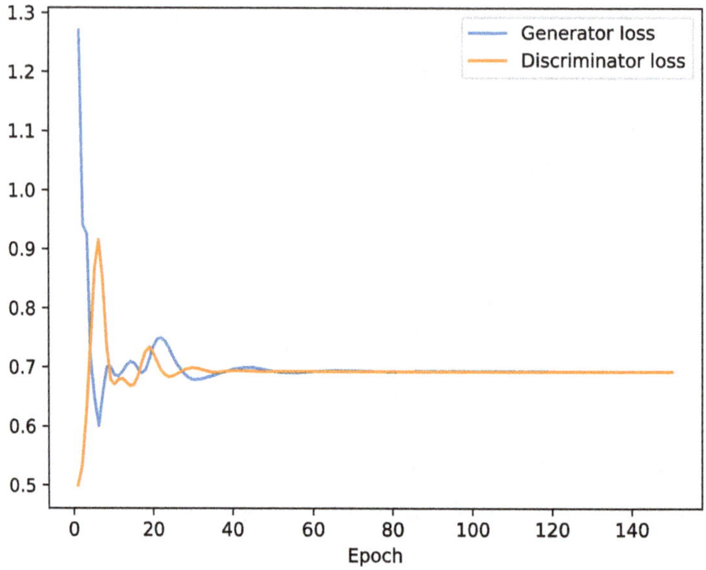

Figure 12.6: Evolution of the losses of the discriminator and the generator along the training process

how to compute them with an exercise that we hope you will find interesting — although, admittedly, is slightly orthogonal to this chapter.

Exercise 12.2

Prove that the Bloch sphere coordinates of a one-qubit state are the expectation values of the observables given by the three Pauli matrices X, Y, and Z.

We can prepare two quantum nodes that return these expectation values for both $|\psi_1\rangle$ and the state returned by the generator after the training. This can be done as follows:

```
@qml.qnode(dev, interface="torch")
def generated_coordinates(weights_gen):
    Generator(weights_gen)
    return [qml.expval(qml.PauliX(0)), qml.expval(qml.PauliY(0)),
    qml.expval(qml.PauliZ(0))]
```

```
@qml.qnode(dev, interface="torch")
def true_coordinates():
    PrepareTrueState()
    return [qml.expval(qml.PauliX(0)),
            qml.expval(qml.PauliY(0)),
            qml.expval(qml.PauliZ(0))]

print("Bloch coordinates")
print("Generated:", generated_coordinates(weights_gen))
print("True:", true_coordinates())
```

And the output that we get is the following:

```
Bloch angles
Bloch coordinates
Generated: tensor([0.3536, 0.6124, 0.7071], dtype=torch.float64,
grad_fn=<MvBackward0>)
True: tensor([0.3536, 0.6124, 0.7071], dtype=torch.float64)
```

The outputs are identical, so we can safely say that our training has been a huge success!

In order to bring this section to an end, we will visually explore how the state created by the generator has evolved throughout the training. We can do this using the array of weights `log_weights` and the `generated_coordinates` function that we have just defined. This function takes the weights of the generator as input, so we can get the Bloch coordinates of the generated states at any point in the training using the saved weights.

We can accomplish this as follows:

```
true_coords = true_coordinates()
def plot_coordinates(cycle):
```

```
coords = generated_coordinates(log_weights[cycle - 1])

plt.bar(["X", "Y", "Z"], true_coords, width = 1,
    color = "royalblue", label = "True coordinates")
plt.bar(["X", "Y", "Z"], coords, width = 0.5,
    color = "black", label = "Generated coordinates")

plt.title(f"Training cycle {cycle}")
plt.legend()
```

This function will plot, for any training cycle, a representation of the Bloch coordinates of the generated states superposed to the coordinates of the state that we want our QGAN to learn. In *Figure 12.7* you can see the plots corresponding to a wide range of cycles.

Exercise 12.3

Try to replicate this example on a different state (you may need to increase the number of training cycles to reach convergence in some cases).

That brings this example to an end. Let's now consider a different QGAN, this time implemented in Qiskit.

12.3 Quantum GANs in Qiskit

An early proposal of a QGAN was introduced by IBM researchers Zoufal, Lucchi, and Woerner [95] to learn a probability distribution using a QGAN with a quantum generator and a classical discriminator. In this section, we will discuss how to implement this kind of QGAN with Qiskit, so let's put everything in more precise terms.

This type of quantum GAN is given a dataset of real numbers that follow a certain probability distribution. This distribution may potentially be continuous, but it could be discretized to take some values $m, m+1, m+2, \ldots, M-1, M$ with $m < M$; this will usually be done by fixing the values m and M, rounding the samples and ignoring those that are smaller than

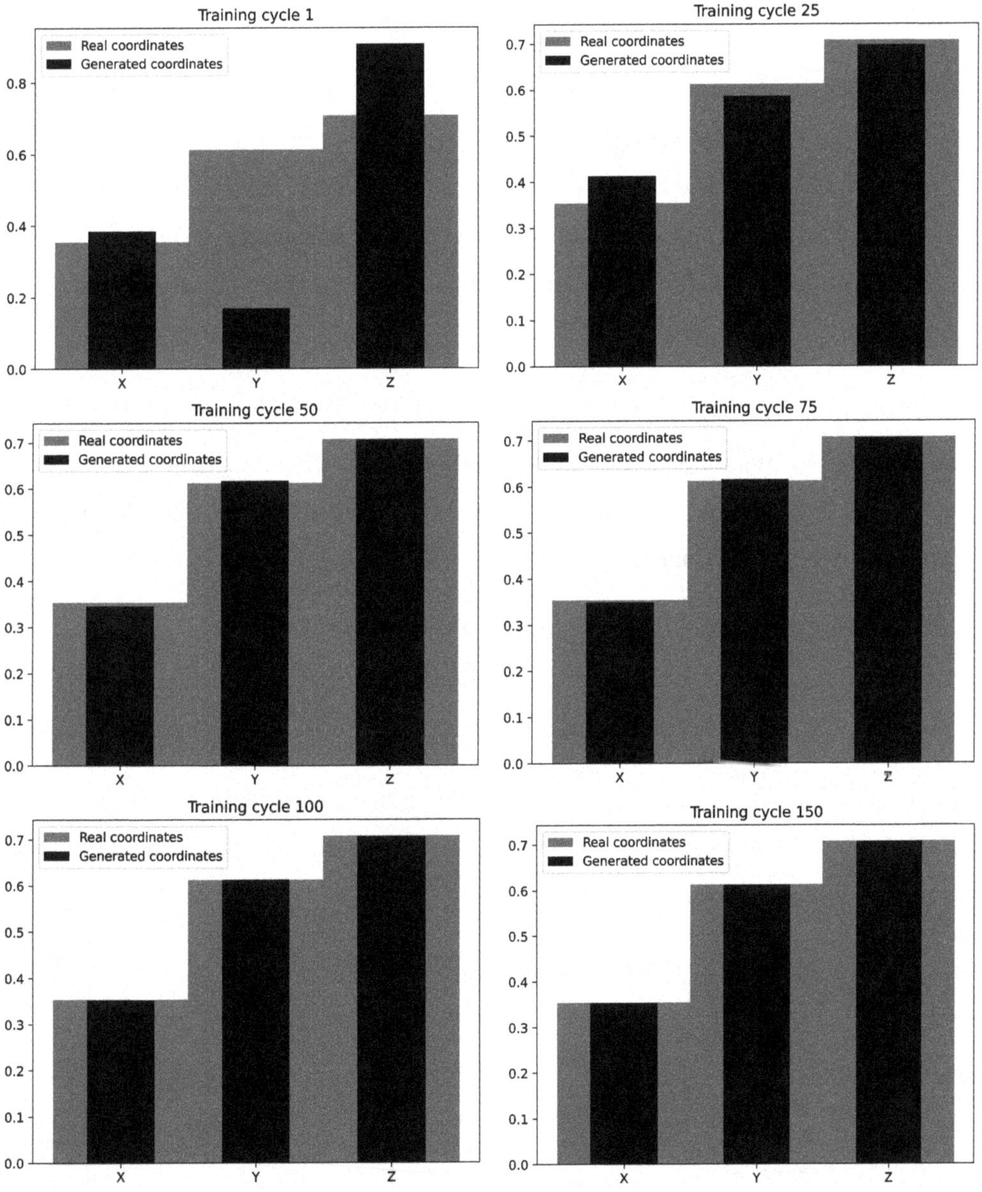

Figure 12.7: Evolution of the Bloch coordinates of the generated state as the training progresses

m or bigger than M. Each of the resulting labels $j = m, \ldots, M$ will have a certain probability p_j of appearing in the dataset. That is the distribution that we want the generator in our QGAN to learn.

And what does the generator of these QGANs look like? It is a quantum generator that is dependent on some classical parameters. It needs to be designed to have n qubits in such a way that $M - m < 2^n$, so that we may assign, to each possible outcome r after a measurement in the computational basis of the generator, a label $\alpha(r)$ in $m, \ldots M$. Thus, the goal of the training will be for the state returned by the generator to be as close as possible to

$$\sum_r \sqrt{p_{\alpha(r)}} \left|r\right\rangle .$$

In this way, measuring samples from the trained generator should be equivalent to extracting more data samples from the original distribution, because the probability of measuring $\left|r\right\rangle$ (which is associated to label $\alpha(r)$) is exactly $\left|\sqrt{p_{\alpha(r)}}\right|^2 = p_{\alpha(r)}$.

The discriminator that enables the training of this QGAN is a classical neural network tasked with distinguishing whether an input datum belongs to the original dataset or has been generated by the discriminator.

So that's the QGAN that we are going to work with: a hybrid QGAN in which the generator is quantum and the discriminator is classic. Sounds interesting? Let's see how we can implement it and train it using Qiskit.

In order to get started, let's import NumPy and Qiskit while setting some seeds to ensure the reproducibility of our results:

```
import numpy as np

from qiskit import *
from qiskit.utils import algorithm_globals

seed = 1234
```

```
np.random.seed(seed)
algorithm_globals.random_seed = seed
```

We will consider a particular example of the general problem that we outlined previously. We will take a dataset with 1000 samples generated from the binomial distribution with $n = 3$ trials and probability $p = 1/2$. These distributions can only take $4 = 2^2$ possible values $(0, 1, 2, 3)$, so will have to use 2 qubits in our generator. We may generate the samples of our dataset using NumPy as follows:

```
N = 1000
n = 3
p = 0.5

real_data = np.random.binomial(n, p, N)
```

The Qiskit framework already incorporates a `QGAN` class that can create and train the QGAN architecture that we discussed previously — it's almost tailor-made for this problem! We may import the class from the `qiskit_machine_learning.algorithms` module and define our QGAN as follows:

```
from qiskit_machine_learning.algorithms import QGAN
from qiskit.utils import QuantumInstance

ncycles = 3000 # Number of training cycles.
bsize = 100 # Batch size.

# Quantum instance on which the QGAN will run.
quantum_instance = QuantumInstance(
    backend=Aer.get_backend('statevector_simulator'))

# Create the QGAN object.
qgan = QGAN(data = real_data,
```

```
num_qubits = [2],
batch_size = bsize,
num_epochs = ncycles,
bounds = [0,3],
seed = seed,
tol_rel_ent = 0.001)
```

In the call to the `QGAN` initializer, we had to specify the dataset whose distribution we want to learn, the bounds at which we want to "cut" the dataset (in this case, we just specified the actual bounds of our distribution), an array containing the number of qubits of the generator circuit, the batch size, the number of training cycles that we want our QGAN to run for, the quantum instance on which the QGAN will run and, lastly, an optional seed.

You may be confused by the fact that we've had to send the number of qubits of the quantum generator in an array. That's because this `QGAN` class could support generating samples of any dimension d (using d generators); in our case, we have $d = 1$, hence we only need to pass an array with a single element.

This QGAN object already comes with a default implementation for the generator and the discriminator, and we will rely on them.

To learn more...

In this default implementation, the discriminator is a dense neural network having two consecutive intermediate layers with 50 and 20 neurons each; the activation function in these intermediate layers is the leaky ReLU function and that of the output layer is the sigmoid function. The generator uses a variational form consisting of a layer of Hadamard gates applied on each qubit followed by the two-local variational form with one repetition and circular entanglement.

These details are not specified in the documentation, but they can be found in the source code.

In order to train the QGAN, we may run the following instruction:

```
result = qgan.run(quantum_instance)
```

The training will take a few minutes to complete, depending on the hardware configuration of your computer. In order to plot the evolution of the generator and discriminator losses throughout the training process, we may run the following code:

```
import matplotlib.pyplot as plt
plt.title("Loss function evolution")
cycles = np.array(range(len(qgan.g_loss))) + 1
plt.plot(cycles, qgan.g_loss, label = "Generator")
plt.plot(cycles, qgan.d_loss, label = "Discriminator")
plt.xlabel("Cycle")
plt.legend()
```

This yields the plot shown in *Figure 12.8*. We can see how both losses are approaching $-\log 1/2$, which can give us hope for the success of our training.

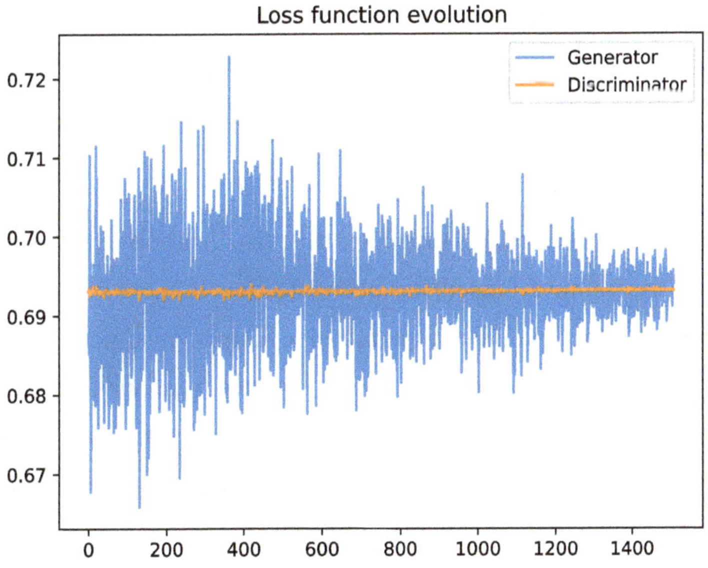

Figure 12.8: Evolution of the generator and discriminator losses during the QGAN training, learning a distribution

In order to check if our training has been successful, we will plot the distribution of the measurement outcomes of our generator against the original distribution. We may generate the data for this plot as follows:

```
samples_g, prob_g = qgan.generator.get_output(qgan.quantum_instance,
                                              shots=10000)


real_distr = []
for i in range(0,3+1):
    proportion = np.count_nonzero(real_data == i) / N
    real_distr.append(proportion)
plt.bar(range(4), real_distr, width = 0.7, color = "royalblue",
        label = "Real distribution")
plt.bar(range(4), prob_g, width = 0.5, color = "black",
        label = "Generated distribution")
```

In this piece of code, we have first asked our QGAN to generate a sample with the distribution it has learned. Then, we have created an array `real_distr` with the relative frequencies of the values in the distribution (entry `j` corresponds to the relative frequency of the value j). Lastly, we have plotted the real distribution against our generated distribution. The output can be found in *Figure 12.9*.

Of course, for the purposes of this example, this visualization is more than enough to convince us that the training, indeed, has been effective. In more sophisticated examples, one may instead want to rely on more quantitative metrics of success. One such metric is the **relative entropy** or **Kullback–Leibler divergence** from one distribution to another. In layman's terms, this entropy measures how "different" two distributions are in a way that if two distributions P_0 and P_1 are identical, the relative entropy from P_0 to P_1 is 0. As P_1 becomes more different from P_0, the relative entropy increases.

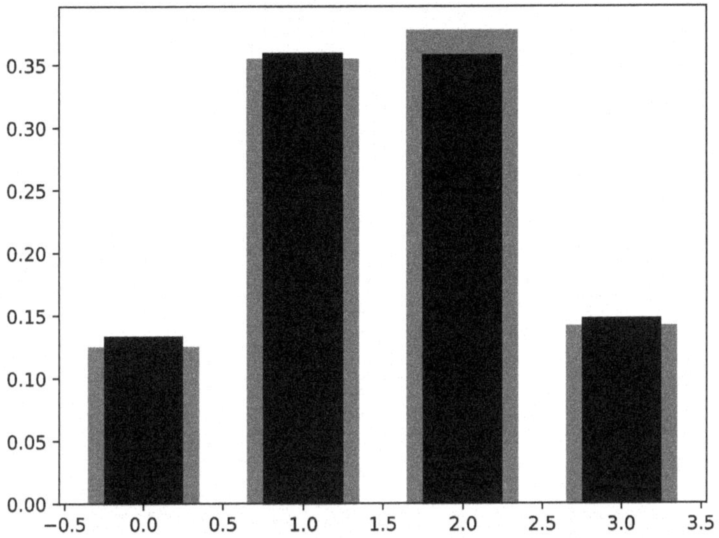

Figure 12.9: Histogram comparing the real distribution (thicker bar) with the one generated by the QGAN (thinner bar)

To learn more...

When you are given two discrete probability distributions P_0 and P_1 over a space X, the relative entropy from P_0 to P_1 can be defined as

$$D(P_1 \| P)0) = \sum_{x \in X} P_1(x) \log\left(\frac{P_1(x)}{P_0(x)}\right).$$

Qiskit's QGAN implementation logs the values of the relative entropy throughout the QGAN training. In this way, we may plot the evolution of the relative entropy over the training process of our QGAN with the following instructions:

```
plt.title('Relative entropy evolution')
plt.plot(qgan.rel_entr)
plt.show()
```

The output is shown in *Figure 12.10*.

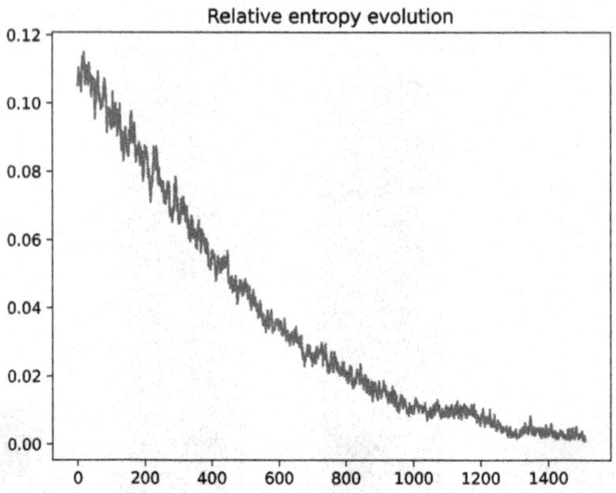

Figure 12.10: Evolution of the relative entropy over the training of our QGAN, learning a distribution

Here it can be clearly shown that the relative entropy approaches 0 as the training progresses, just as we expected. This concludes our example. It's time to wrap up!

Summary

In this chapter, we have explored a whole new kind of quantum machine learning models: quantum GANs. Unlike the models we had considered before, these are used primarily for generation tasks. And, unlike our previous models, they are trained in a fully unsupervised manner.

After understanding what GANs are in general, we introduced the general notion of a QGAN, and then we learned how to implement a couple of QGAN models using PennyLane and Qiskit.

With this, we also conclude our study of quantum machine learning for this book. We hope that you have had a good time learning about all these ways of making quantum computers learn! But your quantum journey does not need to end here. Please, keep on reading for a sneak peek of what you can expect in the near future in the quantum computing field.

Afterword and Appendices

In this part, we provide an afterword that wraps up everything that we have studied in the book as well as a series of appendices that cover some basic mathematical concepts and necessary technical details. We also share some notes on how the book was produced.

This part includes the following contents:

- *Chapter 13, Afterword: The Future of Quantum Computing*
- *Appendix A, Complex Numbers*
- *Appendix B, Basic Linear Algebra*
- *Appendix C, Computational Complexity*
- *Appendix D, Installing the Tools*
- *Appendix E, Production Notes*

13

Afterword: The Future of Quantum Computing

I am not throwing away my shot!

— Alexander Hamilton

This has been a long and (hopefully) interesting journey. In the 12 chapters of this book, we've covered a lot of topics on quantum computing, both from a theoretical and a practical point of view, so maybe it's time to take a look back and see what we have learned.

We started by laying the foundations. We studied the most important mathematical concepts underlying the theory of quantum computing, including how information is stored on qubits, how we can transform their states with quantum gates, and how we can obtain results by measuring them. Then, we explored some of the software tools currently available to implement quantum algorithms, with a special emphasis on the two main software libraries used in this book: Qiskit and PennyLane. We learned how to implement quantum circuits with both frameworks and how to run them on simulators and on actual quantum hardware.

Then, we began to uncover how quantum algorithms can be used to solve optimization problems. To that end, we studied different ways of formulating combinatorial optimization problems as the search for ground states of certain Hamiltonians. This is how we got to know the QUBO and Ising formalisms, that we later used with different types of quantum computers. First, we used quantum annealers, with their heuristic implementation of adiabatic quantum computing, and we learned different ways of solving optimization problems with them. Then, we turned to gate-based quantum computers, and studied the Quantum Approximate Optimization Algorithm and Grover's Adaptive Search.

After that, we widened the scope of our optimization tasks and studied situations in which our interest was set on finding ground states of observables related to problems in different fields, including chemistry and physics. In this case, our focus was on the Variational Quantum Eigensolver. At the same time that we studied its mathematical definition and its practical implementation, we also learned about noisy simulations and readout error mitigation.

Then, we changed gears and started studying the application of quantum computing in machine learning. We first briefly reviewed *classical* machine learning and explained what makes quantum machine learning different. Then, we started studying several quantum machine learning model architectures. The first was that of Quantum Support Vector Machines, which we derived from classical SVMs by introducing quantum kernels.

After that, we focused our attention on how to implement a quantum analog of neural networks that are called – not very imaginatively, we concede – quantum neural networks. For this, we used parametrized circuits in two different ways:

- First, using the free parameters of a part of the circuit to embed classical data into the space of quantum states (as we had already done with QSVMs)
- Then, using the rest of the values as trainable parameters to minimize a cost function

We learned of different ways of computing the gradients needed for the training and we implemented some examples in both PennyLane and Qiskit.

Then, we decided to play a little bit with our new, shiny quantum neural networks and we mixed them with classical neural networks to form hybrid neural networks. We also learned a thing or two about hyperparameter optimization with Optuna and about building and training classical and hybrid networks with PyTorch. All this hard work paid off soon, because it allowed us to implement a very interesting model architecture: that of Quantum Generative Adversarial Networks. We learned a little bit about the theory behind them and we showed how to implement and train a couple of simple examples to learn quantum states and probability distributions.

Most of the methods that we have studied are what we may call modern quantum algorithms: they have been proposed in the last 10 years or so, mainly with current, noisy intermediate-scale quantum computers in mind. And they are the subject of intense development and study by the quantum computing community as we are writing these lines. You may very well say that, with this book, you now know a good deal about what the *present* of quantum computing looks like.

But this should be just the dawn of quantum computing. As wonderful as the ideas behind the methods and devices that we have studied may be (and they truly are!), they still have some limitations (that sometimes are painfully obvious). Due to the reduced scale and resilience to the noise of current quantum computers and due to our incomplete understanding of the capabilities of modern quantum algorithms such as QAOA, VQE, or the different flavors of quantum neural networks, it is difficult to demonstrate quantum computational advantage. In fact, in the cases in which this has been achieved (the most famous one being the quantum supremacy experiment developed by Google researchers [10]), the problems are mainly academic and of little practical application.

Does this mean that we are pessimistic about the future of quantum computing? On the contrary! There is still much work to do, but we think that we are experiencing the beginning of what could very well be a revolution in the way we compute and process information. The famous science fiction writer Arthur C. Clarke formulated three adages that are popularly known as **Clarke's laws**. They go as follows:

1. When a distinguished but elderly scientist states that something is possible, he is almost certainly right. When he states that something is impossible, he is very probably wrong.
2. The only way of discovering the limits of the possible is to venture a little way past them into the impossible.
3. Any sufficiently advanced technology is indistinguishable from magic.

Funnily enough, the three of them could be applied to the field of quantum computing — and, probably, the third is the one that most often has been quoted when talking about quantum computers! But we are especially interested in the second one. We need to keep on investigating, further pushing the frontiers of our knowledge, and exploring new techniques for processing information with quantum devices. To that end, in the next few years, we expect to see developments on different fronts.

Of course, we expect to see the introduction of more capable and reliable quantum computers. In fact, while we were writing this book, IBM announced Osprey, a quantum processor with 433 qubits [97]. But that is not all. They plan to introduce quantum computers with thousands of qubits in the next few years and other big companies such as Google and Honeywell have similar roadmaps. Size is not the only thing that matters, though, and a lot of effort is being put into decreasing gate and readout errors and in increasing coherence times. This will be crucial to reach the ultimate goal of having fault-tolerant, scalable quantum computers.

Thanks to these new, improved quantum computers, we expect to see new demonstrations of quantum advantage: practical experiments in which a quantum computer solves a task much faster than what is possible with the most powerful classical supercomputer available. Probably, the concrete problems tackled in these demonstrations will be not very useful from a practical point of view. However, we think that pursuing these advantages is still very relevant. The more diverse the range of techniques used to achieve quantum supremacy, the better for the field of quantum computing.

But, as you know, this book is mainly about quantum algorithms. So, the developments that we are most excited to see are those related to the development of new quantum methods, the increase of the applicability of quantum techniques, and the deepening of our understanding of the properties of quantum algorithms. In fact, as we have tried to convey throughout this book, this is a field of intense study. To name just a couple of recent highlights, in the last few months, we have witnessed a new quantum exponential speedup [98] and the proof that quantum neural networks may need much less data than their classical counterparts in certain situations [83].

Developing new quantum algorithms, finding new applications of existing quantum techniques, or mathematically proving quantum advantage are, by no means, easy tasks. But we are confident that, as more and more people learn about quantum computing and as more and more quantum computers become available to run experiments, our knowledge and understanding of the power of quantum algorithms will only grow bigger. And with that, new, exciting, and powerful applications will eventually become a reality.

This is why we wrote this book: to invite you on this exciting and wonderful journey. A journey that is just beginning. A journey into the future of quantum computing.

We hope you stay for the ride. And do not throw away your (quantum) shots!

A

Complex Numbers

$$e^{i\pi} + 1 = 0$$

— Leonhard Euler

The set of complex numbers is the set of all numbers of the form $a + bi$ where a and b are real numbers and $i^2 = -1$. This might not be the most formal way of presenting them, but it will do for our purposes!

The way you operate with complex numbers is pretty straightforward. Let a, b, x, and y be some real numbers. We add complex numbers as

$$(a + bi) + (x + yi) = (a + b) + (x + y)i.$$

Regarding multiplication, we have

$$(a + bi) \cdot (x + yi) = ax + ayi + bix + byi^2 = (ax - by) + (ay + bx)i.$$

In particular, when $b = 0$, we can deduce that

$$a(x + yi) = ax + (ay)i.$$

Given any complex number $z = a + bi$, its **real part**, which we denote as $\operatorname{Re} z$, is a, and its **imaginary part**, which we denote as $\operatorname{Im} z$, is b. Moreover, any such number z can be represented in the two-dimensional plane as a vector $(\operatorname{Re} z, \operatorname{Im} z) = (a, b)$. The length of the resulting vector is said to be the **module** of z, and it is computed as

$$|z| = \sqrt{a^2 + b^2}.$$

If $z = a + bi$ is a complex number, its **conjugate** is $z^* = a - bi$. In layman's terms, if you want to get the conjugate of any complex number, all you have to do is flip the sign of its imaginary part. It is easy to check that, given any complex number z,

$$|z|^2 = zz^*,$$

which shows us, incidentally, that zz^* is always a non-negative real number.

One of the most well-known formulas involving the use of complex numbers is Euler's identity, which reads that, for any real number θ,

$$e^{i\theta} = \cos\theta + i\sin\theta.$$

This formula can be easily derived by extending the exponential functions from the usual series that defines it. In particular, according to Euler's identity and using the usual properties of exponentiation, we must have, for any real numbers a and b,

$$e^{(a+ib)} = e^a e^{ib} = e^a(\cos\theta + i\sin\theta).$$

Just to conclude this appendix, let us share with you some fun trivia about our beloved complex numbers:

- Every polynomial of degree n with complex coefficients has exactly n roots, if we account for multiplicity
- Any complex-differentiable function $\mathbb{C} \longrightarrow \mathbb{C}$ is smooth and analytic

To learn more...

If you would like to learn more about complex numbers, we invite you to read the same book that both of us – with a gap of a few years in the middle – used in the complex analysis course of our undergraduate studies: Bak and Newman's *Complex Analysis* [99].

B

Basic Linear Algebra

Algebra is generous. She often gives you more than is asked of her.

— Jean le Rond d'Alembert

In this chapter, we will present a very broad overview of linear algebra. More than anything, this is meant to be a refresher. If you would like to learn linear algebra from the basics, we suggest reading Sheldon Axler's wonderful book [100]. If you are all-in with abstract algebra, we can also recommend the great book by Dummit and Foote [101]. With this out of the way, let's do some algebra!

When most people think of vectors, they think of fancy arrows pointing in a direction. But, where others see arrows, we mathematicians — in our tireless pursuit of abstraction — see elements of vector spaces. And what is a vector space? Simple!

Vector spaces

Let $\mathbb{F}$ be the real or the complex numbers. An $\mathbb{F}$-vector space is a set V together with an "addition" function (usually represented by $+$, for obvious reasons) and a "multiplication by scalars" function (denoted like usual multiplication). Addition needs to take any two vectors and return another vector, that is, + needs to be a function $V \times V \longrightarrow V$. Multiplication by scalars, as the name suggests, must take a scalar (an element of $\mathbb{F}$) and a vector, and return a vector, that is, it needs to be a function $\mathbb{F} \times V \longrightarrow V$. Moreover, vector spaces must satisfy, for any arbitrary $\alpha_1, \alpha_2 \in \mathbb{F}$ and $v_1, v_2, v_3 \in V$, the following properties:

- Associativity for addition : $(v_1 + v_2) + v_3 = v_1 + (v_2 + v_3)$
- Commutativity for addition: $v_1 + v_2 = v_2 + v_1$
- Identity element for addition: there must exist a $0 \in V$ such that, for every vector $v \in V, v + 0 = v$
- Opposites for addition: there must exist a $-v_1 \in V$ such that $v_1 + (-v_1) = 0$
- Compatibility of multiplication by scalars with multiplication in $\mathbb{F}$: $(\alpha_1 \cdot \alpha_2) \cdot v_1 = \alpha_1 \cdot (\alpha_2 \cdot v_1)$
- Distributivity with respect to vector addition: $\alpha_1(v_1 + v_2) = \alpha_1 v_1 + \alpha_1 v_2$
- Distributivity with respect to scalar addition: $(\alpha_1 + \alpha_2)v_1 = \alpha_1 v_1 + \alpha_2 v_1$
- Identity for multiplication by scalars: $1 \cdot v_1 = v_1$

To learn more...

If you, like us, love abstraction, you should know that vector spaces are usually defined over an arbitrary **field** — not just over the real or complex numbers! If you want to learn more, we suggest reading the book by Dummit and Foote [101].

These are some examples of vector spaces:

- The set of real numbers with the usual addition and multiplication is a real vector space.
- The set of complex numbers with complex number addition and multiplication is a complex vector space. Moreover, it can be trivially transformed into a real vector

space by restricting multiplication by scalars to multiplication of complex numbers by real numbers.

- The set $\mathbb{R}^n$ with the usual component-wise addition and multiplication by scalars (real numbers) is a vector space. If we fix $n = 2, 3$, that's where we can find those fancy arrows everyone is talking about!
- Most importantly for us, the set $\mathbb{C}^n$ with component-wise addition and scalar multiplication by complex numbers is a vector space.
- Just to give a cute example, the set of all smooth functions on a closed finite interval of the real numbers is a vector space. You can try to define addition and multiplication by scalars of functions yourself.

When we refer to a vector space on a set V with addition $+$ and multiplication by scalars $\cdot$, we should denote it as $(V, +, \cdot)$ in order to indicate what function we are considering as the addition function and what function we are taking to be the multiplication by scalars. Nevertheless, in all honesty, $(V, +, \cdot)$ is a pain to write, and we mathematicians – like all human beings – have a natural tendency towards laziness. So we usually just write V and let $+$ and $\cdot$ be inferred from context whenever that is reasonable to do.

Bases and coordinates

Some $\mathbb{F}$-vector spaces V are **finite-dimensional**: this means that there is a finite family of vectors $\{v_1, \dots, v_n\} \subseteq V$ such that, for any vector $v \in V$, there exist some unique scalars $\alpha_1, \dots, \alpha_n \in \mathbb{F}$ for which

$$v = \alpha_1 v_1 + \cdots + \alpha_n v_n.$$

The scalars $\alpha_1, \dots, \alpha_n$ are said to be the **coordinates** of v with respect to the basis $\{v_1, \dots, v_n\}$. The natural number n is said to be the dimension of the vector space, and it is a fact of life that any two bases of a vector space need to have the same number of elements, so the dimension is well-defined. If you want proof (which you should want!), check your favorite linear algebra textbook; either of the two that we have suggested should do the job.

Two examples of finite dimensional vector spaces are $\mathbb{R}^n$ and $\mathbb{C}^n$ (with the natural addition and multiplication operations). For example, a basis of $\mathbb{C}^3$ or $\mathbb{R}^3$ would be

$$\{(1, 0, 0), (0, 1, 0), (0, 0, 1)\}.$$

To further illustrate this, if we considered the vector $(i, 3 + 2i, -2)$ in $\mathbb{C}^3$, we would have

$$(i, 3 + 2i, -2) = i \cdot (1, 0, 0) + (3 + 2i) \cdot (0, 1, 0) + (-2) \cdot (0, 0, 1),$$

and this representation in terms of these basis vectors is, clearly, unique. What is more, this basis is so natural and common that it has a name, the **canonical basis**, and its vectors are usually denoted as $\{e_1, e_2, e_3\}$. An analogous basis can be defined on $\mathbb{R}^n$ and $\mathbb{C}^n$ for any n.

To learn more...

We use the canonical basis extensively in this book, but with a different notation. We refer to it as the **computational basis**.

When you have a vector in a finite-dimensional vector space, sometimes it is handy to work with its coordinates with respect to some basis of your choice rather than working with its "raw" expression. In order to do this, we sometimes represent a vector v with coordinates $\alpha_1, \ldots, \alpha_n$ by a column matrix having the coordinates as entries. For example, in the previous example, the vector $(1, 3 + 2i, -2)$ would be represented by the column matrix of coordinates

$$\begin{pmatrix} 1 \\ 3 + 2i \\ -2 \end{pmatrix}$$

with respect to the canonical basis $\{e_1, e_2, e_3\}$.

Important note

It is very important to remember that the column matrix of coordinates of a vector is always defined with respect to a certain basis.

If we considered, for instance, the basis $\{e_1, e_3, e_2\}$, then the coordinates of the aforementioned vector would be

$$\begin{pmatrix} 1 \\ -2 \\ 3 + 2i \end{pmatrix}.$$

And, yes, order matters.

Linear maps and eigenstuff

Now that we know what vector spaces are, it is natural to wonder how we can define transformations $L : V \longrightarrow W$ between some $\mathbb{F}$-vector spaces V and W. In fairness, you could define any such transformation L however you wanted – we are not here to set boundaries on your mathematical freedom. But, if you want L to play nicely with the vector space structure of V and W, you will want it to be linear. That is, you will want to have, for any vectors $v_1, v_2 \in V$ and any scalar $\alpha \in \mathbb{F}$,

$$L(v_1 + v_2) = L(v_1) + L(v_2), \qquad L(\alpha \cdot v_1) = \alpha L(v_1).$$

Keep in mind that the addition and multiplication by scalars on the left-hand side of these expressions is that of V, while the operations on the right-hand side of the expressions are those of W.

Linear maps are wonderful. Not only do they have very nice properties, but they are also very easy to define. If $v_1, \ldots, v_n$ is a basis of V and you want to define a linear map $L : V \longrightarrow W$, all you have to do is give a value – any value – to $L(v_k)$ for every $k = 1, \ldots, n$. Then, by linearity, the function can be extended to all of V as

$$L(\alpha_1 v_1 + \cdots + \alpha_n v_n) = \alpha_1 L(v_1) + \cdots + \alpha_n L(v_n)$$

for any scalars $\alpha_1, \ldots, \alpha_n \in \mathbb{F}$. Furthermore, if we let $\{w_1, \ldots, w_m\}$ be a basis of W and we let $a_{k,l} \in \mathbb{F}$ be the unique scalars such that

$$L(v_k) = a_{1k}w_1 + \cdots + a_{nk}w_n,$$

then the coordinates of $L(v)$ for any $v = \alpha_1 v_1 + \cdots + \alpha_n v_n \in V$ with respect to $\{w_1, \ldots, w_m\}$ will be

$$\begin{pmatrix} a_{11} & \cdots & a_{1n} \\ \vdots & \ddots & \vdots \\ a_{n1} & \cdots & a_{nn} \end{pmatrix} \begin{pmatrix} \alpha_1 \\ \vdots \\ \alpha_n \end{pmatrix}.$$

To put it in perhaps more schematic terms,

$$\begin{pmatrix} | \\ L(v) \\ | \end{pmatrix} = \begin{pmatrix} a_{11} & \cdots & a_{1n} \\ \vdots & \ddots & \vdots \\ a_{n1} & \cdots & a_{nn} \end{pmatrix} \begin{pmatrix} | \\ v \\ | \end{pmatrix},$$

where the column matrices represent the coordinates of the vectors with respect to the bases $\{v_1, \ldots, v_n\}$ and $\{w_1, \ldots, w_m\}$. We say that the matrix $(a_{kl})_{kl}$ is the **coordinate matrix** of L with respect to these bases. If $V = W$ and we have a map $L : V \longrightarrow V$, we say that L is an **endomorphism** and, usually, we consider the same basis everywhere.

There is a very special kind of endomorphism that can be defined on any vector space: the **identity**. This is just a function id that takes any vector v to $\mathrm{id}(v) = v$. If $L : V \longrightarrow V$ is an endomorphism, we say that a function L^{-1} is the **inverse** of L if both $L \circ L^{-1}$ and $L^{-1} \circ L$ are equal to the identity — actually, checking either of the two conditions is already sufficient when working with endomorphisms on finite-dimensional vector spaces. The coordinate matrix of the inverse of a map with coordinate matrix A is just the usual inverse matrix A^{-1}. What is more, a linear map is invertible if and only if so is its coordinated matrix.

When you have an endomorphism $L : V \longrightarrow V$, there may be some vectors $0 \neq v \in V$ for which there exists a scalar λ such that $L(v) = \lambda v$. These vectors are said to be **eigenvectors** and the corresponding value λ is said to be their **eigenvalue**. In some cases, you will be able to find a basis of eigenvectors $v_1, \ldots, v_n$ with some associated eigenvectors $\lambda_1, \ldots, \lambda_n$. With respect to this basis, the coordinate matrix of L would be a diagonal matrix

$$\begin{pmatrix} \lambda_1 & & \\ & \ddots & \\ & & \lambda_n \end{pmatrix}.$$

Inner products and adjoint operators

On an $\mathbb{F}$-vector space V, we may wish to define an **inner product** $\langle - | - \rangle$. This will be an operation taking any pair of vectors and returning a scalar, that is, a function $V \times V \longrightarrow \mathbb{F}$, satisfying the following properties for any $u, v_1, v_2 \in V$, and $\alpha_1, \alpha_2 \in \mathbb{F}$:

- **Conjugate symmetry**: $\langle v_1 | v_2 \rangle = \langle v_2 | v_1 \rangle^*$. Of course, if the vector space is defined over $\mathbb{R}$, then $\langle v_2 | v_1 \rangle^* = \langle v_2 | v_1 \rangle$, so $\langle v_1 | v_2 \rangle = \langle v_2 | v_1 \rangle$.
- **Linearity**: $\langle u | \alpha_1 v_1 + \alpha_2 v_2 \rangle = \alpha_1 \langle u | v_1 \rangle + \alpha_2 \langle u | v_2 \rangle$.
- **Positive-definiteness**: If $u \neq 0$, $\langle u | u \rangle$ is real and greater than 0.

It is easy to check that the following is an inner product on $\mathbb{C}^n$:

$$\langle (\alpha_1, \ldots, \alpha_n) | (\beta_1, \ldots, \beta_n) \rangle = \alpha_1^* \beta_1 + \cdots + \alpha_n^* \beta_n.$$

When we have a vector space with an inner product — which is commonly said to be an **inner product space** — two vectors v and w are said to be orthogonal if $\langle v | w \rangle = 0$. Moreover, a basis is said to be orthogonal if all its vectors are pairwise orthogonal.

With an inner product, we can define a **norm** on a vector space. We won't get into the details of what norms are but, very vaguely, we can think of them as a way of measuring the length of a vector (don't think about arrows, please, don't think about arrows...). The norm induced by a scalar product $\langle \cdot | \cdot \rangle$ is

$$\|v\| = \sqrt{\langle v | v \rangle}.$$

We say that a basis is **orthonormal** if, in addition to being orthogonal, the norm of all its vectors is equal to 1.

When we are given a matrix $A = (a_{kl})$, we define its **conjugate transpose** to be $A^\dagger = (a_{kl}^*)$, that is

$$\begin{pmatrix} a_{11} & \cdots & a_{1n} \\ \vdots & \ddots & \vdots \\ a_{n1} & \cdots & a_{nn} \end{pmatrix}^\dagger = \begin{pmatrix} a_{11}^* & \cdots & a_{n1}^* \\ \vdots & \ddots & \vdots \\ a_{1n}^* & \cdots & a_{nn}^* \end{pmatrix}.$$

The following identities can be easily checked for square matrices and, therefore, for linear maps:

$$(A+B)^\dagger = A^\dagger + B^\dagger, \qquad (AB)^\dagger = B^\dagger A^\dagger.$$

Here, AB denotes the usual matrix multiplication.

If $L : V \longrightarrow V$ is an endomorphism on a finite-dimensional vector space V, we can define its **Hermitian adjoint** as the only linear map $L^\dagger : V \longrightarrow V$ that has as coordinate basis with respect to some basis the conjugate transpose of the coordinated matrix of L with respect to that same basis. It can be shown that this notion is well-defined, that is, that you always get the same linear map regardless of your choice of basis.

To learn more...

The definition that we have given is, well, not the most rigorous one. Usually, when you have a pair of inner product spaces V and W with inner products $\langle\cdot|\cdot\rangle_V$ and $\langle\cdot|\cdot\rangle_W$, the adjoint of a linear map $L : V \longrightarrow W$ is defined to be the only linear map $L^\dagger : W \longrightarrow V$ such that, for every $v \in V$ and $w \in W$,

$$\langle w|L(v)\rangle_W = \left\langle L^\dagger(w)\middle|v\right\rangle_V.$$

We invite you to check that, for the particular case that we have considered ($V = W$ finite dimensional), both definitions agree.

We say that an endomorphism L is **self-adjoint** or **Hermitian** if $L = L^{\dagger}$. And it is a fact of life (again, we encourage you to check your favorite linear algebra textbook) that every Hermitian operator has an orthonormal basis of real eigenvalues.

Also, we say that an endomorphism U is **unitary** if $U^{\dagger}U = UU^{\dagger} = I$, where I denotes the identity matrix.

Matrix exponentiation

Every calculus student is familiar with the exponential function, which is taken to be $\exp(x) = e^{x}$. If you dive deeper into the wonders of mathematical analysis, you'll learn that the exponential function is actually defined as the sum of a series, namely

$$\exp(x) = \sum_{k=1}^{\infty} \frac{x^k}{k!}.$$

As it turns out, this definition can be extended far beyond the real numbers. For instance, Euler's formula – which we introduced in *Appendix A, Complex Numbers* – is the result of extending the definition of the exponential function to every $x \in \mathbb{C}$.

Most importantly for our purposes, the exponential function can be extended to... matrices! In this way, the exponential of a square matrix is defined, rather unsurprisingly, as

$$\exp(A) = \sum_{k=1}^{\infty} \frac{A^k}{k!}.$$

What is more, this definition also works for endomorphisms. If the coordinate matrix of an endomorphism L is A (with respect to a particular basis), we can define the exponential of L to be the endomorphism that has coordinate matrix $\exp(A)$ with respect to the basis under consideration. It can be checked that this notion is well-defined: we always get the same endomorphism regardless of the basis we consider.

Of course, setting out to compute the exponential of a matrix just by summing up an infinite series might not be the best of ideas. Thankfully, there is an easier way. If a matrix is diagonal, it can be shown that

$$\exp\begin{pmatrix} \lambda_1 & & \\ & \ddots & \\ & & \lambda_n \end{pmatrix} = \begin{pmatrix} e^{\lambda_1} & & \\ & \ddots & \\ & & e^{\lambda_n} \end{pmatrix}.$$

As we mentioned in the previous section, when an endomorphism is Hermitian, one can always find a basis with respect to which the coordinate matrix of the endomorphism is diagonal (a basis of eigenvectors), so this enables us to compute the exponential of Hermitian operators. In general, it is always possible to compute the exponential of a matrix [101, Section 12.3], but we won't discuss how to do that here.

Just to bring this appendix to an end, we will briefly touch upon a fairly unrelated topic that we will nevertheless use in some parts of the book: modular arithmetic.

A crash course in modular arithmetic

If your watch says it's 15:00 and we ask you the time, you will say that it is 03:00. But you would be lying, wouldn't you? Your watch says it's 15:00 but you've just said that it is 3:00. What is wrong with you? Well, probably nothing. It turns out that, when you were telling us the time, you were subconsciously working in arithmetic modulo 12.

Vaguely speaking, when you work with numbers modulo n all you are doing is assuming that n and 0 represent the same number. In this way, when you work in arithmetic modulo 4, for example,

$$0 \equiv 4 \equiv 8 \equiv 12 \equiv 16 \pmod{4},$$

$$1 \equiv 5 \equiv 9 \equiv 13 \equiv 17 \pmod{4},$$

$$2 \equiv 6 \equiv 10 \equiv 14 \equiv 18 \pmod{4},$$

and so on, and so forth. Notice how we have written $\equiv$ rather than $=$ to denote that those numbers are not, well, equal on their own, but just that they are equal modulo 4 — that's also why we have that cute (mod 4) on the right.

In this modular arithmetic setting, you can compute additions and multiplications as usual. For example, when working modulo 4,

$$2 \times 3 = 6 \equiv 2 \pmod 4.$$

Ha! Look at what we have done! Now you can tell all your friends that 2 times 3 is 2 (you can then silently whisper "modulo 4" and still be technically correct). But, wait, here comes our favorite one:

$$1 + 1 \equiv 0 \pmod 2.$$

In the end, all those people who claimed that "one plus one doesn't necessarily equal two" had a point, huh? They surely were talking about modular arithmetic. We have no doubt.

To learn more...

Can't get enough of modular arithmetic? Dummit and Foote have you covered! Have fun. [101]

C

Computational Complexity

An algorithm is a finite answer to an infinite number of questions

— Stephen Kleene

Computational complexity theory is the branch of theoretical computer science that is concerned with quantifying the resources needed to solve problems with algorithms. It asks questions such as "How much time is needed to multiply two integer numbers of n bits each?", "Do you need more memory space to solve a problem than to check its solution?", or "Is randomness useful in computational tasks?".

In this brief introduction to computational complexity, we will focus mainly on the concepts involved in estimating how much time is required to solve certain problems. For a thorough treatment of this and other topics (including space or memory complexity, the role of randomness in computation, approximation algorithms, and other advanced matters), you can check standard computational complexity books such as the ones by Sipser [26], Papadimitriou [102], or Arora and Barak [103].

To study the kind of questions posed in computational complexity theory, we need first to introduce a computational model that allows us to measure computation time, memory, and other resources. The usual choice is that of **Turing machines**. It is beyond the scope of this book to mathematically define what Turing machines are (for the details, check the books cited in the previous paragraph), but let us at least give an informal description so you can understand how we can use them to model computational tasks and to measure the resources involved in solving problems with them. Please notice that different textbooks use slightly different definitions of Turing machines, but it is straightforward to show that they are all equivalent in power.

A few words on Turing machines

A Turing machine is a (theoretical) device that has a (potentially infinite) **tape** divided into **cells**. Each of these cells can store a symbol from a finite and fixed number of possibilities (usually, 0, 1, and a "blank" symbol to denote an empty cell). The machine also has a **head** that, at any given moment, is scanning one of the tape cells. Additionally, the machine is in a **state** (also from a finite number of fixed options) at any step in the computation.

The machine has a list of instructions that, depending on the machine's state and the content of the cell that the head is scanning, tell the machine what it should do next. This can involve changing the machine state, writing a different symbol on the cell that is being scanned, and moving the head one cell to the left or to the right. For instance, one such instruction could be "If the state is q_2 and the symbol being read is 1, change the state to q_5, change the symbol to 0, and stay in the same cell," while another could be "If the state is q_0 and the symbol is 0, change the state to q_1, leave the symbol unchanged, and move the head one cell to the right."

Important note

A Turing machine is a (theoretical) device that has an unbounded tape divided into cells and a head that scans one of those cells. At any given moment, the machine is in an internal state from a finite number of possibilities. The instructions of the machine specify, depending on the machine state and the content of the cell that the head is scanning, what the next state is, the new content of the cell, and the action of the machine (move left, move right, or stay, for instance).

In order to perform a computation, the input is given as a finite string of symbols on the tape (the rest are left blank). Then, the Turing machine operates in the following way: it starts in a predefined initial state and with its head scanning the first symbol of the input; then, it changes its state, tape content, and head position following its instructions in discrete steps. Eventually, the machine can stop because it reaches a predefined, halting state. If the machine stops, the output of the computation is the string of symbols written on the tape.

To learn more...

It is not guaranteed that a Turing machine will stop for all its inputs. In fact, it can be proved that determining whether a Turing machine will eventually stop with a given input (what is usually called the **halting problem**) is unsolvable in a very precise way: there is no algorithm that can give the correct answer for every possible Turing machine and every possible input. Check the book by Sipser [26] for a proof of this amazing fact.

Turing machines may seem like too simple a model, but it can be proved that any computation that can be carried out with any other reasonable computational model can also be carried out with a Turing machine (maybe with some slowdown). For instance, it is rather straightforward to prove that if we extend Turing machines by giving them multiple tapes (**multi-tape Turing machines**) or the possibility of non-deterministically choosing among several instructions for the same state-symbol situation (**non-deterministic Turing machines**), the new devices aren't more powerful than our original single-tape, deterministic Turing machines (again, see the book by Sipser [26] for all the details). The same happens if we consider models that are much closer to the actual architecture of modern computers, such as the **Random-Access Machines** model (see Section 3.4 in the book by Savage [104]), or even models, such as that of **while-Programs** (see the book by Kfoury, Moll, and Arbib [105]) that are based on common programming languages.

This has led to the firm belief that Turing machines indeed formally capture the informal notion of what an algorithm is. This fact is usually known as the **Church-Turing thesis**.

Measuring computational time

We can say that the Church-Turing thesis is simply stating that, if you are only interested in identifying which tasks can be solved algorithmically and which cannot, you can just use any of a wide number of equivalent models: single-tape Turing machines, multi-tape Turing machines, non-deterministic Turing machines, Random-Access Machines, while-Programs, and many, many others. Each of them will give you exactly the same power.

But be cautious! If you care about the resources needed to carry out the computations (and that is what computational complexity is all about), then the choice of the model can be important. So let's fix, for now, the single-tape Turing machines (the ones that we have described informally in the previous section) as our computational model. In this way, we can easily measure the time needed to carry out a certain computation with one of these Turing machines as the number of steps that it must take to complete it.

That works well for a fixed Turing machine with a particular input, but we are usually more interested in analyzing how the running time grows with the size of the input than we are in finding concrete running-time values for concrete problem instances. For example, we could be interested in knowing whether the time needed for a certain task grows so rapidly that it quickly becomes unfeasible to solve the problem when the input size becomes moderately big.

For this reason, we will define the running time of a Turing machine as a function of the input length, not as a function of the particular input. Namely, the running time of a Turing machine M is a function T that takes as input a non-negative integer n and returns the maximum number of steps that M performs with an input x of n bits before it stops. Notice that this is a worst-case definition of running-time: it is defined in terms of the string that needs the most time in order to be processed. Note also that, if a machine does not stop for some inputs, its running time for inputs of those lengths will be infinite. This is not a problem for our purposes, because we will only consider machines that always stop.

Important note

The running time of a Turing machine M is a function T such that $T(n)$ is the maximum number of steps that M performs when given an input of length n.

For other computational models, running times can be defined in analogous ways. For instance, for multi-tape Turing machines, the running time is again measured as the maximum number of steps performed on inputs of size n. For computational models that use idealized programming languages (the while-Programs model, for instance) or abstract architectures (the Random-Access Machines model), running time can be defined as the maximum number of basic instructions (setting a variable to zero, incrementing a variable, comparing the value of two variables...) executed with inputs of size n.

Asymptotic complexity

In order to compare different running times associated with different Turing machines, it is convenient to perform some simplifications. We usually do not care about whether the running time of a Turing machine is exactly $T_1(n) = 4321n^2 + 784n + 142$ or, rather, $T_2(n) = n^3 + 3n^2 + 5n + 3$. In fact, we are more interested in whether $T(n)$ grows roughly like n^3 or like n^2, because this implies a qualitative difference: for values of n that are big enough, any polynomial of degree 3 grows more rapidly than any polynomial of 2. In the context of computational complexity theory, we would always prefer a $T(n)$ that grows as n^2 over one that grows as n^3, because its behavior for big inputs (its asymptotic growth, in other words) is better.

This intuitive idea is captured by the famous **Big O notation**. Given two time functions $T_1(n)$ and $T_2(n)$, we say that $T_1(n)$ is $O(T_2(n))$ (and we read it is as "$T_1(n)$ is Big O of $T_2(n)$") if there exist an integer constant n_0 and a real constant $C > 0$ such that for all $n \geq n_0$ it holds that

$$T_1(n) \leq CT_2(n).$$

For instance, you can check that $4321n^2 + 784n + 142$ is $O(n^3 + 3n^2 + 5n + 3)$.

The main idea behind this definition is that if $T_1(n)$ is $O(T_2(n))$, then the growth of T_1 is not worse than that of $T_2(n)$. For example, it is easy to prove that n^a is $O(n^b)$ whenever $a \leq b$ and that n^a is $O(2^n)$ for any a. But, on the other hand, n^b is not $O(n^a)$ and 2^n is not $O(n^a)$. See *Figure C.1* for an example with linear, quadratic, cubic, and exponential functions. Notice how the exponential function eventually dominates all the others despite having 10^{-4} as its coefficient.

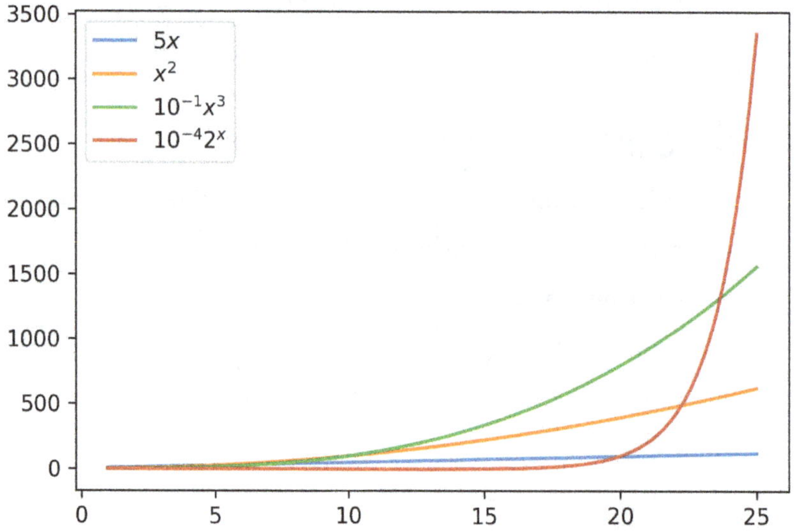

Figure C.1: Growth of linear, quadratic, cubic, and exponential functions

Important note

Given two non-negative functions $T_1(n)$ and $T_2(n)$, we say that $T_1(n)$ is $O(T_2(n))$ if there exist n_0 and $C > 0$ such that

$$T_1(n) \leq CT_2(n)$$

for every $n \geq n_0$.

Big O notation is extremely useful to estimate the behavior of running times without having to focus on small, cumbersome details. If the running time of a Turing machine is $4321n^2 + 784n + 142$, we can just say that it is $O(n^2)$ and forget about the particular coefficients in the time function. This is also the reason why we can abstractly think about the number of steps and not, for example, milliseconds. The particular amount of time that each step takes is a constant that will be "absorbed" by the Big O notation.

However, this comes at a price. A running time such as $10^{100}n^2$ is certainly $O(n^2)$. But it is not preferable to n^3 unless $n > 10^{100}$, something that will never happen in practical situations, because 10^{100} is much, much bigger than the number of atoms in the visible universe. So use this notation wisely: with Big O comes Big Responsibility.

P and NP

As we mentioned at the beginning of this appendix, computational complexity theory studies the amount of resources needed to solve problems with algorithms. So far, we have focused on how to mathematically define the notion of algorithm with the help of Turing machines and on how to measure the time needed to perform computations with them. Now, we turn our attention to defining computational problems and classifying them according to the time they take to be solved. That is, we will think in terms of their inherent complexity and not in terms of specific algorithms.

In computational complexity theory, a **problem** consists of an infinite number of instances or inputs for which an output value needs to be returned. For example, we may be given two natural numbers and asked to compute their product. Or we may be given a graph and asked to check if it has a Hamiltonian path or not. In both cases, the number of possible inputs is infinite and there is a well-defined output or answer associated with each such input.

Problem instances are usually encoded as binary strings in some way. For example, we can represent a natural number by its binary expansion or a graph by (the concatenation of the rows of) its adjacency matrix. In the same way, outputs can also be represented by binary strings. Consequently, a problem can be identified with a function that takes a binary string as its input and returns a binary string as its output. But a Turing machine does exactly that: it receives binary strings as inputs and returns binary strings as outputs. This allows us to study which problems can be solved with Turing machines and how much time is needed to solve them.

In computational complexity, the simplest category of problem that we can consider is that of **decision problems**, in which the output is a single bit (we usually identify 1 with "true" and 0 with "false"). Examples of decision problems include determining whether a natural number m is prime, determining whether a graph has a Hamiltonian path, and determining whether a Turing machine stops for all its inputs.

We say that a Turing machine is a **decider** for a decision problem if, given as input a binary string representing an instance of the problem, it eventually stops and returns the correct output (0 or 1) for that instance. In that case, we also say that the Turing machine **solves** or **decides** the problem. There exist deciders for the problems of determining whether a number is prime and of determining whether a graph has a Hamiltonian path, but not for the problem of determining whether a Turing machine stops for all of its inputs (this is a consequence of the unsolvability of the halting problem that we mentioned earlier).

Once we know that a problem has a decider, we can try to further refine its classification by taking into account the resources used by the decider. This leads, for instance, to the definition of the famous P (short for "polynomial time") class. We say that a decision problem A is in P if there exists a decider for A that runs in polynomial time. That is, there exists a Turing machine D that decides A and whose running time $T(n)$ is $O(n^a)$ for some non-negative integer a. Notice that, for a problem to be in P, it is enough to find one polynomial-time decider for it. However, in order to show that a decision problem A is not in P, we need to prove that no Turing machine running in polynomial time is able to decide A. This is usually much, much harder to do.

As an example, a celebrated result by Agrawal, Kayal, and Saxen [106] shows that the problem of determining whether a natural number is a prime is indeed in P. Other, simpler examples of problems in P include checking whether a number is a perfect square or checking whether a binary string is a palindrome (that is, it reads the same from left to right and from right to left). However, for the problem of determining whether a graph has a Hamiltonian path, we do not know whether it is in P or not. We very strongly believe that it is not in P, but despite the best efforts of thousands of mathematicians over several decades, we still can't prove it.

Important note

We define P as the class of decision problems that can be solved with Turing machines in polynomial time.

Actually, P is interesting for several reasons. First, it is quite robust. We have defined it in terms of the computation time required by deciders that are single-tape Turing machines. However, if we had chosen another computational model such as, for instance, multi-tape Turing machines, then we would have arrived at exactly the same set of problems. This is so because it is possible to simulate a multi-tape Turing machine with a single-tape Turing machine with just a polynomial overhead in running time. The same is true for any other reasonable (classical) computational model, so although the particular running time might differ from one model to another (say $O(n^4)$ with single-tape Turing machines and $O(n^2)$ with 2-tape Turing machines), one will be polynomial if and only if the other is.

What is more, P seems to capture quite well the notion of a problem being efficiently solvable. It is true that in P we allow running times such as n^{1000}, which can hardly be deemed as efficient. However, the running time of naturally-occurring problems that we can prove to be in P is typically much more tame, such as $O(n^2)$ or $O(n^3)$. Moreover, if a decision problem is not in P, then the running time of any of its deciders will grow faster than any polynomial (at least, for an infinite number of its inputs). And that is something that we can unequivocally classify as not efficient at all.

Another central class of problems in computational complexity is NP. It is, again, a class of decision problems. But, in this case, the defining property is not that we can solve them efficiently (as in the case of P) but that we can check their solutions with an efficient algorithm. To make this idea formal, we say that a problem A has a **polynomial-time verifier** if there exists a Turing machine V that runs in polynomial time and a polynomial q with the two following properties:

- If x is an instance of problem A of size n for which the answer is "true," then there exists a binary string y of length at most $q(n)$ such that V on input (x, y) returns 1. The string y is usually called a **witness**, a **certificate**, or a **proof** for x.
- If x is an instance of problem A of size n for which the answer is "false," then for every binary string y of length at most $q(n)$, V on input (x, y) returns 0.

This definition is a little bit convoluted, so let's analyze it in detail. The idea here is that for an instance x of A whose answer is positive, we can find a certificate y that is not long (its length is polynomial in the size x) and that we can check when we are given y together with x, with an efficient algorithm. However, for instances whose answer is negative, there is no such certificate. Note also that the total running time of V on (x, y) is polynomial in the length of x, because V runs in polynomial time in its whole input and y has a length that is polynomial in x. Hence, this definition really captures the notion of checking that the answer to x is positive (through certificate y) with an efficient algorithm.

With this notion at our disposal, we can now define NP as the class of decision problems for which there exists a polynomial-time verifier.

To learn more...

An alternative, but equivalent, definition of NP can be given in terms of non-deterministic Turing machines. In fact, NP is short for "non-deterministic polynomial time." You can find all the details in Sipser's book [26].

Let's discuss an example to illustrate this definition. The problem of determining whether a graph has a Hamiltonian path is in NP. The certificate y can, in this case, be just a Hamiltonian path in the graph. Indeed, it is easy to write a program (in Python, for example) that, given a graph represented by x and a sequence of vertices represented by y, checks whether y is a path in x that visits all the vertices in the graph. Moreover, we can easily do this computation in polynomial time and the certificate is always of size linear in the number of graph vertices. As required, for graphs that have a Hamiltonian path, there exists at least a certificate. However, for graphs without Hamiltonian paths, no y will make the verifier output 1. If needed, we could translate our algorithm into Turing machine instructions; it is a tedious process, but it has no real difficulty.

Important note

NP is the class of decision problems whose solution can be verified with Turing machines in polynomial time.

Similar arguments can be given to prove that many important problems are in NP, including determining whether a Boolean formula is satisfiable, determining whether a graph is 3-colorable, or determining whether a graph has a cut of size bigger than a given integer k. The certificates for them can, of course, be a satisfying assignment, a 3-coloring of the graph, and a cut of size bigger than k. All of them are of a size comparable to the problem instances they certify and can be checked efficiently with obvious procedures.

Additionally, any problem in P is also in NP. This is easily proved. By definition, a problem A in P has a decider. But we can directly use this decider to obtain a verifier for A: we only need to ignore the candidate certificate y and compute the answer with the decider itself.

If the machine knows how to solve the problem in polynomial time on its own, it does not need any external help!

So, we know that P is contained in NP. And it seems like we should be able to prove that they are different, because there must be problems whose solutions we can check efficiently, but for which it is impossible to find those same solutions in a reasonable amount of time, right? Well, it turns out that this is by no means an easy task. In fact, it is literally the million-dollar question!

Determining whether $P = NP$ is one of the seven Millennium Problems selected by the Clay Mathematics Institute in 2000 as the most important open questions in all of mathematics (for an accessible account of the Millennium Problems, check the book by Keith Devlin [107]). Whoever is able to give proof showing that $P \neq NP$ or to show that every problem in NP is also in P, will receive a one-million-dollar prize and will become world-famous.

Important note

Every problem in P is also in NP. The question of whether there are problems in NP that cannot be solved in polynomial time is one of the most important open questions in all of mathematics.

Almost every expert in computational complexity believes that, in fact, $P \neq NP$. All the evidence points in that direction. And it certainly seems logical that *checking* a solution should be easier in general than *finding* a solution. However, no one has yet succeeded in proving that there are problems in NP that are not in P, and the most natural proof techniques have been shown to be insufficient (see *Section 6.5* in the epic book by Moore and Mertens [108]).

Hardness, completeness, and reductions

Although our current mathematical tools are not powerful enough to give satisfactory lower bounds on the resources needed by computational problems, we do know a good deal more about comparing the relative hardness of problems. The main concept used for that kind of comparison is what we call a **reduction**.

Intuitively, a reduction is a procedure to solve a problem from the solution to a different problem. We could say that we reduce solving problem A to solving problem B. So if we know how to solve B with an algorithm, we can use that algorithm and some additional computation to also solve A.

To put it more formally, consider two problems A and B, and imagine that we have an algorithm M_B that solves B. M_B is usually called an **oracle** for B. We say that A is **reducible** to B if we can solve A given an oracle for B. For instance, multiplying two numbers is reducible to adding two numbers: if we are given an oracle that adds numbers, we can use it to multiply by repeated addition.

Of course, when studying computational classes such as P and NP, we are interested in reductions that take a polynomial amount of time. But how can we capture that idea formally? Well, we can simply count each call to the oracle as just another step in the computation. Then, we say that a problem A is **polynomial-time reducible** to a problem B if, given an oracle M_B for B, we can solve any instance x of A with a total number of computational steps plus calls to M_B that is polynomial in the size of x. Another way of seeing this is imagining that we extend our Turing machines with the capability of computing M_B in a single step (these new devices are unsurprisingly called **oracle Turing machines**). Then, showing that A is polynomial-time reducible to B is the same as finding an oracle Turing machine (with an oracle for B) that solves A in polynomial time.

Notice that A being polynomial-time reducible to B has important consequences. The first one is that if B is in P, then A is also in P. This is so because, if B is in P, we can replace every call to M_B with an actual Turing machine that solves B and runs in polynomial time, making the total time involved in solving A also polynomial. This also implies that if A is not in P, then B cannot be in P either, because it would lead us to a contradiction.

Now, we say that a problem B is NP-**hard** if every problem A in NP is polynomial-time reducible to B. This means that B is at least as hard as any problem A in NP, because if we knew how to solve B efficiently, then we would also know how to solve A efficiently. And if at least one problem in A cannot be solved in polynomial time, that implies that B cannot be solved in polynomial time either.

Important note

A problem is NP-hard if every problem in NP is polynomial-time reducible to it.

Being NP-hard seems like a very strong property. Is it really possible for *every* problem A in NP to be reduced to a single problem B? As surprising as this may seem, we know of hundreds (if not thousands) of problems that occur naturally in practice and that are indeed NP-hard. A notable example is the problem of determining whether a Boolean formula is satisfiable or not, also called SAT. That SAT is NP-hard is the content of the famous Cook-Levin theorem (see the book by Sipser for a proof [26]). In *Chapter 3, Working with Quadratic Unconstrained Binary Optimization Problems*, we work with many NP-hard problems. For many other examples and much more on the concept of NP-hardness, you can check the classical book by Garey and Johnson [109].

In fact, it turns out that we can prove that SAT and other decision problems in NP have a property that is a bit stronger than NP-hardness known as **NP-completeness**. In order to discuss it, we first need to talk about a special type of reduction that is very useful when studying decision problems. We say that a decision problem A is **many-one reducible** to a decision problem B if there exists an algorithm F that transforms an instance x of A into an instance $F(x)$ of B with the property that the answer to x in A is positive if and only if the answer to x in B is positive.

Note that, in this case, we indeed have a reduction in the more general sense that we were discussing earlier. If we are given an oracle M_B for B, we can solve any instance x of A by computing $F(x)$ and applying M_B to $F(x)$. Here, we are using only one call to M_B, but in a general reduction, we can use M_B as many times as we see fit. Thus, a many-one reduction is a special case of a reduction. Additionally, in the case in which the transformation F can be computed in polynomial time, we say that we have a **polynomial-time many-one reduction**.

Important note

A polynomial-time many-one reduction of a decision problem A to a decision problem B is a polynomial-time algorithm F that takes instances x of A to instances $F(x)$ of B with the property that the answer to x in A is "true" if and only if the answer to $F(x)$ in B is "true."

Now, we can actually define that subclass of NP-hard problems that we talked about before: the class of NP-**complete** problems. We say that a problem is NP-complete if it is both in NP and every problem in NP is polynomial-time many-one reducible to it. As we mentioned before, SAT, for example, is NP-complete. Other NP-complete problems include determining whether a graph is 3-colorable, determining whether the constraints of a binary linear program can be satisfied, determining whether a graph has a cut of size bigger than a given integer k, and many other natural decision problems.

NP-complete problems are central to the study of the $P \stackrel{?}{=} NP$ question because $P = NP$ if and only if at least one NP-complete problem is in P. So, you can focus on, say, just studying SAT. If you find a polynomial-time algorithm for it, then $P = NP$. If, on the contrary, you show that it is impossible to solve SAT in polynomial time, you have found a problem in NP that is not in P and then, immediately, you can conclude that $P \neq NP$.

Important note

A problem B is NP-complete if it is in NP and every other problem A in NP is polynomial-time many-one reducible to B.

There are, of course, NP-hard problems that are not NP-complete. This is the case, for instance, if you have an NP-hard problem that is not a decision problem (and, hence, cannot be in NP). Many problems that we study in *Chapter 3, Working with Quadratic Unconstrained Binary Optimization Problems*, fall under that category. For instance, finding a minimal coloring for a graph is clearly NP-hard. If you knew how to solve this problem efficiently, then you could also determine whether a graph is 3-colorable (you just need to compute the minimal coloring and check whether its number of colors is at most 3). But checking whether a graph is 3-colorable is NP-hard and, thus, finding a minimal coloring is also NP-hard.

Many other examples of problems that are optimization versions of NP-complete problems are also NP-hard, including determining the maximum number of clauses that can be simultaneously satisfied in a Boolean formula in conjunctive normal form (the MAX-SAT problem), finding a maximum cut in a graph (the Max-Cut problem), finding a minimum-cost solution of a binary linear program, or solving the Traveling Salesperson problem. However, none of them is NP-complete because they are not in NP: they are not decision problems to start with and, moreover, it is far from clear that you could check efficiently that a candidate solution is, indeed, an optimal solution!

A very brief introduction to quantum computational complexity

So far, we have focused only on measuring time complexity with classical models. However, this is a book on quantum computing, so it is natural to ask what will change if we consider quantum computational models instead. This is studied in **quantum computational complexity theory**, a fascinating topic that is totally beyond the scope of this book.

Let us, however, say a few words on the kind of concepts that arise when quantum models are considered instead of classical Turing machines. This is not at all needed to understand any other part of the book, so feel completely free to skip it. We will need to be brief, but you can refer to the survey by Watrous [110] for more details.

It turns out that it is possible to define a class of problems that can be seen as a quantum analogous to P. This class is known as BQP, and it contains those decision problems that can be solved with bounded error in polynomial time with a quantum algorithm.

There are a couple of things that we need to clarify here. The first one is that quantum algorithms being probabilistic, we cannot expect the correct answer to a decision problem to always be obtained. Instead, we impose that this correct answer is returned, for each input, with high probability. Formally, the requirement is that for every positive instance x, the probability of obtaining 1 when the input to the algorithm is x should be at least $2/3$; similarly, for every negative instance x, the probability of obtaining 0 when the algorithm runs on x should be at least $2/3$. In this way, we can repeat the procedure with the same input several times and take the majority result. If the number of repetitions is big enough (but fixed), we can make the probability of error arbitrarily small while still having a total running time that is polynomial.

To learn more...

BQP is not exactly analogous to *P* but to another (classical) computational class called *BPP*. The class *BPP* contains those decision problems that can be solved with bounded error in polynomial time with a probabilistic Turing machine (that is, a Turing machine with multiple instructions for certain state-symbol situations and that can decide which instruction to execute based on a sequence of random bits). *BPP* stands for **bounded-error probabilistic polynomial time** while *BQP* stands for **bounded-error quantum polynomial time**.

The other thing that needs to be clarified about our definition of *BQP* is what we exactly understand by a quantum algorithm. In the classical case, we have identified this notion with a (single-tape) Turing machine. It is possible to define a quantum version of Turing machines (see, for instance, the paper by Bernstein and Vazirani [111]) and use it in our definition. But since our primary model for quantum computations throughout this book is the quantum circuit model, a natural question is whether we can also use it to formalize the notion of quantum algorithm.

In fact, we can give a definition of what is a quantum algorithm in terms of quantum circuits, and this definition is equivalent in computational power to the one in terms of quantum Turing machines (and polynomially equivalent with respect to running time). However, there exist several subtleties that need to be confronted.

The first one is related to being able to consistently measure the execution time of a quantum circuit. To do that, we need to fix a finite set of gates and express every circuit using only those gates. Then, we can assign a cost of one unit to each of those gates and measure the running time of a circuit as its total number of gates. Otherwise, if we allow arbitrary gates, then we could argue that any circuit is just a single unitary gate (plus some measurements), something that is clearly meaningless in terms of analyzing its complexity. Notice that fixing a finite set of permitted gates also allows us to describe every circuit as a finite binary string, for instance, giving a list of the gates that we use and the qubits on which we apply them.

The finite set of gates needs to be chosen in a way that we can approximate any given quantum circuit to arbitrary precision. A possible way of doing this is explained in the survey by Watrous [110].

A second technical problem that we need to tackle is that, while a Turing machine can process inputs of any size, every quantum circuit has a fixed number of qubits and, hence, only admits inputs of a fixed size. As a consequence, we cannot represent a full algorithm (that needs to be able to solve every possible instance of a problem) with just one quantum circuit: we need to consider an infinite family of circuits, one for each input size. So, a quantum algorithm is not a single quantum circuit, but a collection $\{C_n\}$ of circuits, one for each natural number n, so that C_n admits n qubits as its input.

The final issue that we need to address is related to the way in which we select that infinite family of circuits. If we allow any collection of circuits to represent a quantum algorithm, then we can end up in pathological situations such as being able to solve (a problem equivalent to) the Halting problem, which we know to be uncomputable! This is because we could just select a different, totally unrelated quantum circuit for each size in a way that the quantum circuit already "knows" the answer to the Halting problem for its input size. This is not something particular to just quantum circuits. The same happens with classical Boolean circuits (as we mentioned, this is a subtle point; see Section 2.2 in the book by Kitaev et al. [112] or Chapter 6 in the book by Arora and Barak [103], especially what is said there about the $P/poly$ class of problems).

The solution to this issue is to specify all the quantum circuits in the family in a **uniform** way. For instance, we can impose that there exists a (classical) Turing machine that, given a natural number n, generates the circuit for input size n in polynomial time (in n). In this way, we can't hide any additional complexity in the selection of the quantum circuits. Remember that we can represent our quantum circuits as finite binary strings (because we have fixed a finite number of allowable quantum gates), so it makes sense to obtain them as the output of a Turing machine. Moreover, every circuit will have a polynomial size (a polynomial-time Turing machine can only output a polynomial number of bits, after all) and hence a polynomial running time.

Important note

BQP is the class of decision problems that can be solved with bounded error by polynomial-time uniform families of quantum circuits.

Now that we have defined *BQP*, it is natural to ask about its relationship with *P* and *NP* in order to be able to assess the power of quantum computers when compared to that of classical ones.

It is easy to show that $P \subseteq BQP$, that is, that every problem in *P* is also in *BQP*. This follows directly from the fact that we can simulate any classical Boolean circuit with a quantum circuit (as we show in *Section 1.5.2*) and from the fact that polynomial-time uniform families of classical circuits are equivalent to polynomial-time Turing machines (see Section 6.2 in the book by Arora and Barak [103]). But this is not surprising at all, because we expect quantum computers to be at least as powerful as classical computers.

So the question that we should really ask is whether there are problems in *BQP* that are not in *P*. The short answer is that...we don't know. Proving it would imply a major breakthrough not only in quantum computational complexity but also in classical computational complexity theory. It can be proved that *BQP* is contained in *PSPACE*, the class of decision problems solvable in polynomial space. Showing that *P* is different from *BQP* would also imply that *P* is different from *PSPACE*, which is a major open question in computational complexity (although it should be easier to solve than the *P* versus *NP* problem, because *NP* is also contained in *PSPACE*).

That being said, we have good reasons to believe that there are problems in *BQP* that are not in *P*. In fact, we have a very good candidate: the factoring problem (given natural numbers m and k, check whether m has a factor $l \neq 1$ that is less than k) is in *BQP* thanks to Shor's algorithm [6], but it would be really, really surprising if it were in *P*. In fact, many cryptographic protocols currently in use rely on the assumption that factoring is not in *P*. So, every time that you buy something online and you send your credit card number over the internet, you are implicitly trusting that *P* and *BQP* are not equal (and that nobody owns a powerful enough quantum computer!).

And what about *BQP* and *NP*? The situation there is a little bit more complicated. The evidence that we have seems to imply that there are problems in *BQP* that are not in *NP* (one of the strongest results in this direction can be found in a recent paper by Raz and Tal [113]). But we also have some evidence that seems to suggest that there are problems in *NP* that are not in *BQP*, due to results by Bennett, Bernstein, Brassard, and Vazirani [114] that show that Grover's algorithm is, in a certain sense, optimal among quantum algorithms for search tasks.

If all this is true, it would imply that there are problems that we can solve efficiently with quantum algorithms that we couldn't solve efficiently even with non-deterministic machines. But, contrary to what can be read sometimes in the media, it also would imply that not every problem in *NP* could be solved efficiently with a quantum computer, even if it were fault-tolerant. In particular, it would imply that no *NP*-complete problem could be solved efficiently with quantum algorithms (we have represented all these relationships in *Figure C.2*).

Does this mean that quantum computers are not useful at all for optimization problems? Not necessarily. The methods that we describe in *Part 2* of this book may not be able to give the optimal solution to every optimization problem out there. But they provide approximation algorithms that might beat whatever is possible with just classical algorithms. For instance, the QAOA algorithm that we study in *Chapter 5, QAOA: Quantum Approximate Optimization Algorithm*, is considered a possible candidate for that kind of advantage (for some recent results in this direction, see the papers by Basso et al. [115] and by Farhi et al. [116], but

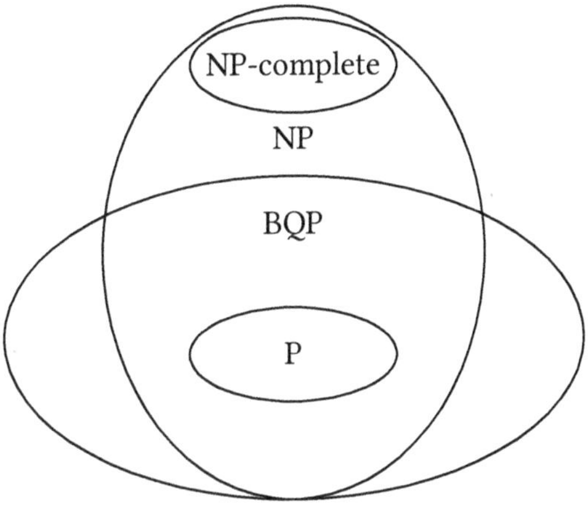

Figure C.2: Possible relationships between P, NP, BQP, and NP-complete problems according to the available evidence and the most accepted conjectures. Be warned: some of these classes might end up being completely equal!

also check the response by Hastings [117]). And even if that were not the case, methods such as quantum annealing (described in *Chapter 4, Adiabatic Quantum Computing and Quantum Annealing*) or QAOA may provide good heuristics that are useful in practice, in the same way that genetic algorithms, simulated annealing, or particle-swarm optimization are used to solve practical problems in many different fields.

D

Installing the Tools

Man is a tool-using animal. Without tools he is nothing, with tools he is all.

— Thomas Carlyle

In this appendix, we will give you all the instructions needed to run the code examples provided in the main text. We will start by guiding you through the process of installing the software that we will use, then we will learn how to access the real quantum computers on which we will run our code, and finally, we will also show you how to accelerate some of the executions by using a GPU.

Getting Python

All the quantum programming libraries that we use in this book are based on Python, so you need to have a working Python distribution. If your operating system is Linux or macOS, you probably have one already. If your Python version is at least 3.7, then you are ready to go.

However, even if you already have Python installed on your system, we recommend that you consider following one of these two options:

- **Installing Anaconda**: Anaconda is a data science software distribution that includes, among other things, Python and many of its scientific libraries. In addition, it also includes Jupyter, an extremely useful web-based interactive computing platform that allows you to run code, write text and formulas, and visualize graphics, all organized into notebooks. For convenience, we provide all the code of the book in Jupyter notebooks that you can download from `https://github.com/PacktPublishing/A-Practical-Guide-to-Quantum-Machine-Learning-and-Quantum-Optimization`.

  If you install Anaconda, you will have most of the non-quantum software libraries that we use in the book already on your system, plus some additional ones that you may find convenient for other, related projects.

  There is a version of Anaconda called **Anaconda Distribution** that is free to download from `https://www.anaconda.com/products/distribution`. It is available for Windows, Linux, and Mac. Anaconda Distribution provides a graphical installer, so it is super easy to set up. In case of doubt, you can always check the installation instructions at `https://docs.anaconda.com/anaconda/install/index.html`.

  Once you install Anaconda, we recommend that you launch it and run JupyterLab. This will open an IDE in your web browser that you can use to manage Jupyter notebooks and start running code right away. For a quick introduction to how to use JupyterLab, you can check this overview of its interface included in the JupyterLab documentation: `https://jupyterlab.readthedocs.io/en/stable/user/interface.html`.

- **Using Google Colab**: If you prefer not to install anything on your own computer, we also have an option for you. Google Colab is a web-based environment provided by Google in which you can run Jupyter notebooks with Python code. In fact, its interface its very similar to that of Jupyter and can be used to run all the code in this book (we know because we did it ourselves!) in addition to many other projects, especially those related to machine learning and data science.

The main difference between using Jupyter and Google Colab is that Colab does not run on your computer but is cloud-based: it uses hardware owned by Google. They provide you with a (usually modest) CPU, some amount of RAM, and some disk space, and you also have the chance to request a GPU to accelerate the training of your machine learning models.

The basic version of Google Colab is free to use: you only need a working Google account to start using it at `https://colab.research.google.com/`. And should you ever need more computational power, you can upgrade to a paid version (see more details at `https://colab.research.google.com/signup`).

By the way, the tutorials at `https://colab.research.google.com/` are really helpful, so you will be running your projects in almost no time.

Each of these options has its pros and cons. With Anaconda, you have perfect control over what you install, you get to use your own hardware (which probably is more powerful than that available at Google Colab, maybe with the exception of those sweet GPUs), and you can work offline. But you need to install everything yourself, keep it up to date, and solve any version conflicts that may arise.

With Google Colab, you can start running code right away from any computer connected to the internet, without the burden of having to install Python and many other libraries, and you can use quite powerful GPUs for free. However, you need to be online all the time, there are some restrictions on the number of projects that you can run simultaneously (at least, with the free version), and the CPU speed is not that great.

The good thing is that any of these possibilities (or any other that gets you a running Python distribution) works perfectly well for the purpose of running the code in this book. Moreover, they are perfectly compatible with each other, so you can start writing a notebook on Google Colab and complete it with Anaconda or vice versa. Since both are free, you can try them both and use the one that better suits your needs at any given moment.

Of course, we don't want to be too prescriptive. If you don't feel like relying on Anaconda or on a cloud service, you can use your local machine without any add-ons and everything will work just fine as long as you have the right versions of the packages that we will use.

Installing the libraries

Although both Anaconda and Google Colab come with a lot of data science, visualization, and machine learning libraries already installed by default, they do not yet include any of the quantum computing libraries that we use in this book.

However, getting them set up and running is a breeze with **pip**, a package manager that comes bundled with Python – you don't need to install Anaconda or access Google Colab to use it. In order to install a new library with pip, you just need to run the following instruction on your terminal:

```
pip install name-of-library
```

If you are using a Jupyter notebook to run your code, you can use exactly that same instruction, but you need to write it in a cell of its own, with no additional code. If you need to install several different libraries and you do not want to create a different cell for each pip instruction, then you can put them all together in the same cell but you need to use the escape symbol `!`. So, for instance, you can install three libraries in the same cell of your Jupyter notebook like this:

```
!pip install first-library
!pip install second-library
!pip install last-library
```

Sometimes, you need to install a particular version of a library. This is the case with some of the examples in this book. Don't worry, because pip has your back in this too. You just need to run the following instruction:

```
pip install name-of-library==version-number
```

For example, to install version 0.39.2 of Qiskit, which is the one that we use in this book, you need to run the following instruction:

```
pip install qiskit==0.39.2
```

Of course, the same comments that we just made about escape symbols in Jupyter notebooks apply to this case.

Important note

If you run a `pip install` command on a Jupyter notebook to install a different version of a library that was already present on the system, you will probably need to restart the kernel (if you are running a Jupyter notebook on your local machine) or the runtime (in Google Colab) for the changes to take place.

In *Table D.1* we have collected all the libraries needed for the code in this book, in the order they appear in the main text, together with the version that we have used to create the examples. The second column specifies the name of each library in pip, so that is the one that you need to use with the `pip install` command.

You may have noticed that there are a couple of libraries in the list that we never explicitly imported into our code. However, they are used by other packages to be able to plot circuits (Pylatexenc) and to obtain Hamiltonians for molecular problems (PySCF), so they need to be present in your system.

Some of the libraries already come with Anaconda and Google Colab. In fact, it is very likely that the code in this book works with whatever version is included in those distributions, so installing the exact version we mention in the table should not be especially important.

The only exceptions are PyTorch and TensorFlow: for them, the versions that you should use are the ones listed in the table.

Library name	Pip name	Version number
Qiskit	qiskit	0.39.2
Pylatexenc	pylatexenc	2.10
Numpy	numpy	1.21.6
Qiskit Aer GPU	qiskit-aer-gpu	0.11.1
PennyLane	pennylane	0.26
PennyLane Qiskit plugin	pennylane-qiskit	0.24.0
Ocean	dwave-ocean-sdk	6.0.1
Qiskit Optimization	qiskit-optimization	0.4.0
Qiskit Nature	qiskit-nature	0.4.5
Scipy	scipy	1.7.3
Matplotlib	matplotlib	3.2.2
PySCF	pyscf	2.11
scikit-learn	scikit-learn	1.0.2
TensorFlow	tensorflow	2.9.1
Qiskit Machine Learning	qiskit-machine-learning	0.5.0
Optuna	optuna	3.0.3
PyTorch	torch	1.13
Qiskit IBM Runtime	qiskit-ibm-runtime	0.7.0

Table D.1: Libraries used in the book and their version numbers

For the libraries that do not come with Anaconda and Google Colab, it is highly recommended to stick to the versions listed in the table. This is especially important for Qiskit and all its modules, which tend to change their APIs quite frequently.

In any case, for convenience, in the book notebooks that you can download from `https://github.com/PacktPublishing/A-Practical-Guide-to-Quantum-Machine-Learning-and-Quantum-Optimization`, we have explicitly included the installation commands of those libraries with the exact version that we have used to create the examples. If you're running the code on a local Python installation, you just need to install these libraries once, so you can remove the `pip install` commands after the first execution. However, if you're using Google Colab, you will need to run those commands every time you create a new runtime, because there is no persistence of data from one session to another.

Accessing IBM's quantum computers

In order to be able to run circuits on IBM's quantum computers from your Python programs, you first need to create an IBM account. This can be done through the IBM Quantum login page located at `https://quantum-computing.ibm.com/login`, and it is completely free.

Then, you need to obtain your API token. You can do this by going to `https://quantum-computing.ibm.com/account`, logging in if necessary, and finding the field titled **API token** (see *Figure D.1*). Then, you can click the icon with the two rectangles next to the string of asterisks to copy the token to your clipboard. Should you need it, this is also the page where you can generate a new API token by clicking on **Generate new token**.

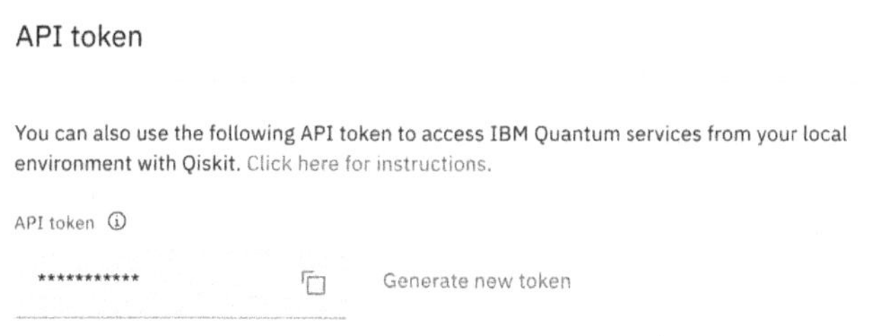

Figure D.1: Getting your IBM Quantum API token

Once you have the token, if you want to access IBM's devices from your Qiskit programs, you need to run the following instructions:

```
from qiskit import IBMQ
IBMQ.save_account("TOKEN")
```

where, of course, you should replace `TOKEN` with your actual token. Then, you can obtain access to the IBM provider by using `IBMQ.load_account()`, as we do in the main text.

If you are using your local Python installation, you only need to save your account once (and, additionally, whenever you change your API token). However, if you are using Google Colab, you need to save your account in each new runtime. We have prepared the notebooks that you can download from `https://github.com/PacktPublishing/A-Pra`

ctical-Guide-to-Quantum-Machine-Learning-and-Quantum-Optimization so that you only need to write your actual token in the ibm_token = instruction.

If you need to access IBM quantum computers from PennyLane, the process is almost the same. The only difference is that you need to, additionally, install the PennyLane-Qiskit plugin as seen in the previous section.

Accessing D-Wave quantum annealers

In order to access D-wave quantum annealers from your code, you first need to create a free D-Wave Leap account at https://cloud.dwavesys.com/leap/signup/. This will give you 1 minute of free access to run your problems on actual quantum devices, as explained in *Chapter 4, Adiabatic Quantum Computing and Quantum Annealing*. If you want to extend this access to get one additional free minute per month recurringly, you can provide your GitHub username and repository by going to https://cloud.dwavesys.com/leap/plans/#Custom and clicking **Get Developer Access**.

In any case, as with IBM quantum computers, you now need to get your API token. You can achieve this by going to https://cloud.dwavesys.com/leap/, signing in if needed, and finding the field titled **API Token**. This is usually located on the left part of the page, under your name and account type (see *Figure D.2*). There, you can click on **COPY** to copy the token to your clipboard and on **RESET** to generate a new token.

Then, you need to configure your access by running dwave config create. You can do this either on a terminal or on a Python notebook or program, but in this latter case, you need to use the escape symbol ! before the command. Then, you will be prompted to enter some options for your configuration. You just need to go with the default values (by pressing *Enter*) on all the questions with the exception of **Authentication token**, for which you need to provide the API token that you copied from the D-Wave Leap website.

If you are using a local Python installation, this only has to be done once and, afterward, you can access D-Wave's quantum annealers as we describe in *Chapter 4, Adiabatic Quantum*

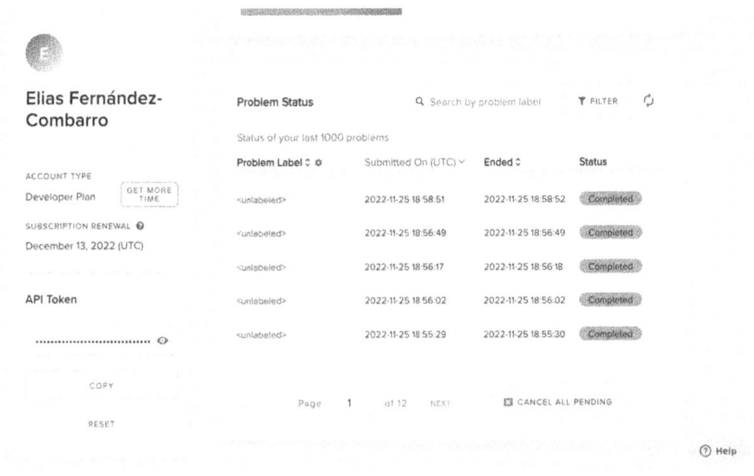

Figure D.2: Getting your D-Wave API token

Computing and Quantum Annealing. If you are using Google Colab, you need to run the configuration step every time you use a new runtime.

Using GPUs to accelerate simulations in Google Colab

As we mention in *Chapter 2, The Tools of the Trade in Quantum Computing*, using a GPU to simulate quantum circuits can offer, in some cases, a noticeable speedup in computation time. In general, the process of setting up a GPU to work with quantum libraries such as Qiskit depends heavily on your hardware configuration and your GPU model (although, in principle, only Nvidia GPUs are supported).

However, if you are using Google Colab, you have the chance of requesting a GPU to run your circuits. The advantages of this approach are two-fold. Not only do you not need to buy the GPU yourself, but you also don't have to set it up.

To request a GPU for one of your Google Colab notebooks, you need to select the **Change runtime type** option in the **Runtime** menu. Then, you need to select the **GPU** option (see *Figure D.3*) and click **Save**. If there is availability, you will be assigned a GPU. To check the status of the GPU, you can run `!nvidia-smi -L`. You will get an output like the following (the model of GPU may vary from one session to another):

```
GPU 0: Tesla T4 (UUID: GPU-a6c87248-f520-fbc1-d604-309890a20713)
```

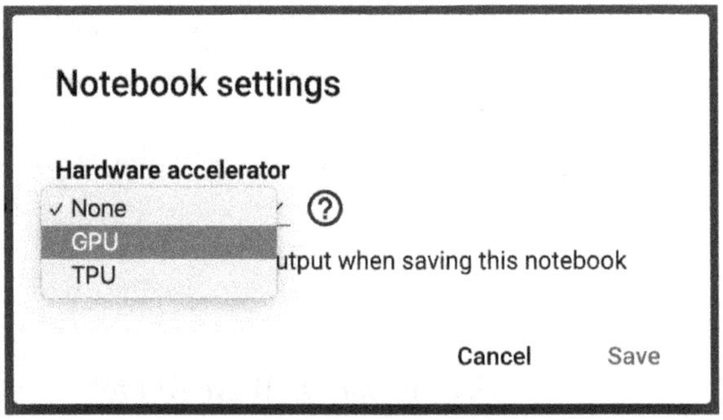

Figure D.3: Requesting a GPU

If this command executes without error, it means that you have access to a GPU. Now, to use it in Qiskit, you need to install the Qiskit Aer GPU package by running the following instruction:

```
pip install qiskit-aer-gpu==0.11.1
```

Notice that this will replace the usual Qiskit Aer module (the one that works with CPUs only), so you may need to restart your runtime if you had already run some Qiskit code. Now, you can try the GPU simulation by executing, for instance, the following instructions:

```
from qiskit import *
from qiskit.providers.aer import AerSimulator

sim = AerSimulator(device = 'GPU')
qc = QuantumCircuit(2, 2)
qc.h(0)
qc.cnot(0,1)
qc.measure(range(2), range(2))
```

```
job = execute(qc, sim, shots = 1024)
result = job.result()
counts = result.get_counts()
print(counts)
```

You will get an output exactly as if you were running the simulation on your CPU. It would be something like this:

```
{'00': 489, '11': 535}
```

E

Production Notes

There are two things nobody should ever have to watch being made, sausage and laws.

— Mark Twain

This book was written in LaTeX by the two of us in three different countries (Ireland, Spain, and Switzerland) and in a wide variety of places: in offices at Maynooth University, the University of Oviedo and CERN; in an apartment in Oviedo, a university dorm in Maynooth, and an apartment in Geneva; in a sports pavilion; in the waiting rooms of emergency departments of two different hospitals; near the beach; near the mountains; on the backseat of a car; on some commuter trains; at several different airports; at a hotel in Almería; and, probably, in some other locations that we do not remember now.

All this wouldn't have been possible without adequate tools and apps. The main one was Overleaf (`https://www.overleaf.com`), which allowed us to collaborate and work simultaneously even while we were thousands of kilometers away from each other.

To help us write formulas, draw circuits, and format code in LaTeX, we used quite a lot of useful packages such as quantikz, physics, siunitx, and listings. To create the figures, we used TikZ, Graphviz (`https://graphviz.org/`), and the Graphviz Visual Editor (`http://magjac.com/graphviz-visual-editor/`). To write the code examples and run them, we used both Anaconda and Google Colab (as described in *Appendix D, Installing the Tools*).

All of them are excellent tools that made writing this book a much more pleasant and easy experience.

Assessments

Chapter 1, Foundations of Quantum Computing

(1.1) The probability of measuring 0 if the state of a qubit is $\sqrt{1/2}\,|0\rangle + \sqrt{1/2}\,|1\rangle$ is exactly

$$\left|\sqrt{1/2}\right|^2 = 1/2.$$

In the same way, the probability of measuring 1 is also 1/2. If the state of the qubit is $\sqrt{1/3}\,|0\rangle + \sqrt{2/3}\,|1\rangle$, the probability of measuring 0 is

$$\left|\sqrt{1/3}\right|^2 = 1/3$$

and the probability of measuring 1 is

$$\left|\sqrt{2/3}\right|^2 = 2/3.$$

Finally, if the qubit state is $\sqrt{1/2}\,|0\rangle - \sqrt{1/2}\,|1\rangle$, the probability of measuring 0 is

$$\left|\sqrt{1/2}\right|^2 = 1/2$$

and the probability of measuring 1 is

$$\left|-\sqrt{1/2}\right|^2 = 1/2.$$

(1.2) The inner product of $\sqrt{1/2}\,|0\rangle + \sqrt{1/2}\,|1\rangle$ and $\sqrt{1/3}\,|0\rangle + \sqrt{2/3}\,|1\rangle$ is

$$\sqrt{1/2}\sqrt{1/3} + \sqrt{1/2}\sqrt{2/3} = \sqrt{1/6} + \sqrt{1/3}.$$

The inner product of $\sqrt{1/2}\,|0\rangle + \sqrt{1/2}\,|1\rangle$ and $\sqrt{1/2}\,|0\rangle - \sqrt{1/2}\,|1\rangle$ is

$$\sqrt{1/2}\sqrt{1/2} - \sqrt{1/2}\sqrt{1/2} = 0.$$

(1.3) The adjoint of X is X itself and it holds that $XX = I$. Hence, X is unitary and its inverse is X itself. The operation X takes $a\,|0\rangle + b\,|1\rangle$ to $b\,|0\rangle + a\,|1\rangle$.

(1.4) The adjoint of H is H itself and it holds that $HH = I$. Hence, H is unitary and its inverse is H itself. The operation H takes $|+\rangle$ to $|0\rangle$ and $|-\rangle$ to $|1\rangle$. Finally, it holds that $X\,|+\rangle = |+\rangle$ and that $X\,|-\rangle = -\,|-\rangle$.

(1.5) It holds that

$$Z\,|0\rangle = HXH\,|0\rangle = HX\,|+\rangle = H\,|+\rangle = |0\rangle$$

and that

$$Z\,|1\rangle = HXH\,|1\rangle = HX\,|-\rangle = -H\,|-\rangle = -\,|1\rangle\,.$$

It also holds that

$$\begin{pmatrix} \frac{1}{\sqrt{2}} & \frac{1}{\sqrt{2}} \\ \frac{1}{\sqrt{2}} & -\frac{1}{\sqrt{2}} \end{pmatrix}\begin{pmatrix} 0 & 1 \\ 1 & 0 \end{pmatrix}\begin{pmatrix} \frac{1}{\sqrt{2}} & \frac{1}{\sqrt{2}} \\ \frac{1}{\sqrt{2}} & -\frac{1}{\sqrt{2}} \end{pmatrix} = \begin{pmatrix} \frac{1}{\sqrt{2}} & \frac{1}{\sqrt{2}} \\ \frac{1}{\sqrt{2}} & -\frac{1}{\sqrt{2}} \end{pmatrix}\begin{pmatrix} \frac{1}{\sqrt{2}} & -\frac{1}{\sqrt{2}} \\ \frac{1}{\sqrt{2}} & \frac{1}{\sqrt{2}} \end{pmatrix} = \begin{pmatrix} 1 & 0 \\ 0 & -1 \end{pmatrix}.$$

(1.6) Since $e^{i\frac{\pi}{4}}e^{i\frac{\pi}{4}} = e^{i\frac{\pi}{2}}$, it is apparent that $T^2 = S$. Also, we have $e^{i\frac{\pi}{2}}e^{i\frac{\pi}{2}} = e^{i\pi} = -1$ by Euler's identity, so $S^2 = Z$. As a consequence, $SS^3 = S^2S^2 = ZZ = I$, so $S^\dagger = S^3$. Also, $TT^7 = T^2T^2T^2T^2 = S^4 = I$, and it follows that $T^\dagger = T^7$.

(1.7) By the definition of R_X we have that

$$R_X(\pi) = \begin{pmatrix} \cos\frac{\pi}{2} & -i\sin\frac{\pi}{2} \\ -i\sin\frac{\pi}{2} & \cos\frac{\pi}{2} \end{pmatrix} = \begin{pmatrix} 0 & -i \\ -i & 0 \end{pmatrix} = -iX.$$

Analogously,

$$R_Y(\pi) = \begin{pmatrix} \cos\frac{\pi}{2} & -\sin\frac{\pi}{2} \\ \sin\frac{\pi}{2} & \cos\frac{\pi}{2} \end{pmatrix} = \begin{pmatrix} 0 & -1 \\ 1 & 0 \end{pmatrix} = -iY$$

and

$$R_Z(\pi) = \begin{pmatrix} e^{-i\frac{\pi}{2}} & 0 \\ 0 & e^{i\frac{\pi}{2}} \end{pmatrix} = \begin{pmatrix} -i & 0 \\ 0 & i \end{pmatrix} = -iZ.$$

Also,

$$R_Z\left(\frac{\pi}{2}\right) = \begin{pmatrix} e^{-i\frac{\pi}{4}} & 0 \\ 0 & e^{i\frac{\pi}{4}} \end{pmatrix} = e^{-i\frac{\pi}{4}}S$$

and

$$R_Z\left(\frac{\pi}{4}\right) = \begin{pmatrix} e^{-i\frac{\pi}{8}} & 0 \\ 0 & e^{i\frac{\pi}{8}} \end{pmatrix} = e^{-i\frac{\pi}{8}}T.$$

(1.8) From the definition of $U(\theta, \varphi, \lambda)$, we have that

$$U(\theta,\varphi,\lambda)U(\theta,\varphi,\lambda)^{\dagger} = \begin{pmatrix} \cos\frac{\theta}{2} & -e^{i\lambda}\sin\frac{\theta}{2} \\ e^{i\varphi}\sin\frac{\theta}{2} & e^{i(\varphi+\lambda)}\cos\frac{\theta}{2} \end{pmatrix} \begin{pmatrix} \cos\frac{\theta}{2} & e^{-i\varphi}\sin\frac{\theta}{2} \\ -e^{-i\lambda}\sin\frac{\theta}{2} & e^{-i(\varphi+\lambda)}\cos\frac{\theta}{2} \end{pmatrix} = I$$

and, analogously, $U(\theta,\varphi,\lambda)^{\dagger}U(\theta,\varphi,\lambda) = I$. Then, $U(\theta,\varphi,\lambda)$ is unitary.

Also, we get that

$$U(\theta,-\pi/2,\pi/2) = \begin{pmatrix} \cos\frac{\theta}{2} & -i\sin\frac{\theta}{2} \\ -i\sin\frac{\theta}{2} & \cos\frac{\theta}{2} \end{pmatrix} = R_X(\theta).$$

Analogously, it holds that

$$U(\theta, 0, 0) = \begin{pmatrix} \cos\frac{\theta}{2} & -\sin\frac{\theta}{2} \\ \sin\frac{\theta}{2} & \cos\frac{\theta}{2} \end{pmatrix} = R_Y(\theta)$$

and that

$$U(0, 0, \theta) = \begin{pmatrix} 1 & 0 \\ 0 & e^{i\theta} \end{pmatrix} = e^{i\frac{\theta}{2}} R_Z(\theta).$$

(1.9) Since $\theta = 2 \arccos \sqrt{p}$, it holds that the state before measurement is

$$\cos\frac{\theta}{2} |0\rangle + \sin\frac{\theta}{2} |1\rangle = \sqrt{p} |0\rangle + \sqrt{1-p} |1\rangle .$$

As a consequence, the probability of measuring 0 is p and the probability of measuring 1 is $1-p$.

(1.10) The probability of obtaining 1 will be $|a_{10}|^2 + |a_{11}|^2$. Upon that measurement result, the state will collapse to

$$\frac{a_{10} |10\rangle + a_{11} |11\rangle}{\sqrt{|a_{10}|^2 + |a_{11}|^2}}.$$

(1.11) It holds that

$$(U_1 \otimes U_2)(U_1^\dagger \otimes U_2^\dagger) = (U_1 U_1^\dagger) \otimes (U_2 U_2^\dagger) = I \otimes I.$$

Analogously, $(U_1^\dagger \otimes U_2^\dagger)(U_1 \otimes U_2) = I \otimes I$. Hence, the inverse of $U_1 \otimes U_2$ is $U_1^\dagger \otimes U_2^\dagger$.

Also, from the definition of tensor product of two matrices we get that, for every matrix A and B (even in they are non-unitary), it holds that

$$A^\dagger \otimes B^\dagger = \begin{pmatrix} a_{11}^* & a_{21}^* \\ a_{12}^* & a_{22}^* \end{pmatrix} \otimes \begin{pmatrix} b_{11}^* & b_{21}^* \\ b_{12}^* & b_{22}^* \end{pmatrix} = \begin{pmatrix} a_{11}^* \begin{pmatrix} b_{11}^* & b_{21}^* \\ b_{12}^* & b_{22}^* \end{pmatrix} & a_{21}^* \begin{pmatrix} b_{11}^* & b_{21}^* \\ b_{12}^* & b_{22}^* \end{pmatrix} \\ a_{12}^* \begin{pmatrix} b_{11}^* & b_{21}^* \\ b_{12}^* & b_{22}^* \end{pmatrix} & a_{22}^* \begin{pmatrix} b_{11}^* & b_{21}^* \\ b_{12}^* & b_{22}^* \end{pmatrix} \end{pmatrix}$$

$$= \begin{pmatrix} a_{11}^*b_{11}^* & a_{11}^*b_{21}^* & a_{21}^*b_{11}^* & a_{21}^*b_{21}^* \\ a_{11}^*b_{12}^* & a_{11}^*b_{22}^* & a_{21}^*b_{12}^* & a_{21}^*b_{22}^* \\ a_{12}^*b_{11}^* & a_{12}^*b_{21}^* & a_{22}^*b_{11}^* & a_{22}^*b_{21}^* \\ a_{12}^*b_{12}^* & a_{12}^*b_{22}^* & a_{22}^*b_{12}^* & a_{22}^*b_{22}^* \end{pmatrix} = (A \otimes B)^\dagger.$$

(1.12) The matrix for $X \otimes X$ is

$$\begin{pmatrix} 0 & 0 & 0 & 1 \\ 0 & 0 & 1 & 0 \\ 0 & 1 & 0 & 0 \\ 1 & 0 & 0 & 0 \end{pmatrix}.$$

The matrix for $H \otimes I$ is

$$\frac{1}{\sqrt{2}} \begin{pmatrix} 1 & 0 & 1 & 0 \\ 0 & 1 & 0 & 1 \\ 1 & 0 & -1 & 0 \\ 0 & 1 & 0 & -1 \end{pmatrix}.$$

(1.13) In the circuit

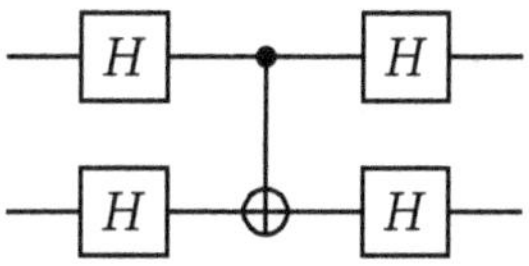

the states $|00\rangle$ and $|10\rangle$ are left unchanged, while $|01\rangle$ and $|11\rangle$ are mapped to each other. This is exactly the action of a CNOT gate whose control is the bottom qubit and whose target is the top one.

The matrix for the circuit is

$$\begin{pmatrix} 1 & 0 & 0 & 0 \\ 0 & 0 & 0 & 1 \\ 0 & 0 & 1 & 0 \\ 0 & 1 & 0 & 0 \end{pmatrix},$$

which is exactly the matrix for the CNOT gate from bottom qubit to top qubit.

On the other hand, the circuit

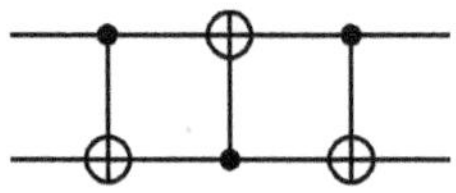

leaves $|00\rangle$ and $|11\rangle$ unchanged, while it maps $|01\rangle$ and $|10\rangle$ to each other. This is exactly the action of a SWAP gate.

Alternatively, the matrix for the circuit is

$$\begin{pmatrix} 1 & 0 & 0 & 0 \\ 0 & 0 & 1 & 0 \\ 0 & 1 & 0 & 0 \\ 0 & 0 & 0 & 1 \end{pmatrix},$$

which is, again, the matrix of the SWAP gate.

(1.14) The state $\sqrt{1/3}(|00\rangle + |01\rangle + |11\rangle)$ is indeed entangled. However, $\frac{1}{2}(|00\rangle + |01\rangle + |10\rangle + |11\rangle)$ is a product state because it can be written as $|+\rangle\,|+\rangle$.

(1.15) If the matrix of U is $(u_{ij})_{i,j=1}^{2}$, then $|00\rangle$ and $|01\rangle$ are taken to themselves by CU. What is more, $|10\rangle$ is taken to $|1\rangle(u_{11}|0\rangle + u_{21}|1\rangle)$ and $|11\rangle$ is taken to $|1\rangle(u_{12}|0\rangle + u_{22}|1\rangle)$.

Hence, the matrix of CU is

$$\begin{pmatrix} 1 & 0 & 0 & 0 \\ 0 & 1 & 0 & 0 \\ 0 & 0 & u_{11} & u_{12} \\ 0 & 0 & u_{21} & u_{22} \end{pmatrix}.$$

The adjoint of CU is $CU^\dagger$ and it holds that $CUCU^\dagger = CU^\dagger CU = I$. Hence, CU is unitary.

(1.16) The equivalence follows directly from the fact that $HXH = Z$.

(1.17) We can prepare $\sqrt{1/2}\,(|00\rangle - |11\rangle)$ with the following circuit:

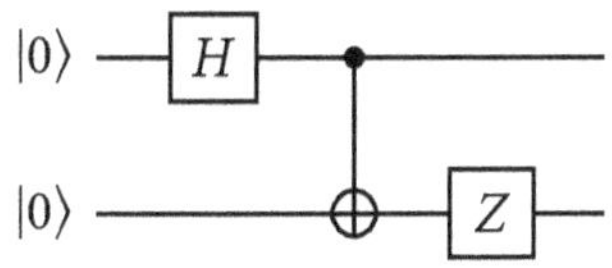

We can use the circuit

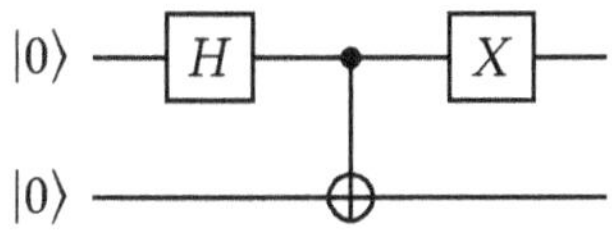

to prepare $\sqrt{1/2}(|10\rangle + |01\rangle)$.

Finally, the circuit

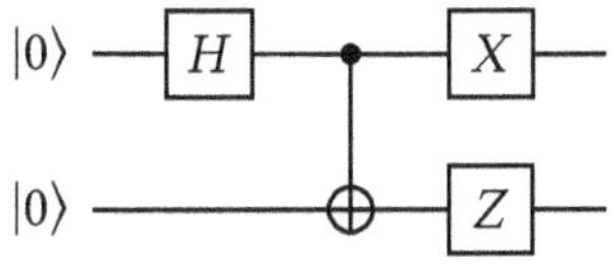

can be used to obtain $\sqrt{1/2}(|10\rangle - |01\rangle)$.

Notice that, to prepare these states, we are only using tensor product gates appended to the circuit that we used to obtain the original Bell state $\sqrt{1/2}(|00\rangle + |11\rangle)$. For instance, we have that

$$\sqrt{1/2}(|10\rangle - |01\rangle) = (X \otimes Z)\sqrt{1/2}(|00\rangle + |11\rangle)$$

and, then, it also holds that

$$(X \otimes Z)\sqrt{1/2}(|10\rangle - |01\rangle) = \sqrt{1/2}(|00\rangle + |11\rangle).$$

If $\sqrt{1/2}(|10\rangle - |01\rangle)$ were a product state, then $\sqrt{1/2}(|00\rangle + |11\rangle)$ would also be a product state. But that is impossible, because we know that $\sqrt{1/2}(|00\rangle + |11\rangle)$ is entangled.

(1.18) We can prove it by induction. We know that the result is true for $n = 1$. Now, assume that it is true for $n > 1$ and consider a basis state $|\psi\rangle$ of $n + 1$ qubits. If $|\psi\rangle = |0\rangle |\psi'\rangle$, then the column vector for $|\psi\rangle$ will start with the elements of the column vector of $|\psi'\rangle$ and then it will have 2^n zeroes. But the column vector for $|\psi'\rangle$ is, by the induction hypothesis, exactly of the form that we are interested in. It follows that $|\psi\rangle$ also has the desired structure. The case when $|\psi\rangle = |1\rangle |\psi'\rangle$ is analogous.

On the other hand, since every n-qubit state can be written as a normalized linear combination of basis states, it follows that its vector representation is a unit length column vector with 2^n coordinates.

(1.19) If we measure the j-th qubit of a generic multi-qubit state, the probability of obtaining 1 is given by

$$\sum_{l \in J_1} |a_l|^2,$$

where J_1 is the set of numbers whose j-th bit is 1. The state after the collapse will be

$$\frac{\sum_{l \in J_1} a_l |l\rangle}{\sqrt{\sum_{l \in J_1} |a_i|^2}}.$$

(1.20) The probability of getting 0 when we measure the second qubit of $(1/2)|100\rangle + (1/2)|010\rangle + \sqrt{1/2}|001\rangle$ is

$$\left|\frac{1}{2}\right|^2 + \left|\frac{1}{\sqrt{2}}\right|^2 = \frac{1}{4} + \frac{1}{2} = \frac{3}{4}.$$

The result after measuring the second qubit and obtaining 0 would be

$$\frac{1}{\sqrt{3}}\left|100\right\rangle + \frac{\sqrt{2}}{\sqrt{3}}\left|001\right\rangle.$$

(1.21) Let's denote $x = x_1 \ldots x_n$ and $y = y_1 \ldots y_n$, where x_i is the i-th bit of x and y_i is the i-th bit of y. Then, it holds that

$$\left\langle y|x\right\rangle = \left\langle y_1|x_1\right\rangle \ldots \left\langle y_n|x_n\right\rangle.$$

As a consequence, $\langle y|x\rangle = 1$ if $x = y$ and $\langle y|x\rangle = 0$ if $x \neq y$. From this, it follows that the elements in $\{|x\rangle\}_{x\in\{0,1\}^n}$ are orthonormal. Since the cardinality of this set is 2^n, which is the dimension of n-qubit states, we can conclude that the set forms a basis.

(1.22) It holds that

$$\begin{aligned}
&\frac{1}{\sqrt{2}}\left(\langle 000| + \langle 111|\right)\frac{1}{2}\left(|000\rangle + |011\rangle + |101\rangle + |110\rangle\right) \\
&\quad = \frac{1}{2\sqrt{2}}(\langle 000|000\rangle + \langle 000|011\rangle + \langle 000|101\rangle + \langle 000|110\rangle + \\
&\qquad \langle 111|000\rangle + \langle 111|011\rangle + \langle 111|101\rangle + \langle 111|110\rangle) \\
&\quad = \frac{1}{2\sqrt{2}},
\end{aligned}$$

because all the inner products are 0 except $\langle 000|000\rangle$, which is 1.

(1.23) From its action of the basis states, we deduce that the matrix for the CCNOT gate is:

$$\begin{pmatrix} 1 & 0 & 0 & 0 & 0 & 0 & 0 & 0 \\ 0 & 1 & 0 & 0 & 0 & 0 & 0 & 0 \\ 0 & 0 & 1 & 0 & 0 & 0 & 0 & 0 \\ 0 & 0 & 0 & 1 & 0 & 0 & 0 & 0 \\ 0 & 0 & 0 & 0 & 1 & 0 & 0 & 0 \\ 0 & 0 & 0 & 0 & 0 & 1 & 0 & 0 \\ 0 & 0 & 0 & 0 & 0 & 0 & 0 & 1 \\ 0 & 0 & 0 & 0 & 0 & 0 & 1 & 0 \end{pmatrix}$$

It holds that this matrix is its own adjoint and that its square is the identity. As a consequence, the matrix is unitary.

(1.24) The circuit

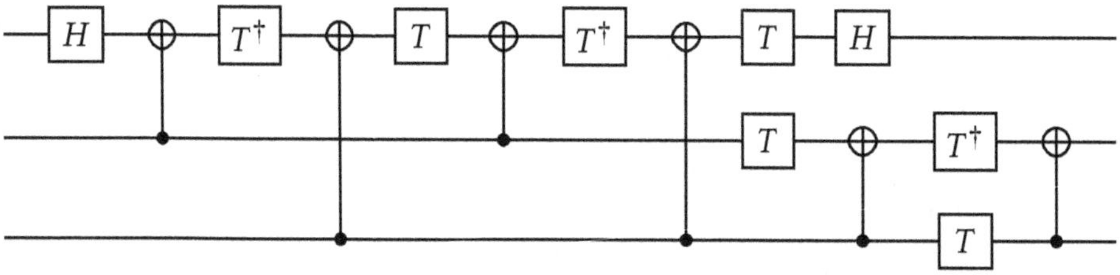

leaves all the states but $|011\rangle$ and $|111\rangle$ unchanged. It also interchanges $|011\rangle$ and $|111\rangle$. That is exactly the action of a CCNOT gate with target on the top qubit.

Chapter 2, The Tools of the Trade in Quantum Computing

(2.1) We already gave you the solution in *Appendix D, Installing the Tools.*

(2.2) In order to construct the circuit in *Figure 2.3b*, you would have to execute the following piece of code:

```
from qiskit import *
import numpy as np
```

```
qc = QuantumCircuit(2)


qc.z(0)
qc.y(1)
qc.cry(np.pi/2, 0, 1)
qc.u(np.pi/4, np.pi, 0, 0)
qc.rz(np.pi/4,1)
```

If you want to visualize the circuit, of course, you can use `qc.draw("mpl")`.

(2.3) You can check IBM's own implementation (`https://github.com/Qiskit/qiskit-terra/blob/5ccf3a41cb10742ae2158b6ee9d13bbb05f64f36/qiskit/circuit/quantumcircuit.py#L2205`) of the method and compare it to your own!

They take some additional steps that we have not considered, such as adding **barriers** in the circuit, but you can ignore those details.

(2.4) You already have the solution in *Appendix D, Installing the Tools.*

(2.5) We have already seen how to construct these circuits in Qiskit. In order to construct them in PennyLane, we would need to run the following piece of code:

```
import pennylane as qml
import numpy as np
dev = qml.device('default.qubit', wires = 2)
@qml.qnode(dev)
def qcircA():
    qml.PauliX(wires = 0)
    qml.RX(np.pi/4, wires = 1)
    qml.CNOT(wires = [0,1])
    qml.U3(np.pi/3, 0, np.pi, wires = 0)
```

```
    return qml.state()


@qml.qnode(dev)
def qcircB():
    qml.PauliZ(wires = 0)
    qml.PauliY(wires = 1)
    qml.CRY(np.pi/2, wires = [0,1])
    qml.U3(np.pi/4, np.pi, 0, wires = 0)
    qml.RZ(np.pi/4, wires = 1)
    return qml.state()
```

If we run this, for instance, for circuit B by executing **print**(qcircB(), we get the following state vector:

```
tensor([ 0.        +0.j        , -0.35355339+0.85355339j,
         0.        +0.j        ,  0.14644661-0.35355339j],
         requires_grad=True)
```

On the other hand, if we simulate that very same circuit with Qiskit, we get this output:

```
Statevector([-5.65831421e-17-3.20736464e-17j,
              2.34375049e-17+1.32853393e-17j,
             -3.53553391e-01+8.53553391e-01j,
              1.46446609e-01-3.53553391e-01j],
            dims=(2, 2))
```

Notice that this is the same result that we got with PennyLane. We first have to take into account the fact that the first two entries are — computationally speaking — zero. And then we have to draw our attention to how Qiskit, following its own conventions, gives us the amplitudes of the basis states in the following order: $|00\rangle$, $|10\rangle$, $|01\rangle$, and $|11\rangle$.

Chapter 3, Working with Quadratic Unconstrained Binary Optimization Problems

(3.1) We can put vertices 0, 1, and 4 in one group, and vertices 2 and 3 in the other. Then, five edges belong to the cut, namely $(0, 2), (1, 2), (1, 3), (2, 4)$, and $(3, 4)$.

(3.2) The optimization problem for the Max-Cut of the graph in *Figure 3.3* is

$$\begin{array}{ll} \text{Minimize} & z_0z_1 + z_0z_2 + z_1z_2 + z_1z_4 + z_2z_3 + z_3z_4 + z_3z_5 + z_4z_5 \\ \text{subject to} & z_j \in \{-1, 1\}, \qquad j = 0, \ldots, 5. \end{array}$$

The value of the cut given by $z_0 = z_1 = z_2 = 1$ and $z_3 = z_4 = z_5 = -1$ is 4. This cut is not optimal, because, for instance, $z_0 = z_1 = z_2 = z_5 = 1$ and $z_3 = z_4 = -1$ achieves a lower value.

(3.3) It holds that $\langle 010| (Z_0Z_1 + Z_0Z_2) |010\rangle = 0$ and that $\langle 100| (Z_0Z_1 + Z_0Z_2) |100\rangle = -2$. This latter value is the minimum possible, because we only have two edges in our graph.

(3.4) We can compute the required expectation values with the following code:

```
from qiskit.quantum_info import Pauli
from qiskit.opflow.primitive_ops import PauliOp
from qiskit.quantum_info import Statevector
H_cut = PauliOp(Pauli("ZZI")) + PauliOp(Pauli("ZIZ"))
for x in range(8): # We consider x=0,1...7
    psi = Statevector.from_int(x, dims = 8)
    print("The expectation value of |",x,">", "is",
        psi.expectation_value(H_cut))
```

If we run it, we obtain the following output:

```
The expectation value of | 0 > is (2+0j)
The expectation value of | 1 > is 0j
```

```
The expectation value of | 2 > is 0j
The expectation value of | 3 > is (-2+0j)
The expectation value of | 4 > is (-2+0j)
The expectation value of | 5 > is 0j
The expectation value of | 6 > is 0j
The expectation value of | 7 > is (2+0j)
```

Thus, we can see that there are two states that obtain the optimal value and both correspond to the cut in which 0 is in one group and 1 and 2 in the other.

(3.5) The QUBO problem would be

$$\begin{aligned} \text{Minimize} \quad & x_0^2 - 4x_0x_1 + 6x_0x_2 - 8x_0x_3 + 4x_1^2 - 12x_1x_2 + 16x_1x_3 + 9x_2^2 \\ & - 24x_2x_3 + 16x_3^2 \\ \text{subject to} \quad & x_j \in \{0,1\}, \qquad j = 0,1,2,3. \end{aligned}$$

The equivalent Ising ground state problem would be

$$\begin{aligned} \text{Minimize} \quad & - z_0z_1 + \frac{3z_0z_2}{2} - 2z_0z_3 + z_0 - 3z_1z_2 + 4z_1z_3 - 2z_1 - 6z_2z_3 + 3z_2 - 4z_3 \\ \text{subject to} \quad & z_j \in \{1,-1\}, \qquad j = 0,1,2,3, \end{aligned}$$

where we have dropped the independent term $\frac{17}{2}$.

(3.6) The binary linear program would be

$$\begin{aligned} \text{Minimize} \quad & - 3x_0 - x_1 - 7x_2 - 7x_3 \\ \text{subject to} \quad & 2x_0 + x_1 + 5x_2 + 4x_3 \leq 8, \\ & x_j \in \{0,1\}, \qquad j = 0,1,2,3. \end{aligned}$$

(3.7) The QUBO problem is

$$\begin{aligned} \text{Minimize} \quad & (x_{00} + x_{01} - 1)^2 + (x_{10} + x_{11} - 1)^2 + (x_{20} + x_{21} - 1)^2 + (x_{30} + x_{31} - 1)^2 \\ & + x_{00}x_{10} + x_{01}x_{11} + x_{00}x_{20} + x_{01}x_{21} + x_{10}x_{30} + x_{11}x_{31} + x_{20}x_{30} + x_{21}x_{31} \\ \text{subject to} \quad & x_{jk} \in \{0, 1\}, \qquad j = 0, 1, 2, 3, k = 0, 1. \end{aligned}$$

(3.8) The expression for the route cost is

$$\begin{aligned} &+2x_{00}x_{11} + x_{00}x_{21} + 3x_{00}x_{31} + 2x_{10}x_{01} + 4x_{10}x_{21} + x_{10}x_{31} + x_{20}x_{01} \\ &+4x_{20}x_{11} + x_{20}x_{31} + 3x_{30}x_{01} + x_{30}x_{11} + x_{30}x_{21} + 2x_{01}x_{12} + x_{01}x_{22} \\ &+3x_{01}x_{32} + 2x_{11}x_{02} + 4x_{11}x_{22} + x_{11}x_{32} + x_{21}x_{02} + 4x_{21}x_{12} + x_{21}x_{32} \\ &+3x_{31}x_{02} + x_{31}x_{12} + x_{31}x_{22} + 2x_{02}x_{13} + x_{02}x_{23} + 3x_{02}x_{33} + 2x_{12}x_{03} \\ &+4x_{12}x_{23} + x_{12}x_{33} + x_{22}x_{03} + 4x_{22}x_{13} + x_{22}x_{33} + 3x_{32}x_{03} + x_{32}x_{13} + x_{32}x_{23}. \end{aligned}$$

Chapter 4, Adiabatic Quantum Computing and Quantum Annealing

(4.1) We first consider a state $|x\rangle = |x_0\rangle |x_1\rangle \cdots |x_{n-1}\rangle$ where each $|x_j\rangle$ is either $|+\rangle$ or $|-\rangle$. The set of all 2^n such states for an orthonormal basis and, hence, any generic state $|\psi\rangle$ can be written as $|\psi\rangle = \sum_x a_x |x\rangle$, where $\sum |a_x|^2 = 1$.

Then, for each j, it holds that

$$\langle x| X_j |x\rangle = \langle x_j| X_j |x_j\rangle .$$

But $\langle x_j| X_j |x_j\rangle$ is 1 if $|x_j\rangle = |+\rangle$ and it is -1 if $|x_j\rangle = |-\rangle$. Hence, it holds that

$$\langle \psi| X_j |\psi\rangle = \sum_x |a_x|^2 \langle x| X_j |x\rangle \leq 1,$$

because $|a_x|^2 \geq 0$ for every x and $\sum |a_x|^2 = 1$.

Then, since $H_0 = -\sum_{j=0}^{n-1} X_j$, by linearity we have

$$\langle\psi| H_0 |\psi\rangle = -\sum_{j=0}^{n-1} \langle\psi| X_j |\psi\rangle \geq -n.$$

On the other hand, if we consider $|\psi_0\rangle = \bigotimes_{i=0}^{n-1} |+\rangle$, by the previous reasoning we have that $\langle\psi_0| X_j |\psi_0\rangle = 1$. Hence, $\langle\psi_0| H_0 |\psi_0\rangle = -n$, which is the minimum possible value and, hence, $|\psi_0\rangle$ is the ground state that we were looking for.

(4.2) We can define the QUBO problem of minimizing $x_0x_2 - x_0x_1 + 2x_1$ with the following code:

```
import dimod
J = {(0,1):-1, (0,2):1}
h = {1:2}
problem = dimod.BinaryQuadraticModel(h, J, 0.0, dimod.BINARY)
print("The problem we are going to solve is:")
print(problem)
```

We can then solve it with the following:

```
from dwave.system import DWaveSampler
from dwave.system import EmbeddingComposite
sampler = EmbeddingComposite(DWaveSampler())
result = sampler.sample(problem, num_reads=10)
print("The solutions that we have obtained are")
print(result)
```

(4.3) For simplicity, we will denote the slack variables as s_0 and s_1. Then, the penalty term is $(y_0 + 2y_1 + s_0 + s_1 - 2)^2$. When you multiply by 5, expand it, and add it to the cost function, you obtain exactly the expression computed by the `cqm_to_bqm` method.

(4.4) For qubits 0 through 7, we have the following connections:

```
{0: {4, 5, 6, 7, 128}, 1: {4, 5, 6, 7, 129},
 2: {4, 5, 6, 7, 130}, 3: {4, 5, 6, 7, 131},
 4: {0, 1, 2, 3, 12}, 5: {0, 1, 2, 3, 13},
 6: {0, 1, 2, 3, 14}, 7: {0, 1, 2, 3, 15}}
```

Clearly, each vertex from 0 to 3 is connected to each from 4 to 7, as we need. Moreover, each vertex from 0 to 3 is connected to one vertex from 128 to 131, which are in the cell below the first one, and each vertex from 4 to 7 is connected to one vertex from 12 to 15, which are in the cell to the right of the first one.

(4.5) You can easily check those values with the following instructions:

```
sampler = DWaveSampler(solver = "DW_2000Q_6")
print("The default annealing time is",
    sampler.properties["default_annealing_time"],"microsends")
print("The possible values for the annealing time (in microseconds)"\
    " lie in the range",sampler.properties["annealing_time_range"])
```

The output, in this case, will be as follows:

```
The default annealing time is 20.0 microsends
The possible values for the annealing time (in microseconds)
    lie in the range [1.0, 2000.0]
```

Chapter 5, QAOA: Quantum Approximate Optimization Algorithm

(5.1) The QAOA circuit for $Z_1Z_3 + Z_0Z_2 - 2Z_1 + 3Z_2$ with $p = 1$ is the following:

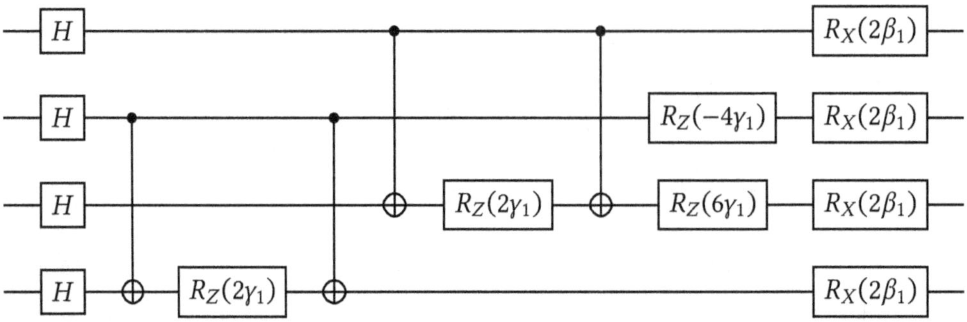

(5.2) It holds that

$$\langle 100| H_1 |100\rangle = 3\langle 100| Z_0Z_2 |100\rangle - \langle 100| Z_1Z_2 |100\rangle + 2\langle 100| Z_0 |100\rangle = -3 - 1 - 2 = -6.$$

(5.3) We can rewrite the problem as

$$\begin{aligned} &\text{Minimize} \quad (1-x_0)(1-x_1)x_2(1-x_3) + x_0(1-x_1)(1-x_2)(1-x_3) + x_0(1-x_1)x_2x_3 \\ &\text{subject to} \quad x_j \in \{0,1\}, \qquad j = 0,1,2,3. \end{aligned}$$

(5.4) The operation can be implemented with the following circuit:

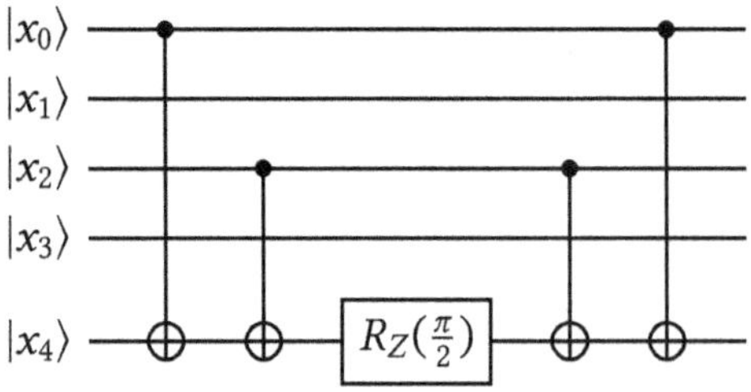

(5.5) It holds that

$$\langle 100| H_1 |100\rangle = \langle 100| Z_0Z_1Z_2 |100\rangle + 3\langle 100| Z_0Z_2 |100\rangle - \langle 100| Z_1Z_2 |100\rangle + 2\langle 100| Z_0 |100\rangle = -1 - 3 - 1 - 2 = -7.$$

(5.6) You can use the following code to obtain reproducible results:

```
from qiskit import Aer
from qiskit.algorithms import QAOA
from qiskit.algorithms.optimizers import COBYLA
from qiskit.utils import algorithm_globals, QuantumInstance
from qiskit_optimization.algorithms import MinimumEigenOptimizer

seed = 1234
algorithm_globals.random_seed = seed
quantum_instance = QuantumInstance(Aer.get_backend("aer_simulator"),
    shots = 1024, seed_simulator=seed, seed_transpiler=seed)
qaoa = QAOA(optimizer = COBYLA(),
            quantum_instance=quantum_instance, reps = 1)
qaoa_optimizer = MinimumEigenOptimizer(qaoa)
result = qaoa_optimizer.solve(qp)
print('Variable order:', [var.name for var in result.variables])
for s in result.samples:
    print(s)
```

(5.7) We can define the $-3Z_0Z_1Z_2 + 2Z_1Z_2 - Z_2$ Hamiltonian using the following instructions:

```
coefficients = [-3,2,-1]
paulis = [PauliZ(0)@PauliZ(1)@PauliZ(2),
    PauliZ(1)@PauliZ(2),PauliZ(2)]
```

```
H = qml.Hamiltonian(coefficients,paulis)
```

We can also use

```
H = -3*PauliZ(0)@PauliZ(1)@PauliZ(2)
    + 2*PauliZ(1)@PauliZ(2) -PauliZ(2)
```

Chapter 6, GAS: Grover Adaptative Search

(6.1) From our definition, O_f always takes one basis state to another basis state. Hence, in matrix representation, its columns are vectors in which exactly one element is 1 and the rest are 0. This means that, in particular, all its entries are real.

What is more, this matrix is symmetric. To prove this, suppose that the matrix has entries m_{jk}. If it is not symmetric, there exist j, k such that $m_{jk} \neq m_{kj}$. We can suppose, without loss of generality, that $m_{jk} = 0$ and $m_{kj} = 1$. We also know that $O_f O_f = I$, so the square of the matrix is the identity. In particular, $\sum_l m_{jl} m_{lj} = 1$, because this is the element in row j, column j of the square of the matrix. But we know that $m_{jk} m_{kj} = 0 \cdot 1 = 0$ and that $m_{lj} = 0$ if $l \neq k$, because there is a single 1 in each column. Nevertheless, then, $\sum_l m_{jl} m_{lj} = 0$, which is a contradiction.

Thus, we have $O_f^\dagger = O_f$ and, since $O_f O_f = I$, it follows that O_f is unitary.

(6.2) We can use the following circuit:

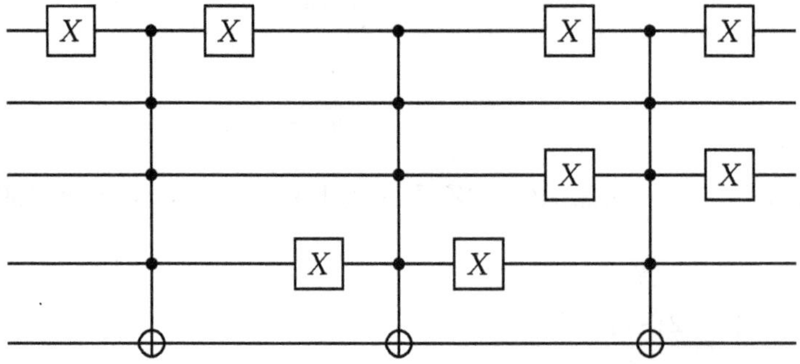

(6.3) The representation of 10 is 01010 and the representation of −7 is 11001. Their addition is 00011, which encodes 3.

(6.4) We can use the following circuit:

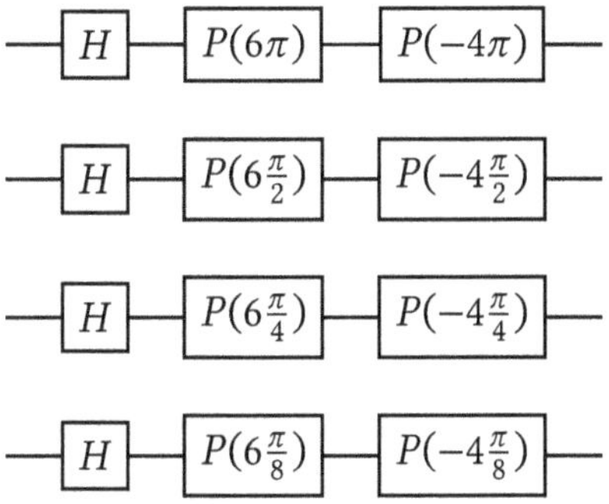

(6.5) We can use the following circuit:

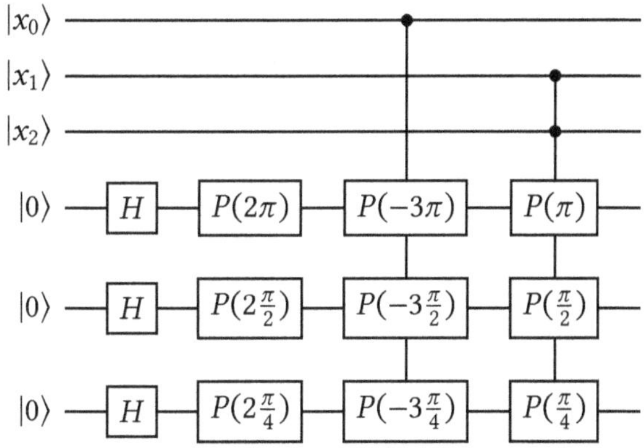

(6.6) We can use the following code:

```
from qiskit_optimization.problems import QuadraticProgram
from qiskit_optimization.algorithms import GroverOptimizer
from qiskit import Aer
from qiskit.utils import algorithm_globals, QuantumInstance
seed = 1234
algorithm_globals.random_seed = seed
```

```
qp = QuadraticProgram()
qp.binary_var('x')
qp.binary_var('y')
qp.binary_var('z')
qp.minimize(linear = {'x':3,'y':2,'z':-3}, quadratic = {('x','y'):3})

quantum_instance = QuantumInstance(Aer.get_backend("aer_simulator"),
    shots = 1024, seed_simulator = seed, seed_transpiler=seed)
grover_optimizer = GroverOptimizer(num_value_qubits = 5,
    num_iterations=4, quantum_instance=quantum_instance)
results = grover_optimizer.solve(qp)
print(results)
```

Chapter 7, VQE: Variational Quantum Eigensolver

(7.1) This all follows from the fact that the matrix is diagonal and all its diagonal entries are different, with eigenvalues corresponding to the actual labels of the measurement outcomes.

Remember that, if the coordinate matrix of an operator with respect to a basis is diagonal, this means that the basis vectors are eigenvectors of the operator and, what is more, the corresponding eigenvalues are found on the diagonal.

(7.2) We know that

$$(A_1 \otimes \cdots \otimes A_n) |\lambda_1\rangle \otimes \cdots \otimes |\lambda_n\rangle = A_1 |\lambda_1\rangle \otimes \cdots \otimes A_n |\lambda_n\rangle .$$

Since $A_j |\lambda_j\rangle = \lambda_j |\lambda_j\rangle$, the result follows directly.

(7.3) It holds that $Z|0\rangle = |0\rangle$, $Z|1\rangle = -|1\rangle$, $X|+\rangle = |+\rangle$, $X|-\rangle = -|-\rangle$, $Y\left(1/\sqrt{2}\right)(|0\rangle + i|1\rangle) = \left(1/\sqrt{2}\right)(|0\rangle + i|1\rangle)$, and $Y\left(1/\sqrt{2}\right)(|0\rangle - i|1\rangle) = -\left(1/\sqrt{2}\right)(|0\rangle - i|1\rangle)$.

Since it is also true that $I|\psi\rangle = |\psi\rangle$ for any $|\psi\rangle$, the result follows.

(7.4) A possible orthonormal basis of eigenvectors of $Z \otimes I \otimes X$ is formed by $|0\rangle|0\rangle|+\rangle$, $|0\rangle|1\rangle|-\rangle$, $|1\rangle|0\rangle|-\rangle$, $|1\rangle|1\rangle|+\rangle$, $|0\rangle|0\rangle|-\rangle$, $|0\rangle|1\rangle|+\rangle$, $|1\rangle|0\rangle|+\rangle$ and $|1\rangle|1\rangle|-\rangle$. The first four eigenvectors are associated with the 1 eigenvalue, and the rest with the -1 eigenvalue.

Let's denote, for simplicity, $|i\rangle = \left(1/\sqrt{2}\right)(|0\rangle + i|1\rangle)$ and $|-i\rangle = \left(1/\sqrt{2}\right)(|0\rangle - i|1\rangle)$. Then, a possible orthonormal basis of eigenvalues for $I \otimes Y \otimes Y$ is $|0\rangle|i\rangle|i\rangle$, $|0\rangle|-i\rangle|-i\rangle$, $|1\rangle|i\rangle|-i\rangle$, $|1\rangle|-i\rangle|i\rangle$, $|0\rangle|i\rangle|-i\rangle$, $|0\rangle|-i\rangle|i\rangle$, $|1\rangle|i\rangle|i\rangle$ and $|1\rangle|-i\rangle|-i\rangle$. The first four eigenvectors are associated to the 1 eigenvalue and the rest, to the -1 eigenvalue.

(7.5) It holds that $H|0\rangle = |+\rangle$ and $H|1\rangle = |-\rangle$. This proves that H takes the computational basis to the eigenvectors of X. Also, $SH|0\rangle = \left(1/\sqrt{2}\right)(|0\rangle + i|1\rangle)$ and $SH|1\rangle = \left(1/\sqrt{2}\right)(|0\rangle - i|1\rangle)$, so SH takes the computational basis to the eigenvectors of Y.

(7.6) This follows directly from the fact that if $\{|u_j\rangle\}_j$ and $\{|v_k\rangle\}_k$ are eigenvector bases of A_1 and A_2, respectively, then $\{|u_j\rangle \otimes |v_k\rangle\}_{j,k}$ is an eigenvector basis of $A_1 \otimes A_2$.

(7.7) The Hamiltonian for our problem is

$$H = Z_0Z_1 + Z_1Z_2 + Z_2Z_3 + Z_3Z_4 + Z_4Z_0.$$

Then, we can use the following code to solve it with VQE:

```
from qiskit.circuit.library import EfficientSU2
from qiskit.algorithms import VQE
from qiskit import Aer
from qiskit.utils import QuantumInstance
import numpy as np
from qiskit.algorithms.optimizers import COBYLA
from qiskit.opflow import Z, I

seed = 1234
```

```
np.random.seed(seed)

H= (Z^Z^I^I^I) + (I^Z^Z^I^I) + (I^I^Z^Z^I) + (I^I^I^Z^Z) + (Z^I^I^I^Z)

ansatz = EfficientSU2(num_qubits=5, reps=1, entanglement="linear",
    insert_barriers = True)
optimizer = COBYLA()
initial_point = np.random.random(ansatz.num_parameters)
quantum_instance = QuantumInstance(backend =
    Aer.get_backend('aer_simulator_statevector'))
vqe = VQE(ansatz=ansatz, optimizer=optimizer,
    initial_point=initial_point,
    quantum_instance=quantum_instance)
result = vqe.compute_minimum_eigenvalue(H)
print(result)
```

(7.8) We can use the following code:

```
from qiskit_nature.drivers import Molecule
from qiskit_nature.drivers.second_quantization import \
    ElectronicStructureMoleculeDriver, ElectronicStructureDriverType
from qiskit_nature.problems.second_quantization import \
    ElectronicStructureProblem
from qiskit_nature.converters.second_quantization import QubitConverter
from qiskit_nature.mappers.second_quantization import JordanWignerMapper
from qiskit_nature.algorithms import VQEUCCFactory
from qiskit import Aer
from qiskit.utils import QuantumInstance
from qiskit_nature.algorithms import GroundStateEigensolver
import matplotlib.pyplot as plt
```

```
import numpy as np

quantum_instance = QuantumInstance(
    backend = Aer.get_backend('aer_simulator_statevector'))

vqeuccf = VQEUCCFactory(quantum_instance = quantum_instance)

qconverter = QubitConverter(JordanWignerMapper())
solver = GroundStateEigensolver(qconverter, vqeuccf)

energies = []
distances = np.arange(0.2, 2.01, 0.01)
for d in distances:
  mol = Molecule(geometry=[['H', [0., 0., -d/2]],
                            ['H', [0., 0., d/2]]])

  driver = ElectronicStructureMoleculeDriver(mol, basis='sto3g',
        driver_type=ElectronicStructureDriverType.PYSCF)
  problem = ElectronicStructureProblem(driver)
  result = solver.solve(problem)
  energies.append(result.total_energies)

plt.plot(distances, energies)
plt.title('Dissociation profile')
plt.xlabel('Distance')
plt.ylabel('Energy');
```

(7.9) We can use the following code:

```
from qiskit import *
```

```
from qiskit.providers.aer import AerSimulator
from qiskit.utils.mitigation import CompleteMeasFitter
from qiskit.utils import QuantumInstance

provider = IBMQ.load_account()
backend = AerSimulator.from_backend(
    provider.get_backend('ibmq_manila'))

shots = 1024

qc = QuantumCircuit(2,2)
qc.h(0)
qc.cx(0,1)
qc.measure(range(2),range(2))

result = execute(qc, backend, shots = shots)
print("Result of noisy simulation:")
print(result.result().get_counts())

quantum_instance = QuantumInstance(
    backend = backend, shots = shots,
    measurement_error_mitigation_cls=CompleteMeasFitter)

result = quantum_instance.execute(qc)
print("Result of noisy simulation with error mitigation:")
print(result.get_counts())
```

Our results when running these instructions were the following:

```
Result of noisy simulation:
```

```
{'01': 88, '10': 50, '00': 453, '11': 433}
Result of noisy simulation with error mitigation:
{'00': 475, '01': 12, '10': 14, '11': 523}
```

We know that the ideal result of running this circuit should not produce any 01 or 10 measurement. These are present quite prominently in the noisy simulation, but not so much when we use readout error mitigation.

(7.10) We can use the following code:

```
from qiskit.opflow import Z
from qiskit.providers.aer import AerSimulator
from qiskit.algorithms import QAOA
from qiskit.utils import QuantumInstance
from qiskit import Aer, IBMQ
from qiskit.algorithms.optimizers import COBYLA
from qiskit.utils.mitigation import CompleteMeasFitter

H1 = Z^Z

provider = IBMQ.load_account()
backend = AerSimulator.from_backend(
    provider.get_backend('ibmq_manila'))

quantum_instance = QuantumInstance(backend=backend,
                        shots = 1024)
qaoa = QAOA(optimizer = COBYLA(), quantum_instance=quantum_instance)
result = qaoa.compute_minimum_eigenvalue(H1)
print("Result of noisy simulation:",result.optimal_value)

quantum_instance = QuantumInstance(backend=backend,
```

```
    measurement_error_mitigation_cls=CompleteMeasFitter,
    shots = 1024)
qaoa = QAOA(optimizer = COBYLA(), quantum_instance=quantum_instance)
result = qaoa.compute_minimum_eigenvalue(H1)
print("Result of noisy simulation with error mitigation:",
    result.optimal_value)
```

The results that we obtained when we ran it were the following:

```
Result of noisy simulation: -0.8066406250000001
Result of noisy simulation with error mitigation: -0.93359375
```

We know that the actual optimal value for our Hamiltonian is -1. We observe, then, that noise has a negative effect on the performance of QAOA in this case and that it is reduced by the use of readout error mitigation.

Chapter 8, What is Quantum machine Learning?

(8.1) Let's proceed by a proof by contradiction. We will assume that there exist some coefficients w_1, w_2, b such that

$$0w_1 + 1w_2 + b = 1, \qquad 1w_1 + 0w_2 + b = 1,$$

$$0w_1 + 0w_2 + b = 0, \qquad 1w_1 + 1w_2 + b = 0.$$

Simplifying, these are equivalent to

$$w_2 + b = 1, \qquad w_1 + b = 1, \qquad b = 0, \qquad w_1 + w_2 + b = 0.$$

The first three identities imply that $b = 0$ and $w_1 = w_2 = 1$, hence the last identity cannot be satisfied.

(8.2) Histograms are usually versatile and powerful options. In this case, however, since our dataset has two features, we could have also drawn a scatter plot using the `plt.scatter` function.

(8.3) The function is clearly strictly increasing, for its derivative is

$$\frac{e^x}{(e^x + 1)^2} > 0.$$

Moreover, it is immediate that $\lim_{x \to \infty} S(x) = 1$ and $\lim_{x \to -\infty} S(x) = 0$.

The ELU function is smooth because the derivative of x at 0 is 1 and so is that of $e^x - 1$ at 0. Both functions are strictly increasing and $\lim_{x \to \infty} x = \infty$ and $\lim_{x \to -\infty} e^x - 1 = -1$.

The image of the ReLU function is, clearly, $[0, \infty)$. It is not smooth because $(x)' = 1$ yet $(0)' = 0$.

(8.4) Without loss of generality, we will assume $y = 1$ (the case $y = 0$ is fully analogous). If $M_\theta(x) = y = 1$, then $H(\theta; x, y) = -1 \log(1) + 0 = 0$ because $1 - y$ is 0 for any value of x. On the other hand, when $M_\theta(x) \to 0$, then $-\log(M_\theta(x)) \to \infty$, hence H also diverges.

(8.5) We can plot the losses using the following piece of code:

```
val_loss = history.history["val_loss"]
train_loss = history.history["loss"]
epochs = range(len(train_loss))
plt.plot(epochs, train_loss, label = "Training loss")
plt.plot(epochs, val_loss, label = "Validation loss")
plt.legend()
plt.show()
```

(8.6) The result is less accurate when we decrease the learning rate without increasing the number of epochs, because the algorithm can't take enough steps to reach a minimum. We also get worse results when we reduce the training dataset to 20, because we have overfitting. This can be identified by looking at the evolution of the validation loss and noticing how it skyrockets while the training loss plummets.

Chapter 9, Quantum Support Vector Machines

(9.1) We will prove that the distance between a hyperplane H_1, characterized by either $\vec{w}\cdot\vec{x}+b=1$ or $\vec{w}\cdot\vec{x}+b=-1$, and H_0, given by $\vec{w}\cdot\vec{x}+b=0$, is $1/\|w\|$. The result will then follow from the fact that $\vec{w}\cdot\vec{x}+b=\pm 1$ are projections of each other over H_0.

Let us consider a point $\vec{x}_0 \in H_0$. The distance between H_0 and H_1 will be the length of the only vector in the direction normal to H_0 that connects $\vec{x}_0$ to a point in H_1. What is more, since $\vec{w}$ is normal to H_0, such a vector needs to be of the form $\alpha\vec{w}$ for some scalar α. Let us find that scalar.

We know that $\vec{x}_0 + \alpha\vec{x}_1 \in H_1$, so we must have

$$\vec{w}\cdot\left(\vec{x}_0 + \alpha\vec{w}\right) + b = 1.$$

But taking into account that $x_0 \in H_0$ and, therefore, $\vec{w}\cdot\vec{x}_0 + b = 0$, this can be further simplified to

$$\vec{w}\cdot\alpha\vec{w} = 1 \iff \alpha = \frac{1}{\|w\|^2}.$$

The length of $\alpha\vec{w}$ will just be $|\alpha|\cdot\|\vec{w}\|$, which is $1/\|w\|$, just as we wanted to prove.

(9.2) Let's assume that the kernel function is defined as $k(a,b) = |\langle\varphi(a)|\varphi(b)\rangle|^2$. The inner product in $\mathbb{C}^n$ is conjugate-symmetric, so we must have

$$k(b,a) = |\langle\varphi(b)|\varphi(a)\rangle|^2 = \left|\overline{\langle\varphi(a)|\varphi(b)\rangle}\right|^2 = |\langle\varphi(a)|\varphi(b)\rangle|^2 = k(a,b).$$

(9.3) Quantum states need to be normalized and, therefore, the scalar product with themselves must be one.

(9.4) The following piece of code would implement that function:

```
from qiskit import *
from qiskit . circuit import ParameterVector


def AngleEncodingX(n):
    x = ParameterVector("x", length = n)
    qc = QuantumCircuit(n)
    for i in range(n):
        qc.rx(parameter[i], i)
    return qc
```

Chapter 10, Quantum Neural Networks

(10.1) It suffices to remember that, while kets are represented by column matrices, bras are represented by row matrices. In this way,

$$|0\rangle\langle 0| = \begin{pmatrix} 1 \\ 0 \end{pmatrix} \begin{pmatrix} 1 & 0 \end{pmatrix} = \begin{pmatrix} 1 & 0 \\ 0 & 0 \end{pmatrix}.$$

Analogously,

$$|1\rangle\langle 1| = \begin{pmatrix} 0 \\ 1 \end{pmatrix} \begin{pmatrix} 0 & 1 \end{pmatrix} = \begin{pmatrix} 0 & 0 \\ 0 & 1 \end{pmatrix}.$$

The results follow trivially from this.

(10.2) The process would be fully analogous. The only differences would be found in the definition of the network, the device, and the weights dictionary, which could be the following:

```
nqubits = 5
dev = qml.device("default.qubit", wires=nqubits)


def qnn_circuit(inputs, theta):
    qml.AmplitudeEmbedding(features = [a for a in inputs],
```

```
        wires = range(nqubits), normalize = True, pad_with = 0.)
    TwoLocal(nqubits = nqubits, theta = theta, reps = 2)
    return qml.expval(qml.Hermitian(M, wires = [0]))

qnn = qml.QNode(qnn_circuit, dev, interface="tf")

weights = {"theta": 15}
```

Also, remember that you should train this model on the original dataset (`x_tr` and `y_tr`), not on the reduced one!

(10.3) On quantum hardware, one may use – for most return types – the finite differences method and the parameter shift rule. On simulators – under certain conditions – one may use these methods in addition to backpropagation and adjoint differentiation.

Chapter 11, The Best of Both Worlds: Hybrid Architectures

(11.1) In order to include the extra classical layers, we would have to execute the same code but define the model in the following manner:

```
model = tf.keras.models.Sequential([
    tf.keras.layers.Input(20),
    tf.keras.layers.Dense(16, activation = "elu"),
    tf.keras.layers.Dense(8, activation = "elu"),
    tf.keras.layers.Dense(4, activation = "sigmoid"),
    qml.qnn.KerasLayer(qnn, weights, output_dim=1)
])
```

This model has similar performance after training. The addition of the classical layers doesn't make a very significant difference.

(11.2) In order to have optimized both the learning rate and the batch size, we could have defined the objective function as follows:

```
def objective(trial):
    # Define the learning rate as an optimizable parameter.
    lrate = trial.suggest_float("learning_rate", 0.001, 0.1)
    bsize = trial.suggest_int("batch_size", 5, 50)

    # Define the optimizer with the learning rate.
    opt = tf.keras.optimizers.Adam(learning_rate = lrate)

    # Prepare and compile the model.
    model = tf.keras.models.Sequential([
        tf.keras.layers.Input(20),
        tf.keras.layers.Dense(4, activation = "sigmoid"),
        qml.qnn.KerasLayer(qnn, weights, output_dim=1)
    ])
    model.compile(opt, loss=tf.keras.losses.BinaryCrossentropy())

    # Train it!
    history = model.fit(x_tr, y_tr, epochs = 50, shuffle = True,
        validation_data = (x_val, y_val),
        batch_size = bsize,
        callbacks = [earlystop],
        verbose = 0 # We want TensorFlow to be quiet.
    )

    # Return the validation accuracy.
    return accuracy_score(model.predict(x_val) >= 0.5, y_val)
```

(11.3) We could try to find a numerical approximation of $f(x) = (x-3)^2$ using Optuna as follows:

```
import optuna
from optuna.samplers import TPESampler


seed = 1234


def objective(trial):
    x = trial.suggest_float("x", -10, 10)
    return (x-3)**2


study = optuna.create_study(direction='minimize',
sampler=TPESampler(seed = seed))
study.optimize(objective, n_trials=100)
```

Of course, Optuna was not designed with (general) function minimization in mind, but it was conceived solely as a hyperparameter optimizer.

(11.4) Let $y = 0, 1$ be the expected label. It suffices to notice that y in one-hot form is $(1-y, y)$ and that the probability assigned by the model to x being of class y is $N_\theta(x)_j$. Hence

$$\begin{aligned} H(\theta; x, (1-y, y)) &= \sum_j -y_j \log\big(N_\theta(x)_j\big) \\ &= -(1-y)\log(N_\theta(x)_0) - y\log(N_\theta(x)_1) \\ &= -(1-y)\log(1 - N_\theta(x)_1) - y\log(N_\theta(x)_1), \end{aligned}$$

where we assume that $N_\theta(x)$ is normalized and, therefore, $N_\theta(x)_0 + N_\theta(x)_1 = 1$. The result now follows from the fact that, in binary cross entropy, the probability that we consider is that of assigning label 1, that is, $N_(\theta)_1$.

(11.5) We just have to use the y targets instead of the y_hot targets when preparing the datasets and then call the sparse categorical cross-entropy loss given in the statement when compiling the model.

(11.6) You can create a suitable dataset with 1000 samples and 20 features with the following instruction:

```
x, y = make_regression(n_samples = 1000, n_features = 20)
```

Then, you can construct the model as follows:

```
nqubits = 4
dev = qml.device("lightning.qubit", wires = nqubits)

@qml.qnode(dev, interface="tf", diff_method = "adjoint")
def qnn(inputs, theta):
    qml.AngleEmbedding(inputs, range(nqubits))
    TwoLocal(nqubits, theta, reps = 2)
    return [qml.expval(qml.Hermitian(M, wires = [0]))]

weights = {"theta": 12}

model = tf.keras.models.Sequential([
    tf.keras.layers.Input(20),
    tf.keras.layers.Dense(16, activation = "elu"),
    tf.keras.layers.Dense(8, activation = "elu"),
    tf.keras.layers.Dense(4, activation = "sigmoid"),
    qml.qnn.KerasLayer(qnn, weights, output_dim=1),
    tf.keras.layers.Dense(1)
])
```

Then it would be trained like any of our previous models using the MSE loss function, which you can access with `tf.keras.losses.MeanSquaredError()`.

Chapter 12, Quantum Generative Adversarial Networks

(12.1) (1) QSVMs, QNNs or Hybrid QNNs (2) QGANs. (3) QSVMs, QNNs or Hybrid QNNs. (4) QSVMs, QNNs or Hybrid QNNs. (5) QGANs.

(12.2) The steps to produce the solution are analogous to what we did in the main text; it's only necessary to change the values of the angles that define the state $|\psi_1\rangle$, and, possibly, to increase the number of training cycles.

(12.3) Let's consider a point (x, y, z) in the Bloch sphere. Its spherical coordinates are (θ, φ) such that

$$(x, y, z) = (\sin\theta\cos\varphi, \sin\theta\sin\varphi, \cos\theta)$$

and a state with those spherical Bloch sphere coordinates is in the state

$$|\psi\rangle = \cos(\theta/2)\,|0\rangle + e^{i\varphi}\sin(\theta/2)\,|1\rangle\,.$$

The expectation value of $Z = |0\rangle\langle 0| - |1\rangle\langle 1|$ in the state $|\psi\rangle$ is

$$\langle\psi|\,Z\,|\psi\rangle = \cos^2(\theta/2) - e^{-i\varphi+i\varphi}\sin^2(\theta/2) = \cos\theta = z.$$

Regarding the expectation value $Y = i\,|1\rangle\langle 0| - i\,|0\rangle\langle 1|$, we have

$$\begin{aligned}\langle\psi|\,Y\,|\psi\rangle &= ie^{-i\varphi}\sin(\theta/2)\cos(\theta/2) - ie^{i\varphi}\sin(\theta/2)\cos(\theta/2)\\ &= i(e^{-i\varphi} - e^{i\varphi})(\sin(\theta/2)\cos(\theta/2))\\ &= \sin\varphi\cdot 2\sin(\theta/2)\cos(\theta/2) = \sin\varphi\sin\theta = y.\end{aligned}$$

Lastly, in regard to the expectation value of $X = |1\rangle\langle 0| + |0\rangle\langle 1|$,

$$\begin{aligned}\langle\psi| X |\psi\rangle &= e^{-i\varphi}\sin(\theta/2)\cos(\theta/2) + e^{i\varphi}\sin(\theta/2)\cos(\theta/2)\\ &= (e^{-i\varphi} + e^{i\varphi})(\sin(\theta/2)\cos(\theta/2))\\ &= \cos\varphi\cdot 2\sin(\theta/2)\cos(\theta/2) = \cos\varphi\sin\theta = x.\end{aligned}$$

Bibliography

[1] E. F. Combarro, S. Vallecorsa, L. J. Rodríguez-Muñiz, Á. Aguilar-González, J. Ranilla, and A. Di Meglio, "A report on teaching a series of online lectures on quantum computing from CERN," *The Journal of Supercomputing*, vol. 77, no. 12, pp. 14 405–14 435, 2021.

[2] C. H. Bennett, G. Brassard, C. Crépeau, R. Jozsa, A. Peres, and W. K. Wootters, "Teleporting an unknown quantum state via dual classical and einstein-podolsky-rosen channels," *Physical review letters*, vol. 70, no. 13, p. 1895, 1993.

[3] D. Bouwmeester, J.-W. Pan, K. Mattle, M. Eibl, H. Weinfurter, and A. Zeilinger, "Experimental quantum teleportation," *Nature*, vol. 390, no. 6660, pp. 575–579, 1997.

[4] C. H. Bennett and G. Brassard, "Quantum cryptography: Public key distribution and coin tossing," in *Proceedings of IEEE International Conference on Computers, Systems, and Signal Processing*, Bangalore, 1984, p. 175.

[5] D. Deutsch and R. Jozsa, "Rapid solution of problems by quantum computation," *Proceedings of the Royal Society of London. Series A: Mathematical and Physical Sciences*, vol. 439, no. 1907, pp. 553–558, 1992.

[6] P. W. Shor, "Polynomial-time algorithms for prime factorization and discrete logarithms on a quantum computer," *SIAM Review*, vol. 41, no. 2, pp. 303–332, 1999.

[7] R. S. Sutor, *Dancing with Qubits: How quantum computing works and how it can change the world.* Packt Publishing Ltd, 2019.

[8] D. R. Simon, "On the power of quantum computation," *SIAM journal on computing*, vol. 26, no. 5, pp. 1474–1483, 1997.

[9] L. K. Grover, "A fast quantum mechanical algorithm for database search," in *Proceedings of the twenty-eighth annual ACM symposium on Theory of computing*, 1996, pp. 212–219.

[10] F. Arute, K. Arya, R. Babbush, *et al.*, "Quantum supremacy using a programmable superconducting processor," *Nature*, vol. 574, no. 7779, pp. 505–510, 2019.

[11] E. Pednault, J. Gunnels, D. M. Maslov, and J. Gambetta, *On "Quantum Supremacy"*, `https://www.ibm.com/blogs/research/2019/10/on-quantum-supremacy/`.

[12] F. Pan, K. Chen, and P. Zhang, "Solving the sampling problem of the Sycamore quantum circuits," *Physical Review Letters*, vol. 129, no. 9, p. 090 502, 2022.

[13] R. L. Rivest, A. Shamir, and L. Adleman, "A method for obtaining digital signatures and public-key cryptosystems," *Commun. ACM*, vol. 21, no. 2, pp. 120–126, 1978.

[14] A. W. Harrow, A. Hassidim, and S. Lloyd, "Quantum algorithm for linear systems of equations," *Physical Review Letters*, vol. 103, no. 15, p. 150 502, 2009.

[15] J. Preskill, "Quantum computing in the NISQ era and beyond," *Quantum*, vol. 2, p. 79, 2018.

[16] M. A. Nielsen and I. L. Chuang, *Quantum Computation and Quantum Information: 10th Anniversary Edition*. Cambridge University Press, 2011.

[17] A. Acín and L. Masanes, "Certified randomness in quantum physics," *Nature*, vol. 540, no. 7632, pp. 213–219, 2016.

[18] A. Einstein and M. Born, *Born-Einstein Letters 1916-1955: Friendship, Politics and Physics in Uncertain Times*. Springer, 2014.

[19] A. Aspect, J. Dalibard, and G. Roger, "Experimental test of Bell's inequalities using time-varying analyzers," *Physical Review Letters*, vol. 49, pp. 1804–1807, 25 1982.

[20] J. F. Clauser, M. A. Horne, A. Shimony, and R. A. Holt, "Proposed experiment to test local hidden-variable theories," *Physical Review Letters*, vol. 23, no. 15, p. 880, 1969.

[21] S. J. Freedman and J. F. Clauser, "Experimental test of local hidden-variable theories," *Physical Review Letters*, vol. 28, no. 14, p. 938, 1972.

[22] M. S. ANIS, Abby-Mitchell, H. Abraham, *et al.*, *Qiskit: An open-source framework for quantum computing*, 2021. DOI: `10.5281/zenodo.2573505`.

[23] V. Bergholm, J. Izaac, M. Schuld, *et al.*, "PennyLane: Automatic differentiation of hybrid quantum-classical computations," *arXiv preprint arXiv:1811.04968*, 2018.

[24] J. Håstad, "Some optimal inapproximability results," *Journal of the ACM (JACM)*, vol. 48, no. 4, pp. 798–859, 2001.

[25] G. Gallavotti, *Statistical mechanics: A short treatise.* Springer Science & Business Media, 1999.

[26] M. Sipser, *Introduction to the Theory of Computation.* Cengage Learning, 2012.

[27] R. M. Karp, "Reducibility among combinatorial problems," in *Complexity of computer computations*, Springer, 1972, pp. 85–103.

[28] R. J. Wilson, *Four Colors Suffice: How the Map Problem Was Solved - Revised Color Edition.* Princeton University Press, 2021.

[29] R. Diestel, *Graph Theory*, 5th. Springer Publishing Company, Incorporated, 2017.

[30] K. H. Rosen, *Discrete Mathematics and its Applications*, 8th ed. New York; McGraw-Hill, 2019.

[31] M. Garey, D. Johnson, and L. Stockmeyer, "Some simplified NP-complete graph problems," *Theoretical Computer Science*, vol. 1, no. 3, pp. 237–267, 1976.

[32] B. Korte and J. Vygen, *Combinatorial Optimization: Theory and Algorithms*, 5th. Springer Publishing Company, Incorporated, 2012.

[33] A. Lucas, "Ising formulations of many NP problems," *Frontiers in physics*, p. 5, 2014.

[34] Ö. Salehi, A. Glos, and J. A. Miszczak, "Unconstrained binary models of the travelling salesman problem variants for quantum optimization," *Quantum Information Processing*, vol. 21, no. 2, pp. 1–30, 2022.

[35] E. Farhi, J. Goldstone, S. Gutmann, and M. Sipser, "Quantum computation by adiabatic evolution," *arXiv preprint quant-ph/0001106*, 2000.

[36] M. Born and V. Fock, "Beweis des adiabatensatzes," *Zeitschrift für Physik*, vol. 51, no. 3, pp. 165–180, 1928.

[37] D. Aharonov, W. van Dam, J. Kempe, Z. Landau, S. Lloyd, and O. Regev, "Adiabatic quantum computation is equivalent to standard quantum computation," *SIAM Journal on Computing*, vol. 37, no. 1, pp. 166–194, 2007.

[38] T. S. Cubitt, D. Perez-Garcia, and M. M. Wolf, "Undecidability of the spectral gap," *Nature*, vol. 528, no. 7581, pp. 207–211, 2015.

[39] C. Carugno, M. Ferrari Dacrema, and P. Cremonesi, "Evaluating the job shop scheduling problem on a D-Wave quantum annealer," *Scientific Reports*, vol. 12, no. 1, pp. 1–11, 2022.

[40] G. Palubeckis, "Multistart tabu search strategies for the unconstrained binary quadratic optimization problem," *Annals of Operations Research*, vol. 131, no. 1, pp. 259–282, 2004.

[41] S. Kirkpatrick, C. D. Gelatt Jr, and M. P. Vecchi, "Optimization by simulated annealing," *Science*, vol. 220, no. 4598, pp. 671–680, 1983.

[42] E. Farhi, J. Goldstone, and S. Gutmann, "A quantum approximate optimization algorithm," *arXiv preprint arXiv:1411.4028*, 2014.

[43] B. Barak, A. Moitra, R. O'Donnell, *et al.*, "Beating the random assignment on constraint satisfaction problems of bounded degree," *arXiv preprint arXiv:1505.03424*, 2015.

[44] A. Biere, M. Heule, and H. van Maaren, *Handbook of satisfiability.* IOS press, 2009, vol. 185.

[45] M. Fernández-Pendás, E. F. Combarro, S. Vallecorsa, J. Ranilla, and I. F. Rúa, "A study of the performance of classical minimizers in the quantum approximate optimization algorithm," *Journal of Computational and Applied Mathematics*, vol. 404, p. 113 388, 2022.

[46] J. Watrous, *Quantum computation lecture notes*, 2005. [Online]. Available: `https://cs.uwaterloo.ca/~watrous/QC-notes/`.

[47] N. S. Yanofsky and M. A. Mannucci, *Quantum computing for computer scientists.* Cambridge University Press, 2008.

[48] M. Boyer, G. Brassard, P. Høyer, and A. Tapp, "Tight bounds on quantum searching," *Fortschritte der Physik: Progress of Physics*, vol. 46, no. 4-5, pp. 493–505, 1998.

[49] C. Durr and P. Høyer, "A quantum algorithm for finding the minimum," *arXiv preprint quant-ph/9607014*, 1996.

[50] A. Gilliam, S. Woerner, and C. Gonciulea, "Grover adaptive search for constrained polynomial binary optimization," *Quantum*, vol. 5, p. 428, 2021.

[51] L. Ruiz-Perez and J. C. Garcia-Escartin, "Quantum arithmetic with the quantum Fourier transform," *Quantum Information Processing*, vol. 16, no. 6, pp. 1–14, 2017.

[52] A. Peruzzo, J. McClean, P. Shadbolt, *et al.*, "A variational eigenvalue solver on a photonic quantum processor," *Nature communications*, vol. 5, no. 1, pp. 1–7, 2014.

[53] J. Preskill, *Lecture notes for physics 229: Quantum information and computation*, 1998. [Online]. Available: `http://theory.caltech.edu/~preskill/ph219/index.html#lecture`.

[54] S. McArdle, S. Endo, A. Aspuru-Guzik, S. C. Benjamin, and X. Yuan, "Quantum computational chemistry," *Reviews of Modern Physics*, vol. 92, no. 1, p. 015 003, 2020.

[55] Y. Cao, J. Romero, J. P. Olson, *et al.*, "Quantum chemistry in the age of quantum computing," *Chemical reviews*, vol. 119, no. 19, pp. 10 856–10 915, 2019.

[56] P. J. O'Malley, R. Babbush, I. D. Kivlichan, *et al.*, "Scalable quantum simulation of molecular energies," *Physical Review X*, vol. 6, no. 3, p. 031 007, 2016.

[57] K. L. Sharkey and A. Chancé, *Quantum Chemistry and Computing for the Curious*. Packt Publishing Ltd, 2022.

[58] O. Higgott, D. Wang, and S. Brierley, "Variational quantum computation of excited states," *Quantum*, vol. 3, p. 156, 2019.

[59] S. Bravyi, S. Sheldon, A. Kandala, D. C. Mckay, and J. M. Gambetta, "Mitigating measurement errors in multiqubit experiments," *Physical Review A*, vol. 103, no. 4, p. 042 605, 2021.

[60] K. Temme, S. Bravyi, and J. M. Gambetta, "Error mitigation for short-depth quantum circuits," *Physical review letters*, vol. 119, no. 18, p. 180 509, 2017.

[61] S. Endo, Z. Cai, S. C. Benjamin, and X. Yuan, "Hybrid quantum-classical algorithms and quantum error mitigation," *Journal of the Physical Society of Japan*, vol. 90, no. 3, p. 032 001, 2021.

[62] R. LaRose, A. Mari, S. Kaiser, *et al.*, "Mitiq: A software package for error mitigation on noisy quantum computers," *Quantum*, vol. 6, p. 774, 2022.

[63] Y. S. Abu-Mostafa, M. Magdon-Ismail, and H.-T. Lin, *Learning from data.* AMLBook New York, 2012, vol. 4.

[64] A. Géron, *Hands-on machine learning with Scikit-Learn, Keras and TensorFlow*, 2nd ed. O'Reilly, 2019.

[65] S. Shalev-Shwartz and S. Ben-David, *Understanding Machine Learning: From Theory to Algorithms.* Cambridge University Press, 2014.

[66] I. Goodfellow, J. Pouget-Abadie, M. Mirza, *et al.*, "Generative adversarial nets," *Advances in neural information processing systems*, vol. 27, 2014.

[67] R. S. Sutton and A. G. Barto, *Reinforcement learning: An introduction.* MIT Press, 2018.

[68] A. Skolik, S. Jerbi, and V. Dunjko, "Quantum agents in the gym: A variational quantum algorithm for deep Q-learning," *Quantum*, vol. 6, p. 720, 2022.

[69] K. Hornik, M. Stinchcombe, and H. White, "Multilayer feedforward networks are universal approximators," *Neural Networks*, vol. 2, no. 5, pp. 359–366, 1989.

[70] M. Ford, *Architects of Intelligence: The truth about AI from the people building it.* Packt Publishing, 2018.

[71] M. Schuld and F. Petruccione, *Machine Learning with Quantum Computers* (Quantum Science and Technology). Springer International Publishing, 2021.

[72] H.-Y. Huang, M. Broughton, J. Cotler, *et al.*, "Quantum advantage in learning from experiments," *Science*, vol. 376, no. 6598, pp. 1182–1186, 2022.

[73] J. Biamonte, P. Wittek, N. Pancotti, P. Rebentrost, N. Wiebe, and S. Lloyd, "Quantum machine learning," *Nature*, vol. 549, no. 7671, pp. 195–202, 2017.

[74] Y. S. Abu-Mostafa, M. Magdon-Ismail, and H.-T. Lin, *Learning from data, e-chapter 8: Support vector machines.* [Online]. Available: `https://amlbook.com/eChapters.html`.

[75] V. Havlíček, A. D. Córcoles, K. Temme, *et al.*, "Supervised learning with quantum-enhanced feature spaces," *Nature*, vol. 567, no. 7747, pp. 209–212, 2019.

[76] M. Schuld, "Supervised quantum machine learning models are kernel methods," *arXiv preprint arXiv:2101.11020*, 2021.

[77] S. Sim, P. D. Johnson, and A. Aspuru-Guzik, "Expressibility and entangling capability of parameterized quantum circuits for hybrid quantum-classical algorithms," *Advanced Quantum Technologies*, vol. 2, no. 12, p. 1 900 070, 2019.

[78] D. Dua and C. Graff, *UCI machine learning repository*, 2017. [Online]. Available: `http://archive.ics.uci.edu/ml`.

[79] R. Orús, "A practical introduction to tensor networks: Matrix product states and projected entangled pair states," *Annals of physics*, vol. 349, pp. 117–158, 2014.

[80] M. Schuld, V. Bergholm, C. Gogolin, J. Izaac, and N. Killoran, "Evaluating analytic gradients on quantum hardware," *Phys. Rev. A*, vol. 99, p. 032 331, 3 Mar. 2019.

[81] J. R. McClean, S. Boixo, V. N. Smelyanskiy, R. Babbush, and H. Neven, "Barren plateaus in quantum neural network training landscapes," *Nature communications*, vol. 9, no. 1, pp. 1–6, 2018.

[82] M. Cerezo, A. Sone, T. Volkoff, L. Cincio, and P. J. Coles, "Cost function dependent barren plateaus in shallow parametrized quantum circuits," *Nature communications*, vol. 12, no. 1, pp. 1–12, 2021.

[83] M. C. Caro, H.-Y. Huang, M. Cerezo, *et al.*, "Generalization in quantum machine learning from few training data," *Nature Communications*, vol. 13, no. 1, p. 4919, 2022.

[84] A. Pérez-Salinas, A. Cervera-Lierta, E. Gil-Fuster, and J. I. Latorre, "Data re-uploading for a universal quantum classifier," *Quantum*, vol. 4, p. 226, Feb. 2020.

[85] M. Schuld, R. Sweke, and J. J. Meyer, "Effect of data encoding on the expressive power of variational quantum-machine-learning models," *Phys. Rev. A*, vol. 103, p. 032 430, 3 Mar. 2021.

[86] D. Kraft, *A software package for sequential quadratic programming*, Technical Report DFVLR-FB 88-28, 1988.

[87] T. Karras, S. Laine, and T. Aila, "A style-based generator architecture for generative adversarial networks," in *Proceedings of the IEEE/CVF conference on computer vision and pattern recognition*, 2019, pp. 4401–4410.

[88] I. Cong, S. Choi, and M. D. Lukin, "Quantum convolutional neural networks," *Nature Physics*, vol. 15, pp. 1273–1278, 2019.

[89] M. Henderson, S. Shakya, S. Pradhan, and T. Cook, "Quanvolutional neural networks: Powering image recognition with quantum circuits," *Quantum Machine Intelligence*, vol. 2, no. 1, pp. 1–9, 2020.

[90] M. Arjovsky, S. Chintala, and L. Bottou, "Wasserstein generative adversarial networks," in *International conference on machine learning*, PMLR, 2017, pp. 214–223.

[91] I. Goodfellow, "NIPS 2016 tutorial: Generative adversarial networks," *arXiv preprint arXiv:1701.00160*, 2016.

[92] S. Wiesner, "Conjugate coding," *ACM Sigact News*, vol. 15, no. 1, pp. 78–88, 1983.

[93] S. Lloyd and C. Weedbrook, "Quantum generative adversarial learning," *Physical review letters*, vol. 121, no. 4, p. 040 502, 2018.

[94] P.-L. Dallaire-Demers and N. Killoran, "Quantum generative adversarial networks," *Physical Review A*, vol. 98, no. 1, p. 012 324, 2018.

[95] C. Zoufal, A. Lucchi, and S. Woerner, "Quantum generative adversarial networks for learning and loading random distributions," *npj Quantum Information*, vol. 5, no. 1, pp. 1–9, 2019.

[96] J. B. Altepeter, E. R. Jeffrey, and P. G. Kwiat, "Photonic state tomography," *Advances in Atomic, Molecular, and Optical Physics*, vol. 52, pp. 105–159, 2005.

[97] IBM, *IBM Unveils 400 Qubit-Plus Quantum Processor and Next-Generation IBM Quantum System Two*, `https://newsroom.ibm.com/2022-11-09-IBM-Unveils-400-Qubit-Plus-Quantum-Processor-and-Next-Generation-IBM-Quantum-System-Two`.

[98] T. Yamakawa and M. Zhandry, "Verifiable quantum advantage without structure," *arXiv preprint arXiv:2204.02063*, 2022.

[99] J. Bak and D. J. Newman, *Complex analysis*, 3, Ed. Springer, 2010.

[100] S. Axler, *Linear algebra done right.* Springer, 2015.

[101] D. S. Dummit and R. M. Foote, *Abstract algebra*, 3rd ed. John Wiley and Sons, 2004.

[102] C. H. Papadimitriou, *Computational complexity.* Addison-Wesley, 1994.

[103] S. Arora and B. Barak, *Computational complexity: a modern approach.* Cambridge University Press, 2009.

[104] J. E. Savage, *Models of computation.* Addison-Wesley Reading, MA, 1998, vol. 136.

[105] A. J. Kfoury, R. N. Moll, and M. A. Arbib, *A programming approach to computability.* Springer Science & Business Media, 2012.

[106] M. Agrawal, N. Kayal, and N. Saxena, "PRIMES is in P," *Annals of mathematics,* pp. 781–793, 2004.

[107] K. Devlin, *The Millennium Problems: The Seven Greatest Unsolved Mathematical Puzzles of Our Time.* Basic Books, 2002.

[108] C. Moore and S. Mertens, *The nature of computation.* OUP Oxford, 2011.

[109] M. R. Garey and D. S. Johnson, *Computers and intractability.* Freeman, 1979, vol. 174.

[110] J. Watrous, "Quantum computational complexity," *arXiv preprint arXiv:0804.3401,* 2008.

[111] E. Bernstein and U. Vazirani, "Quantum complexity theory," *SIAM journal on computing,* vol. 26, no. 5, pp. 1411–1473, 1997.

[112] A. Y. Kitaev, A. Shen, M. N. Vyalyi, and M. N. Vyalyi, *Classical and quantum computation.* American Mathematical Soc., 2002.

[113] R. Raz and A. Tal, "Oracle separation of BQP and PH," *ACM Journal of the ACM (JACM),* vol. 69, no. 4, pp. 1–21, 2022.

[114] C. H. Bennett, E. Bernstein, G. Brassard, and U. Vazirani, "Strengths and weaknesses of quantum computing," *SIAM journal on Computing,* vol. 26, no. 5, pp. 1510–1523, 1997.

[115] J. Basso, E. Farhi, K. Marwaha, B. Villalonga, and L. Zhou, "The quantum approximate optimization algorithm at high depth for maxcut on large-girth regular graphs and the Sherrington-Kirkpatrick model," *arXiv preprint arXiv:2110.14206,* 2021.

[116] E. Farhi, J. Goldstone, S. Gutmann, and L. Zhou, "The quantum approximate optimization algorithm and the Sherrington-Kirkpatrick model at infinite size," *Quantum,* vol. 6, p. 759, 2022.

[117] M. B. Hastings, “A Classical Algorithm Which Also Beats $\frac{1}{2} + \frac{2}{\pi}\frac{1}{\sqrt{D}}$ For High Girth Max-Cut,” *arXiv preprint arXiv:2111.12641*, 2021.

Index

3-CNF, 185

activation function, 323
Adam algorithm, 330
adiabatic evolution, 127
adiabatic quantum computing, 7, 126–129
    discretizing, 172, 174
adiabatic theorem, 128
adjacent vertices, 116
adjoint transpose, 11
Aer simulator
    for PennyLane device, 78
Amazon Braket, 48
amplitude amplification, 219
amplitude encoding, 369, 370
amplitudes, 10
Anaconda, 566
    Distribution version, 566
    installation, 566
analogous quantum model
    data output, 392
    data preparation, 391
    data processing, 391
AND gate, 8
angle encoding, 368, 369
anneal fraction, 157
annealer topologies, 150
annealing parameters
    controlling, 155–160
annealing process, 155
annihilation operator, 269
ansatzs, 261, 427
API token field, 571
arithmetic modulo 4, 540, 541
artificial feed-forward dense neural network, 324
asymptotic complexity, 547–549
attributes, 307
AUC (area under ROC curve), 346
Authentication token, 572
autoencoders, 435
automated machine learning (AutoML), 444
Azure Quantum, 45

backpropagation, 331
barren plateaus, 403
BasicAer, 58
batch gradient descent, 332
Big O notation, 547–549
binary classifier
    performance, 340–347
    recall, 341
binary cross-entropy, 328
binary linear programming problems, 110–113
bipartite graphs, 118
bit (binary digit), 9
Bloch sphere, 18–21
    qubit state, 20
    rotations, 20
BPP, 559
BQP, 558–561
bra-kets, 11
bras, 11

C++, 46
canonical basis, 534
categorical cross-entropy loss, 451
CCNOT gate, 40
change of basis operator, 257
Chimera topology, 150, 152
chromatic number, 116
Church-Turing thesis, 546
circuit engineering 101, 70–77
CircuitQNN class, 426
circuits, QAOA, 178–180
    energy estimation, 181–183
Cirq, 46, 49
Clarke's laws, 524
classical neural network
    data output, 391
    data preparation, 391
    data processing, 391
classical solvers, 165
    SimulatedAnnealingSampler, 167
    SteepestDescentSolver, 165
    TabuSolver, 166
classifier model, 307
clustering, 318
CNOT gate, 28, 29
combinatorial optimization problems, 109
    binary linear programming, 110
    graph coloring problem, 116
    Knapsack problem, 114
    Traveling Salesperson Problem, 119
complex numbers, 527–529
computational basis, 10, 534
    measuring, 11
computational complexity, 543, 544
    decision problems, 550
    problem, 549
computational model, 308

computational problems
completeness, 556–558
hardness, 555
reduction, 555, 556
computational time
measuring, 546, 547
confusion matrix, 340
conjugate transpose, 11, 538
conjunctive normal form (CNF), 185
constrained problems
running, on quantum annealers, 141–145
constrained quadratic models, Ocean, 136–138
solving, with dimod package, 138–140
ConstrainedQuadraticModel object, 136
controlled gates, 33, 34
controlled-NOT (controlled-X) gate, 28
coordinate matrix, 536
coordinates, 533
counterfeit money, 484, 485
coupling strengths
significance, 160, 162–164
couplings, 149
creation operator, 269
cross validation, 316

D-Wave company, 47
D-Wave quantum annealers
accessing, 572
data reuploading technique, 404
decider, 550
dimod package, 138
constrained quadratic models, solving with, 138–140
Dirac notation, 9–12
inner product, calculating, 12
discrete Fourier transform, 225
DiscreteQuadraticModel class, 168
discretization, 173
dissociation profile, 282
Dürr-Høyer algorithm, 234
Dürr-Høyer method, 212

early stopping, 338
eigenvalue, 93, 536
eigenvector, 93, 536
embedding, 152–154
empirical error, 311
empirical risk, 312
encoder network, 435
endomorphism, 536
entangled states
creating, 35
entanglement, 7, 30, 31
epochs, 335
excited state, 127
finding, with VQE, 263–266
expectation value, 96, 252, 253

expectation values, observables
estimating, 253–260
exponential linear unit(ELU) activation function, 325

false positive rate, 343
feasible, 110
feature map, 41, 365
amplitude encoding, 369, 370
angle encoding, 368, 369
constructing, 368
implementing, 367
ZZ feature map, 370, 371
feature space, 365
fermionic Hamiltonian, 267
finite-dimensional vector spaces, 533
forward annealing, 157

GAS in Qiskit
implementing, 236–242
gauge symbol, 8
generalization error, 310
Generative Adversarial Networks (GANs), 318, 483, 486
discriminator neural network, 486
generative neural network, 486
technicalities, 491, 492
training process, 487, 489
generative model
training, 318
global phase, 19
Google, 46
Google Colab, 566, 567
GPU simulations, 573, 574
gradient computation, PennyLane
adjoint differentiation, 421
backpropagation, 421
device gradient computation, 422
finite differences, 421
parameter shift rule, 422
gradient computation, QNNs
automatic differentiation, 401
numerical approximation, 401
parameter shift rule, 402
gradient descent, 328, 329
minibatch gradient descent, 331
gradient descent algorithms, 176, 329–331
gradient descent method, 329
gradient vector, 330
graph, 84
edges, 85
non-optimal cut, 86
optimal cut, 86
vertices, 85
graph coloring problem, 116–119
Graphviz, 578
Graphviz Visual Editor
URL, 578
ground state, 96

Grover Adaptive Search (GAS), 83, 211
Grover's algorithm, 212, 213
circuits, 216–218
minima, finding with, 223
probability, of finding marked element, 219–223
Grover's diffusion operator, 218

H gate, 8, 16
halting problem, 545
Hamiltonian, 97
Hamiltonian function, 89
Hamiltonians, 246–248
QAOA, using with, 188–193
Hartree-Fock state, 281
Hermitian adjoint, 538
Hermitian matrix, 96
Hermitian operator, 249, 539
Hermitian transpose, 11
Higher Order Binary Optimization (HOBO), 184
Hilbert space, 10
hinge loss, 363
HOBO problems
dealing with, 186
hybrid algorithm, 176
hybrid architectures, 432
hybrid architectures in PennyLane, 436
binary classification problem, solving, 438–442
classification problems, solving, 436, 437
model training, 442–449
multi-class classification problem, solving, 449
hybrid architectures in Qiskit, 458
future, 481, 482
hybrid binary classifier, building, 472–476
PyTorch, using, 459
Qiskit QNNs, training with Runtime, 477–480
hybrid models, 432
hybrid networks, 349
hybrid quantum neural networks, 432
example, 433, 434
hybrid solvers, 165, 167
LeapHybridCQMSampler, 168
LeapHybridDQMSampler, 168
LeapHybridSampler, 168
hyperparameters, 315

IBM Quantum
PennyLane, connecting to, 79, 80
using, 64–68
IBM Quantum provider, 50
IBM's quantum computers
accessing, 571
inner product
conjugate symmetry, 537

defining, 537
linearity, 537
positive-definiteness, 537
inner product space, 537
integer linear programming, 113
interference, 7
inverse quantum Fourier transform, 226
inversion about the mean operation, 219
Ising model, 89
example, 89, 90
Ising problem
transforming to QUBO, 105–109

Jordan-Wigner transformation, 269
Jupyter notebook, 53

k-colorable, 116
k-fold cross-validation, 316
Keras, 47
Keras sequential model, 333
kernel functions, 366
kernel trick, 364
kets, 9
Knapsack problem, 114, 115
Kullback–Leibler divergence, 516

labeled dataset, 318
Lagrangian dual, 363
Leap
for solving optimization problems, 146
Leap annealers, 146–149
LeapHybridDQMSampler, 168
LeapHybridSampler
properties, 168
learning rate, 330
libraries
installing, 568–570
Lie-Trotter formula, 174
linear algebra
adjoint operators, 538
bases, 534, 535
coordinates, 533, 534
inner products, 537
linear maps, 535, 536
overview, 531
vector spaces, 531, 533
linear maps, 535
loss function, 311
defining, 327, 328

machine learning
basics, 306
components, 307
computational model, 308
reinforcement learning, 319
supervised learning, 317
trained model, assessing, 312–317
training procedure, 309–311
types, 317
unsupervised learning, 318

Machine Learning (ML) frameworks, 49
macro average, 342
many-one reducible, 556
margin, 358
matplotlib, 53
matrix exponentiation, 539
Max-Cut problem, 84
  cut, 85
  formulation, 86–88
  graph example, 85
  maximum cut, 85
  size of cut, 85
mean squared error (MSE), 327
measurement-based quantum computing, 7
Microsoft's Quantum Development Kit (QDK), 45, 46
mini-batch gradient descent
  fixed batch size, 331
minibatch gradient descent, 331
minus state, 16
mixer Hamiltonian, 205
mode collapse, 490
model
  picking, 322–326
  training, 320–322, 335–339
modular arithmetic, 540, 541
molecular problem
  defining, in Qiskit, 267–270
multi-class classification problem, 449
  general perspective, 450–452
multi-qubit gates, 39, 40
multi-qubit systems, 36–39
multi-tape Turing machines, 545

neural networks, 308, 322–324
  common activation functions, 326
  graphical representation, 323
  working, 323
neurons, 323
  set of biases, 323
NISQ devices, 5, 50
no-cloning theorem, 32, 33
non-deterministic Turing machines, 545
non-parametric models, 309
normalization condition, 10
NOT gate, 8, 15
NP class, 552–554
NP-complete problem, 105, 557
NP-completeness, 556
NP-hard problems, 88, 555, 556

observables, 248, 250–252
  expectation values, estimating, 253–260
Ocean, 47, 49
  constrained quadratic models, 135–138
  optimization problem, formulating , 135

optimization problem, transforming, 135
one-hot encoding, 168
one-hot form, 451
one-qubit quantum gates, 13–17
one-versus-all method, 450
one-versus-one method, 450
one-way quantum computing, 7
optimization problem formulation
classical variables, to qubits, 91, 93–97
expectation value computation, with Qiskit, 97–105
optimization problems
formulating, with Ocean, 135
solving with Leap, 146
transforming, with Ocean, 135
OR gate, 8
oracle Turing machines, 555
orthonormal, 538
overfitting, 313, 314
example, 314

parameter shift rule, 402
parity observable, 400
Pauli matrices, 17
Pegasus, 152
penalty terms, 112
PennyLane, 47, 48
circuit engineering 101, 70–77
circuit, simulating with Aer , 78
connecting to IBM, 80
connecting to IBMQ, 79
Device object, 70
interoperability, 77
QAOA, using with, 203–209
working with, 69
perceptron, 308
phase encoding, 229
phase gate, 226
phase kickback technique, 218
physical observable, 248
pip, 568
plus state, 16
polynomial time class (P), 550, 551
Polynomial Unconstrained Binary Optimization (PUBO), 184
polynomial-time many-one reduction, 556
polynomial-time reducible, 555
polynomial-time verifier, 552
positive predictive value, 341
precision, 341
primitive gates, 40
principal component analysis, 377
principal directions, 378
product state, 30
PyQuest, 69

Python, 566
Python-based simulator (BasicAer), 50
PyTorch, 47, 51, 459
    model training, 463, 465–467, 469–471
    model, setting up, 460–463
    using, 459
PyTorch interface, 47

Q#, 45, 46, 49
QGANs in PennyLane, 495, 496
    model training, 502–509
    QGAN model, preparing, 497–501
QGANs in Qiskit, 510, 514–517
Qiskit, 46, 48
    for expectation value computation, 97–105
    future, 291–294
    overview, 50, 52
    working with, 50
Qiskit Aer, 50
    for simulating quantum circuits, 58–64
Qiskit Dynamics, 52
Qiskit Experiments, 51
Qiskit Finance, 51
Qiskit Machine Learning, 50
Qiskit Metal, 52
Qiskit Nature, 51
Qiskit Optimization, 51
Qiskit Terra, 50
    for building quantum circuits, 52
QNNs in PennyLane, 405, 406
    data preparation, 406–409
    gradient computation, 420–425
    network building, 410–413
    TensorFlow, using, 413–415, 417–420
QNNs in Qiskit, 426–429
QSVMs in PennyLane, 372
    custom feature maps, implementing, 380, 381
    dimensionality of dataset, reducing, 377–379
    kernel, implementing, 375, 376
    model training, 372–374
QSVMs in Qiskit, 382
Quadratic Unconstrained Binary Optimization (QUBO), 84
    combinatorial optimization problems, 109
quantum annealers, 47, 49, 83
    constrained problems, running on, 141–145
    using, 125
quantum annealing, 129–135
    annealing schedule, 130
    annealing time, 130

Quantum Approximate Optimization Algorithm (QAOA), 6, 83, 172, 175, 176, 178
  circuits, 178–180
  using, with Hamiltonians, 188–192, 194
  using, with PennyLane, 203–209
  using, with Qiskit , 188
quantum circuit
  creating, 22
  example, 8
  overview, 7
  simulating, with Qiskit Aer, 58–64
quantum circuit model, 7
  implementation, 83
  measurements, 7
  qubits, 7
quantum circuits, with Qiskit Terra, 52
  initializing, 53, 54
  measurements, 57, 58
  quantum gates, 54–56
quantum classical models, 347–349
quantum computational complexity, 558–562
quantum computing
  defining, 5, 6
  overview, 4
  universal gates, 40
Quantum Exact Simulation Toolkit (QuEST), 46, 49
quantum Fourier transform (QFT), 225, 226
  circuit, 226
quantum GANs (QGANs), 483, 484, 493
  categories, 493, 494
quantum generative adversarial networks, 349
Quantum Neural Networks (QNNs), 6, 41, 349, 389
  building, 390
  gradient computation, 401
  implementing, for ternary classification problem, 452–457
  measurements, 398–400
  usage, 403, 404
  variational forms, 393–397
Quantum Node (QNode), 72
quantum oracles, 214, 216
quantum oracles, for combinatorial optimization, 224
  constructing, 234, 235
  integer numbers, adding, 230–232
  integer numbers, encoding, 227, 228
  polynomial, computing, 232, 233
Quantum Phase Estimation (QPE), 261
quantum processing units, 149

Quantum Support Vector Machines (QSVMs), 6, 349, 351
on IBM quantum computer, 385, 386
on Qiskit Aer, 382–385
working, 366
quantum Turing machines, 7
qubit, 9
collapsed state, 8, 11
normalized state, 10
qubit Hamiltonians, 267
QUBO problems, 106, 184
solving, with QAOA in Qiskit, 195–198, 200–203
transformations, from Ising problem, 107–109
Quirk, 44
Grover search algorithm demo, 45
quota conversion rate, 168

Random-Access Machines model, 545
readout error mitigation methods, 285
Receiver Operating Characteristic (ROC) curve, 343
rectified linear unit (ReLU) function, 325
reduction, 555
registers, 53
regression problem
tackling, 317
regularization, 314
reinforcement learning, 319
agent, 319
environment, 319
environment state, 319
policy, 319
relative entropy, 516
reverse annealing, 158, 159
Runtime, 192

S gate, 18
SAT, 183
Schrödinger equation, 13
scikit-learn, 47
scikit-learn package (sklearn), 320
second quantization, 268
self-adjoint, 539
self-adjoint matrix, 96
sigmoid activation function, 325
simulated annealing, 167
SimulatedAnnealingSampler, 167
slack variables, 110
soft-margin training, 361, 362
softmax activation function, 455
solvers, 146, 148
sparse categorical cross entropy loss, 452
spectral gap, 128
spherical coordinates, 19
statevector method, 61
SteepestDescentSolver, 165
step function, 324
stochastic gradient descent, 332

stratification, 333
strongly entangling layers variational form, 396, 397
Subset Sum problem, 105
superposition, 7
supervised learning, 317
Support Vector Machines (SVMs), 352
  hard-margin case, 356–360
  kernel trick, 364, 365
  simple classifier, 352–355
  soft-margin training, 361–363
  training, 356–360
support vectors, 364

T gate, 18
TabuSolver, 166
tensor product, 24
TensorFlow, 47, 319, 332
  model, defining, 333
ternary classification problem
  QNN, implementing, 452–454, 456, 457
test dataset, 310
TikZ, 578
time-independent Schrödinger equation, 126
Toffoli gate, 40
  decomposition, 41
tool installation, 565
  Anaconda, 566
  Google Colab, 566, 567
  Python, 565
tools
  for quantum computing, 44
  non-exhaustive survey, 44
topology, 149
total ground state energy, 280
training dataset, 310
training procedure, 307
transpilation, 68
Traveling Salesperson Problem, 119–121
tree tensor variational form, 395, 396
Trotterization, 173
true error, 310
true positive rate, 343
true risk, 311
Tseitin transformation, 185
Turing machines, 544, 545
  cells, 544
  head, 544
  state, 544
  tape, 544
two-local variational form, 394
two-qubit states, 24, 25
  inner products, computing, 26
  tensor products, 27, 28
TwoLayerQNN class, 426

U-gate (universal one-qubit gate), 21
UCSSD ansatz, 280

for solving problem, 281
uncomputation technique, 235
underfitting, 315
unitary, 539
unitary matrices, 14
universal function approximators, 308
universal gates, 41
unlabeled datasets, 318
unsupervised learning, 318

validation dataset, 316
validation loss, 317
variational circuits, 392
variational forms, 41, 261, 393
strongly entangling layers, 396, 397
tree tensor variational form, 395, 396
two-local variational form, 394, 395
variational principle, 96
Variational Quantum Deflation (VQD), 276
Variational Quantum Eigensolver (VQE), 6, 83, 245, 260
excited states, finding, 263–266
pseudocode, 262, 263
running, on quantum computers, 288–291
structure, 261
vector spaces, 532, 533
norm, defining, 537
VQE with Hamiltonians
using, 270–272, 274, 275
VQE with PennyLane
implementing, 296, 297
molecular problem, defining in PennyLane, 294–296
running, 297, 298
running, on real quantum devices, 298–300
using, 294
VQE with Qiskit
excited states, finding with Qiskit, 276, 278
molecular problem, defining in Qiskit, 267–270
simulations, with noise, 282–288
using, 267
using, with molecular problems, 278–282

Wasserstein GANs (WGANs), 490
while-Programs, 546
wires, 7

X gate, 8, 15
Xanadu, 47
XOR function, 40

Y gate, 8, 17

Z gate, 8
zero-one linear programming, 110
ZZ feature map, 370, 371